Lecture Notes in Computer Science 7200

Commenced Publication in 1973
Founding and Former Series Editors:
Gerhard Goos, Juris Hartmanis, and Jan van Leeuwen

Services Science

Subline of Lectures Notes in Computer Science

Giuseppe Anastasi   Emilio Bellini
Elisabetta Di Nitto   Carlo Ghezzi
Letizia Tanca   Eugenio Zimeo (Eds.)

Methodologies and Technologies for Networked Enterprises

ArtDeco: Adaptive Infrastructures
for Decentralised Organisations

 Springer

Volume Editors

Giuseppe Anastasi
Università di Pisa
Dipartimento di Ingegneria dell'Informazione
Largo Lucio Lazzarino 1, 56122 Pisa, Italy
E-mail: g.anastasi@iet.unipi.it

Emilio Bellini
Elisabetta Di Nitto
Carlo Ghezzi
Letizia Tanca
Politecnico di Milano
Dipartimento di Elettronica ed Informazione
Piazza Leonardo da Vinci 21, 20133 Milano, Italy
E-mail: emilio.bellini@polimi.it
E-mail: {dinitto, ghezzi, tanca}@elet.polimi.it

Eugenio Zimeo
Università degli Studi del Sannio
Sistemi di Elaborazione dell'Informazione
Via Traiano 1, 82100 Benevento, Italy
E-mail: zimeo@unisannio.it

ISSN 0302-9743   e-ISSN 1611-3349
ISBN 978-3-642-31738-5   e-ISBN 978-3-642-31739-2
DOI 10.1007/978-3-642-31739-2
Springer Heidelberg Dordrecht London New York

Library of Congress Control Number: 2012942116

CR Subject Classification (1998): H.4, I.2, H.3, H.5, C.2, J.1

LNCS Sublibrary: SL 3 – Information Systems and Applications, incl. Internet/Web
and HCI

Typesetting: Camera-ready by author, data conversion by Scientific Publishing Services, Chennai, India

Printed on acid-free paper

Springer is part of Springer Science+Business Media (www.springer.com)

Preface

The needs for flexibility and globalization force enterprises to decentralize their activities and continuously (re)structure their network of relationships regarding both their productive "supply chain" and their design and innovation processes.

The goal of the ArtDeco project (http://artdeco.elet.polimi.it/) was to address these issues by studying the problem and proposing solutions from three main diverse perspectives: the organizational perspective aimed at studying how companies work in a network and how their design processes can benefit from the collaboration with other companies and subjects; the informational perspective focused on how to acquire relevant knowledge from unstructured information and processes and on how to organize and manage such knowledge to make it useful for the networking and innovation objectives of enterprises; finally, the infrastructural perspective has focused on understanding how self-adaptive workflows and software systems can help in supporting the dynamic interconnection of enterprises and on how pheripheral devices such us wireless sensor networks (WSNs) and radio frequency identification (RFID) devices can be useful in acquiring on-the-field data and in communicating such data to all interested parties belonging to the same supply chain.

ArtDeco was funded by the Italian Ministry of Education and Scientific Research under the FIRB program (funding for basic research) and was developed by the following Italian research centers and universities: Centro Nazionale delle Ricerche, Free University of Bolzano, IBM, Politecnico di Milano (project coordinator), Scuola Superiore S. Anna, Università dell'Aquila, Università di Bari, Università dell'Insubria, Università di Pisa, Università del Sannio. The project started in 2006 and ended in 2010.

The objective of the monograph is to collect in a coherent manner an overview of the main results produced by the project and to provide a comprehensive survey of the main organizational, conceptual, and technological challenges raised by networked enterprises. The intended audience for the book is researchers that are interested in getting a multidisciplinary overview of the main issues and challenges concerning networks of enteprises.

The book is structured into four main parts addressing the organizational (Part I), informational (Part III), and infrastructural (Parts II and IV) perpectives we have mentioned. It also includes a case study that is used through the book as the reference example and is developed in detail in the last chapter.

We wish to thank the Italian ministry for research and university (MIUR) for funding the project and Professor Maurizio Zamboni (Politecnico di Torino) for acting as project reviewer. Special thanks go to Professor Paolo Tiberio (University of Modena) for reviewing the whole book. We also wish to thank all students that contributed to different aspects of this project. Finally, we would like to thank all the industries of the wine and fashion areas that participated in our surveys.

February 2012

Giuseppe Anastasi
Emilio Bellini
Elisabetta Di Nitto
Carlo Ghezzi
Letizia Tanca
Eugenio Zimeo

Contents

14   Context Support for Designing Analytical Queries
Cristiana Bolchini, Elisa Quintarelli, and Letizia Tanca

Part IV Management of Peripheral Devices

15   Wireless Sensor Networks for Monitoring Vineyards
Cesare Alippi, Giacomo Boracchi,
Romolo Camplani, and Manuel Roveri

16   Design, Implementation, and Field Experimentation of a
Long-Lived Multi-hop Sensor Network for Vineyard Monitoring
Giuseppe Anastasi, Marco Conti, Mario Di Francesco, and
Ilaria Giannetti

20   Enabling Traceability in the Wine Supply Chain 397
Mario G.C.A. Cimino and Francesco Marcelloni

Part V Case Study

**21   Putting It All Together: Using the ArtDeco Approach in the Wine
Business Domain** 415
Eugenio Zimeo, Valentina Mazza, Giorgio Orsi, Elisa Quintarelli,
Antonio Romano, Paola Spoletini, Giancarlo Tretola,
Alessandro Amirante, Alessio Botta, Luca Cavallaro,
Domenico Consoli, Ester Giallonardo, Fabrizio Maria Maggi, and
Gabriele Tiotto

List of Contributors

Cesare Alippi
Dipartimento di Elettronica e Informazione, Politecnico di Milano, Piazza L. da Vinci 32, 20133 Milano (Italy)
e-mail: alippi@elet.polimi.it

Alessandro Amirante
Ph.D. student at the GII Doctoral School, Boosting Services and Information in Adaptive Networked Enterprise, 2008, L'Aquila (Italy)

Gaetano F. Anastasi
Scuola Superiore Sant'Anna, Piazza Martiri della Libertà, 33, 56127 Pisa (Italy)
e-mail: g.anastasi@sssup.it

Giuseppe Anastasi
Dipartimento di Ingegneria dell'Informazione: Elettronica, Informatica, Telecomunicazioni, Università of Pisa, Largo Lucio Lazzarino 1, 56122 Pisa (Italy)
e-mail: g.anastasi@iet.unipi.it

Rezaei Mahdiraji Alireza
Center for Applied Software Engineering (CASE), Free University of Bozen-Bolzano, Piazza Domenicani 3, 39100 Bolzano (Italy)
e-mail: alireza.rezaei@unibz.it

Teresa Baldassare
Dipartimento di Informatica, Università degli Studi di Bari "Aldo Moro"
e-mail: baldassarre@di.uniba.it

Bruno Basalisco
Scuola Superiore Sant'Anna, Piazza Martiri della Libertà, 33, 56127 Pisa (Italy) and Imperial College Business School, London (UK)
e-mail: b.basalisco@imperial.ac.uk

Emilio Bellini
Università degli Studi del Sannio, Via Traiano, 1, 82100 Benevento (Italy)
Dipartimento di Ingegneria Gestionale, Politecnico di Milano, Piazza L. da Vinci, 32 20133 Milano (Italy)
e-mail: emilio.bellini@polimi.it

Antonia Bertolino
Istituto di Scienza e Tecnologie dell'Informazione "A. Faedo", ISTI-CNR , Via
Giuseppe Moruzzi, 1, 56124 Pisa (Italy)
e-mail: antonia.bertolino@isti.cnr.it

Enrico Bini
Scuola Superiore Sant'Anna, Piazza Martiri della Libertà, 33, 56127 Pisa (Italy)
e-mail: e.bini@sssup.it

Cristiana Bolchini
Dipartimento di Elettronica e Informazione, Politecnico di Milano, Piazza L. da
Vinci 32, 20133 Milano (Italy)
e-mail: bolchini@elet.polimi.it

Giacomo Boracchi
Dipartimento di Elettronica e Informazione, Politecnico di Milano, Piazza L. da
Vinci 32, 20133 Milano (Italy)
e-mail: boracchi@elet.polimi.it

Alessio Botta
Ph.D. student at the GII Doctoral School, Boosting Services and Information in
Adaptive Networked Enterprise, 2008, L'Aquila (Italy)

Nicolò M. Calcavecchia
Dipartimento di Elettronica e Informazione, Politecnico di Milano, Piazza L. da
Vinci 32, 20133 Milano (Italy)
e-mail: calcavecchia@elet.polimi.it

Simone Campanoni
Dipartimento di Elettronica e Informazione, Politecnico di Milano, Piazza L. da
Vinci 32, 20133 Milano (Italy)
e-mail: campanoni@elet.polimi.it

Romolo Camplani
Dipartimento di Elettronica e Informazione, Politecnico di Milano, Piazza L. da
Vinci 32, 20133 Milano (Italy)
e-mail: camplani@elet.polimi.it

Gerardo Canfora
Università degli Studi del Sannio, Via Traiano, 1, 82100 Benevento (Italy)
e-mail: canfora@unisannio.it

Luca Cavallaro
Ph.D. student at the GII Doctoral School, Boosting Services and Information in
Adaptive Networked Enterprise, 2008, L'Aquila (Italy)
e-mail: cavallaro@elet.polimi.it

Mario G.C.A. Cimino
Dipartimento di Ingegneria dell'Informazione: Elettronica, Informatica, Telecomunicazioni, Università di Pisa, Largo Lucio Lazzarino 1, 56122 Pisa (Italy)
e-mail: m.cimino@iet.unipi.it

Domenico Consoli
Ph.D. student at the GII Doctoral School, Boosting Services and Information in
Adaptive Networked Enterprise, 2008, L'Aquila (Italy)

Marco Conti
Istituto di Informatica e Telematica, IIT-CNR, Via Giuseppe Moruzzi, 1, 56124
Pisa (Italy)
e-mail: marco.conti@iit.cnr.it

Guglielmo De Angelis
Istituto di Scienza e Tecnologie dell'Informazione "A. Faedo", ISTI-CNR , Via
Giuseppe Moruzzi, 1, 56124 Pisa (Italy)
e-mail: guglielmo.deangelis@isti.cnr.it

Claudio Dell' Era
Dipartimento di Ingegneria Gestionale, Politecnico di Milano, Piazza L. da Vinci,
32, 20133 Milano (Italy)
e-mail: claudio.dellera@polimi.it

Mario Di Francesco
Dept. of Computer Science and Engineering, Aalto University, Finland,
and CReWMaN, University of Texas at Arlington, USA
e-mail: mariodf@uta.edu

Felicita Di Giandomenico
Istituto di Scienza e Tecnologie dell'Informazione "A. Faedo", ISTI-CNR , Via
Giuseppe Moruzzi, 1, 56124 Pisa (Italy)
e-mail: felicita.digiandomenico@isti.cnr.it

Elisabetta Di Nitto
Dipartimento di Elettronica e Informazione, Politecnico di Milano, Piazza L. da
Vinci 32, 20133 Milano (Italy)
e-mail: dinitto@elet.polimi.it

Daniel J. Dubois
Dipartimento di Elettronica e Informazione, Politecnico di Milano, Piazza L. da
Vinci 32, 20133 Milano (Italy)
e-mail: dubois@elet.polimi.it

William Fornaciari
Dipartimento di Elettronica e Informazione, Politecnico di Milano, Piazza L. da
Vinci 32, 20133 Milano (Italy)
e-mail: fornacia@elet.polimi.it

Carlo Ghezzi
Dipartimento di Elettronica e Informazione, Politecnico di Milano, Piazza L. da
Vinci 32, 20133 Milano (Italy)
e-mail: ghezzi@elet.polimi.it

Ester Giallonardo
Ph.D. student at the GII Doctoral School, Boosting Services and Information in
Adaptive Networked Enterprise, 2008, L'Aquila (Italy)
e-mail: ester.giallonardo@unisannio.it

Ilaria Giannetti
Dipartimento di Ingegneria dell'Informazione: Elettronica, Informatica, Telecomunicazioni, Università di Pisa, Largo Lucio Lazzarino 1, 56122 Pisa (Italy)
e-mail: ilaria.giannetti@iet.unipi.it

Giuseppe Lipari
Scuola Superiore Sant'Anna, Piazza Martiri della Libertà, 33, 56127 Pisa (Italy)
e-mail: g.lipari@sssup.it

Fabrizio Maria Maggi
Ph.D. student at the GII Doctoral School, Boosting Services and Information in
Adaptive Networked Enterprise, 2008, L'Aquila (Italy)
e-mail: maggi@di.uniba.it

Francesco Marcelloni
Dipartimento di Ingegneria dell'Informazione: Elettronica, Informatica, Telecomunicazioni, Università di Pisa, Largo Lucio Lazzarino 1, 56122 Pisa (Italy)
e-mail: f.marcelloni@iet.unipi.it

Eda Marchetti
Istituto di Scienza e Tecnologie dell'Informazione "A. Faedo", ISTI-CNR , Via
Giuseppe Moruzzi, 1, 56124 Pisa (Italy)
e-mail: eda.marchetti@isti.cnr.it

Valentina Mazza
Dipartimento di Elettronica e Informazione, Politecnico di Milano, Piazza L. da
Vinci 32, 20133 Milano (Italy)
e-mail: vmazza@elet.polimi.it

Marcello Montedoro
Business Development Executive, IBM Italia S.p.A., Circonvallazione Idroscalo,
20090 Segrate (Italy)
e-mail: marcello_montedoro@it.ibm.com

Marcello Mura
ALaRI, Faculty of Informatics, University of Lugano (Switzerland)
e-mail: muram@usi.ch

Giorgio Orsi
Dipartimento di Elettronica e Informazione, Politecnico di Milano, Piazza L. da
Vinci 32, 20133 Milano (Italy)
e-mail: orsi@elet.polimi.it

Elisa Quintarelli
Dipartimento di Elettronica e Informazione, Politecnico di Milano, Piazza L. da
Vinci 32, 20133 Milano (Italy)
e-mail: quintare@elet.polimi.it

Guido M. Rey
Scuola Superiore Sant'Anna, Piazza Martiri della Libertà, 33, 56127 Pisa (Italy)
e-mail: g.rey@sssup.it

Antonio Romano
Scuola Superiore Sant'Anna, Piazza Martiri della Libertà, 33, 56127 Pisa (Italy)
e-mail: a.romano@sssup.it

Bruno Rossi
Center for Applied Software Engineering (CASE), Free University of Bozen-
Bolzano, Piazza Domenicani 3, 39100 Bolzano (Italy)
e-mail: bruno.rossi@unibz.it

Matteo Rossi
Dipartimento di Elettronica e Informazione, Politecnico di Milano, Piazza L. da
Vinci 32, 20133 Milano (Italy)
e-mail: rossi@elet.polimi.it

Guido Rota
Dipartimento di Elettronica e Informazione, Politecnico di Milano, Piazza L. da
Vinci 32, 20133 Milano (Italy)
e-mail: guido.rota@gmail.com

Manuel Roveri
Dipartimento di Elettronica e Informazione, Politecnico di Milano, Piazza L. da
Vinci 32, 20133 Milano (Italy)
e-mail: roveri@elet.polimi.it

Antonino Sabetta
Istituto di Scienza e Tecnologie dell'Informazione "A. Faedo", ISTI-CNR , Via
Giuseppe Moruzzi, 1, 56124 Pisa (Italy)
e-mail: antonino.sabetta@isti.cnr.it

Mariagiovanna Sami
Dipartimento di Elettronica e Informazione, Politecnico di Milano, Piazza L. da
Vinci 32, 20133 Milano (Italy)
e-mail: sami@elet.polimi.it

Licia Sbattella
Dipartimento di Elettronica e Informazione, Politecnico di Milano, Piazza L. da
Vinci 32, 20133 Milano (Italy)
e-mail: sbattell@elet.polimi.it

Fabio A. Schreiber
Dipartimento di Elettronica e Informazione, Politecnico di Milano, Piazza L. da
Vinci 32, 20133 Milano (Italy)
e-mail: schreiber@elet.polimi.it

Alberto Sillitti
Center for Applied Software Engineering (CASE), Free University of
Bozen-Bolzano, Piazza Domenicani 3, 39100 Bolzano (Italy)
e-mail: alberto.sillitti@unibz.it

Paola Spoletini
Università dell'Insubria, Via Ravasi 2, Varese (Italy)
e-mail: paola.spoletini@uninsubria.it

Giancarlo Succi
Center for Applied Software Engineering (CASE), Free University of
Bozen-Bolzano, Piazza Domenicani 3, 39100 Bolzano (Italy)
e-mail: gsucci@unibz.it

Letizia Tanca
Dipartimento di Elettronica e Informazione, Politecnico di Milano, Piazza L. da
Vinci 32, 20133 Milano (Italy)
e-mail: tanca@elet.polimi.it

Roberto Tedesco
MultiChancePoliTeam, Politecnico di Milano, Piazza L. da Vinci 32, 20133 Milano
(Italy)
e-mail: roberto.tedesco@polimi.it

Gabriele Tiotto
Ph.D. student at the GII Doctoral School, Boosting Services and Information in
Adaptive Networked Enterprise, 2008, L'Aquila (Italy)
e-mail: gabriele.tiotto@polito.it

Giancarlo Tretola
Università degli Studi del Sannio, Via Traiano, 1, 82100 Benevento (Italy)
e-mail: tretola@unisannio.it

Roberto Verganti
Dipartimento di Ingegneria Gestionale, Politecnico di Milano, Piazza L. da Vinci,
32, 20133 Milano (Italy)
e-mail: roberto.verganti@polimi.it

Eugenio Zimeo
Università degli Studi del Sannio, Via Traiano, 1, 82100 Benevento (Italy)
e-mail: zimeo@unisannio.it

Chapter 1
Introduction

Giuseppe Anastasi, Emilio Bellini, Elisabetta Di Nitto, Carlo Ghezzi,
Letizia Tanca, and Eugenio Zimeo

The goals of flexibility and globalization force enterprises to decentralize their activities and continuously (re)structure their network of relationships, with respect to their productive "supply chain" as well as to their design and innovation processes. The scale of involved companies is often small, but, thanks to the synergies enabled by operating within a network, these companies are able to increase their competitiveness. Currently there are several obstacles to the optimal operation of the networks of enterprises:

- The organizational models that regulate their operation are not always made explicit, and therefore they must be studied, at least by examining a set of "best practices".
- The knowledge shared by the organizations in the network is often kept implicit and tacit, hidden within common practice that is never formalized or within artifacts that are hard to identify. Since collective knowledge evolves dynamically, sometimes it may manifest itself upstream in the supply (or design) chain, sometimes downstream (e.g., at the consumers level). Therefore, there is a need

Giuseppe Anastasi
Dept. of Information Eng., Univ. of Pisa, Italy
e-mail: g.anastasi@iet.unipi.it

Emilio Bellini · Eugenio Zimeo
University of Sannio - Department of Engineering
e-mail: zimeo@unisannio.it

Emilio Bellini
Politecnico di Milano - Dipartimento di Ingegneria Gestionale - Piazza L. da Vinci, 32 20133
Milano, Italy
e-mail: emilio.bellini@polimi.it

Elisabetta Di Nitto · Carlo Ghezzi · Letizia Tanca
Politecnico di Milano -Dipartimento di Elettronica e Informazione - Politecnico di Milano,
Piazza L. da Vinci 32, 20133 Milano, Italy
e-mail: {dinitto,ghezzi,tanca}@elet.polimi.it

G. Anastasi et al. (Eds.): Networked Enterprises, LNCS 7200, pp. 1–5, 2012.
© Springer-Verlag Berlin Heidelberg 2012

to focus on the problems of finding, extracting, representing and formalizing knowledge, in the various forms in which it may be embedded (in particular, natural language) in order to build a semantic model (via ontologies) of the business domains of networked enterprises. Making implicit and hidden knowledge explicit is the starting point to formalize the business relationships within the network and to stimulate the organizational changes required to pursue the improvement goals of the business itself.

- The flexibility requirements of networked enterprises demand new innovative solutions for the technological infrastructure, which must support unprecedented levels of dynamicity. The largely evolutionary nature of the organizations, of their relationships, and of the business as a whole, require that the technological infrastructure on top of which they operate be capable of evolving in a seamless way, rapidly, and in a cost-effective way, adapting to changes, and auto-organizing as much as possible, showing an autonomic behavior. As soon as deviations are discovered from the desired or acceptable quality of service offered by the information system, the latter should be able to reconfigure itself in an autonomic manner. Autonomic behaviors, in turn, require domain-specific knowledge on the business logic to be always available in a fully distributed and pervasive manner. The information system will have to include intelligent peripheral systems which interact with the physical world via pervasive devices such as wireless sensor networks (WSNs) and intelligent tags (RFIDs). This will make the Internet-of-Things a key enabler of the networked enterprises.

This book will contribute to the above issues by presenting the approach we have adopted within the ArtDeco project, where several research areas have offered an integrated assortment of solutions to the above problems. In our view (see Figure 1.1), an explicit organizational model constitutes the foundation for a network of enterprises (in the figure, A, B, and C represent three enterprises), and it is supported by inter-enterprise processes that exploit the processes local to each enterprise, typically packaged into proper services, and that rely on some shared knowledge. Knowledge is built by gathering information from local repositories, sensors of various kinds (devices), and external, often unstructured, sources. The organizational model, the sharing of processes and information are enabled by proper technology that should, in the first place, be able to be adaptable to the dynamics of the network and of the devices used to support it.

This book touches all aspects highlighted in Figure 1.1. The various chapters in parts from I to IV present state-of-the-art and novel methodologies, methods and techniques for the support of the following aspects of dynamic networked enterprises:

- Part I is focusing on the study of the issues and organizational models that regulate the operation of a network of enterprise.
- Part II contributes to the definition of the autonomic infrastructure that supports the execution and adaptation of the local and networked processes. It moreover, provides an overview of the possible approaches to be adopted for verifying, at design time and runtime, that the infrastructure works correctly.

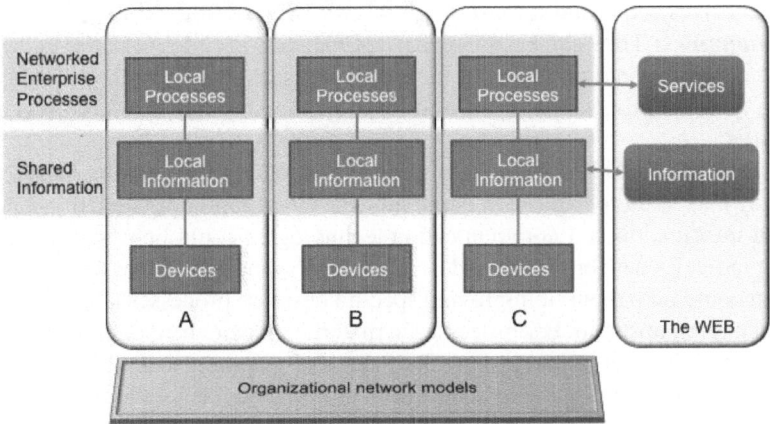

Fig. 1.1 Structure of `ArtDeco` project and book

- Part III focuses on the management of contextualized knowledge and on its extraction from various sources.
- Part IV discusses on how peripheral devices can help in the acquisition of information and in the execution of actions that impact on the physical environment.

In order to analyze the above referring to a practical case, we focus on an application domain that appears to be very relevant for the Italian economy: the wine production sector. This sector is populated by many small-to-medium enterprises focusing on a product that is becoming more and more technology-intensive, but that still maintains strong connections with historical traditions and knowledge accumulated over the past centuries. Such enterprises usually organize themselves into consortia, still tending not to have strict collaborations with one another. A grand challenge in this area is thus to highlight the effectiveness of networks and to offer proper technological infrastructures that enable the creation of such networks and demonstrate their feasibility. In Chapter 2 we present a case study in this domain as this will be referred to and elaborated in most of the chapters in this book. Moreover, in Part V we show how the various `ArtDeco` aspects can be put together to build a comprehensive solution for the case study.

1.1 How to Read This Book

This book aims to be useful to people of widely different backgrounds, from people with no previous experience in networked enterprises to experienced researchers and executives. Depending on their own background, certain chapters may be more or less important to them. The following can be considered a "recommended reading list" for various types of readers:

- **Executives, Technology Managers and Marketing Managers of ICT Companies:** They can easily approach Chapter 2 in order to better understand the organizational problems of their customers, especially if they are developing a competitive strategy focused on the emerging market of IT services for SMEs.
- **Executives, CIOs and Marketing Managers of Manufacturing Companies:** They may want to explore in depth how the evolution of organizational models and information and communication technologies could help them in designing radically new business models. They may want to discover how this evolution could help them in improving specific business processes along their value chain (e.g., upstream key processes in material management or downstream key processes in sales management) They can easily approach all chapters of Part I that focus on the organizational aspects of networked enterprises, in particular, for what concerns the process of conceiving a new product. They might also be interested in the power of data management techniques and read the chapters of Part III. If they are in the wine industry, they could be interested in getting an idea of the capabilities of small sensing devices available for their industry (Part IV). Finally, if they are interested in seeing how the overall ArtDeco approach works, they can read Part V.
- **Researchers on Management, especially on Innovation Management, Organization, Entrepreneurship:** They can approach all the chapters of Part 1 in order to have a general overview on emerging approaches to innovation management in single companies, especially SMEs involved in networked enterprises. If they are interested in identifying specific open problems for further empirical researches comparing different collaborative strategies (e.g. open vs. closed) that are available for a single company or a single networked enterprise, they can read the Chapter 2. If they are interested in identifying a research agenda for further empirical researches comparing different new product development strategies (e.g. design driven approaches vs. market pull approaches), they can read the Chapter 3. If they are interested in a policy-maker perspective, especially concerning SMEs networks, they can read the Chapter 4. Finally they can receive many insights from the integration between organizational issues described in the Part 1 and emerging technological issues resulting from the Part 2, especially reading the Chapter 6 and the Chapter 7. The final Part 5 on the application of the ArtDeco approach to the winemaking industry may offer specific insights on the technological constraints/opportunities for the management of business processes within networked enterprises in many different industries.
- **Researchers on Software Engineering, especially on design and development of adaptable software:** In this book they can find an analysis of the state-of-the-art on software engineering approaches for designing, developing and verifying adaptive large-scale software systems. In particular, they can focus on Part II for an in depth analysis of existing solutions and for the identification of the most important concerns related to the development of adaptive software systems deployed across networked enterprises. In particular, Chapter 6 introduces the most relevant problems of autonomic computing, the impact

of this approach on software engineering and a distributed model for promoting self-composition of pieces of code (capabilities) to satisfy emerging needs in networked systems. Chapter 7 addresses the problem of programming large-scale networked systems by leveraging service oriented architecture and business processes, and taking into account the unpredictable events that influence the execution context of long-running processes that span across the boundaries of several organizations. Finally, Chapter 8 proposes several techniques to verify the correctness of adaptive and autonomic software systems and processes, starting from traditional approaches based on testing and considering more dynamic ones centered on monitoring. They will also find useful to read the chapters of Part I as they give the business context within which the approaches in Part II have to be applied. FInally, Part III and Part IV will be particularly important to understand all problems related to data acquisition and management.

- **Researchers on Database and Knowledge Management:** In this book they will find a systematic framework where several state-of-the-art database and knowledge base researches are applied in an integrated way to the solution of the most cogent contemporary problems of a networked enterprise: information discovery, on-the-fly extraction of knowledge from information sources, dynamic interpretation and integration of the acquired information, analysis and dissemination of such knowledge to all decision levels, information personalization based on the user's function and context. Not only such techniques are integrated among themselves in Part 3, but the other parts provide insight into the various ways in which the information needed to build knowledge can be delivered to the knowledge management framework, and be exploited by the same mechanisms adopted for the more traditional data sources, like internal databases or data warehouses. For example, Chapters 17 and 18 illustrate two different ways to gather data from a wireless sensor network. These data will contribute to the overall information on the status of the enterprise by being included within the framework by means of the techniques illustrated in Chapters 15 and 19.

- **Researchers on Wireless Sensor Networks and Pervasive Sensing Technologies:** This book shows how wireless sensor networks and, more generally, pervasive sensing technologies can help to create the abstraction of networked enterprise in a specific domain (i.e., the wine business domain). Therefore, it allows getting a comprehensive view of problems that need to be solved when using pervasive sensing technologies in a specific practical scenario. Some possible solutions are also proposed. For instance, Chapters 15 and 15 address the problem of energy management in sensor networks for vineyard monitoring and propose solutions based on energy harvesting and energy conservation, respectively. Chapters 17 and 18 present two alternative approaches for data extraction from a wireless sensor network. Finally, a system based on RFID for the traceability of the entire wine supply chain (i.e., from vineyard to retailers) is described in Chapter 19.

Chapter 2
Reference Case Study

Giuseppe Anastasi, Emilio Bellini, Elisabetta Di Nitto, Carlo Ghezzi,
Letizia Tanca, and Eugenio Zimeo

Italian winemaking industry provides a fruitful context for the empirical research on technological and organizational issues concerning the networked enterprise.

The structure of Italian economy is mainly based on networks of Small to Medium-sized Enterprises (SMEs), as the result of a specific historical process. Often these networks have assumed peculiar organizational forms (e.g. Italian industrial districts) inspiring new approaches to the management of inter-firms flows of data and resources.

The specific nature of SMEs, as incomplete organizational entities (e.g. limited investments in information systems, absence of internal technical staff units), forces SMEs entrepreneurs to strengthen relationships with external actors, especially in order to share complementary assets, like trade channels or R&D efforts.

Nevertheless, the winemaking industry shows some peculiar features of typical problems in the networked enterprises, since the complex relationships between all the actors of the supply chain concern not only virtual resources (e.g. marketing services), but also physical resources resulting from the interactions between

Giuseppe Anastasi
Dept. of Information Eng., Univ. of Pisa, Italy
e-mail: g.anastasi@iet.unipi.it

Emilio Bellini · Eugenio Zimeo
University of Sannio - Department of Engineering
e-mail: zimeo@unisannio.it

Emilio Bellini
Politecnico di Milano - Dipartimento di Ingegneria Gestionale - Piazza L. da Vinci, 32 20133 Milano, Italy
e-mail: emilio.bellini@polimi.it

Elisabetta Di Nitto · Carlo Ghezzi · Letizia Tanca
Politecnico di Milano -Dipartimento di Elettronica e Informazione - Politecnico di Milano, Piazza L. da Vinci 32, 20133 Milano, Italy
e-mail: {dinitto,ghezzi,tanca}@elet.polimi.it

G. Anastasi et al. (Eds.): Networked Enterprises, LNCS 7200, pp. 7–13, 2012.

agricultural activities, manufacturing processes and customer relationship management services.

In this chapter we introduce the case study that has been used as a reference to instantiate the ArtDeco platform. It is representative of the typical situation of SMEs in Italy and focuses, in particular, in one of the most relevant sectors in the country, the wine industry. Wine is typically a product with a slow innovation cycle, but it can rely on technological innovations in a number of areas ranging from grapes cultivation and harvesting to wine production, to logistic. Our case has been inspired by a number of discussions we had with Donnafugata (http://www.donnafugata.it) and other companies both in Sicily and in Tuscany. Here we describe the main issues to be addressed in the various phases of cultivation, production, delivery and sale, while in the following chapters we provide an overview of the solutions we propose to address them.

2.1 The GialloRosso Winery

The GialloRosso winery is a hypothetical winery that uses IT technologies in order to increase winemaking variety and quality. It interacts with other companies that provide materials, transportation, sensors and storage and operates in a location where other wineries focusing on similar products are located. GialloRosso cultivates grapevines from some vineyards, but it also purchases some certified quality grapevines. It tries to minimize synthetic chemicals or fertilisers, and principally uses bio-diesel in vineyard tractors, and bluebirds for controlling insects. The strength, colour, and flavour of the wine are controlled during the fermentation process. Its red wines age from seven to ten years before being sold. The quality of wine depends on proper aging, the quality of grapes, when the grapes are picked, the care for handling, and the fermentation process.

In order to manage these critical factors, wineries, delivery companies, retailers, bulk wine distributors, finished goods distributors, exporters demand for efficient and effective information exchange. Effective and well-timed partners' coordination is based upon the accuracy of the information about the products realised by the various supply chain partners.

The main business processes in the company are shown in Figure 2.1. We consider some operators - agronomist, oenologist, quality manager, ... - that use the techniques and the software systems proposed by the ArtDeco project to control and improve wine quality and sales.

Wine is a very delicate product. Temperature alteration, shaking, delays in delivering are all problems that may cause an alteration of the perceived wine quality. Evaluating several distributors, and measuring their quality of services, preserving the data and using them in subsequent deliveries is useful to improve the quality of the process and the perceived quality by final customers. Finally, in order to complete the process life cycle, feedbacks from final customers and from the market trends should be collected. According to the sense and response paradigm such

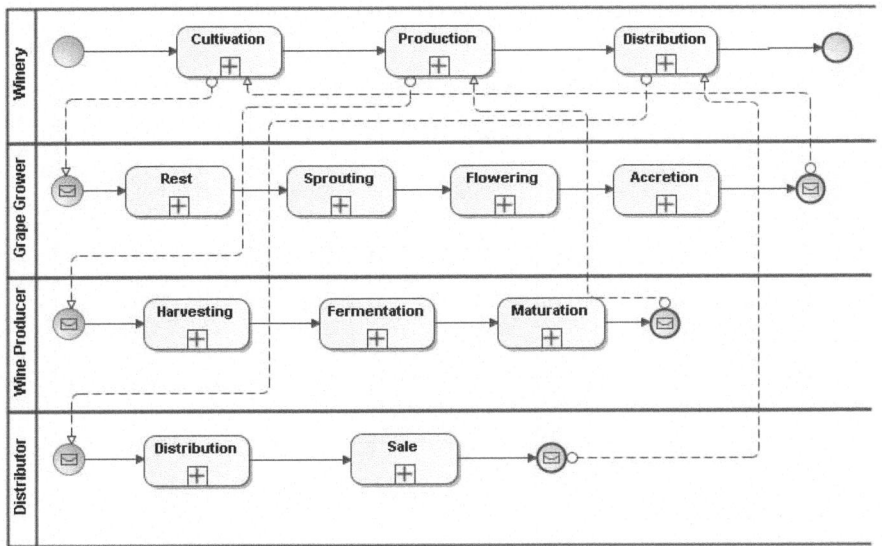

Fig. 2.1 GialloRosso main processes

information is used together with high level objectives in order to improve the business strategy. Feedbacks, in fact, may be used to plan actions for future productions.

In the following sections, each phase of wine production is described in detail to reach a formalized view of data, processes, and sensor parameters.

2.2 Cultivation Process

The agronomist and the oenologist operators handle N vineyards of a winery. Handling means monitoring some significant functional parameters, such as temperature, humidity, light, wind speed, wind direction, etc., in specific months of the year. The monitoring of vineyard provides real-time information about the health of grapes and so an estimation of the quality of the wine produced from that vineyard.

Some possible critical conditions during monitoring could be: (1) 'frost has destroyed all or part of the vineyard'; (2) 'the estimated quality Qs from the monitoring activity seems to be too much different from the quality Q desired from customers'. In both cases, the enterprise can solve the problem by buying from other producers an amount of grapes automatically suggested by the monitoring system by using appropriate rules able to perform specific queries to the internal database.

Vineyard monitoring can be performed either from the process manager of a workflow engine or from a sensor network system, that monitors each useful parameter of a specific zone Z of the vineyard VT, for example utilizing a query layer that offers the ability of programming events generation to deploy directly into the

Table 2.1 Correlation between field parameters and critical conditions of grapes

Adversity	Phenological phase	Temporal interval	Parameters	Critical threshold
Frost	Vegetative rest	October-January	Temperature(T)	$T < 3^oC$
Frost	Green Tips	February	T, wind speed(w)	$T < 5^oC$ $w < 0,5$ m/s
Frost	Sprouting	March	T, w, light(L)	$T < 5^oC$ $w < 0,5$ m/s $L < 100$ lux
Hoarfrost	Expanded leaves	March	T, w, L, Humidity(Hr)	$T < 5^oC$ $w < 0.5$ m/s $L < 250$ lux $Hr > 75\%$
Hoarfrost	Visible bunches	March	T, Hr	$T < 5^oC$ $Hr > 75\%$
Powdery mildew, Downy mildew	Separated bunches	April	T, Hr, w	$T > 30^oC$ $Hr > 75\%$ $w > 9$m/s
Powdery mildew, Downy mildew	Flowering	May	T, Hr, w	$T > 35^oC$ $Hr > 75\%$ $w > 9$ m/s
Powdery mildew, Downy mildew	Berries accretions	June	T, Hr, w	$T > 35^oC$ $Hr > 75\%$ $w > 9$ m/s
Powdery mildew, Downy mildew	Bunches closure	June	T, Hr, w	$T > 35^oC$ $Hr > 75\%$ $w > 9$ m/s
Powdery mildew, Downy mildew	Veraison	June	T, Hr, w	$T > 35^oC$ $Hr > 75\%$ $w > 9$m/s
Powdery mildew, Downy Mildew	Maturation	August-September	T, Hr, w	$T > 35^oC$ $Hr > 75\%$ $w > 9$ m/s

external business environment. In both cases, context variations, mainly in terms of different periods of the year and times of the day, may influence the type of queries that are issued to the sensors.

The identification of particular conditions, of the vineyard or of the vines, has to be handled through the execution of appropriate actions (directly programmed or inferred from a knowledge base). Examples of actions are simple notifications to the agronomist or the oenologist (e.g. SMS notifications), direct interventions on the vineyard (e.g. prolonged irrigations), or execution of more complex processes (e.g. sending a purchase order to an another wine producer). Information about vine species and their specific peculiarities, in terms of their combinations for wine design, is also needed by the oenologist, who, possibly together with the agronomist, is in charge of studying possible relocation of vine species, based on their fitness with the soil conditions.

Table 2.1 shows some relationships among functional parameters identifying critical conditions in the vineyards and possible management actions. It is worth to note that during process execution, in addition to domain-dependent unexpected events (derived from the 'sense-and-respond' business model) also domain-independent anomalies could occur. These are mainly related to the execution infrastructure, such as wireless sensor networks failures. Moreover, dynamic changing of the business environment impacts also onto the execution infrastructure in terms of its reconfiguration: sensor networks could be dynamically reconfigured on the basis of the parameters to monitor with the objective of reducing both energy consumption and data traffic toward the network sinks. The reconfiguration can be suggested by programmed actions in the main workflow or by launching the action of some rules.

2.3 Harvesting, Fermentation and Maturation

The harvesting phase is critical for the wine production process. One important constraint is the necessity to minimize the time between the harvesting and the processing of the grapes (a typical goal is to start processing within one hour). Also, it is important to evaluate the climatic conditions at the time of the harvesting. For some special productions, harvesting is performed at night, in order to take advantage of the lower temperature. In all cases, the trucks used for transportation of the grapes must be thermally isolated.

During fermentation, the wine is periodically subject to chemical analysis to control the evolution. Temperature and humidity of the wine cell are constantly monitored, whereas the analysis of the other parameters is performed in the laboratory by a machine.

Among the many monitored parameters there are the Acetic Acid and the presence of dangerous bacteria that can damage the entire production. In the future, it is envisioned to monitor frequently (i.e. more than once a day) such parameters 'in loco' instead that in a separate laboratory as many of the monitored parameters may be influenced by external factors.

Sensors are also distributed throughout the cellars, where humidity, temperature, light and wind speed must be recorded. The oenologist and the quality manager are interested in such information, which is useful in order to monitor the quality of the produced wine. In the harvesting and transportation phases, it is important to monitor humidity and temperature, and to store the observed data in the database together with the identifier of the phase (e.g. 'harvesting'). An appropriate combination of RFIDs and data loggers on site could solve the problem.

During the fermentation phase, it is important to keep a record of the 'harvesting' identifier in order to be able to track quality dependencies. In addition, by using a combination of WSNs with appropriate sensors, the temporal evolution of the fermentation process can be recorded. It is necessary to signal events that require manual interventions, like presence of dangerous bacteria or percentage of acetic acid outside the limits.

2.4 Distribution

Wine distribution phase includes packing and distributing wine cartons to wholesalers or retailers. The packer is responsible for the receipt, storage, processing, sampling, analysis, packing, dispatch of finished goods, and recording information about what is received and is dispatched. The final distributor is responsible for the receipt of wine bottles, record information about what is received, storage of wine, re-packing and re-labelling, quarantining products, record information about what is received and what is dispatched, and dispatching wine bottles to wholesalers or retailers. Wholesaling is defined as the resale (sale without transformation) of wine bottles to retailers, to industrial, commercial, institutional, or other professional business users, or to other wholesalers.

In general, it is the sale of goods to anyone other than a standard consumer. Wholesalers physically assemble, sort and grade wine bottles in cartons and redistribute them. The wholesaler receipts finished goods and sends cartons, while the Retailer receipts cartons, picks and dispatches them and record information about what is dispatched.

In the Giallorosso winery, the aspect of wine distributing after bottling is examined with respect to humidity and temperature. Wine, in fact, should not experience temperature fluctuations higher than 5 degrees, and it has not to suffer humidity fluctuations higher than 5. These two attributes are used in this case study to substantially influence choice of suppliers made by the winery (main contractor). These parameters, in fact, are measured and exploited for evaluating the quality (through the reputation parameter) of the previous subcontractors' performances. Temperature and humidity information of the package during the distribution time are so recorded in data loggers for subsequent analyses. Measurements are obtained using RFIDs, and a query layer for non functional requirements and pervasive infrastructures has been introduced.

2.5 Sale

The analysis of wine sales helps to deriving information about the success of a wine. The quality manager is interested in the returns in order to compare them with the conditions of the product during all the phases of the lifecycle. The collected information is important to adopt the Sense and Respond model. In fact the model is not aimed to anticipate market but it is settled to quickly react to change and to evolve in order to rapidly adapt the business processes and the activities to the altered situations. This implies receiving events and information that have to be compared with policies and stored data, and using them to alter the business processes. In addition to the higher-level flow of information there is another one deriving from the context. It allows for achieving the required results with the desired quality. These two flows of information can be then used for performing adapting interventions and for design more refined business processes that adapt themselves to the altered conditions of the market.

2.6 Remark

The case study we have presented above results from an elaboration of discussions we had with wine producers, and reflect their point of view that often disregards the presence of other similar companies in their same area of operation. In the following of the book we will show how the ArtDeco approach can address the requirements emerging from this case not only by considering GialloRosso as an isolated company, but also considering the network that it could potentially build with its peers and its providers and suppliers located in neighbor geographical areas.

Part I
Organizational Issues: Methodologies, Empirical Contexts and Policies

Coherently with the aim of this volume, in the first part of the volume we assume a managerial perspective deriving from the joint analysis of organizational challenges and of technological opportunities for networked enterprises acting in wine industry. Since the management literature has consolidated many models on inter-firms networks involved in joint functional activities (e.g supply chain, manufacturing, marketing), the three chapters face, from different points of view, the specific case of networked enterprises involved in the inter-functional process of innovation management. More in detail the authors move, in a progressive way, from a general view on the organizational issues affecting networked enterprises involed in new product development, to the possible strategic options, to the implications for policymakers involved in promoting the adoption of advanced network infrastructures by Small and Medium Enterprises (SMEs).

In the Chapter 3 the author faces the crucial problem of management of organizational boundaries in the new product development process, providing a first definition of Networked Enterprise in Wine Industry (NEWI). The chapter proposes an application to the NEWI of the collaborative innovation model model developed by Pisano & Verganti identifying four collaboration strategies based on different combinations of openness and flatness. This application allows to identify, for each collaboration strategy, some specific organizational challenges and some requirements for balancing orchestration and choreography in the development of an autonomic system.

In the Chapter 4 the authors face the crucial problem of innovation strategy in networked Enterprises, providing a detailed analysis of different options within three different approaches: user-centered, technology push and design-push. The chapter allows to understand the organizational implications deriving from the complex nature of a new product (e.g. a new wine label), that is not only a set of technical functions (e.g. organoleptic features of grapes and wines), but also a set of messages and meanings shaped by the evolution of socio-cultural systems (e.g. new meaning of wine as emotional experience).

In the Chapter 5 the authors focus the role of SMEs in networked enterprises, since their strong role in wine industry (e.g. small wine makers, service SMEs as oenologists and agronomists, small farmers, retail SMEs as restaurants). The authors propose the application of a platform strategy in order to promote the adoption of advanced network infrastructures by SMEs, analyzing the incentives of industrial players in a context where SMEs playing an relevant economic role and where traditional rather than high-tech sectors of activity are predominant with specific reference to the Italian case. The chapter allows to review governance aspect which are paramount to platform functioning given the policy context.

Chapter 3
Which Collaboration Strategy
for the Networked Enterprise in Wine Industry?
Technological and Organizational Challenges

Emilio Bellini

Abstract. The technological dimension and the organizational dimension are the two faces of Information Technology (IT) revolution shaping the life of large and small firms in last 50 years, from the first adoption of mainframe and reusable software to the recent integration of hardware and software technologies within Internet of Things. Over the last few decades, a number of scholars coming from computer science and from organization theory shared a deep confidence on the magic power of networks, especially of technological networks, more and more fast and reliable, and of organizational networks, more and more effective in balancing cooperation and competition among firms. Coherently with the aim of this volume, in this chapter we assume a more critical perspective deriving from the joint analysis of organizational challenges and of technological opportunities for networked enterprises aimed at developing new products/services. The management of innovation within networked enterprises requires a strategic approach to many dimensions. In this chapter we apply at open-closed trade-off the model developed by Pisano & Verganti in the paper Which kind of Collaboration is Right for You published by Harvard Business Review in December 2008. In the Chapter 4 we will focus the strategic models aimed at managing the new product development within the networked enterprise modelled as a design discourse. In the Chapter 5 authors will propose a third strategic perspective, proposing the application of a platform strategy in order to promote the a doption of advanced network infrastructures by Small and Medium Enterprises (SMEs).

3.1 Introduction

This technological dimension and the organizational dimension are the two faces of Information Technology (IT) revolution shaping the life of large and small firms in

Emilio Bellini
University of Sannio - Department of Engineering and Politecnico di Milano -
Department of Management, Economics and Industrial Engineering Piazza L. da Vinci,
32 20133 Milano, Italy
e-mail: emilio.bellini@polimi.it

G. Anastasi et al. (Eds.): Networked Enterprises, LNCS 7200, pp. 17–30, 2012.

last 50 years, from the first adoption of mainframe and reusable software to the recent integration of hardware and software technologies within the Internet of Things era. Over the last few decades, a number of scholars coming from computer science and from organization theory have shared a deep confidence on the "magic" power of networks, especially of technological networks, more and more fast and reliable, and of organizational networks, more and more effective in balancing cooperation and competition among firms.

The global financial crisis of 2007-2010 is freezing many easy and oversimplified enthusiasms about this network miracle both in ICT industry, that showed in 2009 a decline of -5,8% in worldwide sales, and in single firms, who are facing deep revisions of strategies and financial plans. Managers and entrepreneurs are forced to improve their network governance, since they are experiencing that a dogmatic and axiomatic adoption of open organizational forms, such as communities, supply networks, virtual enterprises, cannot always bring to a successful conclusion.

In this chapter we focus on the network governance problem from the perspective of the *networked enterprise*, defined as the evolution of virtual organizations and extended enterprises, as an integrated set of manufacturers, suppliers, business partners, and even customers, where relationships, virtual ties, flows of information and knowledge creation processes play a pivotal role [29, 36, 33].

Within this model, we face a first key question about the governance of a Networked Enterprose in Wine Industry (NEWI): *to what extent the involvement in a networked enterprise is suitable for long term firms' performances?*

To answer this question, we focus on a specific strategic process, the new products development (NPD), where the interactions among different actors of the networked enterprise are characterized by high level of complexity. These interactions usually concern acquiring information (e.g. about the evolution of technologies and of customers needs), creating knowledge (e.g. technical problem solving), managing dynamic configurations both of internal intra-organizational processes (e.g. integration of manufacturing and marketing capabilities) and external inter-organizational processes (e.g. integration of upstream knowledge about raw materials and downstream knowledge about new retail channels). Their complexity calls for a deep comprehension of relations between information technologies and organizational (intra- and inter-organizational) contexts.

Following the structurational model of technology proposed by Orlikowski [22, 23], we reject both the "technological imperative", where technology is an objective external force that would have deterministic impacts on organizational properties, and the "organizational imperative", where technology is a second-order variable, which depends on the outcome of strategic choices and social actions. In the structurational model, technology is neither a totally external variable, out of control for managers, neither an internal resource, fully under the control for managers who could shape it coherently with their strategic goals. Thus, we focus on how the *technologies-in-practice* (rules and resources instantiated in use of technology) are influenced, and at the same time, could influence the ongoing situated use of technology, in terms of facilities (e.g. hardware, software), norms (e.g. protocols, etiquette), interpretive schemes (e.g. assumptions knowledge).

This technological dimension inspires a second key question about the governance of a NEWI involved in a NPD process: *how could IT improve the capabilities in acquiring information, and in creating knowledge in a hyper-dynamic context where both the external environment (e.g. consumer needs, macroeconomic influences, R&D evolutions) and the internal environment (e.g. number and characteristics of firms belonging to the networked enterprise) are constantly changing?*

In order to face both the key questions we have adapted, to the networked enterprise involved in new product development processes, the *collaborative innovation model* proposed by Pisano and Verganti ([24]) who distinguished four collaboration options for companies involved in innovation strategies developed with potential innovation partners.

In the next section we discuss the crucial issue of organizational boundaries in innovation management. In the third and fourth sections we discuss the two key dimensions of the Pisano and Verganti model: 1) the openness (can anyone participate to the networked enterprise, or just selected players?); 2) the flatness (who makes key decisions about, one kingpin participant or all players?). In the fifth and sixth sections we derive specific insights about contingent circumstances of each strategic option (when to use each of the four options?), organizational challenges (what are the trade-off of each option?) and technological opportunities (what are possible technical requirements in order to develop an Autonomic System supporting networked enterprise?).

3.2 Boundaries in the New Product Development

The management of boundaries is crucial for the success of innovation strategies, especially in the current economy, where each firm needs a broad scope, distributing activities in many countries, and integrating relationships with many partners, customers and communities.

Moving from the seminal work of Williamson on transaction cost economics [39], the enduring debate on firms boundaries has driven the management scholars to refer constantly to some metaphors of indistinct boundaries, as in the cases of "the doughnut principle", where, in an overturned logic, the center of the doughnut is the unmodifiable substance, while the extern assures flexibility and balancing citeHandy1992 of *blur economy* [11], of *fuzzy boundary* [28], and of *changing boundaries* [10].

The exceptional resonance of *open innovation paradigm* is confirming that successful firms are able to develop higher capabilities in integrating external and internal knowledge, technological and marketing competences, improvements in R&D results and in business models design ([6, 7]). Many reasons are being identified in order to explain this phenomenon. For example, globalization [35], technological intensity [21], converging technologies [20], new business models and knowledge leveraging [15]. Indeed these forces are shifting the locus of innovation from individual large firms to networks of learning [25]. Exploration and exploitation activities are now to be performed by collaborating with a series of external subjects in order to benefit from the external generated knowledge.

In order to analyze the problem of networked enterprise governance we need to overcome the oversimplified view of *openness* as dogmatic solution for the innovation problem. As showed by several empirical cases, new product development strategies could be successful, also in *closed* networks. As shown by several cases and empirical data, successful new product development can derive both from open and from closed approaches, where a small number of actors share information and decisions about idea, concept, design, development, test, and commercialization.

These successes can be the effect of different strategies, like the *technology-push* approach, in which the success of new product development depends on the internal R&D capabilities of the firm to enhance product functionality and offer new ways of satisfying customers [1, 34, 9], or the *user-centered* approach, shaped by the dominant role played by the comprehension of market needs on the introduction of new technologies [31, 30, 32], or the *design-driven* approach where radical innovation results from integration between new technologies and new product concepts ([26, 27, 4]).

At the same time many firms are experiencing that the adoption of open organizational forms for managing innovation processes could be very risky both for huge time to market, and for low motivation of best innovators in sharing their ideas and solutions within broadest communities.

The organizational theory allows these problems to be framed within the evolution of three basic organizational forms for coordinating knowledge-based processes; Adler [2] argues that each institution or set of networked enterprises involved in new product development processes, is a combination of three ideal-typical organizational forms, corresponding to three coordination mechanisms:

- *Hierarchy*, which allows the coordination through the mechanism of authority, especially through forms of legitimate power attributed to an actor that plans, controls, and distributes the division of activities between different organizations, and different members of teams. Within networked enterprises, good examples of projects where hierarchy is higher than market and community are shown by new product development in automotive industry where the Original Equipment Manufacturer is the formal responsible of heading, coordinating, and of financial controlling).
- *Market*, which allows the coordination through the mechanisms of prices and of contracts, namely, through different forms of exchange/competition among different actors who agree about the economic value of transaction, finding an equilibrium between conflicting interests. Within networked enterprises, good examples of projects where market is prevalent, are showed by new product development in aerospace industry where formal agreements among many manufacturers detail all the flows and exchanges of information, services and money.
- *Community*, which allows the coordination through the mechanism of trust, namely through cooperation forms where each of actor evaluates that other actors fulfill their obligations, even without a direct control or a direct sanction. Within networked enterprises, good examples of projects where community is

higher than hierarchy and trust, are showed by new product development in software industry, where informal and tacit processes allow a fast integration between deliverables carried out by remote and often unknown partners.

It is evident that as tacit knowledge becomes increasingly relevant, we expect higher relevance of the community organizational form. Consequently a successful strategy of openness within networked enterprise involved in a complex knowledge-based process like NPD, depends on the right level of trust between partners, defined as the subjective probability with which actors assess that other actors or groups of actors will perform a particular action, both before they can monitor such action and in a context in which this action affects their own actions [14].

3.3 The First Dimension of "Participation" in the Pisano and Verganti Model: To What Extent Is the Open Participation Suited to a Successful Networked Enterprise?

Moving from the key relevance of trust between partners involved in the innovation processes, Pisano and Verganti ([24]) criticize the fervor around open models of collaboration, describing a number of cases where the success/failure of innovation strategies depends on the right level of openness/closeness in collaborating with external partners.

The authors propose a model of collaborative innovation for choosing the right type of collaboration options within different technological and competitive contexts. They define as *participation* the first dimension of the model, distinguishing between *open mode* and *closed mode* of collaboration.

In the open mode anyone can participate to the network; for example, in the case of a networked enterprise of wine industry involved in a new product development (see Figure 3.1) each of the following partners can freely enter the network giving its own contribution:

- R&D providers, like universities or R&D departments of vineyard equipment suppliers, can propose new technical problems or offer solutions about biological and chemical issues.
- Designers and technical service providers, like oenologists or sommeliers, can propose new wine products or new wine-making processes.
- Raw materials suppliers, like grape-producing or farmer cooperatives, can propose patents for new agricultural processes, or new typologies of grapes, or new industrial uses of vintage grapes.
- Manufacturers, like different wine making companies competing on the same or on different market segments, can propose new wines for emergent markets based on the integration between results of customer needs analysis and results of socio-cultural analysis about impact of regional tastes on wine consumption.

- Retail channels and marketing service providers, like traditional wine importers or innovative food and beverage B2B/B2C e-commerce companies, can propose new business models for launching new products or new services in the wine industry.
- Consumers or lead users, like gourmets, wine lovers and wine experts or restaurants head chefs, can propose insights about changes of tastes and consumers behaviours.

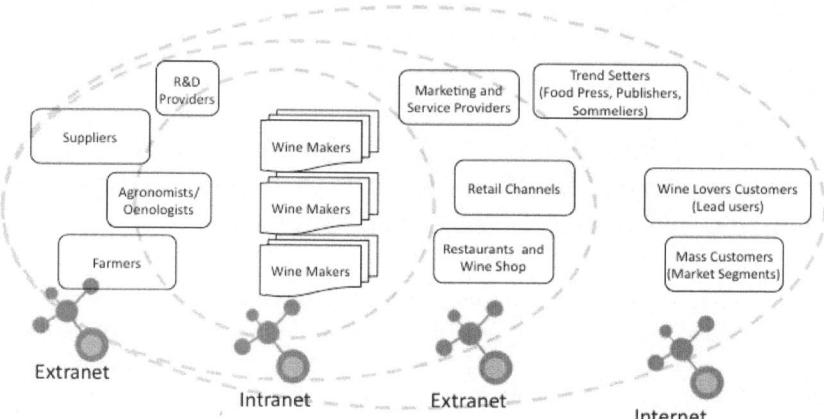

Fig. 3.1 The Networked Enterprise in Wine Industry: Actors involved in New Prouct Development

Clearly, in the closed mode just selected players can be part of the network, for example, reproducing previous trade alliances between a few number of oenologists, wine makers, retail channels, restaurants, where face to face relationships enable a more direct coordination.

Following the Pisano & Verganti model, we expect that the advantage of the open mode is that the networked enterprise could attract a wide range of possible ideas from domains beyond the experience of the single wine maker company or the single oenologist, who are the typical promoters of initial steps in new product development process. When the networked enterprise is designed and managed in the open mode, the community of participants benefits from a very large number of proponents and of sponsors, assuming a potential leadership in shaping trends and future directions of innovation paths. Finally the open networked enterprise could benefit from contributors by unknown emergent actors; for example innovative farmers, young oenologists, wine lovers coming from emergent markets who are nurturing lateral, creative solutions and updated visions about both technologies and consumer needs, that are invisible for well-established traditional market leaders.

Since the advantages of the open mode rely on quantitative effects of the reception of all possible contributors, some disadvantages arise from qualitative side effects. Often the open mode could be perceived as too popular or not much professional from the best players who could be detracted from participating in a networked enterprise where whoever could participate, often in anonymous way, like for example in a innovation initiative in wine industry launched by a forum or a social network.

A second disadvantage is associated with the high costs of searching for, screening and selecting proposals and solutions developed within the open networked enterprise, since we expect to receive a huge number of contributors affected by a low level of homogeneity in terms of focus on the specific domains and in terms of technical expertise.

Finally a third disadvantage depends on the long time required for attracting, interpreting, sharing and using the mass of information and knowledge developed within the open networked enterprise; while you finalize the previous proposals, the context could be radically changed producing new and more effective ideas and solutions. In this case the key problem is due to the nature of innovation processes where technologies and customers needs could evolve in unpredictable ways, and to the dynamic and oscillating behavior of the involved actors, and of the links between them. In other terms an open networked enterprise runs the risk of attracting a lot of ideas and proposals but no solutions to the innovation problem, with a high cost of coordination and selection.

We can make similar but opposite considerations in the case of closed mode networks; in this case, the networked enterprise attracts a low number of ideas and proposals, with a low level of heterogeneity. The key advantage is the high likelihood of attracting the best solutions from a selected knowledge domain, where best players are very interested to participate. Vice-versa the key disadvantage is associated with the verticality of this mode that can produce information and knowledge too much contextualized that cannot be used out of the small group of experts and/or of customers with similar level of expertise, same languages, shared values and common behaviours.

3.4 The Second Dimension of "Governance" in the Pisano and Verganti Model: To What Extent Is Flatness Suited to a Successful Networked Enterprise?

Many contributions on organizational networks and on innovation management have focused on the macro-problem of boundaries, studying how the openness to external relations could impact on long term performances of the single firm. Only a minority of researchers has focused on the micro-problem of political use of knowledge and information developed within networks, especially during decision-making processes about key questions arising from day by day interactions between partners [38, 18, 3].

Therefore, in the case of a networked enterprise in wine industry the problem is not only *how* to attract the right number and the right configuration of partners. A strategy for attaining the goal of a successful innovation requires also the definition of *who* makes decisions about questions arising during day by day activities, for example about initial ideas, creativity and R&D directions, technical choices, marketing options. Pisano & Verganti define as *governance* this second dimension in their model for collaborative innovation, distinguishing between flat and hierarchical mode of collaboration.

In the hierarchical mode one *kingpin* controls the direction and value of innovation; for example, in the case of a networked enterprise in wine industry involved in new product development, one wine maker company defines the initial direction in terms of viticulture methods, organoleptic features, tastes, marketing strategy, and it controls the coherency of contributions of other partners to these initial requirements.

Vice-versa, in the flat mode all players can participate in these decisions with different degrees of influence on the final outcome. Often, a negotiation between partners with the same formal level of power is required.

The inclusion of the governance dimension in the collaborative innovation model has a theoretical relevance, since it allows to highlight the managerial implications of many contributions concerning the political dimension of consensus decision-making in inter-organizational relationships about knowledge acquisition and management practices.

Various authors [5, 8, 19, 3] study the innovation not as a process of acquiring, building and transferring of knowledge as an object, but as a *knowing process*, as a social construction where the truth of the statements is driven not by a presumed representation of objective realities, but by an interpretation (therefore a construction) of practices developed by actors. More precisely, in the case of new product development the decisions about ideas, directions, concept, design, test, commercialization derive from the collective sense-making that drives the actors to connect the truth of knowledge statements to the consensus of a relevant community.

Indeed, as pointed out by Weick [37] in the seminal work on sense-making, the cycles of organizing are "consensually validated grammars of reducing equivocality by means of sensible interlocked behaviors". The peculiar nature of knowing processes developed within networked enterprise makes the decision-making processes pregnant of ambiguity, uncertainty and equivocality, arising from relationships between partners in implementing innovations, through the generation of new meanings. At the same time these new meanings could be connected to the generation of new technical solutions, such as incremental or radical improvements in technical functions (e.g. platforms, architectures, components) of products and services, or in business models and marketing methods.

The political construction realized by several actors involved in the networked enterprise influences the everyday power bargaining games, through several vehicles of social interaction (roles, symbols and rituals, cultural myths, institutions, rules for congruent action, self-fulfilling prophecies) constructing their sense making processes [12, 16, 17]. We argue that the capability of managers in influencing these collective knowing processes in the networked enterprise depends not only on conscious and deliberate plans about openness-closeness, but also on counter-intuitive mechanisms resulting from choices about flatness-hierarchy.

In the Pisano & Verganti model the main advantage of flat mode is the ability to share with others costs, risks, and technical challenges of innovating, since the decisions are either decentralized or made jointly by some or all the partners. The networked enterprise goes well in flat mode when no single player has the necessary bundle of creative vision, technical competences, marketing capabilities. Moreover, the networked enterprise can benefit from flat mode when all the players are interested in influencing some common choices that could become a long term standard for all the competitors. The main disadvantages of flat mode is the cost of continuous permanent negotiation between partners who need arrive at mutually beneficial solutions.

We can make similar but opposite considerations in the case of hierarchical mode; in this case the networked enterprise benefits of an efficient decision making process, since one kingpin player controls the direction and value of the innovation. The key advantage is the high likelihood of responding in fast way to all the proposals and insights coming from all the players, since the communication process is strongly structured, avoiding confusion between data gathering, information retrieval, social exchange, proposal posting, acceptance/rejection. Vice-versa the key disadvantage of a hierarchical networked enterprise is associated with the strong rigidity of this mode that can fail in capturing and using all the dynamic signals coming from informal relationships between the partners and from their genuine commitment in posting ideas and proposals.

3.5 Applying the Pisano and Verganti Model: Organizational Challenges and Technological Requirements from Four Collaborative Strategies in the Networked Enterprise

Using the two dimensions proposed by the Pisano & Verganti model, we can frame some specific organizational challenges for the networked enterprises in wine industry, deriving also some hypotheses about requirements for the technological systems and the methods developed in the other parts of this volume.

As mentioned above, our research questions focus on the strategic decisions about the networked enterprise involved in innovation activities, like new product development. Some of the sub-questions that arise are the following: who can participate in the network? How could an actor become partner of the network? Who can set questions, directions and solutions to the problems faced during the life of the network? How could IT improve the capabilities in acquiring information and in creating knowledge in a hyper-dynamic context where both external environment and internal environment are constantly changing?

The analysis of the literature underlying the two dimensions of participation and governance, has shown how the possible answers to these questions are associated to a contingent perspective. Each option has its own advantages and disadvantages, there is no one unique best way to organize the networked enterprise, and each way is not equally effective under all conditions [13].

Figure 3.2 shows the four ways to collaborate resulting from the application of the Pisano & Verganti model to the Networked Enterprise in Wine Industry (NEWI).

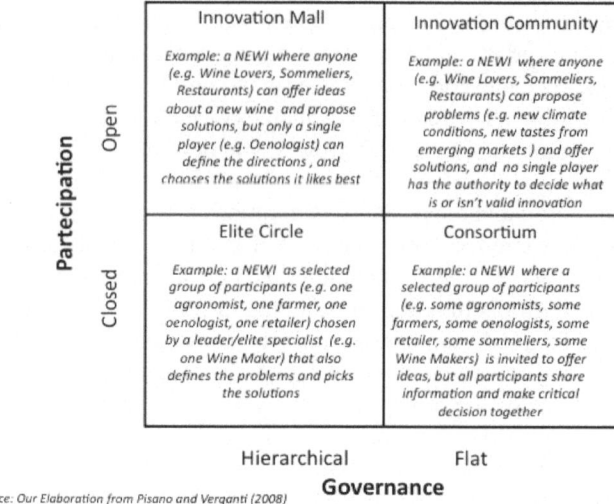

Source: Our Elaboration from Pisano and Verganti (2008)   **Governance**

Fig. 3.2 Applying Collaborative Innovation Model to the Networked Enterprise in Wine Industry (NEWI)

Pisano & Verganti define as Innovation Mall the collaborative strategy resulting from the adoption of open mode and hierarchical mode. On the organizational side this strategy implies a coordination mechanism based on a strong relevance of formal authority (recognized to the leader), a medium relevance of prices (resulting from negotiations about service exchange among partners) and a low relevance of trust (resulting from shared values among partners). This decision implies a good balancing between the interests of a leader (e.g. a wine maker company, leader in a specific market niche) and the interests of many other players (e.g. farmers, oenologists, retailers, restaurants) who accept the subjugation to the innovation directions defined by the leader, because they obtain both some short term advantages (e.g. they sell part of innovation services) and some long term advantages (e.g. they increase their reputation among their own customers, thanks to the qualified partnership with a big player).

The Innovation Community results from the adoption of open mode and flat mode. On the organizational side this strategy implies a coordination mechanism based on a strong relevance of trust, and a low relevance of formal authority and prices. In this case a networked enterprise could benefit from a strong vitality of relationships among a multitude of partners, who share values, curiosity, passion for new products and new services. In this case a large number of farmers, oenologists, winemakers, retailers, who usually compete on the same markets, could be interested both in the short term pleasant social experience, and in long term individual benefits coming from the exposition to qualified information, tacit knowledge, and new emerging technical standards.

The Consortium results from the adoption of closed mode and flat mode. On the organizational side this strategy implies a coordination mechanism based on a strong relevance of prices and of trust between few partners, and a low relevance of formal authority. In this case a networked enterprise could benefit from a good balancing between a control of accesses to the network and a strong integration of contributions coming from different perspectives. In this case a small number of farmers, oenologists, winemakers, retailers, who usually compete on the same markets, could be interested in combining cooperation and competition, realizing some specific joint initial part of an innovation program (e.g. R&D exploration).

The Elite Circle arises from the adoption of closed mode and hierarchical mode. On the organizational side this strategy implies a coordination mechanism based on a strong relevance of formal authority recognized to the leader, and of trust between few partners, with a lower relevance of prices. In this case a networked enterprise could benefit from strategic vision, creativity and organizational capabilities developed in the past by the leader; a big player (e.g. a wine maker or an oenologists who won awards in most prestigious wine competitions) could lead a small number of farmers, oenologists, winemakers, retailers toward radical innovations and new standards in the worldwide competition.

In a contingent perspective, there is no one best way to choose one of these four collaborative strategy, but we need to analyze the particular circumstances faced by the specific networked enterprise.

Dimension	Advantages	When to use	Trade Off	Requirements for Autonomic System
Open	• NE attracts a wide range of ideas and solutions from many perspectives and experiences	• Actors can evaluate proposed solutions cheaply • Problems could be partitioned into small well—defined chunks	• flexibility vs. stability of network configurations	• Dynamic configuration and re-configuration of firms in/out of the NE, and of interactions between them
Closed	•NE receives the best solutions from a select knowledge domain	• Actors know the correct knowledge domain and parties to draw on • Problems require fewer but "better" ideas	• technical depth vs. creative breadth of ideas and solutions	•Enlarging dynamic data gathering and information retrieval from a small number of qualified sources
Hierarchical	• Leaders control the direction and value of the innovation	• Actors Recognize a Leader with higher technical and marketing capabilities •Actors prefer to follow leader's directions /standards	• control vs. motivation of participants	• De-contextualizing and Contextualizing knowledge from high-level Actors to the domains of lower-level Actors
Flat	•Players share the costs, risks, and technical challenges of innovating	• No single player has the necessary breadth of perspective or capabilities •Actors are self-interested in influencing decisions /standards	• numerousness of participants vs. focus on shared benefits	• Reducing information overload in order to quicken collective decision making

Fig. 3.3 Organizational Challenges and Requirements for Autonomic Systems from the four Collaboration Options for Networked Enterprise

As shown in the Figure 3.3, each collaborative dimension is shaped by specific organizational challenges, while the fulfillment of specific requirements for the Autonomic System presented in the other chapters of this volume, could help in order to partially overcome these trade-offs. In the open mode the organizational challenge is associated with the trade-off *flexibility vs. stability* of network configurations. Indeed, the more contributors you attract, from different perspectives and knowledge domains, the more you are able to assume dynamic configurations in order to respond to unpredictable changes in the external environment (e.g. new laws affecting innovation in wine industry, radical results from R&D in biotechnologies, economic crisis in specific local markets, new cultural trends and consumer behaviours), and the more you need to manage instability of constantly changing configurations. An autonomic system can help managers in facing this organizational changes, offering advanced functions in order to manage both dynamic configuration and reconfiguration of firms that continuously enter/exit from the networked enterprise, and dynamic configuration of interactions and links between them.

In the closed mode the organizational challenge is associated with the trade-off *technical depth vs. creative breadth* of ideas and solutions proposed within the networked enterprise. Indeed, the more you receive vertical solutions coming from well-established actors proposing fewer, and, hopefully, better ideas, the more you are locked up within current paradigms (e.g. a new problem in harvesting grapes depending on recurrent weather instability, faced by old experts who exclude possibility of climate changes). An autonomic system can help managers in facing this problem by offering advanced functions to enlarge data gathering and information retrieval from a dynamic configuration of a small number of qualified sources.

In the hierarchical mode the organizational challenge is associated with the trade-off *control of processes vs. motivation of participants* in proposing ideas and creative solutions. Indeed, the more you obtain efficiency reducing times and costs of proposals acceptance-rejection, the more single participants could reduce their eagerness, or, worse, they could feel themselves less responsible for the final outcomes (e.g. a group of young innovative oenologists coming from different countries and cultures who are not able in building consensus about radical innovation that will be exploited some months after by other competitors). An autonomic system can help managers by offering advanced functions in order to help participants in contextualizing, de-contextualizing, and re-contextualizing knowledge from-in different domains.

Finally, in the flat mode the organizational challenge is associated with the trade-off *numerousness vs. focus* of the networked enterprise on shared benefits for all the players. Indeed, the more you enlarge the number of decision-makers coming from different positions in the supply chain, from different cultures, from different languages, the more you risk that the network turns around many ideas that are not clear to all the participants (e.g. a networked enterprise working on a new wine for the emerging market niche of aperitif, where participant are not able in selecting initial ideas). An autonomic system can help managers by offering advanced functions in order to reduce information overload and to quicken collective decision making.

3.6 Conclusions

The adoption of the Pisano & Verganti model to collaborative innovation strategies for the networked enterprise has helped us in answering to both the research questions about the relationships between long term performances and level of organizational involvement in networked enterprises, and about the role of IT in helping decision-making processes in different dynamic strategic contexts.

As suggested by the structurational approach, technology could enable new configurations of organizational problems, and, at the same time the analysis of organizational properties could inspire radical improvements in IT development.

It is not surprising that the four strategic options presented in the previous section can lead to different organizational challenges, and to different technical requirements for an advanced autonomic system. Indeed, the key contribution of these technologies is the improvement of all three ideal-typical forms of coordination within the networked enterprise. Advanced technologies could help managers both in negotiating transparent prices for coordinating exchanges of products and services, and in making the formal authority acceptable as an advantage for all the partners.

Above anything else, good technologies could improve trust between partners of the networked enterprise, because they share the same language and they are more and well-informed, reducing their fear about diversity of cultures, languages, technical background.

References

1. Abernathy, W., Clark, K.: Innovation: mapping the winds of creative destruction. Research Policy (14) (1985)
2. Adler, P.: Market, hierarchy, and trust: the knowledge economy and the future of capitalism. Organization Science (12), 215–234 (2001)
3. Bellini, E., Canonico, P.: Knowing communities in project driven organizations: Analysing the strategic impact of socially constructed hrm practices. International Journal of Project Management (IJPM) (26), 44–50 (2008)
4. Bellini, E., Dell' Era, C.: How can product semantics be embedded in product technologies? the case of the italian wine industry. International Journal of Innovation Management (IJIM) (3) (2009)
5. Blackler, F.: Epliogue: Knowledge, knowledge work and organizations. The Strategic Management of Intellectual Capital and Organizational Knowledge (2002)
6. Chesbrough, H.: Open innovation: the new imperative for creating and profiting from technology. Harward Business School Press, Boston (2003)
7. Chesbrough, H.W., Vanhaverbeke, W., West, J.: Open innovation: Researching a new paradigm. Oxford University Press, Oxford (2006)
8. Choo, C., Bontis, N. (eds.): Sense making, knowledge creation and decision making: organizational knowing and emergent strategy. Oxford University Press, Oxford (2002)
9. Christensen, C., Rosembloom, R.: Explaining the attacker's advantage: technological paradigms, organizational dynamics and the value network. Research Policy (24), 233–257 (2006)

10. Colombo, M.G.: The changing boundaries of the firm (1998)
11. Davies, S., Meyer, C.: Blur: The speed of change in the connected economy (1998)
12. Friedberg, E.: Le pouvoir et la régle. Le Seuil, Paris (1993)
13. Galbraith, J.: Designing complex organizations. Addison-Wesley Publishing Co, Reading (1973)
14. Gambetta, D.: Trust: Making and breaking cooperative relations (1988)
15. Gassman, O.: Opening up the innovation process: towards an agenda. R&D Management 36(3) (2006)
16. Giddens, A.: The constitution of society: Outline of the theory of structure. University of California Press, Berkeley (1984)
17. Goffmann, E.: Frame analyses. Harper and Row, New York (1984)
18. Hertog, J., Huizenga, E.: The knowledge enterprise: Implementation of intelligent business strategies (2000)
19. Itami, H., Numagami, T.: Knowing in practice: Enacting a collective capability in distributed organizing. Organizational Science (3), 119–136 (2002)
20. Kodama, F.: Technology fusion and the new r&d. Harvard Business Review (1992)
21. Miotti, L., Sachwald, F.: Co-operative r&d: Why and with whom? an integrated framework of analysis. Research Policy 32(8) (2003)
22. Orlikowski, W.: The duality of technology: Rethinking the concept of technology in organizations. Organizational Science 3(3) (1992)
23. Orlikowski, W.: Using technology and constituting structures: A practice lens for studying technology in organizations. Organizational Science 11(4) (2000)
24. Pisano, G., Verganti, R.: Which kind of collaboration is right for you? Harvard Business Review (12), 78–86 (2008)
25. Powell, W., Koput, K., Smith-Doerr, L.: Interorganisational collaboration and the locus of innovation: Networks of learning. Administrative Science Quarterly (41) (1996)
26. Verganti, R.: Innovating through design. Harvard Business Review (12), 114–122 (2006)
27. Verganti, R.: Design, meanings and radical innovation: A metamodel and a research agenda. Journal of Product Innovation Management (25), 435–456 (2008)
28. Raffa, M., Zollo, G.: Sources of innovation and professionals in small innovative firms. International Journal of Technology Management 9(3-4) (2009)
29. Schonsleben, P., Buchel, A.: Organising the extended enterprise. Journal of Product Innovation Management (2008)
30. Seybold, P.: Get inside the lives of your customers (2001)
31. Stein, E., Iansiti, M.: Understanding user needs (1995)
32. Thomke, S., Von Hippel, E.: Customers as innovators: a new way to create value. Harvard Business Review, 47–56 (2002)
33. Tonchia, S., Tramontano, A.: Process management for the extended enterprise: organizational and ict networks (2004)
34. Tushman, M., Anderson, P.: Technological discontinuities and organizational environment. Administrative Science Quarterly (3), 439–465 (1986)
35. Tushman, M., Anderson, P.: Technological discontinuities and dominant designs: a cyclic model of technological change (35) (1990)
36. Venkatraman, N., Henderson, J.: Real strategies for virtual organizing. Sloan Management Review (Fall 1998)
37. Weick, K.: The social psychology of organizing. Addison- Wesley, MA (1979)
38. Wenger, E., Snyder, W.: Communities of practice: The organizational frontier. Harvard Business Review (78) (2000)
39. Williamson, O.: Comparative economic organization: The analysis of discrete structural alternatives. Administrative Science Quarterly (36) (1991)

Chapter 4
A Design-Driven Approach for the Innovation Management within Networked Enterprises

Emilio Bellini, Claudio Dell' Era, and Roberto Verganti

Abstract. The intensification of the international competition and the necessity to compete in a global context push companies to introduce new organizational forms; the continuous evolutions of the contexts both in terms of market demand and available technologies, the progressive reduction of the time-to-market, the necessity to personalize the offering according to the expectations of the single customer favour the development of new organizational paradigms such as the "networked enterprise". Supply-chain organization constituted by clients and suppliers has to consider also a series of actors with very different competences and specializations. If the concept of innovation is interpreted as the output of a new recombination of existing factors, nets of companies appear proper and coherent with the development activities. In many industries the access to tacit and distribute knowledge constitutes one of the principal critical success factor: for example, in the textile industry, the interaction with actors such as research centre, designer, design services, etc. allows to access and to interpret the diffused knowledge along the entire development chain. In this kind of scenario the ICTs cover a critical role. The identification of a model able to describe potential interrelations and synergies between ICTs and organizational structures has to consider a double aspect: (i)On the one hand the diffusion of new organizational models based on the networked enterprise constitutes a continuous stimulus for the creation of new technologies able to improve the effectiveness and the efficiency of the knowledge flows; (ii)On the other hand the availability of new information technologies offers the opportunity to create new organizational models. The increasing presence of international competition as well as the importance of socio-cultural and contextual settings are transforming the future of industrial and social scenarios. The increasing pressures of international competition are

Emilio Bellini
University of Sannio - Department of Engineering,

Emilio Bellini · Claudio Dell' Era · Roberto Verganti
Department of Management, Economics Industrial Engineering Piazza L. da Vinci, 32 20133 Milano, Italy
e-mail: {emilio.bellini,claudio.dellera,roberto.verganti}@polimi.it

G. Anastasi et al. (Eds.): Networked Enterprises, LNCS 7200, pp. 31–57, 2012.
© Springer-Verlag Berlin Heidelberg 2012

challenging companies that find themselves unable to reach international sources of supply of components and knowledge, as well as unable to offer their products and services worldwide. Moreover, historical, social, demographic and industrial factors are also forcing companies to develop a system of offerings that merge profitability with social and cultural values. Couple these factors with an ever-increasing standard of living means that the development of products and industrial processes in order to survive must not only be profitable for the manufacturers, but also socially constructive. For this reason companies must develop the knowledge and new product development processes that can support them in designing and developing socially, culturally and physically constructive products. The obstacles that obstruct the development of this kind of products are related to two fundamental capabilities: the capability to develop new technologies, and the capability to understand and design new lifestyles. While several methods exist that support companies in understanding the needs from a technically functional basis (such as Quality Functional Deployment) or in managing the explicit knowledge of organisations (Knowledge Management Systems), a methodology for the identification, capturing and interpretation of value intensive information related to socio-cultural trends and latent needs is less formalized.

4.1 Introduction

The intensification of the international competition and the necessity to compete in a global context push companies to introduce new organizational forms; the continuous evolutions of the contexts both in terms of market demand and available technologies, the progressive reduction of the time-to-market, the necessity to personalize the offering according to the expectations of the single customer favour the development of new organizational paradigms such as the "networked enterprise". Supply-chain organization constituted by clients and suppliers has to consider also a series of actors with very different competences and specializations. If the concept of innovation is interpreted as the output of a new recombination of existing factors, nets of companies appear proper and coherent with the development activities. In many industries the access to tacit and distribute knowledge constitutes one of the principal critical success factor: for example, in the textile industry, the interaction with actors such as research centre, designer, design services, etc. allows to access and to interpret the diffused knowledge along the entire development chain.

In this kind of scenario the ICTs cover a critical role. The identification of a model able to describe potential interrelations and synergies between ICTs and organizational structures has to consider a double aspect:

- On the one hand the diffusion of new organizational models based on the networked enterprise constitutes a continuous stimulus for the creation of new technologies able to improve the effectiveness and the efficiency of the knowledge flows;

- On the other hand the availability of new information technologies offers the opportunity to create new organizational models.

New ICTs introduction has not be analyzed only in terms of intra and inter organizational interchanges of information and knowledge, but also in terms of comprehension of the interactions with the market: for example, in the textile industry, the use of RFID technologies in the shops allows to develop dataset of customers behaviours and consequently, by datamining techniques, it allows to identify recurrent couplings of some items.

Collaboration is a word too much used to label all kinds of web applications. Technology is often identified as the driving force when speaking in general of Internet's potential to enable business-to-business transactions. In reality, collaboration is not an add-on to the information technology architecture of a firm, putting a new face on traditional buyer/seller relations. It is a synergistic approach to handle evolving business processes, driven by important changes in the economic scenario:

- *Global competition*, favoured by the free movement of capital and information. Companies must measure their growth in a continually changing international market. Competition is no longer between individual companies, but between entire supply chains or supply networks.
- *Speed and flexibility*, necessary to ensure adequate service levels to ever more demanding customers. This means re-thinking, or extending, the integrated corporate model which characterised the 80's and 90's. The downstream demand (customers, distributors) needs to be linked to the upstream suppliers in a single flow that goes from forecasting, to purchase and production planning, to order fulfilment.
- *Virtualisation of business*, imposed by the need to reduce costs and concentrate on key competences. Companies ever more tend to focus on a few high added value processes (finance, research, strategic marketing), while delegating those activities affected by economies of scale (logistics, non-strategic production, distribution) to external partners.

By effect of these changes, processes traditionally carried out within the enterprise "four walls" acquire a broader scope, calling for direct involvement of key customers and suppliers. The ultimate goal is to move from linear supply chains, where communication is limited to buyer/supplier links, to supply networks where each node visibility extends as far as it benefits the overall network performance.

Collaborative-commerce is a framework on which to build deep, rich, mutually beneficial relationships with multiple stakeholders. Stakeholders can be inside the organization as well as trading partners. (Gartner Group)

According to the leading analysts and experts of business processes, the following phenomena will characterize the near future:

- A decrease in traditionally managed relationships (purchase order);
- A wider use of "open sourcing" via Internet;
- An ever-increasing use of "collaborative"-type relationships.

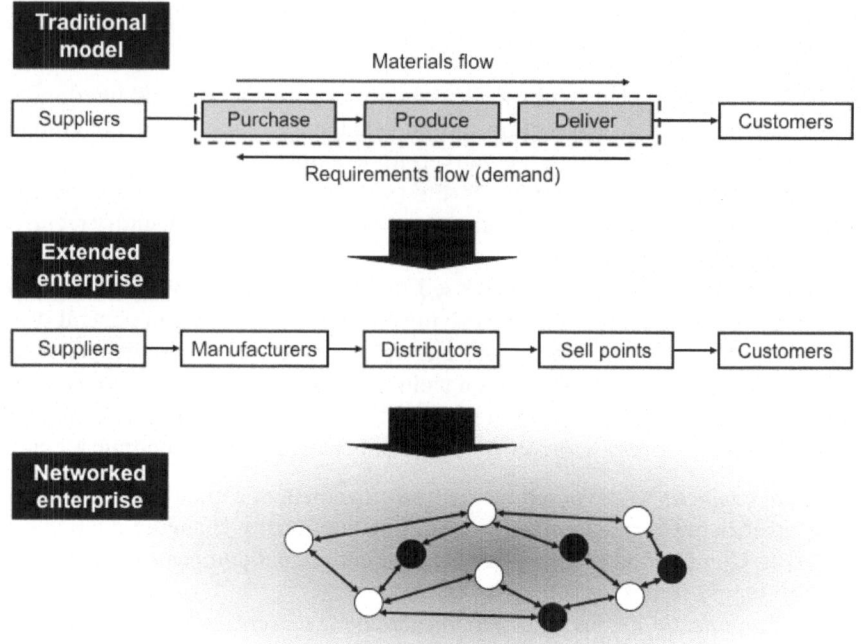

Fig. 4.1 From traditional model to networked enterprise

What defines a collaborative relationship often collides with common enterprise practice: collaboration main benefit is visibility into the customer/supplier/competitor data and strategies. This can be the key to higher supply chain performance:

- *Visibility in the development chain* allows to share several inputs for the new product development process;
- *Visibility in the production chain* allows real-time propagation of requirements and engineering changes, as well as the optimisation of materials and capacity throughout the whole chain;
- *Visibility in the distribution chain* is a fundamental prerequisite for the most advanced replenishment techniques: from vendor-managed inventory (VMI) to collaborative forecasting.

At the same time, visibility can hardly be extended to any business counterpart. In the end it is necessary to devise a collaboration strategy for long-term relationships where mutual benefit is evident and trust motivated. Other relations will be handled by a conventional buyer/seller approach, with several shades of collaboration between these two extremes. Despite acknowledging the benefits, the development of collaborative-type relationships is still fraught with obstacles that prevent the concept spreading even among innovative companies. The barriers that are found on the field may be grouped into two categories:

- *Cultural barriers*, deriving from practices which have become ingrained, such as:

  - Difficult relations with suppliers, to whom internal inefficiencies and costs are often transferred;
  - Resistance to the sharing of planning data;
  - Planning only within the four walls of the division or factory.

- Organisational and technological barriers, specifically:

  - "Waterfall" processes, inflexible and excessively formal;
  - Independent and disconnected, yet irreplaceable, information systems over the various parts of the chain.

4.2 R&D Organization: An Historical Analysis

In this paragraph we analyze the state of the art of the literature about R&D organisation in order to identify main trends. R&D includes a series of activities that have changed many times in the last 50 years in terms of management styles and organisation: first of all it is possible to affirm that they passed from an exclusively technology push model to a progressive integration with a market pull approach ([49]).

In literature five principal R&D generations are described ([56, 55, 46, 11, 49]).

The first generation is characterized by the low level of interaction with the rest of the company and strong focalization on future technologies; in this sense it is possible to affirm that the management style corresponds to the technology push paradigm ([49]) and the R&D unit acts as a sort of ivory tower where employees do not have any interactions with other business functions. In the '50s the R&D units were centralized at the corporate level and for this reason they had great autonomy and possibility to identify the research fields more interesting. Corporate research labs had to flourish technological innovations and to stimulate scientific advances.

The second generation saw to prevail market pull approach, characterized by an increasing importance of the Marketing function in the innovation process ([49]). This historical period (mid-1960s to early 1970s) was characterized by the necessity to increase the market shares ([55]) and the main goal was not to propose a new technological standard, but to obtain maximum margins in the short period. For this reason R&D units were incorporated into single business units and turned their attention to the customers' needs, consequently the Marketing function assumed the integration role between several business functions.

In the '70s the interpretation of technology push and market pull paradigms shifted from alternative solutions to complementary ones. Two approaches can be applied in a synergic way giving life to a more interactive research and development paradigm rather than linear as in the first generations. In this historical period, R&D units were characterized by a strong resource rationalization, that brought to the

Table 4.1 Description of five generations of R&D processes (developed and adapted from [56, 55, 46, 11, 49])

R&D Generations	Context	Process Characteristics	Company Response	Managerial Approaches
First generation	Black hole demand (1950 to mid- 1960s)	R&D as ivory tower, technology-push oriented, seen as an overhead cost, having little or no interaction with the rest of the company or overall strategy. Focus on scientific breakthroughs.	Corporate research labs	-Stimulating scientific advances -Choosing location after competencies
Second generation	Market shares battle (mid-1960s to early 1970s)	R&D as business, market-pull oriented, and strategy-driven from the business side, all under the umbrella of project management and the internal customer concept.	Business unit development	-Appointing internal customers - Ideas gathered from market
Third generation	Rationalization efforts (mid-1970s to mid-1980s)	R&D as portfolio, moving away from individual projects view, and with linkages to both business and corporate strategies. Risk-reward and similar methods guide the overall investments.	R&D projects	- Structuring R&D processes - Evaluating long-term technology strategies -Integrating R&D and marketing
Fourth generation	Time-based struggle (early 1980s to mid-1990s)	R&D as integrative activity, learning from and with customers, moving away from a product focus to a total concept focus, where activities are conducted in parallel by cross-funcional terms	Cross-functional projects	- Parallelizing activities -Involving suppliers and lead customers - Integrating R&D and manufacturing
Fifth generation	Systems integration (mid- 1990s onward)	R&D as network, focusing on collaboration within a wider system -involving competitors, suppliers, distributors, etc. The ability to control product development speed is imperativem separating R from D.	Cross-boundary alliances	- Involving company network -Focusing integration - of systems -Separating/linking R and D.

adoption of portfolio logics in the project activation. The main objective consisted on the development of a balanced portfolio of R&D projects in terms of risk and temporal horizon of investments. For this reason, this R&D generation favoured a great diffusion of project management techniques with the purpose to get all consequential advantages to operate for projects. At the same time it was also possible to assist to a progressive integration among R&D and Marketing, shifting from a sequential collaboration to a parallel and cyclical interaction in order to anticipate in the first stages of the innovation process all possible constraints.

The fourth generation considered the R&D as an activity of integration between client and enterprise, it introduced the concept of lead user and foresaw the early involvement of suppliers in the innovation process in order to compare different perspectives and consequently to increase the cross-functionality; the fourth generation emphasized the role of feed-backs and the point that innovation is by definition cross-functional. In the '80s, the economic context was characterized by the introduction of new functionalities and new services associated to the existing products as a new kind of differentiation. For example great Japanese companies such as Toyota, Sony and Honda introduced some products in the European and America markets with a series of additional services never proposed by competitors ([55]).

The fifth generation of R&D proposed to manage the innovation process as a network, in which all those people that could bring useful know-how had to participate: suppliers, clients, distributors and even competitor. The role of teams shifted from cross-functional and cross-disciplinary to cross-firms; the necessity to share the huge technological investments and to divide the research from the development, with consequent increase of costs and resources, spontaneously suggested to cooperate with other subjects.

In literature it is possible to identify several contributions about this kind of phenomenon: Allen ([2]) spoke about collective invention, Morrison et al. ([48]) deepened the role of lead user in the Information Technology industry, [28] analysed the knowledge dynamics in the case of open source software and Von Hippel ([72]) proposed the concept of open innovation. An example of this type of collaboration could be identified between Toyota and PSA Group that has brought in 2004 to the development of a new city car: Toyota Aygo, Peugeot 1007 and Citroen C1 are three cars that directly compete on the market and share a common research base.

Finally, the fifth generation of R&D foresaw the separation between research and development: they were considered completely different, the first one totally devoted to the exploration of new technological solutions and the second one with a strong engineering component (Kodama, 1995). This separation allowed to use two different organizational models, with the purpose to leave the maximum freedom to the researchers and to formalize the development process.

4.3 Innovation and Networked Enterprise

The increasing presence of international competition as well as the importance of socio-cultural and contextual settings are transforming the future of industrial and social scenarios. The increasing pressures of international competition are challenging companies that find themselves unable to reach international sources of supply of components and knowledge, as well as unable to offer their products and services worldwide.

Moreover, historical, social, demographic and industrial factors are also forcing companies to develop a system of offerings that merge profitability with social and cultural values. Couple these factors with an ever-increasing standard of living means that the development of products and industrial processes in order to survive must not only be profitable for the manufacturers, but also socially constructive. For this reason companies must develop the knowledge and new product development processes that can support them in designing and developing socially, culturally and physically constructive products.

The obstacles that obstruct the development of this kind of products are related to two fundamental capabilities: the capability to develop new technologies, and the capability to understand and design new lifestyles. While several methods exist that support companies in understanding the needs from a technically functional basis (such as Quality Functional Deployment) or in managing the explicit knowledge of organisations (Knowledge Management Systems), a methodology for the identification, capturing and interpretation of value intensive information related to socio-cultural trends and latent needs is less formalized.

Analysing the Italian lighting industry, [67] observe that companies developing design-driven innovations have to be able to furnish the necessary flexibility, opening and dynamism. The project resources involved in the innovation process have to be organized so that to favour the creativeness and the possibility of exploration and recombination. It is not possible to think that a design firm possesses or wants to develop internally all competences necessary to innovate; rather, the project resources have to be select outside.

There are many studies about Italian districts ([8, 39, 42, 60]), but they evidence the same weakness: the industrial districts find great difficulties to face radical changes because the local nature of the relationships limits the possibility to introduce innovations produced out of the system.

The amplification and the extension of the relationships out of the system are possible solutions to these limits; the concept of "milieu innovateur" introduced by Camagni ([7]) can guarantee a greater ability of innovation. Also the dynamic nets ([45]) or the innovative nets ([53]) are based on the critical role of some firms that, despite being strongly connected to the local productive system, extend their collaboration to external enterprises and act as intermediary of the knowledge transmission.

Cagliano et al. ([6]) analyse differences and similarities in managing technological collaborations in research, development and manufacturing: more in detail for each phase of the innovation process they identify some characteristics of the t echnological collaborations (content, motivations, typology of partners) and corresponding organizational forms (number of partners, contractual formalization, structure of control, time horizon, density of relationship).

Iansiti and Levien ([29]) introduce the concept of "business ecosystem" and sustain that drawing the precise boundaries of an ecosystem is impossible, but stand-alone strategies cannot work when the company's success depends on the collective health of the organizations that influence the creation of the product.

Other researches identify some relationships between organizational variables (internal competences, access to networks of innovators, etc.) and innovativeness of the company ([4, 21, 61]).

4.4 Strategies of Innovation

Models and classifications traditionally proposed in the innovation management literature focus mainly on technological innovations, but interpreting the new product development as a process of generation and integration of knowledge, the literature identifies two principal sources: the knowledge about the availability of new technologies and the knowledge about explicit customers' needs.

Starting from this basic reflection, Dosi ([16]) analyzes two antithetic approaches to innovation: market-pull and technology-push. On the one hand, the market-pull approach is primarily characterized by the dominant role that the comprehension of market needs plays over the introduction of new technologies. Consequently, in this particular approach, the main source of innovation is the market and the new product development is a direct consequence to explicit needs manifested by the consumers. However, the primary assumption of this approach is that user needs are explicit elements that can be identified, captured and translated in new products able to satisfy the consumer needs ([62, 59, 64, 9]).

On the other hand, the technology-push approach looks at the innovation process from a completely different perspective; in fact, this approach does not believe in a process driven by the market. Instead, it believes that the source of innovation stems from the research and development activities of the company that, through the identification and development of new technologies, allow to realize new products. If in the market-pull approach the central role is covered by the market and the consumer, in the technology-push approach it is given to the company and the development of new technologies that subsequently drive the company's innovation processes ([1, 27, 13]).

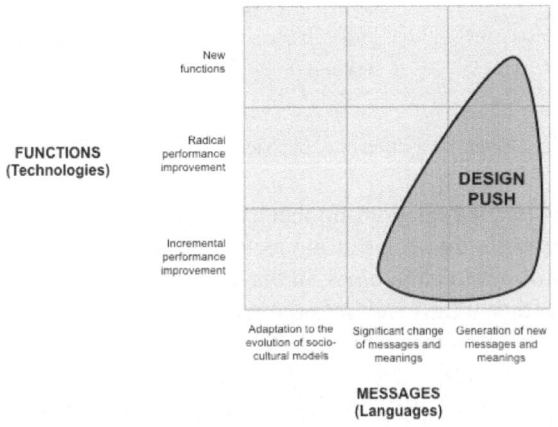

Fig. 4.2 Design-push approach to innovation (adapted from [66])

In more recent literature, a third approach to innovation allows to enrich the possible innovation approaches that a company can adopt. This approach focuses on the importance of design in innovation and can be defined as the design-push approach. Verganti ([66, 68]) describes this approach as complementary to the previously described market-pull and technology-push approaches. In fact, he emphasizes that it may require the development of new technologies or the satisfaction of explicit market needs. In this approach, it is the semantic dimension that drives the innovation process rather than the market or the technology.

More in detail, Verganti ([66, 68]) proposes a third source of knowledge to add the knowledge about user needs and the knowledge about technological opportunities: the knowledge about product languages "is the knowledge about the signs that can be used to deliver a message to the user and about the socio-cultural context in which the user will give meaning to those signs" (see Figure 4.2).

In all approaches to innovation three types of knowledge (knowledge about user needs, knowledge about technological opportunities, knowledge about product languages) are present, but their relative importance is different: in the case of design-push strategy the driver of innovation, the starting point, is not the technology or the customer needs, but is the capability to understand, anticipate and influence the emergence of new product meanings. Verganti defines the radical design driven innovation as "an innovation where novelty of message and design language is significant and prevalent compared to novelty of functionality and technology" (see Figure 4.3).

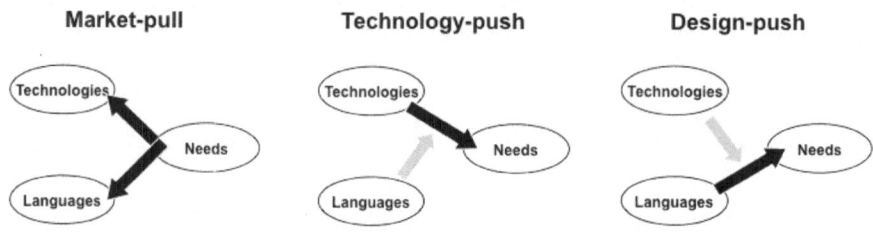

Fig. 4.3 Knowledge drivers in a different approaches to innovation ([66])

It becomes evident the relationship that exists between the technology-push and design-push approaches: whereas in the technology-pushed approach the driver is the development of new technologies, in the design-push approach the driver is the meaning or semantics of the resulting product, but product languages and messages can be modified acting on the technologies.

Plastic pieces of furniture developed by Kartell in the sixties can be considered meaningful examples: Castelli, founder and past president of Kartell, gave a new sense of modernity to furniture products through the use of plastic materials. The adoption of the plastic was reinterpreted to the point that it assumed the meaning of "noble" material, breaking the dominant cultural models that foresaw the diffusion of other materials such as wood, steel, marble and glass in the furniture industry.

For what concerns the relationship between the market-pull and design-push approaches, it lies mainly in the fact that a consumer can manifest explicit needs from a semantic perspective only when the innovation is incremental. A radical design driven innovation however drives towards the development of new meanings that change the socio-cultural context. Consequently, as long as the degree of innovation is incremental, the two approaches can coincide and operate complementarily.

This, however, cannot be the case when the degree of innovation tends towards radical since the market and/or consumer is not able to manifest coherent needs that can stimulate the company in developing new product meanings that break with the past. More in detail it is presumable that the role of market factors in the design-push approach changes according to the level of novelty of product meanings.

Market drivers become remarkable in the case of incremental innovation of product meanings, where, in other words, incremental adaptations of product meanings are determined by the continuous and natural evolution of explicit cultural models adopted by the customers. On the opposite, market factors lose importance in the case of radical innovation of product meanings, where innovations originate from a cultural scenario developed through the collaboration between companies and designers. In other words radical innovations of product meanings require the comprehension of possible or latent social dynamics that can influence consumers lifestyles and behaviours. The above considerations can be summarized in the following two propositions (see Figure 4.4):

- The market becomes less capable to propose semantic innovations the more radical the change in meaning. Consequently, the more radical the design driven innovation, the less it can be generated by a market-pull approach;
- The more the technology represents a mean to generate a change in meaning, the more the technology-push approach to innovations can be considered design driven.

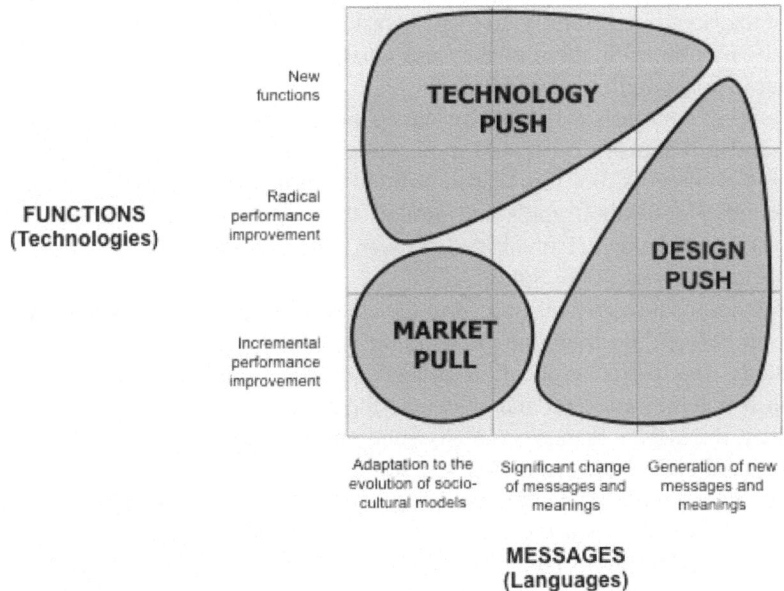

Fig. 4.4 Mapping different Approaches to innovation (adapted from [66])

In order to adopt the design-push approach it is necessary to develop a particular sensibility towards the evolution of socio-cultural contexts; it is indispensable to recognize "weak" signals that will be fundamental characteristics of the future scenarios.

A correct management of radical design driven innovation allows companies to interpret new lifestyles and subsequently to propose a coherent system of values to the market. The considerations mentioned above evidence the necessity to access to a series of stakeholders connected one each other in order to comprehend the evolution of socio-cultural contexts and then to introduce radical design driven innovations.

The following paragraphs describe in detail two approaches to innovation: the user centered design is based on the market-pull approach, while the design driven innovation starts from the frameworks associated to the design-push approach.

4.5 User Centered Design

A specific approach to design, usually referred to as "user centered design", is the basic topic of several recent researches ([9]). The success of major design firms such as IDEO (Kelley, 2001) or Continuum (Lojacono and Zaccai, 2004) underlines the potentialities of the user centered design approach; it foresees that product development should start from a deep analysis of user needs. The basic assumption is that a company can successfully innovate by asking users about their needs or, more effectively, by observing them as they use existing products and by tracking their behaviour in consumption processes.

This kind of approach abandons the classic and common interpretation of design as style and provides a deeper and more valuable interpretation of design as an organizational process. For this reason in the recent literature it is possible to find several contributions about tools and models that support the application of user centred design approach (Patnaik and Becker, 1999; Kumar and Whitney, 2003; Rosenthal and Capper, 2006; Sutton, 2001).

Some theoretical references related to the classification of the needs are particularly useful before describing the methodologies mostly used in the analysis of the users' needs. The basic classification of needs in explicit and latent depends on the degree in which they are clear and evident to the subject. Obviously the more needs are explicit, the more it is easy to satisfy such needs, while it could be particularly arduous to understand and satisfy the i rrational feelings that cannot be explicit.

Maslow's theory (1974) classifies human needs according to a hierarchical structure that goes from the most urgent to the less pressing (see Figure 4.5).

The classification proposed by Maslow describes a broad range of needs: from primary needs, to the safety desire, to the needs linked to the affective-sentimental sphere up to those of respect and self-realization. According to the structure proposed by Maslow, product attribute assume different importance and consequently a company can design them in a precise way: at the lowest levels, the set of

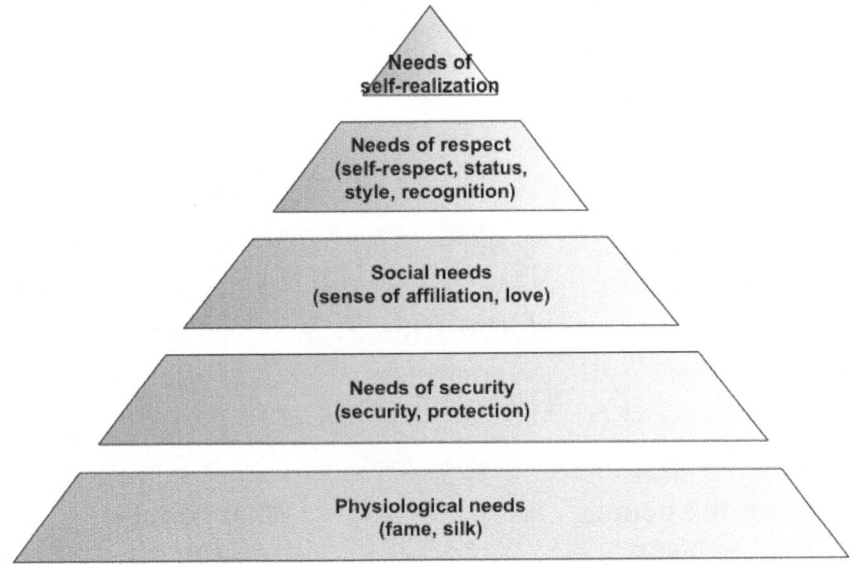

Fig. 4.5 Hierarchy's needs (Maslow, 1974)

functional characteristics respond to the hygienic and safety regulations, while, rising in the structure, the needs gradually move towards the semantic, symbolic and cultural dimensions: sense of affiliation, love, status, style, recognition, etc. In this sense design plays an important role in the development of basic and crucial product aspects that allow to satisfy needs and desires connected with the last three levels of the pyramid.

Despite the classification provided by Maslow is particularly interesting and stimulating, it does not consider the role played by the context of use. For this reason recent literature about new product development process tries to better describe the interconnections between users' needs and context of use going from the "design for users" paradigm to the "design with users" one (Sanders [57]).

During the first years of the 1980s many designers used to collaborate with sociologists and anthropologists in the process of user needs analysis, but this kind of approach showed very soon several limits: it didn't gather many aspects connected to the emotions, to the memories of the consumer, to his actual and ideal experiences. With this intention in mind, it was necessary to introduce the philosophy of triangulation that analyzes the three following different degrees of knowledge about the consumer (Sawhney et al., 2003): What people say, What people do, What people make (see Figure 4.6).

Explicit needs can be identified by listening what the customer; at the same time, it is necessary to consider that customer reveals only what he wants, and consequently he determines the direction of development of the analysis. To analyze what the customer does and uses can be insufficient because it underlines only the

Fig. 4.6 The phylosophy of triangulation

observable needs skipping an unexplored area connected to what the customer knows, feels and dreams (Sanders).

The investigation related to knowledge and convictions of the customer provides to the researcher some indications about perception of the reality and different experiences of the customer; moreover the comprehension of feelings and sentiments can allow to increase the empathy with the analyzed subject, underlining the tacit knowledge (Polanyi, 1983). What the customer dreams, exemplifies how he/she would like the future to be. However, to gather such information it is necessary to actively involve the studied subject in the development process in order to observe the kind of solutions that he/she proposes without knowing the need from which they originate. For example the paradigm of the Experience Design focuses on the creation of an experience for the customer: the emotional aspect of the interaction with the product become the fulcrum of the entire project (Figure 4.7).

It is possible to observe that the analyses traditionally effected by the marketing are directed to the identification of the explicit needs (what people say), while the observation on the contexts of use, typical of the designers, are more directed to identify what the customers do with the products (what people do and what people make). Anthropologists mainly develop analyses about the category what people make using tools such as the collage ([57]), the narration (Lerdhal, 2002), the construction of metaphors ZMET (Zaltman, 1997)

The three categories of needs can be also reinterpreted according to the temporal horizon : what people do underlines the actual situation; what people say reveals

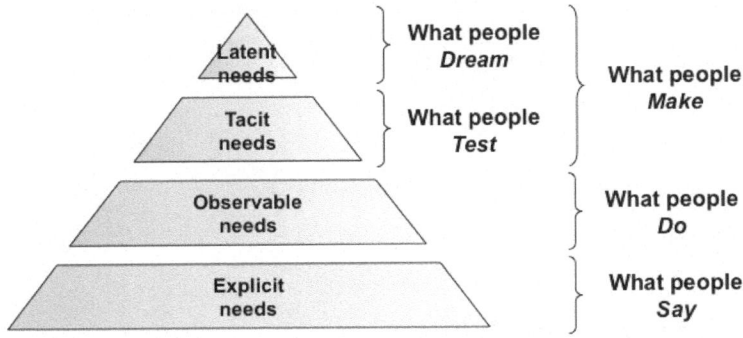

Fig. 4.7 Typologies of needs

the past and the immediate future; what people make stimulates the researcher to investigate in the remote past (memory) and in the most distant future (dreams).

The category of needs that a company wants to analyze has obviously also an impact on the more appropriate technique of investigations to apply. In the research field a first distinction consists in qualitative and quantitative investigations; having said this, it must be clear that we are dealing with a binary variable in which the analysis can either be quantitative or qualitative, in reality we have to manage a continuum of methods that range from purely quantitative analyses to purely qualitative ones.

The quantitative investigation uses formalized systems of data collection; its more typical tool is the questionnaire with multiple-choice answers, it allows to implement statistical analysis. It is typically very useful for the detailed verification of well-circumscribed hypothesis (eventually consequential from a first qualitative investigation).

Instead the qualitative research uses tools that don't present a rigid formalization; the most typical and common tool is the free interview. This kind of analysis is very detailed, and it allows to examine also variables initially not anticipated. The qualitative investigation foresees a strong interaction between the researcher and the observed subject.

Naturally the given definitions are very general and they do not refer specifically to methodologies developed for investigating the needs. Market investigations (from the analysis of the sales of a product, to the investigation of the percentage of stock-out, from the observation of the unsold up to benchmark with the direct competitors) and questionnaires to be submitted to statistic revision can be considered quantitative investigations.

Instead, the qualitative researches go from the focus-groups, to the analysis of the lead-users, from the ethnographic to socio-cultural researches. As shown in Figure 4.8, moving from the left side to the right one the capability to interpret the influence of context of use on user needs and his/her active role increase.

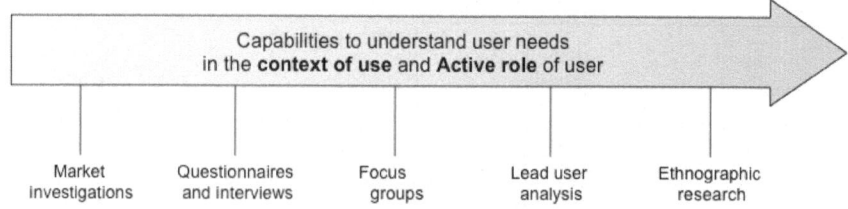

Fig. 4.8 Methodologies of user needs analysis

4.6 Design Driven Innovation

The interpretation of the meaning that a customer will give to the product is essential in the field of design driven innovation; it cannot be unilaterally imposed by the company or the designer because the interpretation depends on the cognitive schemes of the customer and socio-cultural context in which he lives.

The meaning attributed to a product descends from its languages and the customer's interpretation filtered through individual, social and cultural models. In the case of design driven innovation it is necessary to develop a great capability to interpret the social groups in which the user lives because the contextual cultural factors have a strong influence; the culture has to be considered as one of the fundamental factors in the identification of the perceived needs (Kotler and Scott, 1999). Kotler and Scott (1999) propose a model that supports the analysis of the factors influencing consumer needs (Figure 4.9).

The investigation of a target market needs requires to analyze the characteristics of the group that individuals belong and the socio-cultural models that can influence determined personal and psychological aspects. Literature states that a individual's buying choices are influenced by four major psychological factors: motivation, perception, learning and beliefs. Several psychologists have developed numerous theories of human motivation. However, given that the scope of this chapter is not related to the psychology of motivation, it focuses on social and cultural factors.

Every person presents some peculiarities: from the age to the employment, from the economic condition to the lifestyle, etc. The term personality represents the set of characteristics that distinguish the individual and bring him to answer in coherent and constant way to the environment; in this field it is possible to introduce concepts like self-realization, dominance, autonomy, ability to socialize. A consumer's behaviour is influenced by such social factors as reference groups, family and roles and status.

A person's reference group consists in all the groups that have a direct or indirect influence on a person's attitudes or behaviour. Groups having a direct influence on a person are called membership groups (e.g. family, friends, neighbourhoods, co-workers etc.), whereas groups having an indirect influence on a person are called aspiration groups. Groups to which a person aspires to belong (i.e. a teen-ager may hope one day become a football player for the F.C. Juventus). Consequently,

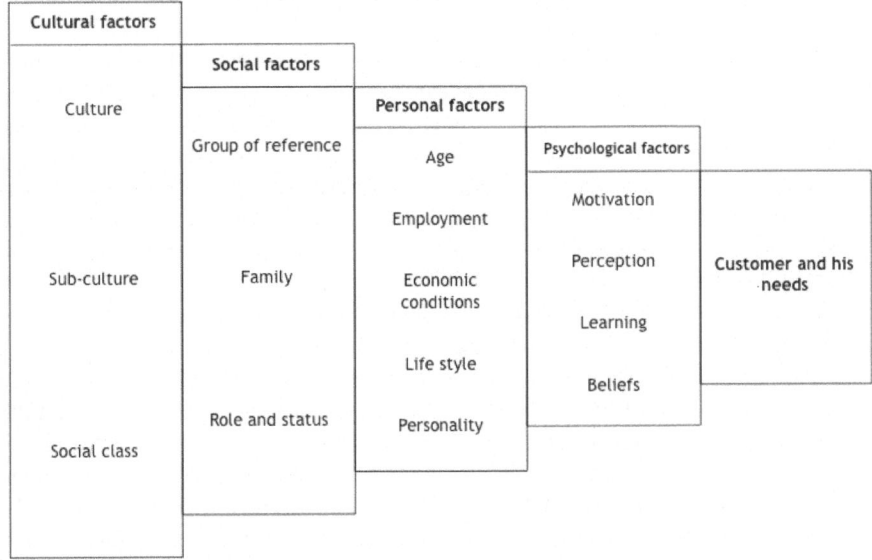

Fig. 4.9 Influencing factors

marketers must try to identify their consumers' reference groups and their attitudes towards the product that they are trying to sell in order to understand how reference groups can influence brand and/or product choice.

Manufacturers of products and brands where group influence is strong must determine how to reach and influence the opinion leaders in these reference groups. Marketers try to reach opinion leaders by identifying demographic and psychological characteristics associated with opinion leadership, identifying the media read by opinion leaders and directing messages towards these people in order for them to then directly or indirectly influence people that are part of their reference group. A person's position in society can be also defined in terms of role and status. A role consists in the activities that a person is expected to perform. Each role carries a status. People choose products and brand to communicate their role and status in society. Consequently, for marketers, this means that they must are aware of the status symbol potential of their products and brands.

In addition to social factors, cultural factors represent the broadest and deepest influence on consumer behaviour and can be divided into the two following sublevels: culture, subculture and social class. Culture can be summarized as the most fundamental determinant of a person's desires and, consequently, behaviours. In fact, we as individuals acquire sets of values, perceptions, preferences and behaviours from our external environment that we internalize into our personality.

Each culture consists of smaller subcultures that provide more specific identification and socialization for its members. Typical subcultures include nationalities, religions, racial groups, and geographical regions. These sub-cultural strata predominantly influence or consumption patterns (I.e. food and clothing preferences carrier aspirations etc).

All human societies exhibit social stratification. More frequently, stratifications take the form of social classes, which can be defined as hierarchically ordered groups of individuals whose members share similar values, interests and behaviour. Social classes do not reflect income alone, but also other indicators such as occupation, education, area of residence etc. Social classes differ in their dress, speech patterns, recreational preferences, media preferences where, for example, upper class segments prefer magazines and books and lower class segments prefer television (Geertz, 1973).

A product innovation can be communicated acting on the product signs and languages; it can be understood by the customer if the company knows the social and cultural models that surround the customer and embeds an opportune lexicon in the product. The criticism of the analysis resides in the different speed with which the models evolve: the evolution of a new reference model or paradigm has to face a sort of inertia that increases with the number of involved subjects. For this reason the individual interpretative models evolve more rapidly than socio-cultural models.

The society employs a significant amount of time to align the behaviours of different subjects to the dominant models until they become interpretative standards. The same phenomenon occurs at the level of relationships among different societies until an interpretative scheme is spread and becomes part of the culture of reference. The whole phenomenon was described by Flichy (1996) through three concentric circles of different dimension, in movement with different speed but with the same quantity of motion (Figure 4.10).

The comprehension of cultural models is particularly complicated, but provides results characterized by a long term validity. The development of radical design driven innovations requires to capture and interpret cultural aspects because they aim to strongly change the meaning attributed to the object. This type of innovation is particularly risky because it implies deepened analysis that require time and have a remarkable impact on the final result; the slow dynamics of the cultural models explains the difficulty connected to the innovation in this field: a change at cultural level is a sum of rapid movements at individual and social level.

In the last years many scholars have developed several studies about the semantic dimension of a product recognizing and underling its importance ([14, 36, 30, 43, 22, 24, 31, 40, 20, 50]).

Different terminologies have been used to describe the communicative qualities of a product. McDonagh-Philp and Lebbon use the term *emotional domain, soft design* or *soft function*, while Desmet et al. write about *added emotional value, emotional fit* and *product emotions*; finally Durgee (2001) refers to *product soul* while Marzano (2000) speaks about *product experience*. The semantic dimension allows to innovate creating new meanings and acting on the socio-cultural models or, in less radical way, modifying the already incorporated meanings in the products.

According to the approach of these scholars, design deals with the meanings that people give to products, and with the languages that can be used to convey those

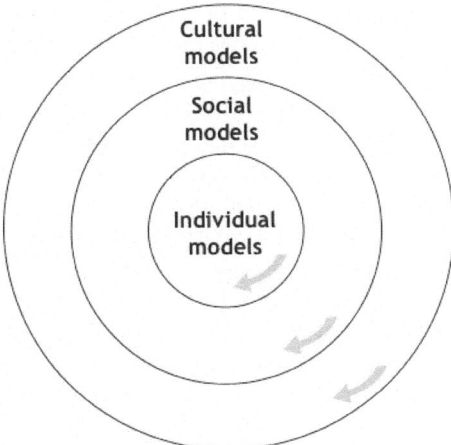

Fig. 4.10 Individual, social and cultural models

meanings. In this sense the following definition proposed by Klaus Krippendorff on Design Issues in 1989 ([36]) results particularly meaningful:

The etymology of design goes back to the latin de + signare and means making something, distinguishing it by a sign, giving it significance, designating its relation to other things, owners, users or gods. Based on this original meaning, one could say: design is making sense (of things).

Gotzsch ([24]) introduces the concept of symbolic value in opposition to the concept of functional value and proposed the following working definition: *some products include a symbol or a message that is subconsciously recognised by its customer. This subconsciously recognised symbol adds important value to the product as its user feels that the product really suits him.*

Sanders ([57]) underlines the relationships between the success of a product and three specific factors: usefulness, usability and desirability; the latter represents the communicative aspects of the product and the psychological aspects linked to the interpretation of the customer.

Some interesting considerations come also from scholars of brand management: in particular Holt (2003) speaks about "icons" as myths created by leader companies and able to propose charismatic visions of the socio-cultural contexts; moreover Holt (2002) interprets the brand as the "culture of the product", in his opinion the innovation process has to borrow inputs from anthropology, history and sociology in order to propose products as *cultural artefacts*.

In order to underline the increasing interest to the semantic dimension of a product, Gotzsch ([24]) describes the different philosophies in product design of the 20th century (see Figure 4.11). She tries to demonstrate how the attention of practitioners and academics has been moved progressively from the functional dimension of a product to the semantic aspect.

In this sense at the beginning of the twentieth century the paradigm "form follows function" became the dominant approach in the USA and later it was reinterpreted and refined by the German Bauhaus and the Ulm School. But, partially in the 30s, and, above all, in the 70s, new approaches move their interests to the aesthetic dimension of the product.

Last breakthrough in terms of design philosophy was probably introduced by Ettore Sottsass who leaded the Italian design movement in the 80s: he proposed to shift the attention to the product meanings. In his eyes, the design has to participate to social and cultural phenomena and for this reason design products became a sort of media to transfer values, emotions, meanings and messages. The dialectic of "Function versus Form" is a classic that very often leads to confine the interpretation of the term "Form" to the aesthetic appearance of products. However, a product can bring messages to the market in several ways; styling is just one of these ways. While the functionalities of a product aim to satisfy the operative needs of the customer, the product meanings aim to satisfy the emotional and socio-cultural needs of the customer.

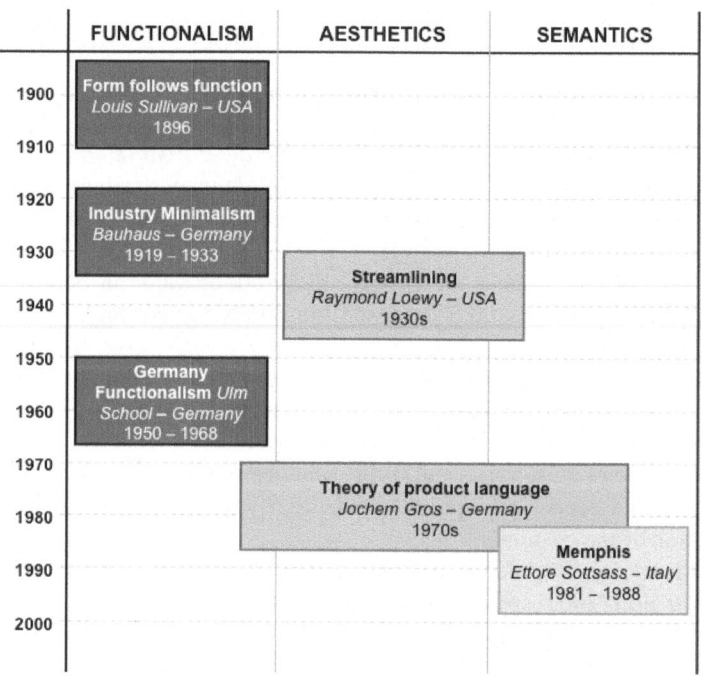

Fig. 4.11 Philosophies in product design of the 20th century

As mentioned before, successful Italian companies such as Alessi, Artemide, Kartell or B&B Italia base their innovation strategy on the development of new product meanings in terms of sense, personality, identity, emotion and values. What consumers are looking for more and more in consumer products are new forms of psychological satisfaction that go beyond normal and simple consumption.

Today, more than ever, products define their own presence not only through their attributes, but also through the meanings that they assume, through the dialogue that they establish with the user, and also through the symbolic nature that they emanate. They become magnets of meaning. Therefore, adopting a design-push approach means having the right set of skills to understand, anticipate and influence the emergence of new product meanings and messages.

Design driven innovations allow to communicate new meanings and values to the customers and for this reason they can be defined as semantic innovations. As a functional innovation can improve technical performances in an incremental or in a radical way, an innovation of product meanings may be more or less radical.

Many "fashionable" products are often the result of incremental innovations; they adopt design languages that match existing aesthetical frameworks and definitions of beauty, and propose messages in line with the existing socio-cultural models. Contrarily, as far as a radical semantic innovation is concerned, embedded languages and delivered messages need significant reinterpretation of meanings.

In many luxury industries (fashion, jewels, watches, etc.) the rapidity to capture and react to the emergence of new customers' needs can be considered one of the principal critical success factors because these sectors are characterized by continuous incremental improvements. Contrarily radical semantic innovations sometimes are not immediate: they take time to diffuse and achieve acclaimed success.

The comprehension of new radical product meanings and messages requires that users find new connections to their socio-cultural context, establish new patterns of interaction with the products and explore new symbolic values. If radical technological innovations require profound changes in the technological regimes, radical semantic innovations ask for profound changes in the socio-cultural models.

Design driven innovations allow to develop products that are completely different from "fashionable" or stylish products, they may contribute to the definition of new aesthetic standards proposing new interpretations of socio-cultural models. Differently from "fashionable" or stylish products, radical innovations of meanings can become icons and can contribute to the definition of new aesthetic parameters.

In other words design driven innovations cannot be considered answers to user needs, but the result of a dialogue with consumers in order to satisfy not only utilitarian needs but also symbolic end emotional meanings. Successful Italian manufacturers in design-intensive industries such as Kartell, Artemide, Alessi, B&B Italia, etc. have demonstrated unique capabilities to understand social needs and develop systems of offering with higher value for the socio-cultural environment. They have superior capability to understand, anticipate and influence the emergence of new product meanings. Their product portfolio is characterized by many incremental innovations and few strategic projects developed with the purpose to introduce breakthrough changes of product meanings. Analyzing the innovation process of these companies, it is possible to observe that they do not apply user needs analysis such as ethnographic research or in general user centered methodologies. Contrarily, entrepreneurs of leading design driven companies demonstrate a completely different approach to innovation:

Market? What Market? We do not look at market needs. We make proposals to people.[Ernesto Gismondi, Chairman of Artemide]

Working within the meta-project transcends the creation of an object purely to satisfy a function and necessity. Each object represents a tendency, a proposal and an indication of progress which has a more cultural resonance. [Alberto Alessi, CEO of Alessi]

These sentences underline that the adoption of a design-push approach does not start from users' insights and for this reason customers have to considered a limited source of knowledge to introduce radical innovations of product meanings.

The diffusion of socio-cultural models and consequently their impacts on the interpretation of design languages depend on many interactions between several stakeholders; customers' interpretations are in line with what is happening today and for this reason they can rarely provide interesting indications in terms of radical changes.

Today socio-cultural context in which consumers are immersed and the current values in the society shape users needs and interpretations. Both managers claim that their companies make proposals to people rather than develop answers to customers. They are able to introduce radical innovations of product meanings by looking at long-term phenomena with a broader perspective.

In this sense the design-push approach seems more similar to the technology-push rather than the market-pull one because, as suggested by Dosi ([16]), changes in technological paradigms are mainly technology-push while incremental innovations within existing technological paradigms are mainly market-pull.

Each product has particular language and meaning and they can be innovated; besides to consider product functionalities, the framework proposed by Verganti ([66]) expands and elaborates the concept of form, in order to consider the symbolic and emotional values of a product (see Figure 4.12).

As mentioned before the radical design driven innovation is defined as *an innovation where novelty of message and design language is significant and prevalent compared to novelty of functionality and technology.* Design can be defined as a set of activities that aim to create meanings ([36]).

Product languages developed by designers allow products to speak and convey precise meanings. For this reason, on the one hand, the designer has to understand the socio-cultural context that surrounds customers and, on the other hand, he/she must also be able to translate this inputs into concrete product languages. Consequently, it is particularly important to apply the knowledge and techniques developed in the disciplines of semantics and semiotics.

Many studies analyse the potential advantages and disadvantages of the external acquisition of knowledge and technology ([25, 3, 15, 26, 71, 11]) as well as many others propose several typologies of co-operation with external sources of knowledge and technology ([5, 34, 10, 17, 47, 65, 12]).

Most of these studies deepens the technological collaborations, but it is necessary to underline that the knowledge about the dynamics of socio-cultural models shows different specificities (for example, two peculiarities of how knowledge related to

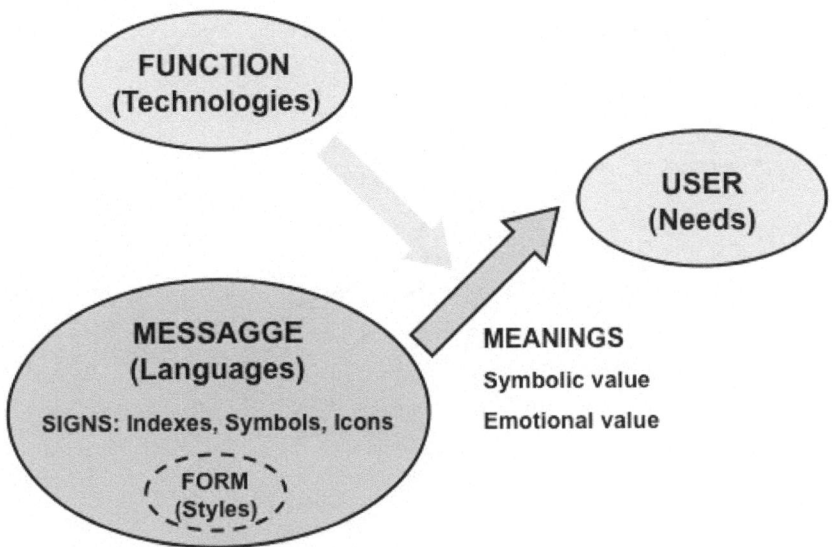

Fig. 4.12 Design driven innovation

socio-cultural models differs are the elevated degree of dispersion of knowledge among several actors, and the relative implicit structure of this knowledge).

Designers can support companies in the creation of breakthrough product meanings acting on the semantic dimension; they capture, recombine and integrate knowledge about socio-cultural models and product semantics ([36]), but they have not to be considered the only possible support.

Unfortunately there is not a centralized repository where it is possible to retrieve all information about future scenarios; the affirmation of socio-cultural models and consequently their impact on the interpretation of product languages depends on many interactions among several stakeholders: users, firms, designers, communication media, cultural centres, schools, etc. Verganti ([68]) sustained that this knowledge is diffused within our environment in a design discourse (see Figure 4.13).

For this reason, in order to introduce radical design-driven innovations it is necessary to develop several channels to access tacit and distributed knowledge about socio-cultural trends; it is necessary to be part of a network in which several stakeholders interact and share their knowledge.

All these actors are not only interested in understanding possible future domestic scenario, but they have also the possibility to influence user needs and desires with their actions and outputs (products, projects, reports, artworks, shows, etc.).

For this reason Italian leading companies consider these actors as key interpreters of the evolution of future scenarios and for this reason they develop continuous dialogue about possible visions of the future, exchange and compare information about emerging trends, verify the robustness of their assumptions.

In this way they can, on the one hand, to be part of design discourse and, on the other hand, to influence its evolution. The networked research process can be

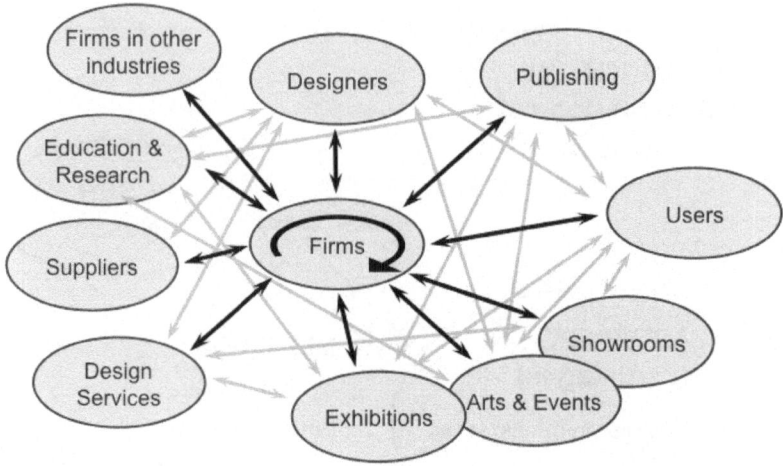

Fig. 4.13 Design discourse

Fig. 4.14 User centered design vs. Design driven

considered one of the principal difference between design driven innovation and user centered design (see Figure 4.14):

- Design driven innovation is based on a huge amount of relationships that span widely outside the boundaries of the firm, including users, but also and mainly several other interpreters;
- Design driven innovation is based on sharing and exchanging of knowledge about socio-cultural models, product languages and meanings;
- Design driven innovation aims not only to capture and interpret socio-cultural models, but also to influence and modify meanings and values in the society thanks to the collaboration with the interpreters;
- Differently from the user centered design that considers fundamental to get as close as possible to users, support the identification of implicit needs and finally be creative in finding solutions, the key capability in design driven innovation is accessing and sharing knowledge about socio-cultural models and emerging values and meanings in the society.

References

1. Abernathy, W., Clark, K.: Innovation: mapping the winds of creative destruction. Research Policy (1985)
2. Allen, R.: Collective invention. Journal of Economic Behavior and Organization (1), 1–24 (1983)
3. Bidault, F., Cummings, T.: Innovating through alliances: expectations and limitations. R&D Management 24(1), 33–45 (1994)
4. Bougrain, F., Haudeville, B.: Innovation, collaboration and smes internal research capacities. Research Policy, 735–747 (2002)
5. Brockhoff, K.: R&d cooperation between firms: a classification by structural variables. International Journal of Technology Management (6), 361–373 (1991)
6. Cagliano, R., Chiesa, V., Manzini, R.: Differences and similarities in managing technological collaborations in research, development and manufacturing: a case study. Journal of Engineering and Technology Management, 193–224 (2000)
7. Camagni, R.: Innovation networks: Spatial perspective. Belhaven Press (1991)
8. Celaschi, F., Collina, L., Simonelli, G.: Design for district. progetti per un distretto. POLI design Editore, Milano (2001)
9. Chayutsahakij, P., Poggenpohl, S.: User-centered innovation: The interplay between user-research and design innovation. In: Proceedings of The European Academy of Management 2nd Annual Conference on Innovative Research in Management EURAM, Stockholm, Sweden (2002)
10. Chesbrough, H., Teece, D.: When is virtual virtuous? organizing for innovation. Harvard Business Review 74, 65–71 (1996)
11. Chiesa, V.: R&d strategy and organisation. Imperial College Press (2001)
12. Chiesa, V., Manzini, R.: Organizing for technological collaborations: a managerial perspective. R&D Management (28) (1998)
13. Christensen, C., Rosembloom, R.: Explaining the attacker's advantage: technological paradigms, organizational dynamics and the value network. Research Policy (1990)
14. Csikszentmihalyi, M., Rochberg-Halton, E.: The meaning of things: Domestic symbols and the self. Cambridge University Press, Cambridge (1981)
15. de Brentani, U.: New industrial service development: scenarios for success and failure. Journal of Business Research (32), 93–103 (1995)
16. Dosi, G.: Technological paradigms and technological trajectories. Research Policy, 147–162 (1982)
17. Doz, Y.: The evolution of cooperation in strategic alliances: initial conditions or learning processes. Strategic Management Journal (17), 55–83 (1996)
18. Durgee, J.: Freedom of superstar designers? lessons from art history. Design Management Review (3), 29–34 (2001)
19. Flichy, P.: L'innovazione tecnologica: le teorie dell'innovazione di fronte alla rivoluzione digitale, pp. 412–453 (1996)
20. Forlizzi, J., Mutlu, B., Di Salvo, C.: A study of how products contribute to the emotional aspects of human experience. In: Proceedings of Fourth International Conference on Design and Emotion, Ankara (July 2004)
21. Freel, M.: Sectoral patterns of small firm innovation, networking and proximity. Research Policy, 751–770 (2003)
22. Friedman, K.: Theory construction in design research: criteria: approaches, and methods. Design Studies (24), 507–522 (2003)
23. Geertz, C.: Deep play: Notes on the balinese cockfight. The interpretation of cultures, 412–453 (1973)

24. Gotzsch, J.: Creating products with symbolic value. In: Proceedings of the 3rd European Academy of Design Conference, Sheffield (1999)
25. Hagedoorn, J.: Understanding the rationale of strategic technology partnering: interorganizational modes of cooperation and sectorial differences. Strategic Management Journal 14(5), 371–385 (1993)
26. Hargadon, A., Sutton, R.: Technology brokering and innovation in a product development firm. Administrative Science Quarterly (42), 716–749 (1997)
27. Henderson, R., Clark, K.: Architectural innovation: the reconfiguration of existing product technologies and the failure of established firms. Administrative Science Quarterly (1990)
28. Henkel, J.: Software development in embedded linux: Informal collaboration of competing firms. In: Proceedings der 6 Internationalen Tagung Wirtschaftsinformatik (2003)
29. Iansiti, M., Levien, R.: Strategy as ecology. Harvard Business Review (2004)
30. H. J.: Industrial design. Thames and Hudson London (1990)
31. Karjalainen, T.: Strategic design language: Transforming brand identity into product design identity. In: Proceedings of the 10th International Product Development Management Conference, Brussels (2003)
32. Kelley, T.: The art of innovation. Curreny, New York (2001)
33. Kodama, F.: Emerging patterns of innovation. Harvers Business School (1995)
34. Kotabe, M., Swan, S.: The role of strategic alliances in high-technology new product development. Strategic Management Journal (16), 621–636 (1995)
35. Kotler, P., Scott, W.: Marketing management. Prentice Hall International (1999)
36. Krippendorff, K.: On the essential contexts of artefacts or on the proposition that design is making sense (of things). Design Issues, 9–38 (1989)
37. Kumar, V., Whitney, P.: Faster, deeper user research. Design Management Journal (2), 50–55 (2003)
38. Lerdahl, E.: Using fantasy storywriting and acting for developing product ideas (2002)
39. Lipparini, A., Lorenzoni, G.: Le organizzazioni ad alta intensità relazionale. riflessioni sui processi di learning by interacting nelle aree ad alta concentrazione di imprese. L'Industria (4) (1996)
40. Lloyd, P., Snelders, D.: What was philippe starck thinking of? Design Studies (24), 237–253 (2003)
41. Lojacono, G., Zaccai, G.: The evolution of the design-inspired enterprise. Sloan Management Review, 75–79 (2004)
42. Maffei, S., Simonelli, G.: Il design per i distretti industriali. POLI design Editore, Milano (2000)
43. Margolin, V., Buchanan, R.: The idea of design: A design issues reader. MIT Press, Cambridge (1995)
44. Marzano, S.: Suffusing design through the organisation. Design Management Journal (2000)
45. Miles, R., Snow, C., Coleman, H.J.: Managing 21st century network organisations. Organizational Dynamics, 53–72 (1992)
46. Miller, W., Morris, L.: Fourth generation r&d. Wiley, New York (1998)
47. Millson, M., Raj, S., Wilemon, D.: Strategic partnering for developing new products. Research Technology Management 39(3), 41–49 (1996)
48. Morrison, P., Roberts, J., Von Hippel, E.: Determinants of user innovation and innovation sharing in a local market. Management Science 46, 1513–1527 (2000)
49. Nobelius, D.: Towards the sixth generation of r&d. International Journal of Project Management (22), 369–375 (2004)

50. Norman, D.: Emotional design. why we love (or hate) everyday things. Basic Books (2004)
51. Patnaik, D., Becker, R.: Needfinding: The way and how of uncovering people's needs. Design Management Journal (2), 37–43 (1999)
52. Polanyi, M.: The tacit dimension. MIT Sloan Management Review (1983)
53. Robertson, P., Langlois, R.N.: Innovation, networks, and vertical integration. Research Policy, 543–562 (1995)
54. Rosenthal, S., Capper, M.: Ethnographies in the front end: Designing for enhanced customer experiences. Journal of Product Innovation Management (3), 215–237 (2006)
55. Rothwell, R.: Towards the fifth generation innovation process. International Marketing Review (11), 7–31 (1994)
56. Roussel, P., Saad, K., Erickson, T.: Third generation r&d. Arthur D. Little Inc., Boston (1991)
57. Sanders, E.: Converging perspectives: Product development research for the 1990s. Design Management Journal, 49–54 (1992)
58. Sawhney, M., Prandelli, E., Verona, G.: The power of innomediation. MIT Sloan Management Review (2) (2003)
59. Seybold, P.: Get inside the lives of your customers. Harvard Business Review (2001)
60. Simonelli, G.: Design e innovazione nei distretti industriali italiani. In: Celaschi, F., de Polis, F., Deserti, A. (eds.) Forniture e Textile Design. POLI design Editore, Milano (2000)
61. Soh, P., Roberts, E.: Networks as innovators: a longitudinal perspective. Research Policy, 1569–1588 (2003)
62. Stein, E., Iansiti, M.: Understanding user needs. Harvard Business School Publishing (1995)
63. Sutton, R.: The weird rules of creativity. Harvard Business Review, 95–103 (2001)
64. Thomke, S., von Hippel, E.: Customers as innovators: a new way to create value. Harvard Business Review (2002)
65. Turpin, T., Garrett-Jones, S., Rankin, N.: Bricoleurs and boundary riders: managing basic research and innovation knowledge networks. R&D Management 26(3), 267–282 (1996)
66. Verganti, R.: Design as brokering of languages: The role of designers in the innovation strategy of italian firms. Design Management Journal (3), 34–42 (2003)
67. Verganti, R.: Gestire linnovazione design driven. In: Zurlo, F., Cagliano, R., Simonelli, G., Verganti, R. (eds.) Innovare con il design: Il caso del settore dell'illuminazione in Italia. Il Sole 24 Ore (2002, 2003)
68. Verganti, R.: Innovative through design. Harvard Business Review (2003)
69. Verganti, R.: Design, meanings, and radical innovation: A metamodel and a research agenda. Journal of Product Innovation Management, 434–456 (2008)
70. Verganti, R.: Design driven innovation e changing the rules of competition by radically innovating what things mean (2009)
71. Veugelers, R., Cassiman, B.: Make and buy in innovation strategies: evidence from belgian manufacturing firms. Research Policy (28), 63–80 (1999)
72. Von Hippel, E.: Democratizing innovation (2005)
73. Zaltman, G.: Rethinking market research: Putting people back in. Journal of Marketing Research (23), 424–437 (1997)

Chapter 5
Enhancing the Networked Enterprise for SMEs: A Service Platforms – Oriented Industrial Policy*

Bruno Basalisco and Guido M. Rey

Abstract. This work considers the application of a platform strategy in order to promote the adoption of advanced network infrastructures by Small and Medium Enterprises (SMEs). First, we analyse the incentives of industrial players in a context where SMEs playing an important economic role and where traditional rather than high-tech sectors of activity are predominant – with specific reference to the Italian case. Second, we review the contributions of the economic and managerial literature on platforms or two-sided markets. We leverage this by proposing a policy application of the platform concept, showing its potential benefits in allowing the manufacturing sector to gain from access to the service economy. We classify platforms according to their function and identify those types which can serve the purpose of industrial promotion. This leads us to review governance aspect which are paramount to platform functioning given the policy context. Finally, we discuss the competition policy constraints on national industrial policy (and their evolution in recent times) with a reference to the EU context.

5.1 Introduction

Advances in network technologies and related services can be a key factor supporting firms' innovation and growth. In particular, the networked enterprise business model and the underlying enabling technologies offer a vast potential to firms

Bruno Basalisco
Scuola Superiore Sant'Anna, Pisa and Imperial College Business School, London
e-mail: b.basalisco@imperial.ac.uk

Guido M. Rey
Scuola Superiore Sant'Anna, Pisa

* A previous working paper is at the origin of this study: "Varaldo R., Rey G.M., Ancilotti P., Frey M. (eds.), La Diffusione dei Servizi Innovativi in Rete. Linee Strategiche. Lab In-Sat della Scuola Superiore Sant'Anna di Pisa. Edizioni ETS Pisa, 2006".

G. Anastasi et al. (Eds.): Networked Enterprises, LNCS 7200, pp. 59–81, 2012.
© Springer-Verlag Berlin Heidelberg 2012

willing to invest in this innovation. Notwithstanding the potential benefits that companies can derive from network services, the adoption of these technologies varies
across industries and firms' sizes. As a consequence, economies where SMEs play a
major contribution and where this is confined mainly to low-tech traditional industries, show levels of diffusion of network technologies below international averages.
An empirical analysis of the process of ICT diffusion in Italian SMEs highlights
that these firms fail to realise the expected complementarities between ICT and the
associated investments in human capital and organisational change.

On a similar note, the delay across the whole of Italy in embracing the technical
and organisational change associated to ICTs threatens to foster a split between
Italian businesses and their counterparts in the global economy: Italian SMEs risk
being trapped in a business digital divide. Today, digital, sectorial and geographic
divides are intertwined and compounded; in a "flat" world where competitiveness is
informed by the breadth and depth of each country's connectivity [11], these divides
threaten to weaken the performance of the Italian economy.

Many firms have deployed ICT while switching to a new business model. Scholars who have focused on corporate investment on ICT broadly agree that, for corporations to maximise the benefit from using advanced ICT infrastructure, significant
business transformations must occur.[1]

The extant literature mainly explains what makes a firm's ICT investment successful (or not), which can facilitate and thus encourage ICT deployment strategies.
Nonetheless many firms currently fall short from tapping into the benefits of advanced ICT infrastructures and services, which would be beneficial to society as a
whole. Thus a research gap exists as to the normative issue of how to make sure
that those laggard firms are instead able to fruitfully invest in ICT. We believe that
such line of research is a valid response to the key observation made by Bresnahan and Trajtenberg, who showed that, because of its general purpose technology
(GPT) nature, the development and match of supply and demand of ICT present
peculiar challenges: "in particular, if the relationship between the GPT and its users
is limited to arms-length market transactions, there will be 'too little, too late' innovation in both the GPT and the application sectors" [5, , p. 103]. A policy aimed at

---

[1] There is a wealth of literature in economics and management strategy analyzing how firms
can stand to benefit from investment in their ICT systems. An oft-quoted example is the
clothier Zara, whose business model is highly synergic to the ICT infrastructure embedded throughout its supply chain [10] (Ferdows et al., 2004) – providing the firm with a
strong competitive advantage. Some authors stress that in order to yield benefit to a firm,
ICT investment needs to be matched by some changes in its organization and business. For
instance, a successful organizational adoption of ICT is weaved with the management of
knowledge processes [2, 3]. If the latter does not match the former, the end results on firm
productivity of the new technology may be mitigated or lost completely. This phenomenon
can be generalized by stating that for firms there are complementarities in production between internal reorganization and ICT investment. Finally, taking a techno-economic perspective, it has been argued that ICT follow in the footsteps of the steam engine and the
electric motor in embodying the general purpose technology (GPT) archetype, because of
their extensive breadth of applications and impact on economic growth[5].

ameliorating such technology-dependent market failure is therefore a relevant research subject.

If a given firm's ICT investment can be more successful when matched by the move to a revised business model, we argue that this intuition can be transposed within the policy – rather than corporate – domain. We do so by introducing a policy equivalent of the business model construct, which we will refer to as "policy model". We embrace this novel approach as we believe that it enables both a fruitful positive analysis of the systemic challenges that prevent a silent majority of firms (frequently the smallest ones) to benefit from advanced ICT – as well as normative suggestions as to a potential avenue for a policy of industrial development.

The following section links the context of our analysis – namely the challenges for Italian SMEs to fully exploit the innovative potential of advanced network infrastructures and services – to the policy model which we will analyse: a platform approach.

The platform policy model is in fact seen here as a tool to devise policies aimed at mitigating the structural barriers to the adoption of novel adaptive infrastructural technologies and business models by SMEs. In order to do so, we will draw from the insights of the burgeoning literature on platforms in the fields of industrial organisation and strategic management. This allows us to highlight why and how it is beneficial to induce SMEs to join ICT-enabled services platforms; we will stress the pivotal role that market-consistent policies can play in the promotion of investment in advanced network infrastructures.

5.1.1 The Strategic Links between the SME Industrial Context and the Policy Approach Analysed

The technology adoption story in the economics and business literature is frequently centred on the incentives informing the decision of a single firm (as to whether to invest in a new technology or not), with respect not only to its competitors' actions but specifically when the technology supports complementarity between products and services (downstream or upstream), and when compatibility issues affect the strategic incentives to innovate. More efficient outcomes can be associated to the level of coordination between firms and this, in turn, boils down to trust and to the institutional framework underlying firms' access to markets.

While network effects can drive fast adoption of successful and popular ICT consumer technologies and services, these adoption patterns are not always the case: as well as many take up stories, so have several unsuccessful network products or services entered and exited the markets. This follows from poor coordination in joining networks: an important challenge for Italian SMEs to overcome.

First, the higher the number of actors (consumers, firms or NGOs such as standards bodies) composing a sector is, the more difficult it is for sufficient coordination to arise. This can be associated to the free-riding problem (analysed in game theory): where there are spill-overs from investment (e.g. in network technologies), the incentives for some agents to exploit other agents' investments (without any contribution) increase as the number of members of a community increases. As a result, every time a single firm goes alone in investing in a network technology, other firms may subsequently benefit from the risk reduction and perhaps from connecting to the deployed infrastructure.

Second, on top of the above, an important question that is clear in the mind of every business assessing the benefit of networking technologies is whom can these technologies allow to meet and whether existing and potential clients can be found on the network. Therefore, it is not only the number of participants on a network but also their kind that will matter in adoption decisions and in business investment. This suggests that a platform – rather than simply a network framework – can best represent the issues surrounding business use of ICT products and services. In the light of this coordination challenge, we see that each firm's decision to invest in network technologies and in the organisational change required to leverage them will depend on the counterparts this business faces and on any institutional / regulatory characteristics which may mitigate (or exacerbate) the investment coordination problem,

In countries such as Italy, where a multitude of SMEs compose the backbone of the economy, the presence of both issues (free riding; counterpart uncertainty) makes ICT investment coordination extremely complex and obstructs the network adoption dynamics, thus explaining the lag in the business investment in ICT.

Therefore, there are structural reasons which call for special awareness by Italian policy-makers of the innovation challenge which the networked enterprise business model and technologies pose to Italian SMEs and as a consequence to the whole Italian economy.

The past successes of Italian SMEs have often been associated to an ability to be on the cutting edge of the low-cost dimension and to an entrepreneurial drive to seek new markets. This business model is now mimicked by several emerging countries, which benefit from lower costs of labour and equivalent access to an ever more interconnected global economy. The low-cost concept permeates all cultural barriers and is aptly conveyed via the communication infrastructures (as an inspection of the eBay online marketplace will confirm for the C2C and B2C domains). Notwithstanding their legacy of low-cost production, Italian SMEs are currently striving to transition to a model based on competitive advantage derived from high-quality production, which is less replicable and therefore more sustainable. The journey which must be undertaken today by a firm wishing to sell the quality of its products is significantly different from the venturing salesman journey which may have been sufficient to market low-cost products a few decades ago: quality is intrinsically intertwined with culture, which implies that intangible resources and service provision are key future drivers for Italian SMEs to succeed at marketing quality manufactured goods.

The quality strategy requires a host of complementary services to support not only the production process per se (and the underlying organisation) but also the ability for firms to market their quality goods effectively on a global scale. This is because in a business world where the value paradigm is defined by the services content (even for manufactured goods) the factory – while an important source of value adding activities – is not anymore the central component.

In order to leverage ICT for the purpose of adding value creation to production activities, a fast internet connection (which in any case is not to be taken for granted in many areas of Italy) and access to eBay are not simply enough – as many SMEs are well aware of.

As stressed by marketing scholars, a new service paradigm is now the perspective through which all economic transactions– including those pertaining to manufacturing activities – must be seen in order to understand their value [31]. Because of this, a key challenge for manufacturers is the overcoming of cultural barriers between supplier and customer; this is harder for SMEs to perform without the support of dedicated advanced professional services which at the moment only large companies find available and can afford to tap into.

5.2 The Challenges and Opportunities for SME ICT Services-Enabled Innovation

Modeling innovation is often a tough challenge for large firms, which are frequently encumbered by layers of bureaucracy and have larger opportunity costs descending from the scale of existing business, thus resulting less responsive to market opportunities. On the contrary, the conversion to new business models is likely to be less difficult for smaller firms. Nonetheless, SMEs lack the resources that a large firm can mobilise (in-house strategy management, consultant advice), which can hinder access to professional business services which are fundamental in enabling business innovation. In both cases a vital characteristic which the top managers and/or entrepreneurs must hold for the business transformation to succeed is the capability to evaluate and manage correctly risk in innovative organizations, products, processes and related investments. This is the case whether the assets on which the firm opts to invest are both tangible or (even more importantly) intangible.

In this section we present the two main policy questions which we believe must be addressed in order to successfully promote ICT-enabled innovation in Italian SMEs. This analysis is what leads us to propose in this study a platform policy approach in order to mitigate the resource access challenge structurally associated to the Italian SMEs.

5.2.0.1 Which SME Is Best Suited for ICT-Enabled Innovation?

Firms' innovation relates strongly to how flexible businesses can be in tapping new markets for innovative products (in manufacturing industries) or services or a mix of both. A significant share of Italian manufacturing SMEs are likely to be tied to their markets (for instance to a larger firm which is their sole buyer; or as part of a supply chain) so the choice of their business model is frequently highly interdependent to their industrial milieu rather than independent from it.

Furthermore, another large part of Italian SMEs present a significant industrial legacy of production in low-tech sectors where competition is on the low-cost dimension rather than quality differentiation – an important strategic characteristic [25, 26]. Recovering the costs of capital equipment investments such as ICT can be hard unless firms can be confident to be able to tap into the quality-enhancing potential that the decentralised enterprise business model can deliver. In order to reap the benefits of investments in quality and value creation rather than costs, firms need access to an appropriate range of business services. These services will be most valuable the more suited they are to the firms' scale, sectorial, geographic and perhaps cultural characteristics and their business challenges. This leads us to the following policy question, discussed below.

5.2.0.2 Why Is an SMEs-Focused Service Platform Necessary to Promote Innovation?

The analysis of platform industries would be limited if we were to constrain our view to a framework of upstream/downstream markets (or input/output products or services). Industrial ecosystems are more complex than that: SMEs clusters are a prominent example of forms of industrial organisation where complementarities in products and services play a key role (supported by local proximity).

A related question is how to leverage the platform framework to disentangle the reasons why SMEs lag in exploiting advanced ICT solutions, failing to innovate their business model by investing in technology, human capital and organisational improvements.

5.3 Platforms and Their Functions

The platform perspective suggests that the large benefits of the networked enterprise business model can be placed within the reach of SMEs where institutions and actors face incentives to match markets for products and services. A simple model of SMEs support can start from the promotion of one level of platforms serving business needs, while then envisaging sub-platforms nested at several levels of the value chain. In order for this solution to be viable, policy-makers need to be aware of the issues highlighted in the economics and strategy literature on platforms, from which several insights can be drawn, and on **the economics of multi-sided markets**.

5.3.1 An Alternative Approach to Platform: Acknowledging Different Platform Concepts

Taking into consideration the review of the literature on platforms we have noted that different interpretations of the platform concept are holding ground.

First, platforms can be seen as intermediary agents who enable transactions between two groups of agents, which are mediated by their organisation. Platforms extract value inasmuch as they can support beneficial transaction (which will depend on existing externalities) and can measure (and keep within the platform) the interaction between those on the platforms. This conceptualisation reflects the more static consideration of platforms from the economic point of view. In this case we should refer to platform A.

Alternatively, in a managerial perspective, platforms embody a technology and market coordination device, thus a mighty strategic lever for the organisations who control them. Platforms empower participants to develop new markets based on the provision of platform-complementary services or products. These platforms can facilitate innovation in technology standards by mediating and coordinating the interests of consumers and suppliers. When these aspects are present we will refer to a platform of type B.

Finally, a third type of platform describes a firm which holds a competitive advantage based on a previous innovation effort (i.e. embodied in a set of patents). This firm enhances its potential market by cajoling supplier of complementary services and its clients to enlarge the range of solutions, services and goods which incorporate the initial platform proprietary innovation – thus greatly increasing the diffusion of the latter. We call this Platform C.

Note though that more than one of these aspects may coexist within the same platform, which explains the current ambiguity in the academic conceptualisation of platform; as well as explaining the strong and sustained interest in the platform construct across disciplinary boundaries. For the reasons outlined above, this work transposes the use of the platform construct onto the public policy domain; specifically we argue that, because of their multiple facets, platforms can enable a successful pursuit of public policies aimed at connecting and empowering marginalised economic actors by leveraging their diverse interests, characteristics and capabilities.

A recent flurry of economic literature discusses the incentives of a platform firm which can earn revenues from more than one side of the market. Two-sided markets are those where platforms revenues – coming from two distinct groups of actors – depend on the interactions between the different categories of subscribers to the platform [4, 20, 27, 28, 33]. A platform firm succeeds inasmuch as it can generate value by enabling transactions between users of the platform – and capture part of this value. These benefits depend of the nature of the externalities that the platform is able to internalise. When both sides of the market gain from the interaction with the other side, the platform firm faces a dynamic challenge commonly referred to as the chicken-and-egg problem [6]. This issue occurs in two-sided platforms exactly because they differ from merchants [21]. A new merchant could enter a market by investing part of its capital on stocking supplies; having done that, it could then proceed to resale at a marked-up price. Instead platforms do not "stock up" one side of the market: both sides of the market have to interact simultaneously.[2]

Note that when there are bidirectional externalities, these can potentially be asymmetric (for instance positive in one direction and negative in the other). The strategic management of such externalities is particularly challenging for new platform firms, who have to devise strategies to elicit custom on both sides of the market and may find it beneficial to focus on one side only to start from [9]. Asymmetric prices embodying a subsidy across market sides are likely to be optimal where the platform externalities are not symmetric. This is because the platform has more incentives to enrol those on the side of the market which is relatively more attractive to the other. Non-price factors can play a strong role too. Expectations management and key actors (such as marquee customers) may influence significantly whether a new platform will thrive or fail to attract sufficient interest on both sides of the market.

The two-sided markets framework is used to support rigorous discussions of pricing decisions and welfare implications of different platform structures. Due to their scope, the seminal studies in this field set the issue of innovation aside, while focusing on models where the markets to be analysed are taken as given and the key driver of a platform's success (in terms of adoption) is pricing or any economic return for the services they supply to both sides

On the other hand, inasmuch as platform industries present remarkable rates and patterns of innovation, evidence of market creation (and of the strategic efforts underlying such creation) suggests that the assumption of a given set of markets is untenable. A key driver of business innovation is in fact firms' – whether large or small – aims to expand beyond a given market type.

Pricing is only one side of the innovation story in platform businesses. Firms such as Intel which aim to increase the systemic value of a general purpose technology such as the microprocessor, use a variety of non-price strategies to induce value-enhancing complementary innovation [14]. Platform owners who strive to orchestrate innovation around a proprietary innovative core (Platform C) do not confine themselves to promoting adoption decisions in existing markets which depend on the platform. They seek to engage complementors (firms supplying complementary products or services) into developing new markets – adopting both pricing and non-pricing strategies [16].

The latter platform firms derive their success from careful management of their industrial ecosystem. In the long run, a platform owner (whether a key player

---

[2] Moreover, unlike merchants, platforms take minimal liability for any goods / services provided by one of the sides to the other and intermediated by the platform itself. This is currently the case in many virtual marketplaces (Platform A), although some actors such as eBay / PayPal are introducing procedural safeguards to reassure bidders against potential frauds. Instead, online portals (frequently based on a "walled garden" model) are expected to guarantee the content of the goods and services associated to their environment. This is the case in many mobile internet portals which enable their users to purchase and download applications such as ringtones, etc.

amongst a few competing platforms or the single platform) will grow if the end use of products and services which the platform underlie will expand through time. This will of course have a positive spill-over effect on the firms which have invested so to be able to provide goods and services complementary to the main platform product / service.

The analyses of platforms incentives reveal that different models of competitions can coexist – therefore there is not a one-size-fits-all policy to promote platform development (such as for instance a "the larger the better" platform promotion policy). The economic literature on platforms in fact suggests that competition between several platform agents could endogenously lead to asymmetric equilibria [1]. So, while size (i.e. adopted base across all markets connected via the platform) is a key platform value driver, platforms can successfully specialise and this will result in industries populated by platform of different shapes and sizes. Firms in industries where establishing a platform leadership position is of fundamental importance frequently attempt to leverage usage adoption decisions in their favour, often investing large amount in pricing subsidies to cajole actors from the most valuable side of the market on board [15] .

On the other hand, in many industries there may be room for more than one viable platform – and sometimes the dynamics of pricing incentives and market niches is such that platforms with smaller users bases can be significantly profitable too.

5.3.2 Complementors and Market Creation

The analysis of platform industries would be limited if we were to constrain our view to a framework of upstream/downstream markets (or input/output products or services) as is the case of platform A. Industrial ecosystems are more complex than that: SMEs clusters / districts are a prominent example of forms of industrial organisation where complementarities in products and services play a key role (supported by local proximity). The platform literature so far recognises that in key high-technology industries some firms play a pivotal role in orchestrating innovation throughout their ecosystem – both with downstream and upstream partners. While the former concerns adoption of innovative technologies in user markets (which is fundamental for the underlying platform product to hold a competitive edge), the latter revolves around enhancing the platform value by adding complementary components specifically honed to the platform characteristics (Platform B). A successful platform owner focuses on expanding the overall pie, which depends on all uses of platform-based products and services. Hence, it has an incentive to promote product development in complementors. A platform owner attempting a type C strategy is unlikely to succeed in this pursuit, though, if it sets a one-off transaction as a "procurement" of the complementary products/services development sought. Major ICT firms such as IBM and Philips have mishandled investment into leveraging complementary product developments in the past. This is simply because they failed to consider the need for the complementor to commit to a business plan for the new product. Missing that, the new complementary product developments – dearly paid by platform owners – failed to sustainably contribute to the overall platform development [16].

A platform wishing to promote the development of complementors must avoid simply purchasing the complementary products or services: by doing so it would revert to a supply chain. Smart platform development relies instead in finely crafted inducements for complementors to find their own new markets for their platform-based products / services. By doing so, the value of the platform will be substantially increased in a sustainable way, since the incentives of the complementors, the customers and the platform itself will be aligned (which is not always the case in a supply chain where one value layers could be squeezed from market power downstream in order to extract more rents).

Thus platforms must take care not to foster dependency on any "subsidy" in the firms selling platform-based complementary products / services.[3] Specifically for a platform organisation, promoting (teaching to fish) – rather than simply commissioning (giving the fish) – an innovative product / service requires a joint effort from both platform owner and complementor to create a new market. In order for this to work, solid risk-sharing arrangements must underlie this enterprise, so to provide the correct incentives to both parties.

The relationship between platform firms and complementors can be informed by how firms use hard and soft power [34].[4] Hard power includes threats and direct (financial) incentives to manage complementors. The authors argue this can be effective in the short run but may not be sustainable in the long run – for instance by causing a backlash. Where the influence of a platform firm over a complementor depends on payment for a service, common purpose may be lacking. This will imply a misalignment of incentives, a liability in the long run. On the other hand, soft power – the ability to influence by leveraging shared long-run incentives rather than immediate contractual terms - can hold profound strategic implications where a platform firm and its complementors negotiate investments aimed at market creation.

Moreover, platforms embody not only transactional links but also underlying relational networks which inform the business relationships. While the implications of the nature of relationships within social networks lies beyond the scope of this work, it is the subject of a vast literature which highlights how key gatekeepers of networks otherwise unconnected (such as platform actors may be for different sides of the market) wield power – the so called "strength of weak ties" argument [18].

In conclusion, market creation is a key determinant of platform innovation and is thus different from simple "innovation procurement", where risks are borne entirely by the platform owner.[5] The possibility for firms to add value by providing

---

[3] A Chinese proverb famously states: "Give a man a fish and he will eat for a day. Teach a man to fish and he will eat for the rest of his life."

[4] Yoffie and Kwak draw the notions of soft and hard power from the field of political science [23].

[5] We note that market creation efforts are not confined within platform strategies: many non-platform firms attempt to achieve the same goal by means of engaging and "open" marketing strategies.

complementary services is what defines the potential a platform in enabling innovation beyond mere match-making across market sides.

5.3.3 Governance and Risk Sharing Profitability

Because of the relationship-specific nature of the investment of a complementor in a platform, the latter is in a position of power, when the engagement between actors pans out through time. This is similar to the hold-up problem analysed in the economics literature [22]. In a scenario where a set of complementors can only access one platform (i.e. because of their nature, location, industry), the latter could extract all value derived from the relationship, once this is established. Platform governance can therefore provide the means to address this issue by the adoption of common rules. An example of a platform rule could be that the platform extracts only the positive externalities deriving from its market-matching activities, while redistributing all remaining profits to the complementary agents – a cooperative arrangement.

The concept of governance does not only concern the exercise of market power but also involves political considerations. This is especially poignant where a platform organisation is not exclusively a commercial entity but instead holds characteristics of a governmental or non-governmental organisation (e.g. reflecting a public sector stakeholder or a cooperative charter). By political we refer then to a shared understanding within the community of platform stakeholders – and of the ecosystem of firms transacting with the platform – of the division of power and responsibilities on the strategy and conduct of the platform itself.

As the literature on platform strategy has shown [14], firms who own platforms strive to establish a credible commitment towards supporting the complementor ecosystem. This is done for instance by investing in quasi-independent divisions tasked with supporting innovation in the industry, e.g. via the selective dissemination of IP, engineers / researchers time and the establishment of consultation processes at the stage of innovation design (the Intel case provides a clear illustration of this). In this context of type C platform, the key strategic challenges for the platform are: i) how to resolve the initial make-or-buy trade-off (make of core platform technology assets; delegate/buy of complementary products); but also ii) how to address the incentives within its organisation so that it can both sell (its own core product) and let sell (i.e. have a commercial and legal framework enabling its complementors to develop and maintain viable businesses centred around the platform core). While the above example is set in a commercial platform context, similar strategic challenges will need to be clearly resolved also where platforms have elements of non-governmental or public sector governance, addressing the key questions of who does what and how are competing incentives to appropriate value managed.

For large multinational corporations who seek a platform leadership role, this translated into the opportunity to invest resources in establishing in-house divisions whose aim is tantamount to industrial policy, only outside a governmental framework. In order to do so, these firms need to establish organisational rules and boundaries so to sustain the credibility of their commitment to supporting the whole industry, rather than undertaking exploitative divide et impera strategies. The commitment is embodied by the quasi-regulatory status of the arms-length relationship between commercially-minded units and the industrial-promotion units inside these large firms. A similar principle of division of responsibility and functional separation can ensure platform effectiveness even in contexts where the platform is not corporate.

In Italy, where very large corporations within the national industrial system are scarce, other forms of organisations could still replicate such a platform model, leveraging some of the above strategies. This model can be successful provided that the strategic rationale is focused to a key characteristics of the industrial context, i.e. the pervasive and strong economic contribution of SMEs, which can become the subject of a platform policy.

Given the specificity of the Italian industrial background (above all its fragmentation), platform governance is thus fundamental in ensuring that platform organisations (however established) will pursue the long-run interest of industry promotion, while maintaining a sustainable share of profit. In turn, the trust enabled by the strength of the platform commitment will provide the basis for widespread platform adoption by firms of any size.

The current absence of platforms which aim to serve the interest of SMEs can be associated to the structural characteristics of both the manufacturing industries where SMEs are concentrated in districts and also in the service industries which do not currently include value creation services for SMEs as a relevant stream of business. For instance, while the Italian banking system has introduced network services ("Corporate Banking Interbancario" or CBI) used by SMEs, the services themselves are designed so to reflect the needs of the banks themselves rather than elicit the set of needs which SMEs could satisfy with an advanced tailored e-banking platform. This follows from the joint efforts by multiple parties within the banking system to create a shared infrastructure and electronic services base, funded exclusively by the banks.

Targeted policy intervention embracing all services domains could potentially alter the status quo and bridge the existing market gap / failure. In fact, Gawer (2009) suggests that evolutionary trajectories may define which types of platform characterise an industry: under certain conditions, internal platforms can evolve into supply-chain platform, which on their turn can become industry platforms. Thus, policies aiming to ameliorate the competitiveness of SMEs via the promotion of supply-chain platforms may want to target existing internal platforms (i.e. proprietary to a firm's product / service design) and devise the incentives for these platforms to be shared across the supply chain. Note that this process could be detrimental to the firm currently holding the internal platform (see the analysis of the IBM case in Gawer, 2009). While internal platforms that are based on products rely on design skills and are protected by the system-maker via formal intellectual property such as patents, those based on services – the key target of a policy aimed to enhance the competitiveness of SMEs – will be dependent on "softer" ways (including the

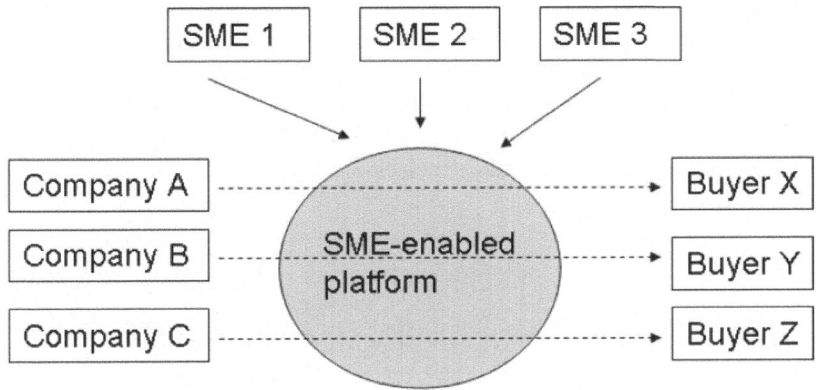

Fig. 5.1 The organisation of industry under a SME-enabled approach

use of existing sectorial and geographic relational networks) to defend competitive advantage at the systemic level, such as trade secrecy, business contacts, etc.

We argue therefore that – by taking into account the characteristics of the diverse range of platforms which already operate across many markets – it is possible to conceive and define a set of platform specifically targeting the SME needs. Specifically, we consider that SMEs can engage in different ways with platforms. A first approach is to consider a platform which allows firms (predominantly large firms) to trade with each other, perhaps facilitating international trade. This platform may compete with an alternative distribution channel, where a merchant will own the suppliers' stock and resale at a margin to the buying firms. Consider an SME that can provide value added services (which could be embodied in a product associated with the products traded by the large firms). This SME will have to overcome several legal and commercial constraints in order to be able to sell its services atop the main goods traded by the merchant and its distribution channel. On the other hand, the same SME – if engaged with a trading platform which incorporates by design the provision of complementary services atop the main traded goods – could draw on the (large) buying firms' incentives to select as trading partner only (large) supplying firms which enable the complementary services. Large suppliers, which benefit from the extra platform trade, have an incentive to allow SMEs services to complement their goods as they may be at a competitive disadvantage against other suppliers on the platform who do so. Finally, this platform may be competing with other trading platforms which may not enable SMEs engagement; if SMEs products/services bring net value added to the platform, then this can provide a source of competitive advantage to those platform who tap into their potential. We will refer to this as the **SME-enabled platform** approach (cf. Fig. 5.1).

A second approach considers a platform where SMEs supply the main goods, seeking trade with buyers (e.g. international buyers). Here, the platform not only intermediates between the SMEs and their (new) target markets, but also enables suppliers of professional services (marketing, communication, finance, ICT consultants) to come in and engage with SMEs. Thus, by aggregating SMEs, the platform provides a mutually beneficial opportunity for professional services firm to tap into underserved SME markets, simultaneously allowing SMEs to invest in these value-adding professional services to improve the service-value of their goods. By similarity with the established wine cooperatives (commonplace in Italy, amongst other countries), we define this as the **services-coop platform** approach (Fig. 5.2).

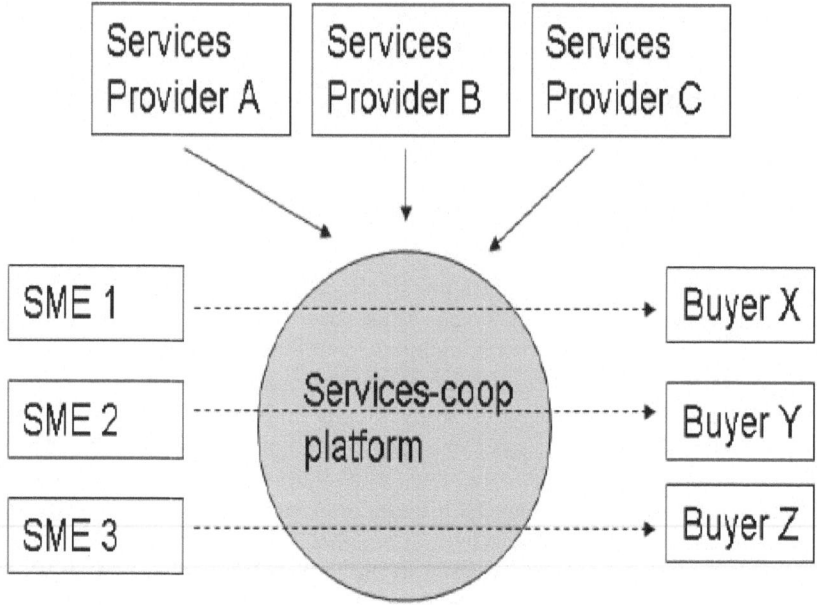

Fig. 5.2 The organisation of industry under a services-coop approach

5.3.4 Innovation in Platforms: The Challenge of Market Creation

One of the major obstacles that Italian SMEs face as they approach renewing their business models along the lines of the networked enterprise is that a relevant component of their competitive advantage derives from strong links with the firm's markets. The organisational, business and human capital changes needed in order to embrace ICT – so to transform the firm into a part of a networked system – can be seen as orthogonal to the tight market link which SMEs often perceive as the key source of their competitive advantage (e.g. supplier relationship to a major buyer, niche position, etc.). Instead, we argue that the benefits that SMEs can gain from investing in network technologies (and the organisational factors which are its complements) can be enhanced by the market creation role of platforms. This is especially relevant to the services-coop platform approach introduced above, namely that which enables professional services firms to complement and add value to SMEs products (e.g. by enhancing their marketing or communication potential).

The economic and strategy literature on market creation frames this issue as a special case of technology adoption / diffusion models or in terms of new product development / marketing issues [17, 19, 24]. The key question tackled is the timing of adoption of new technology, modelled over a continuum of options. The seminal paper of Fudenberg and Tirole (1985) shows how the adoption game can lead to different Pareto-efficient equilibria, which can coexist in a duopoly scenario (although not in a general oligopoly framework). In particular this could be either a delayed joint adoption or a preemptive outcome, where one firm introduces the new technology as first-mover.

Joint adoption will result only when the first-mover imposes a large enough negative externality on the laggard. This follows from the risky nature of investment, as each firm trades-off benefits and risks of its unilateral, uncoordinated action. As Ruiz-Aliseda and Zemsky (2006) point out, this cannot be the case where adoption concerns new markets: preemption will then be the only optimal strategy. Firms who do not embrace the innovation first can be the clear-cut losers and forego entry to new markets. In these models the first-moving firm will be that which benefits from cost/technology advantages and risk.

This is one side of the story. Firms do not only enter markets because of access to a new technology; instead, market creation can be a distinct strategic effort, where new technologies can play an enabling, though not sufficient role. This is because the search for markets is a fundamentally uncertain process – and a pivotal one for Italian SMEs – which cannot be modelled simply as a choice to be made in a decision continuum. Specifically, we stress that SMEs platforms can improve the competitive position of SMEs by realigning the incentives of the large majority of services firms which could benefit from providing platform services; the resulting facilitated interaction between professional services industry and manufacturing SMEs would in turn benefit the latter. As a consequence, an SME-oriented platform could empower SMEs – in conjunction with service providers – to create new markets for their products.

5.3.5 *Catalysts*

Behind the rise of every platform there is a huge entrepreneurial effort and platform strategies play a key role in enabling success. Starting-up a platform requires a keen focus for the cross-market dynamics which a platform owner must manage. Overcoming the chicken-and-egg problem [6] which embodies the essence of market-matching is the aim of any catalyst strategy to build platforms. Evans (2009) describes the strategies which entrepreneurs can choose in order to cajole platform support from actors belonging to different markets.

Price incentives can play a strong role in inducing actors to join the platform – one side of the market at a time. At the same time, platform catalysts can focus on attracting key ("marquee") customers. These are defined as customers who provide relatively larger positive externalities to users from other markets, where these join the platform. Therefore, a first strategy by which an SME-oriented platform can promote adoption is by attracting an "important" player. For instance, a services-coop platform can ensure that a leading professional services supplier commits to engage with SMEs via the platform. This could have a first impact on the number and type of SMEs joining the platform; at a second stage, the presence of the leading service company could induce some of its competitors to join too to serve the SME market. This confirms that for an SME services adoption initiative to achieve maximum impact, exclusivity policies may be counterproductive – which suggests that an open platform approach may be the best policy.[6]

The surveys on the diffusion of ICT amongst SMEs show a systematic gap between a strong SMEs online B2B practice (on purchases) and a significant reluctance to use online channels for their sales. This is explained by SMEs themselves as a result of several factors:

- No trusted third parties can provide an alternative retail channel and the SMEs' customers prefer direct face-to-face contact;
- The trust between buyer and seller is important to obtain the demanded product and ensure agreement on the exact price;
- The product is not standardised enough and must be customised on a case-by-case basis.

Moreover, where a transaction is performed online, it takes place on the website of one of the two parties, rather than on a dedicated portal. In practice, what any e-commerce arrangement fails to convey are all the transactions related to the exchange of information aiming to define the object of the transaction itself.

We consider that asymmetric information and knowledge barriers pose a strong barrier for the take up of professional services by SMEs which would benefit from them. Simply put, SMEs do not hold sufficient information on technologies and managerial knowledge to be able to assess the merits of innovative services which could positively complement their business expertise and competencies – thus enabling greater competitive advantage. For instance, services such as the leasing of adaptive infrastructures could enable some SMEs to transform into a more cost efficient networked enterprise, which could be more effective at capturing demand and create new markets (or market niches); nonetheless, this type of solutions are currently almost exclusively the preserve of larger enterprises.

Taking this into account, we argue that a second catalyst strategy for an SME-oriented platform is to develop a key focus on initiatives aiming to develop SMEs knowledge and confidence in the opportunities offered by professional services. This role is to some extent similar to the functions of the technology / business advocates (or "evangelists" as common parlance in the US West coast), which disseminate and promote the concept and benefits of innovative products or services on behalf of high-tech firms. The key step here is realising that even in a low-tech industry

---

[6] This could have implications on governance. An open platform may be more likely to succeed where multiple stakeholders share risks, oversight responsibilities and rewards from the platform. On the contrary, a platform captured by a dominant player (even if a public sector organisation) may end up trading-off the benefits of openness in favour of a particular interest of the dominant stakeholder.

context, several SMEs face a de facto knowledge barrier as to the applications of high-tech solutions (products and services) and their potential to improve the productivity of firms in low-tech industries.

Therefore, apart from carefully crafting market incentives on both side of the markets, we envisage that successful SMEs-oriented platforms will engage in information dissemination and persuasive confidence building strategies (e.g. pilot projects) to catalyse platform adoption. This "evangelist" catalyst role can be pivotal at the start of a platform take up; when sufficient momentum has been gathered, this catalyst effort can be turned to another platform with a different remit (e.g. a different type of services) which is in the early stages of adoption. One way to fund this catalyst function would be by allocating a share of the platform transaction commissions resulting from its actions – as well as perhaps a policy subsidy devised as an incentive for the platform to fully develop and internalise the cross-market externalities (rather than leave some SMEs sub-markets under-served).

Finally, we note that a competitive advantage for an SME platform may lay specifically in harnessing trust from its subscribers and define itself as an "honest broker", which can be relied upon to clarify the interests of the parties. For this reason, a platform of type A could be successful by enabling interaction between the parties by means of specialised and standardised transactions solutions (products / services). At the same time, a platform C – because of its incentives to "enlarge the pie", i.e. increase the diffusion of its core platform product by enhancing the range of solution, services and goods which incorporate the core product – could also be seen as a balanced mediator of the interests of parties located on different sides of the market. On the other hand, platform B may be a setup less conducive to guarantee SME trust in the transaction domain.

These considerations will inform our discussion of a proposed policy approach for SMEs services platforms in the next section.

5.4 A Proposed Policy of ICT-Based Service Platforms for SMEs

Building upon the analysis presented throughout this work, we now proceed to elucidate a proposal to leverage the platform industrial organisation in order to achieve a policy objective: promote the adoption of advanced ICT solutions (such as, for instance, adaptive infrastructures) amongst SMEs. Specifically, we aim to fine-tune this proposed policy approach so to target SMEs which achieved competitive specialisation in legacy low-tech industries – such as the case of Italian SMEs and those in many low-tech clusters, districts or filières.

We have highlighted how SMEs can benefit from a platform organisation of industry in different ways. First, SMEs can tap into the transactions between large firms occurring within an SME-enabled platform. Here SMEs are the players providing the complementary, value-adding services to the suppliers of the principal product.

Alternatively, SMEs – which engage in direct trade with their buyers – can benefit from services platforms. These, defined above as services-coop platforms (cf. Fig. 5.2), provide a medium for SMEs to engage with professional services suppliers, which can greatly enhance the service value component of the SMEs offering (i.e. manufacture). For instance, a set of SMEs belonging to the same industry or in close geographic proximity may leverage the platform to share access to an adaptive infrastructure, maintained by the ICT services provider. By doing so, each of the SMEs could innovate around its existing product and competencies, without having to suddenly overcome a huge ICT knowledge and management barrier. In turn the service provider can benefit from the demand aggregation function performed by the platform and fine-tune an array of solutions for SMEs.

We scope our proposed policy development around this latter type of platform. What we would like to advocate is for policy-makers to consider promoting the advent of services-centric platforms targeted at SMEs. In order to do so, a policy could be envisaged so to provide incentive for professional services firms (services beyond banking) to coordinate and engage with SMEs.

As discussed above, the key to a platform's success in gathering sufficient momentum to become a sustainable business is how effective it is at catalysing the latent interest amongst parties at all sides of the market (SMEs, professional services suppliers, to some extent also buyers of the final product which may be drawn by a platform certification scheme or the higher profile associated to the input of the professional services firms). Therefore policy intervention, rather than attempting to devise a catch-all subsidy (unlikely to be as effective as expected and perhaps legally incompatible with EU laws, as considered below) could focus on support for a set of agents to mobilise SMEs interest in the professional services supplied on the platform and garner trust in the overall remit and functioning of the platform - thus providing the essential catalyst role.

Another pivotal objective of a public policy aimed at promoting platform based services (including ICT) for SMEs is the creation of a set of appropriate laws and regulations which are sufficiently clear and flexible so to elicit effective governance regimes within platforms. In fact, both during and after the catalyst stage, trust in the platform (especially by the smaller and weaker players) will depend on how transparent is the division of responsibilities, risks and rewards within the platform. Therefore a platform-specific legal framework, adapted perhaps from that which disciplines cooperative enterprises, could be an essential policy input for a fast and sustained establishment of services centric platforms for the benefit of SMEs.

The following sections provide a set of remarks on the context within which a set of policies aiming to promote platform-based adoption of advanced ICT infrastructures and services by SMEs must be set.

5.4.1 The Scope for a Platforms–Oriented Industrial Policy under EU Rules

A remark for any national-level policy-maker in the EU is that any measure aimed to promote industrial competitiveness must comply with the State aid regulations (even though recent exceptional measures aimed at tackling the recession seem to signal a change in trends and views on this matter).[7] These regulations are in place to ensure that competition inside the Single market is not distorted by the activity of member States.[8] The State aid discipline aims to achieve dynamic efficiency in competition within the EU single market.

This is because in the short-run a set of disparate national measures to shore up national champions can have positive consequences on employment and the national economies. On the other hand, if any national intervention results in the more efficient firms being penalised, then only a fully competitive dynamics across the EU single market can ensure that in the long-run social welfare is maximised, which is the aim of State aid regulations.

Nonetheless, some sectors of public interest are to some extent placed beyond the standard framework of competition policy rules (e.g. broadcasting, health), because of the belief that unfettered competition across member states may not be the best means to ensure the provision of services of public interest. Furthermore, as mentioned, the recent economic circumstances have led the EU to waive measures in support of national industries in sectors other than services of public interest (such as banking, automotive), instead choosing to establish some coordination in these actions across member States.[9]

5.4.2 ICT-Based Service Platforms for SMEs

Industrial policy measures geared to promote platforms for SMEs therefore needs to carefully tread along the lines of State aid regulations. On the other hand, horizontal policies aimed at supporting businesses across industry (and country) boundaries are a part of key EU-level initiatives led by the European Commission Directorate-General for Enterprise and Industry. Alongside this, the DG Information Society and Media promotes actions targeted at the adoption of network technologies, as part of its Information Communication Technologies Policy support Programme. Moreover, the EU-level debate on competition in telecommunications, initially concerned with issues of pricing and access, recently resonates with calls for wider access to innovation-enabling technologies. Policy-makers and businesses alike are increasingly aware that the establishment of a truly single market in the field of ICT is not only about cheaper broadband to (as many as possible) residential customers but also depends on the elimination of any barriers that may prevent telecommunications suppliers wishing to serve the advanced ICT needs of large corporations

---

[7] In turn, some of the regulations preventing anti-competitive state intervention are embodied in the international agreements associated to the World Trade Organisation, where EU member states are represented by the European Commission.

[8] A part of the State aid discipline is the "One time, last time" doctrine, which has been used as a condition in exceptional circumstances.

[9] More details about this "temporary State aid framework" are available at:
http://ec.europa.eu/enterprise/newsroom/cf/itemlongdetail.cfm?item_id=2567&lang=en.

as well as diverse industrial ecosystems [32]. This theme has several regulatory implications, which are beyond the scope of this work. In the light of the above developments, it is evident that a scheme designed to promote a platform whose goal is to allow SMEs to fully benefit from the adoption of ICT is therefore in the spirit of EU policies. Its scope may have to be set beyond the national level and include several countries – which can be a source of added benefit to the SMEs taking part to it.

We will thus now proceed to present a policy framework based on service platforms targeted to SMEs, which will be described in greater detail in the following section.

5.4.3 Which Policy and for Which SME?

Public policy needs not disrupt industry equilibria but may instead align itself with industrial dynamics trends. In the general domain of industrial policy promotion, the literature on state subsidies shows how SMEs subsidies could have had unintended negative effects due to adverse selection and moral hazard [7, 8, 30]: firms who rely on state aids and subsidies without more than a passing interest in viable commercialisation do not contribute to sustainable innovation.

Carefully crafted public policies could thus ensure that the positive economic externalities from this process can be reaped in part by the internal platform firm, so to rebalance its otherwise conflicting proprietary incentives. Public intervention can enable a healthy platform governance if it provides the right rules to prevent anti-competitive behaviour within platform. This could be the case if one of the transaction parties on the platform can act as a monopolist and appropriate in its entirety the value generated by other parties (such as SMEs).

For instance, it is possible to argue that a platform-wise legal framework could specify that under all circumstances:

- SMEs serving different products on different markets can participate to the same platform; and
- a stakeholder mechanism ensures SME representation on the platform management board; and
- a defined share of the platform-generated revenues filters back to the SMEs.

5.5 Concluding Remarks

This work builds upon a review of the concept of platform within the economics and technology and management strategy literature in order to identify a positive policy approach to ameliorate shortcomings in ICT investment by SMEs. We have identified across this literature three definitions of platform, each of which stresses the importance of their role in:

- Managing transactions across different market sides (linked by cross-market externalities and platform prices): Platform A – Economics perspective.
- Enabling the creation of new services and markets by platform participants: Platform B – Strategic management perspective.
- Promoting the adoption of a core platform technology by coordinating technological development and the necessary standards: Platform C – Technology innovation perspective.

We have highlighted how SMEs can benefit from a platform organisation of industry in different ways. First, SMEs can tap into the transactions between large firms occurring within an SME-enabled platform. In this case, SMEs provide value added products and services which are complementary to the main product exchanged via the platform. This is similar to the function already performed by many SMEs under currently established industrial structures, where SMEs deal exclusively via a pivot firm, the large supplier of the main product. Note that the latter can appropriate a share of the value provided by the SME if supplied under a vertical rather than platform industrial structure. At the same time a pivot firm may be missing out on the benefits of having its product complemented by products / services provided by SMEs other than those in close proximity to the pivot firm's cluster or filière. Therefore both large and small firms can potentially benefit from a platform approach, given its win-win nature.

As a next step we have focused on services-coop platforms and analysed how this approach to industrial organisation can enable SMEs to benefit from access to professional services inputs in order to enhance their products and capabilities. By way of example, a services-coop platform could enable specialists in the installation and strategic management of ICT (such as adaptive infrastructures) to engage with multiple SMEs, by aggregating their demand.

Moreover, we stress that a services-coop platform would require efforts to be directed towards catalysing interest, support and trust in the platform and in the technology and services provided. This is an area where strong and timely public policy support could make the difference and generate sustainable platforms for SMEs.

References

1. Ambrus, A., Argenziano, R.: Asymmetric networks in two-sided markets. American Economic Journal: Microeconomics 1(1), 17–52 (2006)
2. Antonelli, C.: A regulatory regime for innovation in the communication industries. Telecommunications Policy 21(1), 35–45 (1997)
3. Antonelli, C.: New Information Technology and the Knowledge-Based Economy. The Italian Evidence. Review of Industrial Organization 12(4), 593–607 (1997)
4. Armstrong, M.: Competition in two-sided markets. RAND Journal of Economics 37(3), 668–691 (2006)
5. Bresnahan, T.F., Trajtenberg, M.: General purpose technologies as 'engines of growth'? Journal of Econometrics 65, 83–108 (1995)

6. Caillaud, B., Jullien, B.: Chicken & egg: competition among intermediation service providers. RAND Journal of Economics 34(2), 309–328 (2003)
7. Colombo, M., Grilli, L.: Supporting high-tech start-ups: Lessons from Italian technology policy. The International Entrepreneurship and Management Journal 2(2), 189–209 (2006)
8. Colombo, M.G., Grilli, L.: Technology policy for the knowledge economy: Public support to young ict service firms. Telecommunications Policy 31(10/11), 573–591 (2007)
9. Evans, D.S.: How catalysts ignite: The economics of platform-based start-ups. In: Gawer, A. (ed.) Platforms, Markets and Innovation, ch. 5. Edward Elgar, Cheltenham (2009)
10. Ferdows, K., Lewis, M., Machuca, J.: Rapid-fire fulfillment. Harvard Business Review 82(11), 104–110 (2004)
11. Friedman, T.L.: The World is Flat. The globalized world in the twenty-first century. Penguin books, London (2006)
12. Fudenberg, D., Tirole, J.: Preemption and rent equalization in the adoption of new technology. Review of Economic Studies 52(170), 383 (1985)
13. Gawer, A.: Platform dynamics and strategies: From products to services. In: Gawer, A. (ed.) Platforms, Markets and Innovation. Edward Elgar, Cheltenham (2009)
14. Gawer, A., Cusumano, M.A.: Platform Leadership: How Intel, Microsoft, and Cisco Drive Industry Innovation. Harvard Business School Press Books, Cambridge (2002)
15. Gawer, A., Cusumano, M.A.: How companies become platform leaders. Sloan Management Review 49(2), 28–35 (2008) 0019-848X
16. Gawer, A., Henderson, R.: Platform owner entry and innovation in complementary markets: Evidence from intel. Journal of Economics & Management Strategy 16(1), 1–34 (2007)
17. Ghemawat, P.: Market incumbency and technological inertia. Marketing Science 10(2), 161–171 (1991)
18. Granovetter, M.: The strength of weak ties. American Journal of Sociology 78(6), 1360 (1973)
19. Gupta, S., Jain, D.C., Sawhney, M.S.: Modeling the evolution of markets with indirect network externalities: An application to digital television. Marketing Science 18(3), 396–416 (1999)
20. Hagiu, A.: Pricing and commitment by two-sided platforms. RAND Journal of Economics 37(3), 720–737 (2006)
21. Hagiu, A.: Merchant or two-sided platform? Review of Network Economics 6(2), 115–133 (2007)
22. Joskow, P.: Contract duration and relationship-specific investments: empirical evidence from coal markets. The American Economic Review 77(1), 168–185 (1987)
23. Nye, J.: Soft Power: The Means to Success in World Politics. Perseus Books Group (2006)
24. Parker, G.G., Van Alstyne, M.W.: Two-sided network effects: A theory of information product design. Management Science 51(10), 1494–1504 (2005)
25. Porter, M.M.E.: Clusters and the new economics of competition. Harvard Business Review 76(6), 77 (1998)
26. Porter, M.M.E., Ketels, C.H.M.: UK Competitiveness: moving to the next stage. Technical report, DTI (2003)
27. Rochet, J.-C., Tirole, J.: Platform competition in two-sided markets. Journal of the European Economic Association 1(4), 990–1029 (2003)
28. Rochet, J.-C., Tirole, J.: Two-sided markets: a progress report. RAND Journal of Economics 37(3), 645–667 (2006)

29. Ruiz-Aliseda, F., Zemsky, P.B.: Adoption is Not Development: First Mover Advantages in the Diffusion of New Technology. SSRN, SSRN eLibrary (2006)
30. Santarelli, E., Vivarelli, M.: Is subsidizing entry an optimal policy? Industrial and Corporate Change 11(1), 39–52 (2002)
31. Vargo, S.L., Lusch, R.F.: Evolving to a new dominant logic for marketing. Journal of Marketing 68(1), 1–17 (2004)
32. Various Authors: The economic benefits from providing businesses with competitive electronic communications services. Technical report, BT group plc, London (2007)
33. Weyl, E.G.: A price theory of multi-sided platforms. American Economic Review 100(4), 1642–1672 (2010)
34. Yoffie, D.B., Kwak, M.: With friends like these. Harvard Business Review 84(9), 88–98 (2006)

Part II
Software Methodologies and Technologies

Networked enterprises cannot perform well without a software infrastructure supporting data exchange and coordination. In the current practice, as argued in Part I, these activites are often performed in an ad hoc manner, often, without a specific or with limited software support. This part focuses on some of the challeges that arise when defining a software infrastructure for the networked enterprise.

As a networked enterprise is tipically a dynamic entity where specific organizations can enter and exit over time, the underlying software infrastructure has to self-adapt dynamically to these changes. For instance, while at a certain time the services offered by a company, let us call it A, are available, these same services could be unavailable at a later time, or they could become less convenient than others offered by a new company B that has just entered the market. Moreover, while in a situation of high demand for its products the organization could find convenient to rely on a personalized logistic service, this same service might not be convenient in a situation with a small size demand. In this last case, it could be more convenient to coordinate with other companies with the same needs to share the logistic support. As another example, if we consider the case of a small winery relying on its own production of grapes, this winery might decide to go for the grapes available on the market if the weather conditions appear to be critical for the in house production.

The need for self-adaptation of software infrastructures is one of the problems addressed by the autonomic computing field. Thus, in Chapter 6 we review this research field and discuss on how it addresses the needs of networked enterprises. Moreover, we propose our autonomic framework called the *SelfLets* that focuses on enabling self-adaptation in the cooperation among organizations.

The need for being adaptable is not only a requirement concerning the inter-companies infrastructure, but it also concerns the infrastructure in charge of managing the processes that are local to a specific company. In this case, processes may evolve to account for new situations that have occurred in the company and that are sensed by proper, physical or logical sensors. This problem is addressed by the autonomic workflow management system proposed in Chapter 7, while the issues concerned with the design, deployment, and management of physical sensors and actuators is the focus of Part IV, in this chapter we focus on the interaction between these sensors and actuators and the core processes of each company, as well as on the interaction between these same processes and the data that are acquired and conceptualized in the company repositories according to the techniques and guidelines described in Part III.

Autonomic infrastructures and workflows need to be verified as any other piece of software. While verifying traditional software is tipically an off-line activity performed before the system is deployed in its operation environment, verifying self-adaptable software means being able to check it while it is running. This activity is called on-line verification and exploits specifically tailored analysis and testing tecniques that do not significantly interfer with the execution of the system. A literature review on these aspects is conducted in Chapter 8. This chapter also offers an overview of the approach we have taken to verify the evolving workflow.

Chapter 6
Complex Autonomic Systems for Networked Enterprises: Mechanisms, Solutions and Design Approaches

Nicolò M. Calcavecchia, Elisabetta Di Nitto, Daniel J. Dubois, Carlo Ghezzi, Valentina Mazza, and Matteo Rossi

Abstract. The goals of flexibility and globalisation force enterprises to both decentralise their activities and continuously (re)structure their network of relationships, to support their business processes: "supply chain" manufacturing, design and innovation. Highly dynamic business processes must be supported by new technological infrastructures that can easily evolve to face continuous changes in the requirements and in the environment in which the application is contextualised. Increasingly, such changes would occur in a self-managed autonomic manner, as applications are running. This chapter focuses on achieving autonomic behaviours in the informative systems for supporting networked enterprises. The chapter starts with motivations; it provides an introduction to the autonomic systems literature and then focuses on the SelfLets framework developed as part of the ArtDeco project to support the autonomicity of the applications.

6.1 Introduction

The needs for flexibility and globalisation force enterprises to both decentralise their activities and continuously (re)structure their network of relationships, both regarding their productive processes, their "supply chain", and their design and innovation processes.

This demands for new innovative solutions for the technological infrastructure, which must support unprecedented levels of dynamism. In particular, the overall architecture of the information system must evolve as requirements and context change. Typical cases of context changes occur when services become available, existing services become less competitive, or even they are discontinued. As soon as deviations from the desired or acceptable quality of service are discovered, or

Nicolò M. Calcavecchia · Elisabetta Di Nitto · Daniel J. Dubois · Carlo Ghezzi ·
Valentina Mazza · Matteo Rossi
Politecnico di Milano, via Golgi 42, Milano
e-mail: {calcavecchia,dinitto,dubois,ghezzi,vmazza,rossi}@elet.polimi.it

G. Anastasi et al. (Eds.): Networked Enterprises, LNCS 7200, pp. 85–113, 2012.
© Springer-Verlag Berlin Heidelberg 2012

new opportunities are devised, the information system should be able to reconfigure itself in an autonomic manner.

The information system integrates not only the business-level software usually installed within the various enterprise domains, but also intelligent peripheral systems which interact with the physical world via pervasive devices, such as wireless sensor networks (WSNs) and intelligent tags (RFIDs). These devices are used to *sense* contextual information in a pervasive setting and to *react* by adapting the behaviour in a context-aware manner (this correspond to the paradigm *sense-and-respond*). The goal of this chapter is twofold: first, we provide an introduction to the autonomic systems literature; second, we present the specific autonomic framework we developed as a part of the `ArtDeco` project, the `SelfLets` framework, and explain how this could be beneficial to support the development and operation of networked enterprises.

Consistently with these aims, the chapter is structured as follows. Section 6.2 provides an overview of the literature about autonomic systems. Section 6.3 contextualises the usage of such systems within networked enterprises. Section 6.4 presents our `SelfLets` framework and Section 6.5 instantiates it on the `ArtDeco` reference case study. Finally, Section 6.6 provides a discussion and some conclusions.

6.2 Autonomic Systems: Literature Overview

Dealing with complexity of software is seen as one of the most critical future concerns since it usually results in an increase of costs in terms of human time and expertise needed to manage and evolve software. The term *autonomic computing* has been adapted by IBM to the context of software systems as a way to handle systems complexity through proper self-management mechanisms [23] that, as a consequence, limit the need for human intervention. Automation of software evolution, however, is not only dictated by the need to reduce costs. Self-management capability is in fact increasingly demanded by systems that must be continuously operating. In such cases, solutions that do not require off-line human intervention to evolve the application are needed. Rather, the application should be able to achieve adaptation in an autonomic manner. On the other side, in some cases, due to particular business goals, human presence could not be always substituted using an autonomic behaviour: in such cases, autonomicity guides and supports the human in the decision making when some events occur.

In the remainder of this section we give a brief introduction to autonomic computing followed by a presentation of the most common platforms and approaches that have been identified so far.

6.2.1 Autonomic Computing

The autonomic computing paradigm has been introduced to reduce the complexity of software systems. Typical examples of cases that require such reduction of complexity include the following situations: need to make decisions on how to recover from a hardware or software failure without causing service interruptions; need to optimise the use of resources in a wide network characterised by thousands of nodes to reduce the waiting times of the users; need to guarantee that the interactions within a network of enterprises fulfil certain Service Level Agreements (SLAs).

The main idea is that a system should be able to self-adapt to environmental changes and to individual goals in a seamless way, similarly to what happens to the Autonomic Nervous System in a human body, which regulates the breathing rate, blood pressure, and many other parameters essential to life without the need for each person to be aware of that. This concept of self-adaptation is not new in computer science. For example, adaptation problems are often addressed in the area of control theory [13], while decision problems are addressed by the operation research community [18]. The reason for identifying autonomic computing as a specific discipline is to group all these competences together, focusing on the problem of assigning autonomic properties to an existing system. The autonomic properties identified in the IBM Manifesto are the following [21]:

- *self-configuration*: given a goal, the system should be able to configure itself to reach such goal;
- *self-optimisation*: given a utility function that measures the quality of the achieved goal, the system should be able to tune its working parameters to improve the quality of the achieved goal;
- *self-healing*: if the system, due to any reason, reaches a state that prevents reaching its goal, some corrective actions must be taken to bring the system back to its normal working condition;
- *self-protection*: the system should be able to detect anomalies and security attacks that may mine its future capability to achieve the system goal, and, whenever possible, it should also take defensive actions in advance.

All these properties are often called *self-* properties* since all of them have in common the fact that they are achieved by the system itself in an autonomous way.

From the definitions above it is possible to notice that a generic autonomic system needs a way to understand the goal to be reached, a method to assess the quality of the goal, a way to distinguish a good system state from a bad system state, and a set of possible actions that the system may take at any time. Thus, the autonomic system should be able to read its internal working parameters and the external parameters gathered from the environment (*monitor* phase), process and interpret such data (*analyse* phase), take decisions on how to react based on such data (*plan* phase), and finally apply such decisions (*execute* phase). These four phases are executed in a loop that is supported by an *internal knowledge* used to store the data processed during the four phases: for this reason the loop is often called MAPE-K (Monitor, Analyse, Plan, Execute, Knowledge).

Figure 6.1 shows in detail how this control loop is executed by the system. The "brain" that coordinates these phases is called *Autonomic Manager*, while the environment is modelled as a set of managed resources. Resources provide information

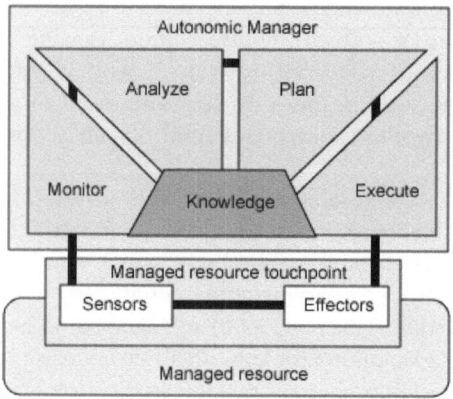

Fig. 6.1 IBM architecture of an Autonomic Element

to the monitor through (abstract) *sensors* and execute the decisions communicated by the autonomic manager through (abstract) *actuators* (or *effectors*). The whole approach is also called *sense and response* and is based on a loop where past decisions may have a feedback effect on future ones. The definition provided by IBM is only a guideline for the engineering of an autonomic system. Most of the existing systems do not implement all the phases of the MAPE-K loop in an individual way, moreover, in some cases, they exploit the model in a compositional way in which multiple individual autonomic components with their own MAPE-K loops are able to interact with each other using sensors/actuators interfaces.

6.2.2 Autonomic Toolkits

Since the introduction of autonomic computing, many different toolkits have been proposed. Some toolkits are optimised for enhancing existing legacy systems with self-* properties, while others offer new architectural models for developing autonomic applications from scratch. Also, some toolkits feature a centralised or hierarchical architecture suitable for small-scale static environments, while some others offer a completely decentralised and strongly decoupled approach suitable in more dynamic settings. Finally, some toolkits assume that self-adaptation is applied in a top-down fashion by executing triggers and policies defined within a management layer, while others adopt a bottom-up approach by exploiting some *emergent* properties [20] of each system component, as it happens in biological systems such as insect colonies and swarms. In the following we describe some of the most important autonomic toolkits along with their main characteristics. For the sake of simplicity we divide the list into three categories: approaches that base their autonomicity on top-down self-adaptation, approaches that base it on bottom-up self-adaptation, and finally hybrid approaches that may self-adapt in both ways.

6.2.2.1 Architectures Based on Top-Down Adaptation

As we have anticipated above, in top-down approaches we have an explicit management layer. This layer is responsible for collecting from the managed resources extensive information on what is going on in the system and therefore it first reasons on the collected data, and then takes the actions to ensure a correct system behaviour. Having a global view of the system makes it easy to apply global policies for reaching global goals. However in some situations top-down approaches have some scalability issues. Think, for example, of a system composed of thousands of components with limited computational resources situated in a dynamic environment. In such situations having global information about the system for implementing a control loop may not be feasible due to the possible difficulties in managing a distributed autonomic manager. The main architectures based on the top-down approach are discussed below.

The *IBM Autonomic Toolkit* [11] is the first autonomic toolkit following the IBM model explicitly. It offers a Java infrastructure for building an autonomic manager. Using this toolkit it is possible to implement the various phases of the MAPE-K loop and the interaction with the managed resources. This approach can be applied when the managed resource is a legacy system. Its drawback is that the autonomic manager is modelled as a centralised "reasoning" engine, which uses the legacy system as a managed resource.

ABLE [8] is another toolkit from IBM with a different architectural style. It extends the IBM Autonomic Toolkit model described above, but exploiting a hierarchical architecture: it does not have a single autonomic manager, but many different autonomic managers running at the same time with their own loops organised in a hierarchy. The final purpose is to separate different autonomic aspects of the managed systems into different agents.

Kinesthetics eXtreme [22] (KX) is similar to the IBM Autonomic Toolkit since it focuses on assigning autonomic properties to legacy systems. The innovation of the approach stems from the strong decoupling between the autonomic manager and the managed resource. Monitored data are collected by the managed resource and translated and processed using *Probes* and *Gauges*, which produce the high-level information that is needed by the autonomic manager. Similarly, the corrective actions are sent by the autonomic manager in the form of high-level *Worklets* that are executed by the managed resource using host-specific adaptors. One limitation of this approach is that it lacks the definition of the architectural style of the autonomic manager, and only provides facilities for monitoring and execution.

ACME [16] is another toolkit that supports the MAPE-K loop in a completely decoupled and distributed fashion. ACME includes the definition of a language for monitoring the system and another language for taking corrective actions. With respect to KX this work not only decouples sensors/effectors from the managed resource, but also from the monitoring and adaptation parts of the control loop.

The *Rainbow* architecture-based self-adaptation approach [15] is a modular infrastructure for realising an autonomic control loop. With Rainbow it is possible to implement an autonomic manager that is composed of four different parts: an abstract *system-layer infrastructure* that provides all the interfaces to read system data and to apply adaptation primitives, an abstract *architecture-layer* infrastructure that comprises any information needed to take decisions about triggering run-time adaptations, a *translation infrastructure* that is needed to map the abstract level of system-layer and architecture-layer infrastructures to concrete system-level operations, and finally a system-specific adaptation knowledge that includes all monitoring and adaptation aspects that cannot be captured by the previous abstract infrastructures. The main advantage of Rainbow is the fact that it is possible to define some *Architectural Styles* that group commonalities of different systems, thus keeping separate system-specific features from more abstract/general reusable ones, therefore this approach may be preferable with respect to the others when the reusability of some components in different parts of the system as well as in different contexts is important.

6.2.2.2 Architectures Based on Bottom-Up Adaptation

The bottom-up approaches, also called self-organising approaches, provide a solution to the problem of giving self-* properties to systems that are characterised by a large number of components and by a level of dynamism that is not easily managed using a top-down approach. In these systems the coordination mechanism will be distributed among the components, therefore each component should contribute to take the whole system to the desired state by relying on local policies and local information only. The advantage is that the coordination mechanism is difficult to break since it is distributed among all the components, moreover the computational resources that are needed for achieving the global goal are scattered among all the components, thus scaling well with respect to the system size. The main disadvantage is the fact that, starting from the desired global properties, it is difficult to find the correct local policies. Another disadvantage is that most of the properties cannot be reached in a deterministic way since the typical algorithms used in this kind of systems have a stochastic behaviour. The following list shows the main self-organising architectures.

The purpose of *Autonomia* [17] is to provide a collection of tools that can be used to deploy autonomic agents. The framework separates its functionalities into the following two modules: the *Application Management Editor* (AME) and the *Autonomic Middleware Service* (AMS). AME features a set of control and management policies for specifying and maintaining the needed self-* properties for each task, and thus supports the creation of the execution environment that implements a full MAPE-K cycle. AMS is used instead to build an application execution environment that can be used to maintain the application requirements during its execution. This environment is then composed of a set of agents that achieve self-management using the information provided by AME.

AutoMate [30] is another toolkit that tries to solve the scalability/centralisation problems of the IBM solution. It is based on a multi-layered architecture optimised for scalable environments such as decentralised middleware and peer-to-peer applications. In this system there is not a global level MAPE-K cycle, but each system component takes decisions based on its own MAPE-K cycle.

BIONETS [12] is an architecture based on two types of nodes: *Tiny-nodes*, which are nodes with limited capabilities such as sensors and tags, and *User-nodes*, which are terminal user nodes with plenty of capabilities such as laptops and smartphones. User-nodes may host services and interact with the environment using the Tiny-nodes and other User-nodes in their proximity. The final purpose of this architecture is not to support legacy systems, but to create a network of services with self-organising behaviour inspired by the natural world. The work focuses on the context of pervasive systems.

AntHill [6] is a framework that takes inspiration from ant colonies. In this framework the peers are interconnected entities called *nests*, characterised by the capability of handling task execution requests originated by local users. These task execution requests (that may contain single tasks, or a workflow of tasks) are called *ants* and are sent in a form of autonomous agents that perform their tasks by traveling across the network. These agents may leave and read information in nests and use such information to evolve overtime to better fulfil the tasks/workflow contained in the initial request. This is another example of self-organising approach and differs from the previous ones by the fact that this model explicits self-organising aspects in terms of interaction between small entities, while the others feature more complex architectures that rely more on coordination and on the explicit management of some resources.

6.2.2.3 Hybrid Architectures

The following are some architectures that may be used to obtain both a top-down and a bottom-up adaptation.

The *CASCADAS Toolkit* [19] supports the development of highly distributed autonomic applications. Using the CASCADAS toolkit the autonomic system is composed of a set of autonomic elements called ACEs (Autonomic Communication Elements). Each element has one or more goals, a list of capabilities, a policy, and the possibility to communicate with other elements. This way elements may exchange information about needed goals and achievable goals, and ask the remote execution of the local requests as needed. Other characteristics of these elements include the possibility to modify all their features (such as goals, capabilities, policies) at run-time, migrate from a machine to a different one, and self-organise themselves into groups. Unlike the previous architectures this one may be used indifferently to easily express centralised solutions with top-down adaptation as well as decentralised solutions with bottom-up (self-organising) adaptation.

In another work [24], the authors define an architectural model by exploiting the characteristics of modern robotic systems in the context of self-managed systems. This architecture consists of a set of interconnected components supported by three different layers: a component layer that is in charge of self-adapting single components in a bottom-up way, a change-management layer that is in charge of reconfiguring different components at the same time in a decentralised manner (it exhibits both bottom-up and top-down behaviour with respect to the other two layers), and finally a goal-management layer that is responsible for giving planning capabilities to the whole system for reaching high-level system goals in a top-down manner. Even in this architecture maximum effort has been given to separate autonomic aspects at different architectural levels: this simplifies the engineering processes of the system and allows the reuse of different components in similar situations without having to start building them again from scratch.

6.3 Autonomic Systems in the Context of Networked Enterprises

Autonomic systems so far have been proposed to manage software systems and to avoid human intervention as much as possible. They, however, have the potential to address the need of networked enterprises as well since they could support the creation and evolution of flexible relationships among enterprises. This has always been a difficult task since usually networks of enterprises are decided a-priori and a human support is of course necessary for a correct management and to correct errors. For this reason we can think of using an autonomic system not only to manage the technical infrastructure of the networked enterprises, but also to trigger human actions among them. This may be considered against the traditional view of autonomic systems where the purpose is to remove the human presence from the control loop, but there are situations in which it may be useful, as explained in [14], such as when dealing to the reorganisation of the roles in a network of enterprises. In this situation it is possible to launch some alerts to trigger human interventions by giving the autonomic infrastructure the capability to analyse the business processes and propose a good network transformation. These manual interventions may involve actions that do not require ICT technical expertise such as the resolution of a contract.

In this area studies are focusing on the extensible exploitation of intangible value exchanges (such as the collection and evaluation of feedbacks and usage reports) since by definition they are able to produce value for the whole network without requiring monetary transfers [5].

More precisely each organisation in a network has one or more goals either compatible or complementary to the ones of other organisations. Compatibility and complementarity help in choosing the current interactions among different organisations. Consider for example a network where some organisations are specialised in logistic services, such as the delivery of goods, and others are specialised in the production of goods. In this case, producing organisations can rely on one or more logistic organisations selected on the fly to deliver their goods, and, in turn, logistic organisations can opportunistically cooperate in order to optimise the costs for movement of goods and increasing their capacity. In this context, relationships among enterprises can be established, managed, and terminated with the support of *Open Autonomic Systems* where different subcomponents may be owned and controlled by different organisations [7]. One of such systems could, for instance, make suggestions on which shipping company to use, based on the feedback obtained from the previous interactions. Even if the proposed suggestions cannot be applied in an automatic way, they can still be used to support human decisions.

By analyzing the approaches described in Section 6.2 we have decided to build our experiments on top of the concepts developed in the CASCADAS project [19]. CASCADAS supports both normal application logic and the autonomic aspects in the same toolkit still keeping them conceptually separated. This feature helps to develop the different parts of a complex autonomic system in a separate way and makes it possible to assign specific human expertise to different specific parts and aspects of the system. The approach, hovewer, has also some limitations. In particular, it does not allow expressing global system goals as a composition of other goals. This decreases the chances to enable collaborative behaviours between different entities in the system. Moreover CASCADAS supports only the capability for an entity to achieve goals in a local way (if possible), or to delegate them to other system elements. In the networked enterprises scenario we want to give the system elements also the capability to share their know-how with others that can then learn how to achieve goals autonomously.

With these new ideas in mind we propose a new framework called SelfLet that is based on the CASCADAS concepts, but that solves the issues identified above. This new framework will be described and used as our reference architecture in the remainder of this chapter.

6.4 The SelfLet Approach

6.4.1 The Concept of SelfLet

SelfLets [9] are the ArtDeco solution to the need of providing Networked Enterprises with a *sense-and-respond* infrastructure based on autonomic services. In this section, we briefly introduce the key elements of SelfLets. To gain a better understanding of the mechanisms on which SelfLets are based, (a subset of) their elements have been formalised through the ArchiTRIO [31] language. ArchiTRIO, whose features are briefly described in the appendix of this chapter, is a UML-compatible formal language based on temporal logic, which is well-suited to describing the dynamic nature of the properties of SelfLets. A complete description of the features of ArchiTRIO and of the formal model of the SelfLet is outside of the scope of this section; nevertheless, to provide readers a quick glance of the issues at hand, we will present snippets of the formal model and some ArchiTRIO details necessary for their comprehension. The interested reader can refer to [28] [27] [31] for further information.

In general, a SelfLet is a self-sufficient piece of software, which operates in some logical or physical network, where it can interact and communicate with other SelfLets. Figure 6.2 shows, through a class diagram annotated with ArchiTRIO stereotypes, the main elements of autonomic services realised as SelfLets.

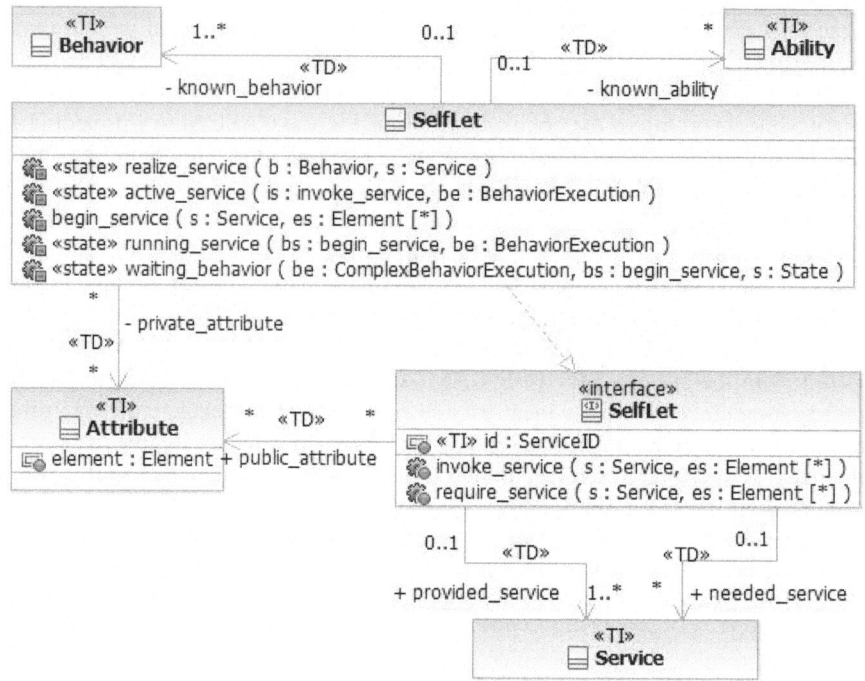

Fig. 6.2 The key elements of SelfLets

Each SelfLet realises the *services* it provides. A service can be invoked by other SelfLets; to provide its service, a SelfLet may require the use of a service provided by the other SelfLets. The way a SelfLet behaves to realise its services, as well as the services it offers or requires may change over time according to:

- its internal state;
- the information coming from the other SelfLet;
- the actions defined in the autonomic policies.

In Figure 6.2 this is represented by marking the associations with the ArchiTRIO «TD» stereotype, which states that the former are "time-dependent".

Services are generic concepts representing high-level, user-defined objectives. Being high-level objectives, there are many different ways to achieve them; in other words, for any given service there are many possible ways to realise it. Each service can then be associated with none, one or more implementations, where each implementation consists of a *behaviour* (see Figure 6.2). For each service it provides, a SelfLet must have at least one behaviour realising that service, as defined by the following ArchiTRIO formula:

$$\text{Alw}(\forall s : \text{Service}($$
$$s \in \text{provided\_service} \rightarrow \qquad\qquad (6.1)$$
$$\exists b : \text{Behavior}(b \in \text{known\_behavior} \wedge \text{realize\_service}(b, s))))$$

Formula (6.1) states that in all instants (Alw is an ArchiTRIO temporal operator that stand for *always*), if *s* is a service provided by the SelfLet in that instant[1], then, at that same time, the SelfLet must have (i.e., know) a behaviour *b* that realises *s*, where *realize_service* is a state-like (i.e., time-dependent) property of the SelfLet that is suitably formally defined by formulae not shown here.

Services provided by a SelfLet can be offered in different ways. A SelfLet can run a service and return the result to the caller: in this case the service is offered in a *Can Do* way. In such a situation, each service can specify a set of parameters, including the arguments it requires for its achievement and the output it produces if the proper parameters are provided. Alternatively, the SelfLet can be available to teach the service, so that the requester is able to execute the service by itself from then on, and this is called the *Can Teach* way. A SelfLet can offer a service even if it does not know it directly, by using the *Knows Who Can Do* and *Knows Who Can Teach* ways. In these cases, the SelfLet will give the requester information about SelfLets able to either offer the service or provide directions for the execution of the service. Every combination of these offer modes is allowed. Notice that, if a service is invoked in a *Can Do* way, there has to be at least one behaviour in the SelfLet that realises the service.

SelfLets can respond to requests for a specific needed service or spontaneously advertise their provided services to inform the other SelfLets about them. Moreover, when issuing a needed service, a SelfLet can specify the offer mode it would prefer. Each SelfLet maintains a list of known providers for each service it requires and that it cannot offer itself. Then, when a need for such a service arises, the SelfLet selects one of these providers according to a given policy, and directly asks it for the service. The SelfLets in a network cooperate to keep such lists always up-to-date with information about the availability of providers.

SelfLets realize services through *behaviours*, which can be seen as workflows realised by the SelfLets. As depicted in Figure 6.4, a behaviour is characterised by a set of traces, which correspond to the possible executions of the behaviour. More precisely, as shown in Figure 6.3, a trace is a sequence of *tuples* of elements (which, in turn, are simply values of any type): since the value of attribute *tuple* of a trace is time-dependent («TD»), a trace is associated with a tuple for each different instant.

Fig. 6.3 Traces and tuples

The traces are used to define the service realised by the behaviour through predicate *realize* (see Figure 6.4). In fact, in a behaviour, predicate *realize* is true for a given service *s* if and only if each trace *bt* of the behaviour is such that its

---

[1] The "current time", i.e., the instant in which time-dependent predicates such as *provided_service* are evaluated, is implicit in ArchiTRIO.

projection over the parameters of the service is in fact a trace of the latter, as defined by ArchiTRIO formula (6.2), which belongs to class *Behavior* of Figure 6.4.

$$\forall s : \text{Service}(\text{realize}(s) \leftrightarrow \\ \forall bt : \text{Trace}(bt \in \text{traces} \rightarrow \\ \exists st : \text{Trace}(st \in s.\text{traces} \wedge \\ \text{Alw}(\forall e : \text{Element}(e \in st.\text{tuple} \rightarrow e \in bt.\text{tuple}))))) \tag{6.2}$$

More precisely, formula (6.2) states that the behaviour (which is implicit in the formula) realises a service s if and only if, for each trace bt of the behaviour, there is a trace st of the service s such that, for each instant (hence Alw) the values of trace st are also values of trace bt (in a sense, trace bt contains st).

When performing a behaviour to realise a service, the SelfLet may be either able to fulfil it directly using *abilities*, which are atomic tasks that may be considered as "portions of a program", or by using other sub-services. These sub-services may then be realised through some other behaviours, and so on in a recursive fashion.

Behaviours are specified as a subset of UML Statecharts [26], and they are characterised by states and transitions. They can be of two types: elementary or complex. An elementary behaviour is composed of only one state, which contains an action that directly invokes an *ability*. A complex behaviour, on the other hand, may be composed of more states and a state may not only be an action that invokes an ability, but also the usage of another service. Since services are defined in the SelfLet framework as high-level goals realised by behaviours, the usage of these sub-services requires the execution of other behaviours.

The formal model includes elements that are necessary to represent the execution of behaviours (either complex or elementary); they are shown in Figure 6.5. An execution essentially corresponds to a sequence of values with a beginning and an end, i.e., to a *bounded trace* (see also Figure 6.3). The execution of a complex behaviour corresponds to a particular path in the state machine of the behaviour; hence, the items of a complex behaviour execution (e.g., *state_begin*) describe when states are entered and exited.

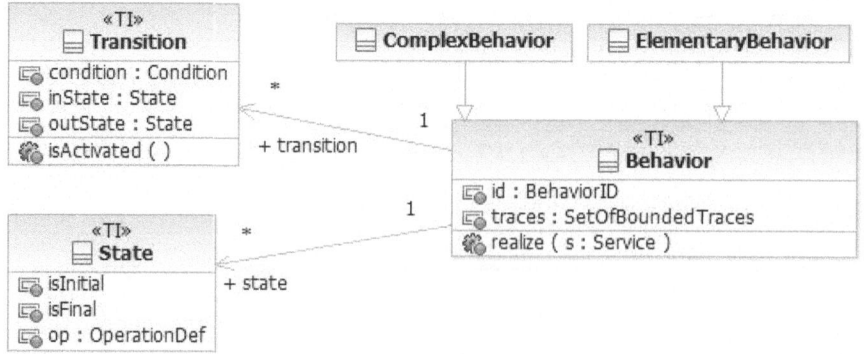

Fig. 6.4 Features of behaviours

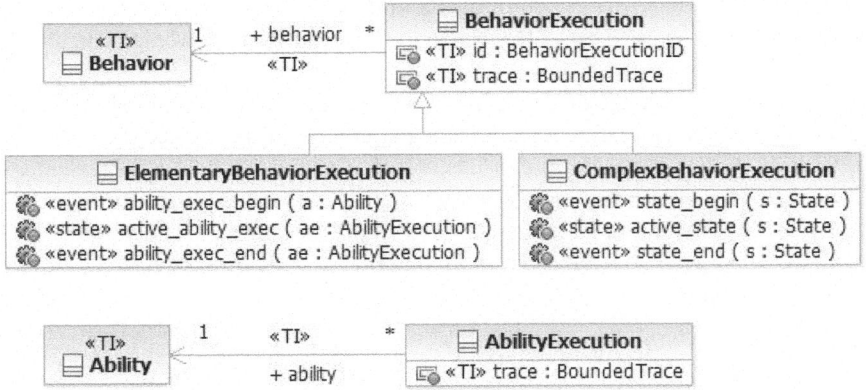

Fig. 6.5 ArchiTRIO classes representing executions of behaviours and abilities

The execution of an elementary behaviour, instead, corresponds to the execution of the ability that realises it. Hence, for example, any time the trace of the execution of an elementary behaviour begins, at that same time the execution of the corresponding ability also starts (which is represented by an occurrence of event *ability_exec_begin*); this is formalised by formula (6.3), which belongs to class *ElementaryBehaviorExecution* of Figure 6.5.

$$\text{Alw}(\text{trace.begin} \to \exists a : \text{Ability}(\text{ability\_exec\_begin}(a))) \tag{6.3}$$

When a behaviour (elementary or complex) is executed, the data used and produced by the abilities are stored in knowledge bases associated with the execution. These knowledge bases are not visible outside the SelfLet, hence they are dubbed "internal" in Figure 6.6.

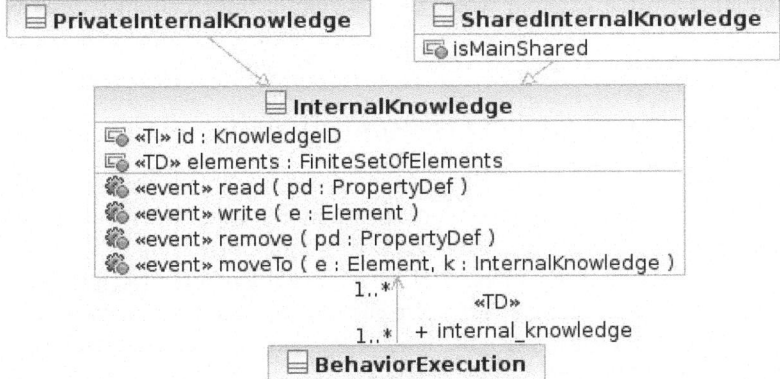

Fig. 6.6 ArchiTRIO classes representing the notion of *internal knowledge*

An internal knowledge basically stores a finite set of values, represented by item *elements* which is marked as «TD», since the contents of a knowledge can change over time.

The internal knowledge of a SelfLet can be accessed and modified during the execution of a behaviour, since it stores the parameters of the service the behaviour implements. For example, formula (6.4), which also belongs to class *Elementary-BehaviorExecution*, defines that, when an ability is launched, there must exist a corresponding execution that starts with the "right" parameters. The value of each one of these parameters is taken from the internal knowledges associated with the behaviour execution.

$$
\begin{aligned}
&\text{Alw}(\forall a : \text{Ability}(\text{ability\_exec\_begin}(a) \rightarrow \\
&\quad \exists ae : \text{AbilityExecution}(\text{active\_ability\_exec}(ae) \wedge \\
&\quad a = ae.\text{ability} \wedge ae.\text{trace.begin} \wedge \\
&\quad \forall pd : \text{PropertyDef}(\\
&\quad\quad pd \in a.\text{known\_property} \wedge pd \text{ instanceOf } ParameterDef \wedge \\
&\quad\quad (pd.\text{direction} = \text{in} \vee pd.\text{direction} = \text{inout}) \rightarrow \\
&\quad\quad \exists ik : \text{InternalKnowledge}(\\
&\quad\quad\quad ik \in \text{internal\_knowledge} \wedge \\
&\quad\quad\quad \exists e : \text{Element}(e \in ik \wedge e.\text{pd} = pd \wedge \\
&\quad\quad\quad\quad e \in ae.\text{trace.tuple.elements}))))))
\end{aligned}
\tag{6.4}
$$

More precisely, formula (6.4) states that, any time the execution *ae* of an ability *a* starts (in an elementary behaviour), for each input parameter *pd* of ability *a* (that is, whose direction is either *in* or *inout*), there is an internal knowledge *ik* associated with the behaviour execution that stores a value (i.e., an *element*) for parameter *pd* which appears in the tuple at the initial instant of the trace of *ae*.

The internal knowledge is implemented as an internal associative memory. Moreover, it can be partially or totally shared among many SelfLets, and it relies on LIME [25] middleware, which implements a distributed tuple space, used to build and maintain such a memory. The internal knowledge could also be integrated with data coming from the external world for example using techniques described in Chapter 17.

6.4.2 Autonomic Rules and Prediction Models

Each SelfLet is assigned a set of autonomic policies, which are responsible for regulating the evolution of services and corresponding behaviours within the same SelfLet, and that can be defined by SelfLet programmers.

Autonomic policies are composed of ECA (Event-Condition-Action) rules [4], that must have a name and can have attributes. Each rule is composed of a LHS (Left Hand Side) and a RHS (Right Hand Side). The LHS contains the event and the condition, the RHS contains the actions, i.e. instructions that should be executed as a reaction when the condition expressed in the LHS is met. The language used to express policies in the SelfLet is Drools [2] and an example of a rule is shown in

Figure 6.10. The kinds of operations that can be performed by policies depend, in general, on the way they are programmed; SelfLets offer to programmers several actions that enable the transformation of several aspects of the SelfLet itself. For example, policies can change the way a service is provided or required and they can also install new services and abilities in the SelfLet; behaviours can also be modified, for example by deleting and replacing their states and transitions.

A good estimate of future internal and external state provides an effective way to promptly trigger autonomic policies. Within a SelfLet-based system, each SelfLet is able to perform local predictions and actuates policies independently from the other ones.

The developer is provided with a framework that allows the definition of the prediction algorithm, its input data and the specific action to be actuated with a prediction. More specifically, the input data is represented by the stream of events generated within the SelfLet such as the update of a knowledge base variable or the dispatch of network messages. It is important to say that the framework is not specifically tied to any prediction technique and the developer can add its own algorithm following a plug-in paradigm.

6.4.3 Design of SelfLet-Based Applications

As previously described, a SelfLet is composed of different parts (i.e. services, behaviours, policies etc.); in order to ease the process of designing a SelfLet based system here we provide some design guidelines. Moreover, the SelfLet autonomic framework is also provided with a prototype of an Integrated Development Environment (IDE), *SelfLetClipse*, supporting the developer during the major design steps.

A developer of a SelfLet-based system has to accomplish the following main operations:

- SelfLet-based systems have a distributed and decentralised nature in which each node (physical or logic) of the system is a SelfLet. As a consequence, the first step to carry out in the design process is the identification of the types of SelfLets involved in the system. Analysing the functionalities of each node, the physical devices and the human actors involved in the system, can help in the achievement of this step.
- Once a clear definition of the involved nodes is available, it is necessary to define services and behaviours associated with each SelfLet. In this case the design can follow a top-down approach: initially, high-level services are defined at each SelfLet then their implementation is provided. It is important to notice that the service implementation is given under the form of a state diagram where each state is in turn an invocation to a service. That is, services are recursively defined in terms of other services; this recursion ends with the abilities which represents the finer level of services in SelfLets.
- After the definition of services, it is necessary to establish their offer strategy (i.e. *CanTeach*, *CanDo*, etc.).

- The last relevant step in the system design involves the definition of possible prediction plug-ins to support autonomic policies and the policies themselves.

The previous described process can be carried out through the *SelfLetClipse* IDE. The tool is implemented as a plug-in for the *Eclipse* development environment [3] and supports the design in different ways:

- *Wizards.* An effective way to support the developer during the design process is represented by wizards[2]. In particular, *SelfLetClipse* offers three wizards

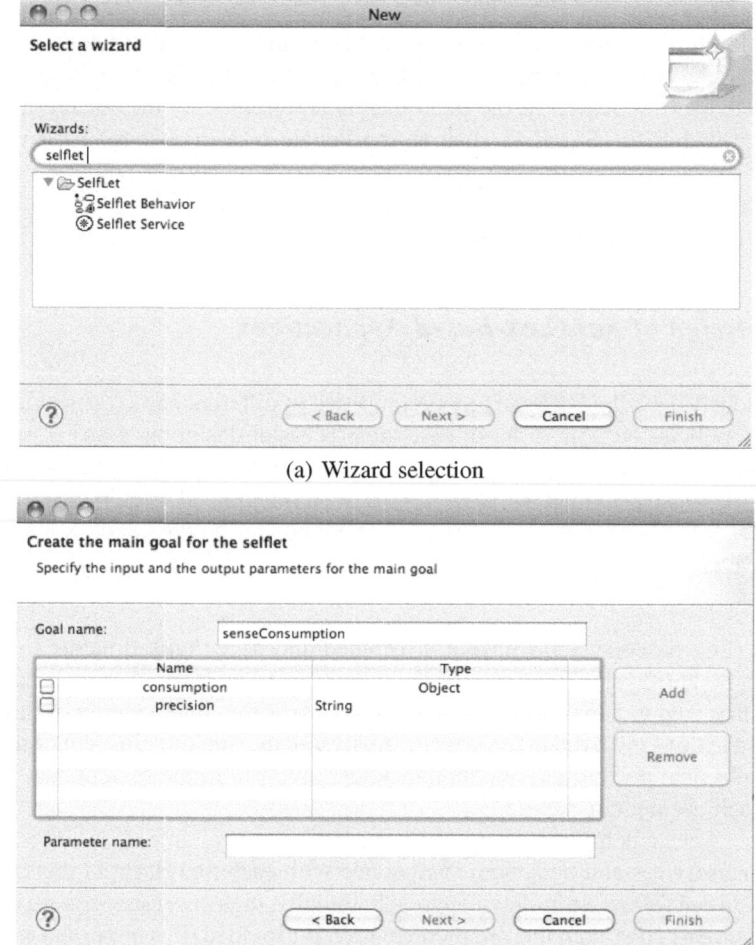

(a) Wizard selection

(b) Part of a wizard procedure in which the parameters of the main goal are specified

Fig. 6.7 Screenshots of the *SelfLetClipse* development environment

---

[2] A wizard is a piece of user interface; it is composed by a sequence of graphical steps (windows) that guide the user in the achievement of a task.

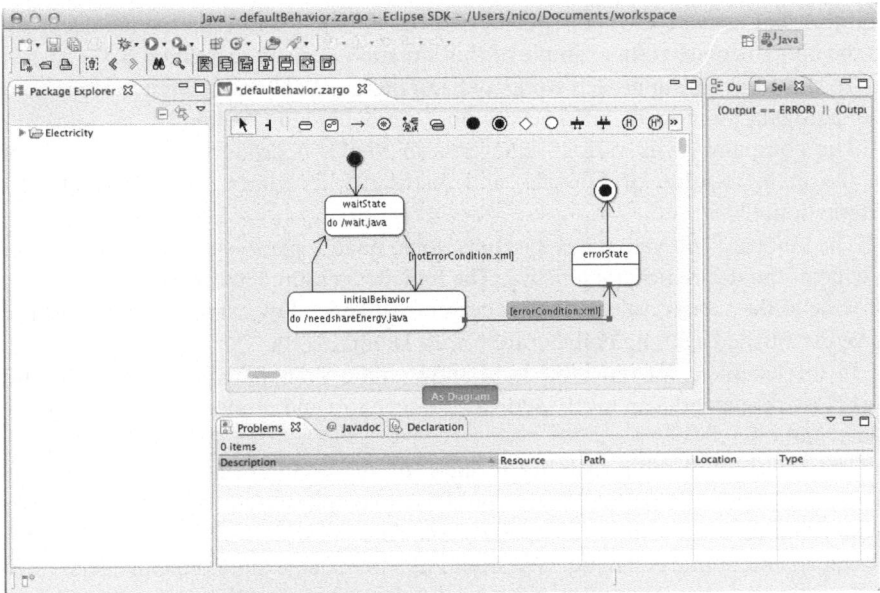

(c) Graphical interface to define behaviours

Fig. 6.7 *(continued)*

supporting the creation of new `SelfLet`, services and behaviours (see Figure 7a and 7b).

- *Behaviour design.* The environment is well integrated with *ArgoEclipse* [1], a plug-in to draw UML diagrams, and it is used to draw behaviour statecharts (see Figure 7c).
- *Error checks.* The IDE also provides a mechanism to perform some verification on the structure of the developed system. Certain properties like well-formed syntax, well-formed statecharts etc, can be automatically verified by the tool itself thus helping the developer during the design process.

6.5 `SelfLets` and Networked Enterprises, an Example Application

When referring to a networked enterprises context, `SelfLets` can be exploited at various levels of abstractions to operate on behalf of various entities. Among the others, a `SelfLet` can represent the way an enterprise interacts with the others within a certain network. While a `SelfLet` can implement a way to automate various aspects of the interaction, it can also be considered as an intermediary toward human operators. In this case, a `SelfLet` behaviour can be seen as the workflow human operators are involved in, that interacts with them through the execution of proper abilities that activate GUIs or specific sensors able to capture the events occurring in the human context. An example of this situation is given in Section 6.5.

To show how the approach works, we refer to the ArtDeco reference scenario of Donnafugata[3]; a more detailed description is provided in Chapter 21.

The company owns various vineyards in Sicily, it produces and stocks wine in the main location of Marsala, and distributes its wines both nationally and internationally.

The vineyard cultivation has to satisfy some quality parameters in order to obtain a wine of the desired characteristics. The logistics required to distributes the bottles of wine to the various sales points is currently managed by external companies that have established a strong collaboration with Donnafugata.

In this chapter, our objective is to show that a system built by exploiting the SelfLet framework can nicely address the scenario and properly react to its possible evolutions.

The focus is on the possibility to have an autonomic behaviour within a networked enterprise. In particular we would like to emphasise the relationships among wine producers and logistic companies, and find possible ways to exploit them to obtain an autonomic behaviour. We would like to manage autonomically such interactions limiting the human intervention. In particular, the application should have some mechanisms to evaluate the goodness of a certain choice when a certain event occurs, and decide for it. On its side, we could image that the wine producer has its own SelfLet-based application to manage (autonomically) its internal processes: in the rest of this section we describe a SelfLet-based application able to manage the status of the vineyard. By means of SelfLets the wine producer is able to monitor the status of the vineyard and react to particular conditions in an autonomous way, without, hopefully, human intervention. In such a way, the reaction time is always reduced, especially when no human intervention is required. In some cases human intervention is needed, and the autonomic application is able to propose to the human a list of suitable choices and the human have to decide what is the best for the desired business goal.

Thanks to this, SelfLets are used for the management of the internal processes of a company, and moreover, for the external processes involving more than one company belonging to a networked enterprise.

In particular, for the case study we are considering, we have conceived a two-level SelfLet architecture. At the low level a SelfLet is associated to each device (sensor or actuator) spread over the vineyard. The purpose is to show that the system is able to properly react to climate changes and degradation of devices performance in order to avoid that these impact on the maturation and harvesting processes. At the high level, SelfLets deal with the networked enterprise. SelfLets describe the behaviour of companies in the network and incorporate the autonomic policies that allow unforeseen changes in the network to be handled.

---

[3] http://www.donnafugata.it/.

6.5.1 The Vineyard *SelfLet* Infrastructure

The purpose of the low level SelfLet infrastructure deployed in the vineyard is to ensure that wine production reaches the required quality standards.

A relevant aspect for agronomists is the constant monitoring of vineyard conditions to ensure that critical situations are avoided (e.g., pests, fires, frosts, etc.). To enable a pervasive control, the vineyard is equipped by a *wireless sensor network* (WSN), which is able to measure vineyard physical parameters such as temperature, humidity, light exposure, wind speed etc. We assume that sensors are placed all over the vineyard and can sense all the relevant physical parameters. An attempt to implement autonomic behaviour on nodes of a WSN can be found in [29]. NesC was used to migrate SelfLet functionalities on the physical nodes composing the network. The WSN-based system we envision is however different from a simple WSN.

Wireless sensors have some limitations for what concerns the duration of their battery and the range of data transmission, since the radio communication range typically covers short distances. Thus, direct transmission of data to a single base station may not be always possible. To deal with this physical constraint, we conceptually divide the vineyard in areas and identify a two-level structure where some devices, the *area leaders*, are able to collect information from the sensors in the area and transmit it to a *vineyard leader* that, in turn, communicates them to the base station. An example of configuration for the vineyard divided in areas is shown in Figure 6.8. Area leaders are not dedicated devices. Instead, they are sensors as the others that, besides their normal sensing tasks, collect data from the others, calculate the average of the collected values and alert the vineyard leader only in case the average values show some potential problem.

The previously described structure represents one of the major differences with respect to a plain WSN approach. In particular, in our case each sensor node is provided with a higher level of intelligence that allows the definition of application specific concepts directly in nodes (i.e. critical conditions) as will be later on discussed.

When dealing with sensor networks, energy represents a strong constraint for the battery life. In our case, the area leader is subject to a higher workload with respect to the other sensor nodes; as a consequence its power consumption is higher. In order to prolong the battery life, whenever the remaining energy at the leader is below a certain threshold, a new area leader is selected among the remaining sensors of the same area.

Each sensor node in the vineyard is implemented through a SelfLet, which behaves according to one of the following roles chosen in the *nodeMainService* behaviour (see Figure 9a):

- A *sensor* represents a simple node able to directly communicate with nodes within its own area. It periodically senses physical parameters and sends them to the area leader (see Figure 9c).

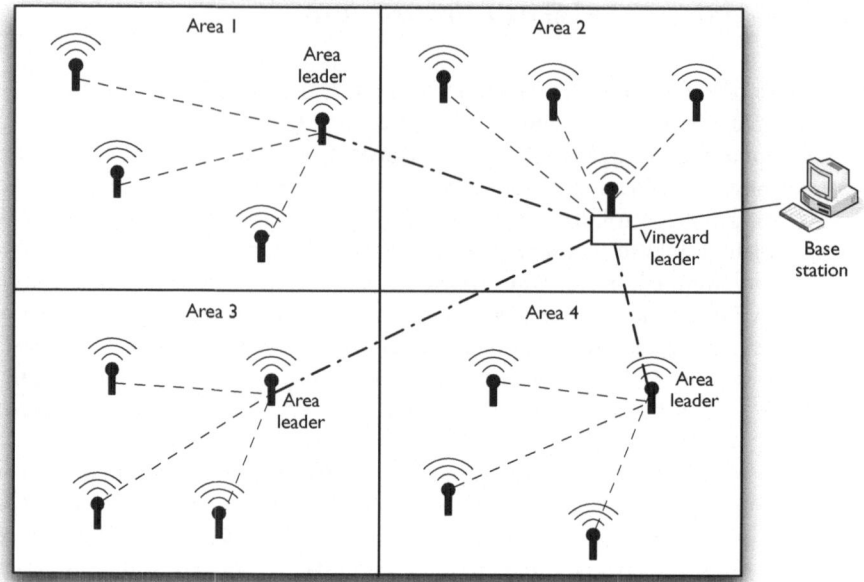

Fig. 6.8 An example of a vineyard equipped with wireless sensors and a base station; the vineyard is divided in four areas

- The *area leader* collects and processes data coming from sensors of its own area and sends this information to the vineyard leader (see Figure 9b). Each area has a single area leader and communication between the vineyard leader and the area leader is achieved through multi-hop routing.
- *Vineyard leader* is unique for the vineyard and is physically connected to the base station, which acts as a main interface for the vineyard operator (see Figure 9d).

Data gathered by the previously described infrastructure contain useful information about the vineyard and can be used to trigger adaptation actions (e.g., recovery actions or notifications of alarms). Through the vineyard leader, the sensor network is linked to the base station, which is dedicated to the analysis of data and to the delivery of notifications and alarms. Data collected by sensors within the network are sent toward the base station that receives a uniform and compact view of the actual vineyard condition. Whenever the base station discovers a critical condition in the vineyard, an adaptation action is executed. To this regard, SelfLet policies represent a flexible mechanism to define such kind of adaptation actions by exploiting the Drools rule manager.

For example the following critical condition:

"Whenever the temperature goes below 5 Celsius degrees and the wind speed goes below 0.5 m/s and the light exposure do not exceed 100 lux during the March month an alarm must be sent to the agronomist"

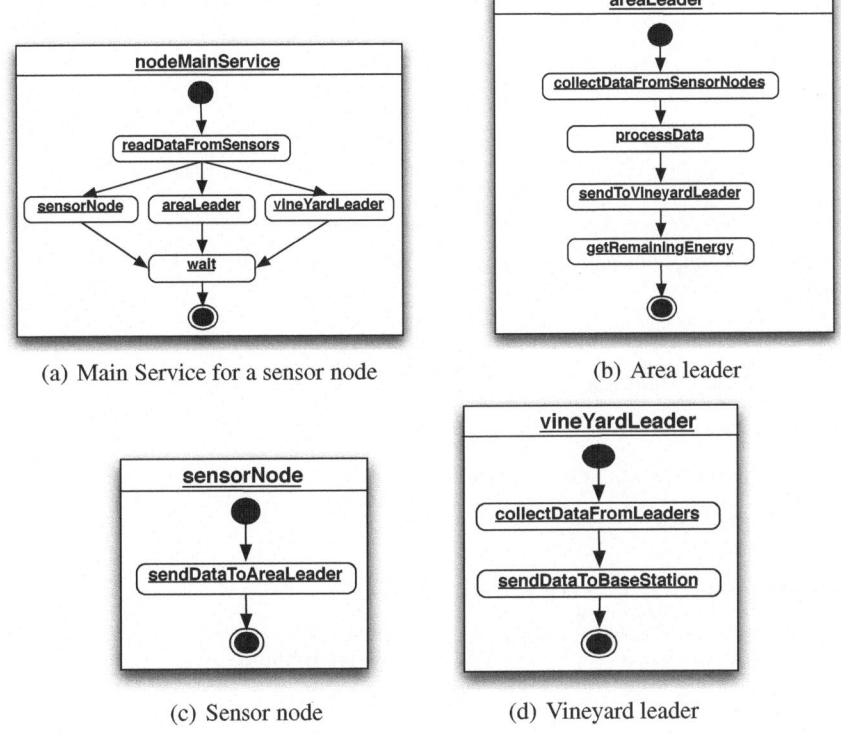

(a) Main Service for a sensor node   (b) Area leader

(c) Sensor node   (d) Vineyard leader

Fig. 6.9 Behaviours implementing services for monitoring physical parameters of the vineyard

can be implemented by the policy depicted in Figure 6.10. In particular, when an event of type *VineYardEvent* is generated internally in the `SelfLet`, the conditions of applicability are checked and in case they are satisfied an alarm is generated.

Figure 6.11 shows the autonomic policy that enables an area leader to abdicate from its role and to pass it to another sensor in the network. In particular, the current area leader selects the sensor with higher remaining energy and then informs of this choice all other nodes within the area; we call this process area leader abdication. While the policy we show does not indicate what happens in the case the selected area leader is not anymore available, another policy is in charge of managing the situation when an area leader does not appear to be alive.

Each node can also optimise the energy consumption by modifying the rate at which messages are periodically sent to the area leader according to the proximity of a critical condition. Indeed, a major source of energy consumption is represented by transmission of message over the network. To motivate such type of optimisation we give an example: suppose that the current temperature is far from the critical threshold in a certain month of the year, that is the vineyard is not close to a be in a critical state. Since there is not an imminent risk, the sensor can slow down the rate at which information are sent without compromising the safety of the vineyard.

```
rule "critical condition 1"
    when
            vineYardEvent: VineYardEvent(month == "March",
                                         temperature < 5,
                                         wind < 0.5,
                                         light < 100)
    then
            sendAlarm();
end
```

Fig. 6.10 An example of policy reacting to a critical condition by sending an alarm to the Agronomist

```
global double THRESHOLD;

rule "abdication policy"
    when
            energyEvent: EnergyLevelEvent(energy <= THRESHOLD)
    then
            abdicate();
end

function void abdicate(){
    askNeighboursEnergy();
    Map<String,Double> neighbours = waitAnswers();
    String newLeader = findNewAreaLeader(neighbours);
    advertiseLeader(newLeader);
}
```

Fig. 6.11 Policy code actuated during the abdication process by the area leader

Dually, whenever a physical parameter is close to a critical threshold, the rate at which messages are sent is increased.

6.5.2 The Networked Enterprise SelfLet Infrastructure

In such scenario, we want to focus our attention to the interactions and cooperations among networked enterprises. The idea here is to consider the possibility of outsourcing some of the phases of the delivery process in situations in which it may be more convenient.

In this context we focus on the need of an autonomic mechanism to monitor shipping requests, take decisions on which actions to take in case of anomaly (e.g., excess in demand, vehicle failures, strikes, and so on), and finally apply such actions taking also into account the risk in trusting external partners. This is an example of problem in which the autonomic policies of the system may trigger human actions instead of pre-determined automatic actions for setting up the opportune decision criteria and behaviour for the current autonomic application, as described in section 6.3. In the remainder of this paragraph we propose the structure and the design of a system based on SelfLets in the context of networked enterprises.

In the wine producers scenario we could imagine that small wine producers at a certain point could analyse their internal processes, such as the shipping one, and restructure/optimise them to increase their profit. A possible way could be the following: when a wine producer needs a truck for a small delivery, he could advertise its need to some other companies that are able to fulfil the request in a more convenient way, these other companies may consist – for example – of close wine producers with an available track going on the same direction, or a specialised shipping company. The requesting company could specify some parameters such as the destination, the amount of product to deliver and the maximum time in which the delivery has to be concluded; after a response is received by other neighbouring producers and external truck companies, benefits/costs are compared to the cost of performing such action internally, and then the best one is chosen. We expect that if an external company is already doing a similar shipment, then it may share the same truck. In such a way the cost of delivery can be reduced and optimised for both. If the wine producer can find no neighbours able to satisfy the request and does not have any available trucks, relying to an external truck rent company may be the only option. However delivery information such as departure, arrival, and maximum time interval for the delivery and the amount of product to deliver will be provided to such company to increase the chance of satisfying more than an incoming request, and therefore, to optimise the deliveries by sharing the same truck among more than one wine producer having compatible logistic requirements.

This solution is however very complex since many parameters and functions need to be defined a priori by a human system manager; another issue is that this solution has unfortunately some problems related to the dynamism of the environment, for example it is difficult to state if a truck-sharing offer is trustable or not, and especially if it actually complies to its advertised quality standards. Due to these problems the system may have another actor with the role of intermediator for receiving advertisements from companies about their availability to cooperate as a truck rental company or as a truck requestor, and — very important — collect feedback information about every interactions. This way it is possible to collect information on the costs of the different options, as well as their risk since future information is influenced by the past one, moreover having this information guides the system (through a market mechanism) to a stable state in which the costs become related to reliability of the provider and to the historical quality of the offered services, thus obtaining a system-wide emergent property that is typical of self-organising autonomic systems.

Summarising, in this high-level scenario we have identified two different autonomic mechanisms. The first one is an autonomic policy that is able to alert the system manager when shipping costs are becoming too high (for example when it is not possible to fulfil all shipping request). When this policy triggers the alert, the system administrator will set up mutual shipping agreements with other companies and modifies a utility function that is used to support the choice whether to send shipping requests using internal trucks or not. The second autonomic mechanism is given by actually introducing the role of intermediator to the system. This intermediator allows to achieve, by collecting feedbacks, as an emergent property the correct evaluation of the reliability of the delivery alternatives that will be used as an additional parameter to the previous utility function. In the following we will show how this scenario may be realised using the SelfLet framework.

Under this architecture we have at least a SelfLet for each enterprise, plus one for the intermediator (that may be a single one for all the network, or more than one, and may be either independent or run by a company that is already part of the network); a deployment example of this system is depicted in Figure 6.12.

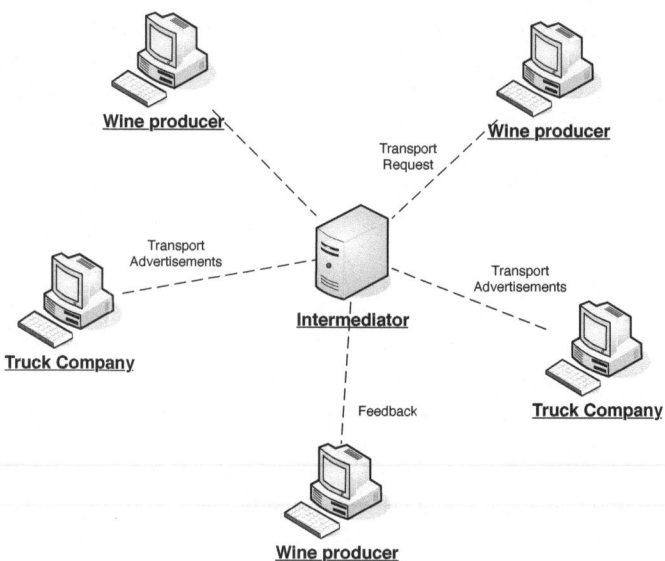

Fig. 6.12 A deployment example of a system composed by 3 wine producers and 2 truck companies; each enterprise (truck or wine producer) has its own SelfLet

Figure 6.13 shows the behaviour implementing the truck sharing features in the enterprise's SelfLets: when a company needs to ship something it will first send its request to the intermediator (which may be the company itself if there are no other intermediators) specifying a filter on the results based on the actual requirements, then it evaluates the answers using a utility function that considers not only the compliance with the request and the maximum affordable costs, but also other parameters such as historical feedback and reputation. This utility function may be very simple at the beginning (it may forward all the request internally unconditionally) and then be updated (with a dedicated ability) while the system is running by the system managers according to the alerts that may be triggered by some autonomic rules that monitor the quality of the shipping services. If the utility value of using internal trucks is greater than the one of using an external service, the request will be managed internally, otherwise it will be managed by an external company. If the maximum utility value obtained both internally and externally does not satisfy the minimum requirements, then the request is suspended and the problem will be reported to the user to manage the exception manually. In addition to this behaviour, every participant may also add to the autonomic policies of their `SelfLets` a rule for sending feedback to the intermediator after every shipment, one for updating the intermediator's information every time the advertised availability of local trucks changes, and another one for automatically managing external requests coming from the intermediator. As we have said before these feedbacks are very important to guarantee the accuracy of the evaluation of the utility function defined before.

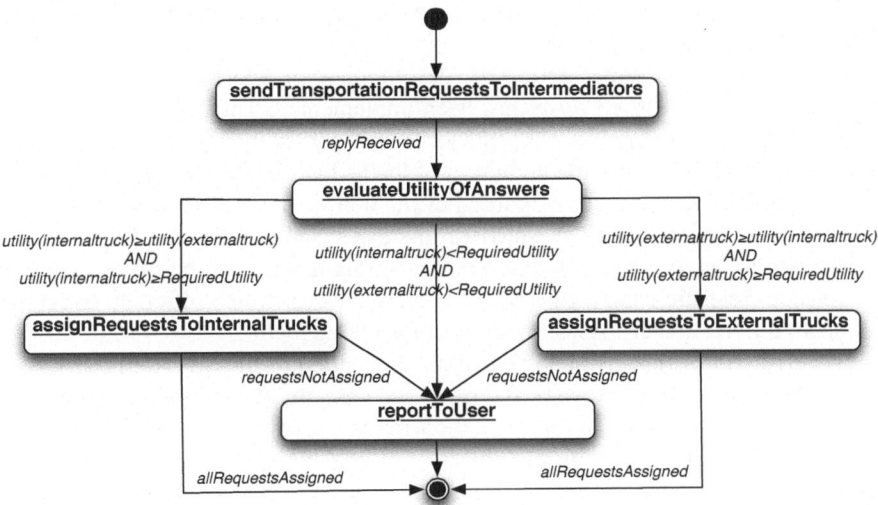

Fig. 6.13 Behaviour implementing the truck sharing service

In conclusion this high-level scenario points out how it is possible to obtain an emergent high-level autonomic behaviour using self-organisation techniques, such as the feedback loops provided above. These feedback loops, using `SelfLets`, are able to support the system evolution by bringing it to a more profitable state; moreover they take into account past information to help deciding future actions.

6.6 Discussion and Conclusions

In this chapter we have provided an overview of existing autonomic systems and shown the details about the `SelfLet` framework developed within the `ArtDeco` project to address the topic of the autonomicity of applications. In particular, the aim of such framework is to provide a mean for enabling the design and the implementation of autonomic behaviours within the context of networked enterprises.

A SelfLet is a self-sufficient piece of software able to provide certain services within an application; it is able to communicate with other SelfLets belonging to the same application in order to achieve a desired goal.

The development of a SelfLet involves two roles: the definition of the business logic in terms of service implementation and the definition of autonomic policies, which dynamically enable changes within the SelfLet structure. More specifically, policies are delegated to offer the autonomic behaviour the SelfLet exhibits and make the SelfLet able to autonomically react to particular conditions.

An attempt to improve the autonomicity was arisen by the intuition that a good estimation of future internal and external states could provide a more efficient way to trigger autonomic policies: some efforts have been spent in this direction in order to have suitable predictive models.

To support the developer during the development of a SelfLet-based system, an Eclipse plugin (*SelfLetClipse*) has been implemented. Thanks to such environment, the development process appears to be friendly enough: it provides a unified environment, which allows the definition of the SelfLet components (behaviours, services, policies); however, a rigorous evaluation of the tool is actually missing and still is to be performed. Related to this aspect, it could be useful to have a mechanism to verify SelfLet-based systems; this feature can be of great help if we consider realistic cases of networked enterprises (in which different SelfLets belong to different enterprises communicating among them).

The final part of this chapter showed an application of the SelfLet framework within the context of a vineyard; the proposed case study supports the process of winemaking at two levels. At low level the wireless sensor network deployed within the vineyard is optimised with the objective of prolonging the battery life of each sensor. At high level, the SelfLet framework is used to enable the interconnection of different wine producers (networked enterprises) in order to share their delivery facilities (e.g., trucks).

Future work will be conducted in order to address issues related to privacy and security aspects; some restrictions could exist and not all the facilities could be allowed every time. For example, we could image that due to some security reason, in some cases, the teach mode could be not possible. Finally, a future work will consists in the development and deployment of the afore described case study into a real context; indeed, this will allow us to effectively evaluate the benefits given by our framework.

6.7 Appendix: A Brief Introduction to ArchiTRIO

ArchiTRIO is a UML-compatible formal language [31], whose underlying philosophy is to approach the problem of modeling complex software-centric systems in a so-called "lightweight" manner. More precisely, the idea behind the ArchiTRIO language is that users should be able to approach the modeling of complex systems with the familiar, widely-used, semi-formal UML notation (or subsets thereof), and introduce formal statements only when (if) and where needed.

ARCHITRIO is based on the TRIO temporal logic [10]. TRIO is a metric logic with a linear notion of time, which provides facilities for constructing formulae that describe the behavior of phenomena, and hence constrain what may happen at particular time instants or over time intervals. In TRIO, the perspective on time is always in terms of the implicit *now*, with other points in time described in terms of their distance from *now* using the Dist operator. For instance, suppose that, at every instant, *temp* represents the temperature of a room that is being monitored. Suppose also that one wanted to state that the temperature does not evolve over time in a completely arbitrary way, but according to the constraint that, for any two instants at a distance of less than ten time units from one another, the difference in temperature cannot be more than two degrees. Such a constraint could be described by the following TRIO formula:

$$\forall \gamma(\text{temp} = \gamma \rightarrow \\ \forall \delta(-10 < \delta < 10 \rightarrow \text{Dist}(\gamma - 2 < \text{temp} < \gamma + 2, \delta))) \tag{6.5}$$

Dist is the only basic temporal operator of the TRIO language; however, a number of derived operators are defined from Dist, through the usual first-order logic constructs. For example, the Alw operator is used to state that a property holds in every instant (i.e., always), while the Lasts (resp. Lasted) operator is used to state that some property will hold (resp. have held) throughout a certain future (resp. past) interval (a comprehensive list of the TRIO operators is available in [10]). For example, if one wanted to state that throughout the interval of length ten from the current instant (endpoints excluded) the temperature will not increase nor decrease more than two degrees from the current value, one could describe this dynamic using the following TRIO formula:

$$\text{Alw}(\text{temp} = \gamma \rightarrow \text{Lasts}(\gamma - 2 < \text{temp} < \gamma + 2, 10)) \tag{6.6}$$

TRIO formulae can be interpreted over various time domains (naturals, integers, rationals, reals or subsets thereof) without modification of their syntax.

ARCHITRIO [31] is a UML-oriented extension of TRIO which allows users to represent the structure of a complex system, its components, and their properties. It uses a subset of UML2 [26] concepts and notations (e.g., structured class, port, and interface) to define structural features of systems, and TRIO formulae to describe their dynamics. From a graphical point of view there is very little difference between UML and ARCHITRIO, and all graphical elements that ARCHITRIO keeps from UML retain their semantics, albeit, in ARCHITRIO, these are defined formally. Thanks to the UML-derived modular features of ARCHITRIO, designers can clearly describe system components and their interfaces; then, modellers can precisely define the properties and behaviours of elements using TRIO-derived formulae. For example, Figure 6.14 shows ARCHITRIO class *Room* providing interface *RoomMeasurable-Phen* (through a UML2 port [26] named *rmp*), which exports item *temp*. *temp* is a logic element representing, at every instant, the actual temperature of the room; the «TD» stereotype indicates that this is a time-dependent value; that is, in different time instants the value associated with *temp* can change. Then, class *Room* could contain the TRIO formula (6.6) presented above describing the regularity constraint on the room temperature.

Fig. 6.14 Example of a simple ArchiTRIO class

ArchiTRIO classes can be composed in more complex models using the UML2 notions of port, interface (provided or required) and connector, as shown also in Section 6.4.1.

References

1. ArgoEclipse, http://argoeclipse.tigris.org/
2. Drools, http://www.jboss.org/drools/
3. Eclipse, http://www.eclipse.org/
4. C. Act-Net Consortium: The active database management system manifesto: a rulebase of adbms features. SIGMOD Rec. 25(3), 40–49 (1996)
5. Allee, V.: A value network approach for modeling and measuring intangibles. In: Proceedings Transparent Enterprise, Madrid (2002)
6. Babaoglu, O., Meling, H., Montresor, A.: Anthill: A framework for the development of agent-based peer-to-peer systems. In: International Conference on Distributed Computing Systems, p. 15 (2002)
7. Baresi, L., Di Nitto, E., Ghezzi, C.: Toward open-world software: Issue and challenges. Computer 39(10), 36–43 (2006)
8. Bigus, J., Schlosnagle, D., Pilgrim, J., Mills, W., Diao, Y.: ABLE: A toolkit for building multiagent autonomic systems. IBM Systems Journal 41(3), 350–371 (2002)
9. Bindelli, S., Di Nitto, E., Mirandola, R., Tedesco, R.: Building autonomic components: The selflets approach. In: 23rd IEEE/ACM International Conference on Automated Software Engineering - Workshops, ASE Workshops 2008, pp. 17–24, 15-16 (2008)
10. Ciapessoni, E., Coen-Porisini, A., Crivelli, E., Mandrioli, D., Mirandola, P., Morzenti, A.: From formal models to formally-based methods: an industrial experience. ACM Transactions on Software Engineering and Methodology 8(1), 79–113 (1999)
11. Cybrynski, J.R.: Abc of the autonomic computing toolkit. Technical report, IBM Autonomic Computing Technical Report (2005)
12. De Pellegrini, F., Miorandi, D., Linner, D., Bacsardi, L., Moiso, C.: Bionets architecture: from networks to serworks. In: 2nd Bio-Inspired Models of Network, Information and Computing Systems, Bionetics 2007, pp. 255–262 (December 2007)
13. Diao, Y., Hellerstein, J., Parekh, S., Griffith, R., Kaiser, G., Phung, D.: Self-managing systems: A control theory foundation. In: 12th IEEE International Conference and Workshops on the Engineering of Computer-Based Systems, ECBS 2005, pp. 441–448 (2005)

14. Dubois, D., Nikolaou, C., Voskakis, M.: A model transformation for increasing value in service networks through intangible value exchanges. In: International Conference on Service Science, ICSS 2010 (2010)
15. Garlan, D., Cheng, S.-W., Huang, A.-C., Schmerl, B., Steenkiste, P.: Rainbow: Architecture-based self-adaptation with reusable infrastructure. Computer 37(10), 46–54 (2004)
16. Garlan, D., Schmerl, B.: Model-based adaptation for self-healing systems. In: WOSS 2002: Proceedings of the First Workshop on Self-Healing Systems, pp. 27–32. ACM, New York (2002)
17. Hariri, X.D., Xue, S.L., Chen, H., Zhang, M., Pavuluri, S., Rao, S.: Autonomia: an autonomic computing environment. In: IEEE International Performance, Computing, and Communications Conference 2003 (2003)
18. Hillier, F.S., Lieberman, G.J.: Introduction to Operations Research. McGraw-Hill Science/Engineering/Math (2005)
19. Hoefig, E., Wuest, B., Benko, B.K., Mannella, A., Mamei, M., Di Nitto, E.: On concepts for autonomic communication elements. In: International Workshop on Modelling Autonomic Communications (2006)
20. Holland, J.H.: Emergence: from chaos to order. Addison-Wesley Longman Publishing Co., Inc, Boston (1998)
21. IBM. Autonomic vision and manifesto. Website, http://www.research.ibm.com/autonomic/manifesto/
22. Kaiser, G., Parekh, J., Gross, P., Valetto, G., Gupta, S., Kaiser, G., Neistadt, D., Grimm, P., Gupta, S.: Kinesthetics eXtreme: An External Infrastructure for Monitoring Distributed Legacy Systems. In: First ACM Workshop on Survivable and Self-Regenerative Systems (2003)
23. Kephart, J., Chess, D.: The vision of autonomic computing. Computer 36(1), 41–50 (2003)
24. Kramer, J., Magee, J.: Self-managed systems: an architectural challenge. Future of Software Engineering, 259–268 (2007)
25. Murphy, A.L., Picco, G.P., Roman, G.-C.: LIME: A coordination model and middleware supporting mobility of hosts and agents. ACM Transactions on Software Engineering and Methodology 15(3), 279–328 (2006)
26. Object Management Group. UML 2.3 superstructure specification. Technical report, OMG, formal/2010-05-05 (2010)
27. Pacifici, S.: Formal modeling and evaluation of selflets supporting the design of dependable autonomic services. Master's thesis, University of Illinois at Chicago (2009)
28. Pacifici, S., Rossi, M.: Towards a formal model of autonomic services based on Self-Lets. In: Proceedings of the 2009 ICSE Workshop on Principles of Engineering Service Oriented Systems, pp. 13–17. IEEE Computer Society (2009)
29. Panzeri, M.: Studio di un approccio per la realizzazione di agenti autonomici in reti di sensori wireless. Master's thesis, Politecnico di Milano (2009)
30. Parashar, M., Liu, H., Li, Z., Matossian, V., Schmidt, C., Zhang, G., Hariri, S.: Automate: Enabling autonomic applications on the grid. Cluster Computing 9(2), 161–174 (2006)
31. Pradella, M., Rossi, M., Mandrioli, D.: ArchiTRIO: A UML-Compatible Language for Architectural Description and Its Formal Semantics. In: Wang, F. (ed.) FORTE 2005. LNCS, vol. 3731, pp. 381–395. Springer, Heidelberg (2005)

Chapter 7
Autonomic Workflow and Business Process Modelling for Networked Enterprises

Gerardo Canfora, Giancarlo Tretola, and Eugenio Zimeo

Abstract. As markets become more and more competitive and dynamic, companies need to increase control over their business processes to quickly adapt them to the changing conditions of the operational environment. Workflow management technology is a means to automate and control business processes, but they need more sophisticated capabilities to cope with highly dynamic execution contexts.

This chapter proposes a novel approach to adaptive workflow management, based on a programming model and a related runtime system. By combining imperative and declarative programming, a specific workflow management system is able to react to events sourced from the business environment by modifying the structure and behaviour of running workflows.

The chapter discusses related work on workflow adaptation, illustrates the proposed autonomic workflow model, the overall architecture of the related management system, the technical motivations and choices for the implementation, and the impact of this kind of workflows onto business modelling.

7.1 Introduction

Dynamic markets and globalization force companies to define new requirements for coordinating business actions among supply chains. ICT plays a fundamental role to build virtual organisations based on electronic collaborations [10]. A special kind of e-collaboration is *Sense and Respond* [4]: supply chains are tuned to collect events and information from the business environment to dynamically alter the configuration and behavior of the supply chain itself according to the modified business requirements.

Gerardo Canfora · Giancarlo Tretola · Eugenio Zimeo
Universitá degli Studi del Sannio
e-mail: {canfora,tretola,zimeo}@unisannio.it

G. Anastasi et al. (Eds.): Networked Enterprises, LNCS 7200, pp. 115–142, 2012.
© Springer-Verlag Berlin Heidelberg 2012

Workflow management systems (WfMSs) are gaining an increasing importance for handling complex business processes, but they often fail in highly dynamic environments, where imperative behaviours could become unsuitable to address the changing conditions of the running context. This is particularly true in service oriented computing, where services are handled in an open world and are provided by different organisations: they can be modified or replaced; they can disappear, and new services with different features may become available.

Workflow execution across geographically distributed organisations, technology, and assets introduces the need for features that are beyond the capability of current enterprise management tools and impact the design, supervision and management of workflows. As a consequence, the complexity of IT infrastructures management increases, as they are forced to accommodate for heterogeneous components [26]. Additional complexity is generated by the need for: (i) analyzing a great amount of data collected during execution to improve and adapt processes, (ii) quickly reacting to the evolution in the operating environment, (iii) being able to produce goods or deliver services that are really required, at the right time and in the right place.

This level of complexity is too high for having manual adjustments of workflows during their execution. Autonomic computing (AC) [16], instead, could be an effective approach to automate workflow handling at runtime, since autonomic systems should not only be able to take automated actions, but they should do this with the support of an innate ability to sense, and respond to changes, by incorporating self-learning and self-managing capabilities. Therefore, by considering a business workflow as a large-scale program, autonomic computing could be a key enabler for ensuring self-* properties to workflows.

This chapter presents the main features of a novel approach for dynamically handling workflows, that we define *Autonomic Workflows*, by exploiting autonomic computing techniques. It provides (i) a programming model that aids users to design autonomic workflows, and (ii) a system that is able to run those workflows with the ability of handling execution anomalies with a limited human intervention. The main objective of the system is to enable workflows management during the entire lifecycle, by collecting and organizing the knowledge from the operating environment. This knowledge, represented in the form of Event Condition Action (ECA) rules, is then used for adapting and improving workflows at run-time.

The rest of the chapter is organized as follows. Section 7.2 discusses research work on workflow adaptation by analyzing in particular the adoption of autonomic computing in the context of workflow management. Section 7.3 presents some typical adaptation situations and defines the conceptual architecture of a workflow management system that is able to handle them. Section 7.4 exploits the model described in the previous section to design a coherent and robust architecture for implementing a workflow management system that is able to seamlessly run autonomic workflows. Section 7.5 shows how the proposed workflow management system can be programmed by using a declarative language based on ECA rules. Finally, Section 7.6 describes how the concept of autonomic workflow can be exploited to model variable business processes; to this end, a step by step example is illustrated. Section 7.7 concludes the chapter by highlighting future enhancements for autonomic workflows.

7.2 Related Work

Automating the management of computing resources is not a new problem [5] [17] [6]. In this direction, Autonomic Computing (AC) has recently emerged as a scientific area to study novel techniques for managing in an automatic fashion the increasing complexity of modern computing environments. Among the different approaches to achieve AC, the most common one is the MAPE model [15]. MAPE stands for *Monitor, Analyse, Plan* and *Execute*. With this model, system resources are continuously monitored in order to identify, through the analysis of data collected, possible anomalies and consequently to plan the actions to execute onto the resources in order to change their state as defined by some policies.

In spite of the interest of many researchers, up to now, AC has been applied with success mainly in system administration, with the aim of freeing administrators from the details of system operation and maintenance, improving robustness of systems, and decreasing the total cost of ownership [16] [11].

More recently, AC has been applied also in the area of process-aware information systems to effectively and efficiently deal with changes in several aspects of these applications [21]. Pautasso et. al. [14] [24] have proposed one of the first solutions to automatically manage a workflow execution engine through the adoption of an autonomic controller that ensures self-configuration, self-tuning, and self-healing properties. In particular, the controller monitors the engine performance and reacts to workload variations by altering the current configuration of the execution environment. This way, the authors use autonomic computing principles to design a sophisticated workflow engine that is able to manage itself to exploit at the best the opportunities arising in the execution environment at run-time. However, the engine has not been designed to react to changes in the business environment.

A preliminary attempt to apply autonomic computing principles to handle the complexity of workflows is by Strohmaaier and Yu [27]. They analyze the problem of autonomy in WfMSs and propose five behaviors: *basic, managed, predictive, adaptive, autonomic*. Each behavior is characterized by a different degree of automation supported by the system. With the basic behavior, the system supports designers only for syntactical correctness. With the managed behavior, the system is able to monitor events and perform statistics about possible changes in the execution flow. With the predictive behavior, the system supports, though simulation, some kinds of prediction based on workflow mining that suggests alternatives in workflow design. The adaptive behavior increases the degree of autonomy by providing the system with the ability of repairing and optimizing a workflow. In the last case, the system exhibits a proactive behavior since it is able to autonomously evolve in the presence of changes by respecting high-level strategies and goals.

Due to the specific focus of this chapter onto adaptation and autonomy, the rest of the survey is limited to these topics. Muller et. al. [22] have introduced the support for dynamic and automatic workflow adaptations though the adoption of a rule-based approach (to specify workflow exceptions), temporal logics and a predictive behavior. Song et al. [13] dealt with managing exceptions in medical workflows. They focused on three topics: representing, handling and analyzing exceptions that are able to re-compose and re-schedule workflows. Dadam et. al. [20] proposed a framework for defining semantic constraints useful for expressing domain knowledge. This knowledge is then used for maintenance and semantic process verification. Starting from approaches already developed to ensure system correctness at syntactical level, they have introduced techniques to ensure semantic correctness of workflow process changes.

The diffusion of service oriented computing and related technologies for implementing large-scale information systems and distributed applications has further stimulated researchers working on workflows. Here, service compositions are typically considered at abstract and concrete levels. The former represents an abstract solution to a problem, whereas the latter considers also the specific services that are used for executing the composition. In [7] the concept of workflow is mapped on Web Services composition and it is executed by exploiting a multi-agent technology. The authors state that enterprises need to redesign their systems to implement novel workflow management systems based on semantically described Web Services obtained by the current *servicisation* process that is characterizing the Web.

An interesting aspect is the ability to bind and re-bind abstract activities to concrete services at run-time, by respecting some local or global constraints. In [2], the authors adopt semantic description of services that are used to identify the semantics of an abstract process. Such information is also used to discover and bind the concrete services to be used in the process instance, by means of a matchmaker component. Zeng et al., 2004 [30] focus on the problem of identifying workflow bindings based on optimum global and local QoS criteria. The global optimization problem, which is shown on the price, response time, availability and reliability QoS attributes, assumes linearity of the constraints and of the objective function, and is solved through integer programming techniques. Rebinding entails the need for capturing information about the running process to estimate the values of QoS parameters. Canfora et al. [8] introduces a QoS-aware binding approach based on Genetic Algorithms. The approach features early run-time re-binding whenever the actual QoS deviates from initial estimates, or when a service is not available. Baresi and Guinea [3] have proposed an interesting approach based on process weaving. In this approach the monitoring policy is defined independently of the process and its definition language. Before executing the process, its monitoring policy is added to the process instance and executed by the workflow. Each monitoring rule is translated into an invocation to the Monitoring Manager which is responsible to evaluate pre and post-conditions to determine the satisfaction of monitoring policies.

The previous papers do not consider the semantics of QoS parameters. Therefore, their approaches could fail in a real business environment where terminological mismatches are very common. Zimeo et. al., [12], present the onQoS ontology, an openly available OWL ontology for QoS, useful to express functions of QoS metrics that are employed to improve QoS-aware, service discovery recall without degrading precision. An improvement for QoS and context monitoring based both onto QoS semantics and ECA rules has been proposed in [28]. With this approach, designers follow an optimistic approach to design workflows, while domain-dependent rules, programmed separately, are used at run-time to monitor running workflows with the aim of planning reactive and corrective actions when anomalies occur.

These works represent isolated solutions to one of the dynamic aspects that characterize service oriented computing on large-scale systems. However, they do not provide global solutions to address the need for adaptation.

Sheth and Verma [29] have perceived the need for addressing adaptation in the new context of SOA by exploiting some concepts of autonomic computing for the definition of *Autonomic Web Processes* (AWPs). AWPs represent the natural evolution of the application of autonomic computing from individual information technology resources to business processes. With this evolution, the canonical self-* properties of autonomic systems are transferred to the workflow and business levels to overcome anomalies that may arise during the execution of workflows onto the Web (*Web processes*), which typically cause the failure of execution. However, the paper mainly focuses on service binding and does not propose concrete solutions to the problems raised.

The adoption of the MAPE model to handle the adaptation of workflows has been proposed in [19]. The paper addresses the problem of workflow adaptation as an AC problem and clearly identifies different kinds of adaptation for scientific, long-running workflows. The anomalies identified are automatically handled by using a functional decomposition of autonomic managers into monitoring, analysis, planning and execution components. Whilst, this approach is similar to our intent, it presents three main gaps to fill: (i) anomalies regard only mapping and scheduling phases of a scientific workflow; (ii) planning consists of selecting corrective actions from a predefined list of actions tied to specific problems identified during the analysis phase; (iii) the architecture proposed is only related to the manager; no details are given about the workflow engine able to use that manager.

Predefined plans to react to particular situations could be an acceptable solution for implementing an autonomic behavior when changes regard the execution environment, however, they are inadequate to handle changes that happen in the business context. Here, a more automated approach is needed to handle the growing number of services that may become available in the Web and consequently to handle the large number of possible solutions when an anomaly occurs [1].

A concrete approach for performing automatic composition of web services is proposed by Klush [18]. Their tool, OWLS-XPlan, converts initial and goal state ontologies in OWL and services in OWL-S 1.1 to equivalent domain and problem descriptions in PDDL 2.1, and returns a composition plan that reaches the given goal state starting from the initial state. Composition of complex services at design time is supported by many other available tools and planners, such as SHOP2 [23]. However, a few results have been obtained as concerning dynamic behaviors of service composition and the adoption of autonomic computing for implementing dynamic compositions based on business knowledge coded by means of semantics and strategies.

7.3 Autonomic Workflow Model

Autonomic Workflow (see figure 7.1) refers to the ability of automatically changing an aspect of a workflow - able to orchestrate both human (i.e. manual) and electronic (i.e. automatic) resources - under some local and global constraints and considering the goal as an invariant.

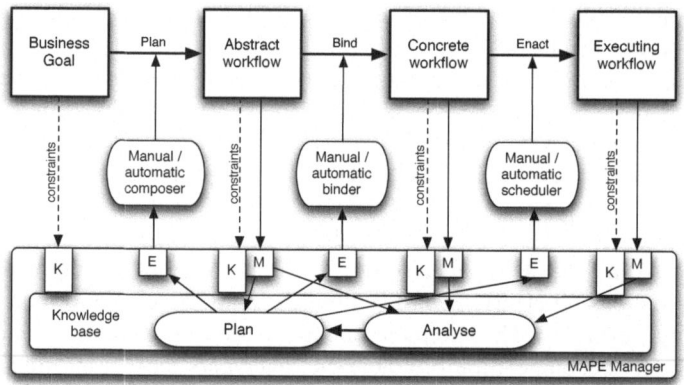

Fig. 7.1 Autonomic Workflow and the conceptual runt-time support

7.3.1 Scenarios

Following the bottom-up sequence of figure 7.1, we initially focus on *executing workflow*. At this level, a workflow is concrete and its activities are scheduled for the execution according to a partial order depending on workflow enactment constraints. During the execution, events generated by the business environment (context events) can influence the scheduling attributes of either the running activities (duration, e.g. through interruption, as in figure 7.2) or the future activities (causal links, start time, duration, etc.) as figure 7.2 shows.

At *concrete workflow* level, a common adaptation is dynamic binding and rebinding, i.e. run-time selection of concrete services on the basis of the abstract description of the desired functionality or the change of this selection on the basis of some conditions identified at run-time.

Generalizing some other approaches mainly focused on QoS constraints at local and global level [25], we consider the binding problem as influenced by global information related to the overall workflow and the business environment that can

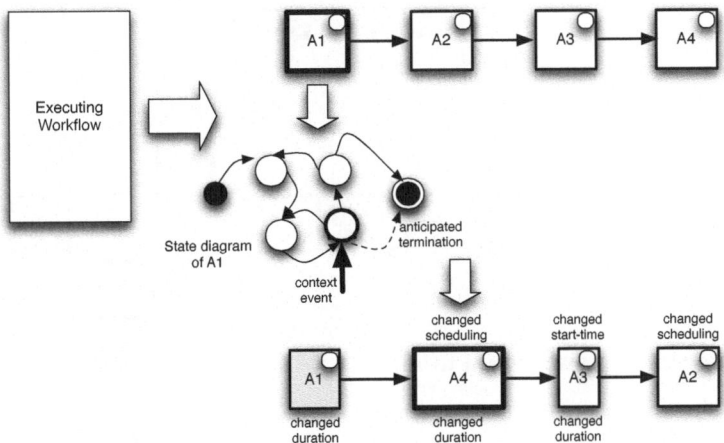

Fig. 7.2 Re-scheduling of workflow activities

impose some restrictions. These restrictions and constraints may also emerge during execution, such as the results of functional or non-functional effects of the services, changing the state of the world and impacting on the execution environment.

If services that implement the activity under focus do not exist, dynamic binding may not be completed. In this case the only possibility is to change the control structure at the *abstract workflow level*, on the assumption that other services, composed together, may perform the same action of the missing service. There is, again, the need for an adaptive intervention for modifying the process description, performing a re-composition (see figure 7.4).

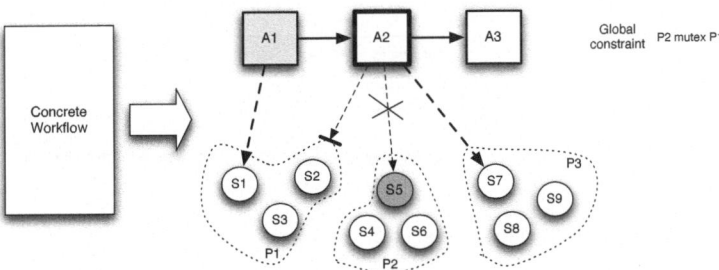

Fig. 7.3 Re-binding of workflow activities

A more abstract level could be considered (as shown in figure 7.1): *business goal*. At this level, monitoring the business environment could suggest to change current business goals and consequently the entire chain of workflows. Even if this ability could be interesting for enterprises working in dynamic markets, the detailed discussion about this kind of adaptation is out of the scope of this chapter.

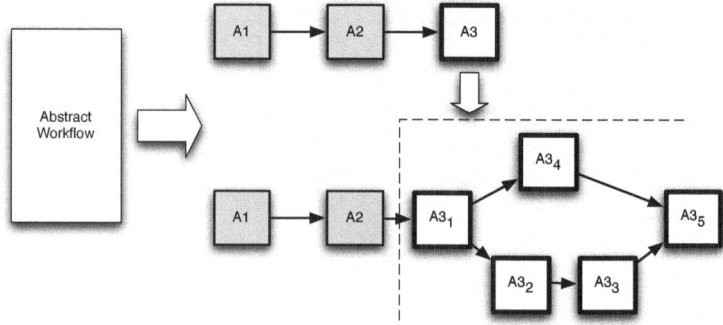

Fig. 7.4 Re-planning of workflow activities

7.3.2 Autonomic Workflow Life Cycle

The underlying logic of an Autonomic Workflow allows for the management of a process (workflow instance) during its entire life cycle: *conception*, *description* or *plan*, *binding*, *enactment*, *execution* and *post-mortem analysis*.

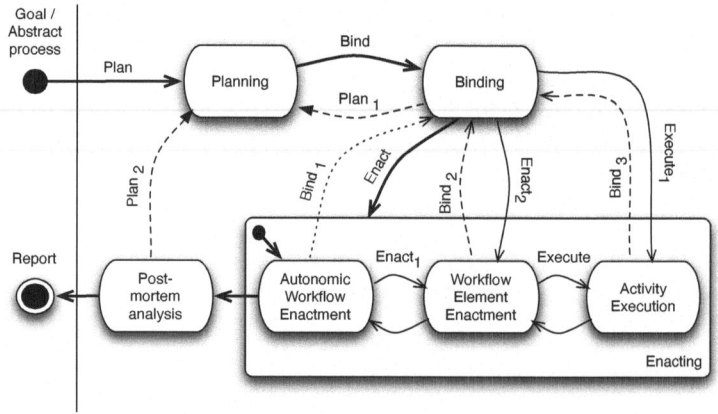

Fig. 7.5 Autonomic workflow life cycle

After the conception, the problem to solve, the goal, the starting state and the domain constraints of a workflow are known. The problem description and the related goal are used for planning (Plan) the activities of the workflow that solve the problem, reaching the goal with a finite sequence of them.

When a planned workflow exists, the next step (Bind) is the *binding* with the concrete resources/services that are able to execute the abstract activities defined in the plan. At this stage, three cases are possible: *static*, *dynamic* or *constrained* binding:

- *Static binding*. It allows for obtaining a concrete workflow (typically at design time), in which each abstract activity is associated with a concrete piece of functionality provided by a service, we refer to it as 'once and for all binding'. The result is one or more directly executable workflow elements.
- *Dynamic binding*. In this case, the concrete workflow is not created at design-time since binding is resolved at run-time. It may happen dynamically (1) prior the starting of the execution, defined early binding or just-in-time binding of the workflow ($Bind_1$), or (2) for each activity that is selected for execution, called late binding or on-the-fly binding ($Bind_2$).
- *Constrained binding*. The problem may have constraints and restrictions that may affect the binding of the services. If the constraints are intensive (not additive), for example "response time has a maximum value that must not be exceed", then each activity binding may be performed respecting this restriction. If the constraints are extensive (additive), for example "total process time must be less than a value", the dynamic binding of a single activity could not fulfill the overall restrictions. In the presence of constraints of the second category, the binding activity may be performed at build-time, but only to check that a set of services that may satisfy the constraints of the workflow exists. When the set of services has been found, the concrete descriptions of the services are used to define the binding restriction for each service in the composition. Each kind of binding is managed from the system in order to use the overall information about the workflow, the service descriptions and the domain knowledge.

After the binding phase, the process is ready for *enactment*. If the workflow has been bound completely at build-time, it is sent to the enactor for the execution, entering in the *Enacting* state. Otherwise the workflow is still abstract and, generally, may have more than one workflow element that must be executed. In this case the workflow enters in the *Enacting* phase but requires binding operations to be performed.

In the *Autonomic Workflow Enactment* sub-state a workflow element is selected for the enactment. The enactment of a workflow element brings the workflow in the *Workflow Element Enactment* sub-state, whose behaviour is similar to the enactment of a traditional workflow; the only difference is that the activities in the workflow element may be still abstract functionalities and so they require a run time binding. This depends on workflow constraints and policies. When information about the concrete implementation of an activity is available, the activity is invoked, and the workflow enters the *Activity Execution* phase.

All these steps characterize the direct action flow inside the lifecycle of the Autonomic Workflow. This flow is associated to a reaction flow that is in charge of the management of the problems that may arise during process enactment. A running activity may be impossible to complete for low level malfunctioning, e.g. network issues or hardware breakdown, or for logical reasons, i.e. the predefined constraints for the service provision are not fulfilled and the interaction is interrupted.

When an anomaly occurs, the system tries to analyse the problem in order to plan a corrective action. Therefore, if a binding violation occurs for a service, the system tries to re-bind the activity ($Bind_3$) with a different service. Then, if another service exists that has the same or a compatible implementation of the failed one, the system may execute a re-binding, changing the mapping of the abstract activity with another service. In this phase, the system should be aware of the process global state and restrictions, and using the contextualized information of the model executes the action to complete the binding, eventually performing also the binding of other activities.

Re-binding failure may cause the process element termination. The process now may continue the execution of another process element or may enact again the failed one. If this is not the case, because the environmental situation is not changed and the required service is still not available, then the process execution is stopped and the binder receives the request for performing another binding for evaluating whether it is possible to perform the process considering alternative binding solutions ($Bind_1$), then, an early binding with the objective of verifying the possibility of completing successfully the enactment. If neither this re-binding may resolve the problem, then the only chance is to force the process back in the planning state ($Plan_1$). In this phase, considering the planner context and the new and updated information available, about the environment and the services, a new planning problem may be generated. The new problem initial state is the last coherent state reached by the failed process, while the objective is still the original goal.

The post-mortem analysis may discover opportunities for process improvement, obtained through a re-planning of the process ($Plan_2$) in which the planning problem has been updated with the new knowledge collected during autonomic workflow enactment.

7.4 Autonomic Workflow Management System

To achieve the behavior described in the previous chapter, a specific system is needed. This comprises not only the engine that is able to drive the execution of autonomic workflows, but also the external environment that needs to provide specific and detailed information to drive the autonomic behavior of the whole system.

In this section a meta model is presented to master the complexity of the autonomic workflows and the system infrastructure needed to run them. The model has a twofold objective: (1) to provide a formalized description of the external environment and of autonomic workflows in order to support designers; (2) to identify the key elements that need specific runt-time support for the autonomic workflow execution.

7.4.1 Meta Model

The meta-model (illustrated in figure 7.6) formalizes the conceptual model underlying our idea of autonomic workflow. The higher level is concerned with the problem description for the autonomic workflow. This description contains also the constraints that must be fulfilled during the enactment, and the domain, that is the set of semantically described services, which may be used. The middle level contains the characterization of the process elements and the functional description of the involved activities, considering input-output parameters and pre- and post-conditions. The lower level regards the concrete description of the activities and the correspondence with deployed services.

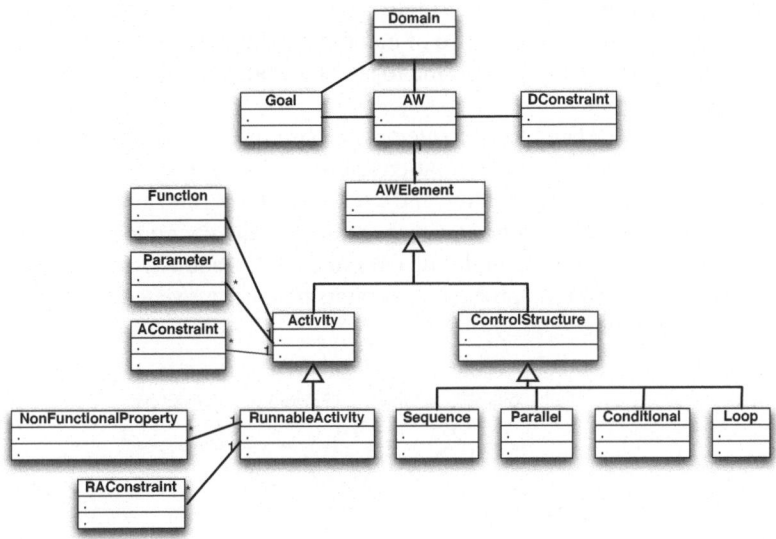

Fig. 7.6 Autonomic workflow meta-model

The central entity of the meta-model is the AW entity, representing the autonomic workflow to be managed. An AW, i.e. a planned solution to the problem, may be expressed with one or more than one AW elements that are represented as abstract processes. Each AWElement is represented as a graph with two kinds of nodes: (1) *simple nodes* are activities to be performed, characterized by their functional semantic description, parameters and constraints. (2) *complex nodes* may contain other nodes, that in turn can be complex or simple nodes, and are used to describe the control structure of the workflow.

The *Goal* and the *Domain* represent the driving descriptions for the management of an autonomic workflow. The Goal contains the description of the initial state and the final state to reach. The Domain contains the available operations that can be considered for solving the problem described by the Goal. *DConstraints* allow for imposing global constraints onto the workflow to manage.

The workflow to manage is in turn composed of one or more *AWElements*, which represent the abstract activities or the control structures. It is worth to note, that AW elements may contain partial solutions if the complete planning is impossible without additional knowledge generated by the execution of the partial solutions, i.e. the problem could be not solved entirely but only partially and the complete solution may be found only after achieving partial results. An *Activity* is an operation related to a service interface and is provided with: functional semantic description (*Function*), inputs and outputs (*Parameter*), and may also have some constraints (*AConstraint*) that represent the non-functional requirements for an activity.

Finally, the description of the concrete implementation of an activity is handled by the *RunnableActivity*. This refers to a concrete service that may be invoked with a specified interaction protocol and that may perform the desired operation. The concrete description is composed also of non functional semantics, that may be used for managing the QoS of the service and to negotiate other parameters or constraints for the service providing.

Other information related to concrete operations are the binding constraints, that are used to take into account the limitations imposed by the process constraints that influence each single activity binding and execution. Therefore, a RunnableActivity is connected with *NonFunctionalProperty*, related to the concrete characterization of the service in term, for example, of QoS and SLAs, and *RAConstraint*, that is information derived from the process constraints that are translated in specific limitations for the particular activity.

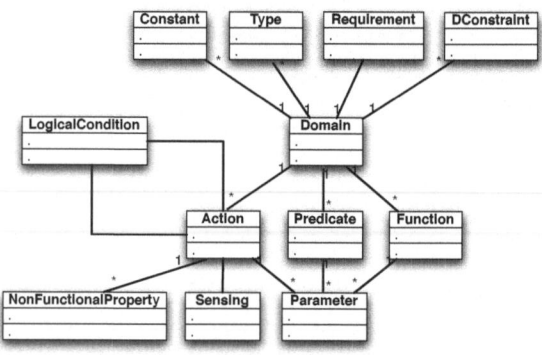

Fig. 7.7 Domain meta model

Figure 7.7 shows a detailed description of the Domain element of the model with related entities. To reduce the gap with the implementation, the meta-model is based on the Planning Domain Definition Language (PDDL), a logic language typically used by AI planners. The basic information needed for defining a problem is related to the Domain entity and should be able to maintain information taking into account the different planning formalisms. The related classes are responsible for modeling every single aspect of the process.

DConstraint is used to specify conditions that may not be violated in the definition of the plan. *Constant* and *Type* are used to handle pre-defined information. *Predicate* and *Function* are used to express information that may be computed during problem solution. *LogicalConditions* are used to express the pre- and post-conditions of the action. Non-functional aspects of the services in the domain are

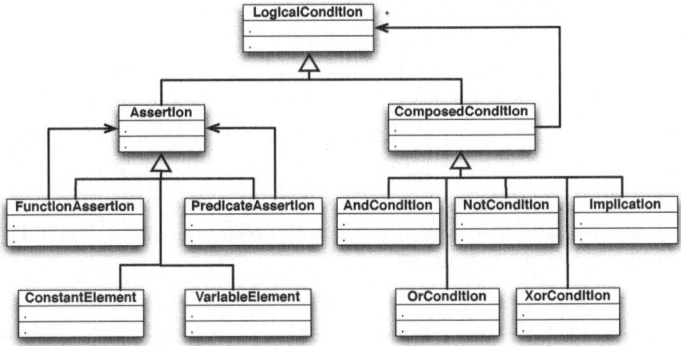

Fig. 7.8 Logical Condition Modeling

modelled with *NonFunctionalProperty*. Finally, *SensingActions*, are actions used at run-time to check variables and logical expressions in the domain.

Figure 7.8 shows the portion of model that represents *LogicalConditions*. They may be simple (*Assertion*), and complex (*ComposedCondition*), able to contain and group together one or more logical conditions, simple or complex, in turn. An *Assertion* may be modeled as a predicate (*PredicateAssertion*), or as function (*FunctionAssertion*). The former is a declaration that may refer to constants and variables. The latter is a function that may refer to input-output variables and constant elements and may be evaluated to obtain a value.

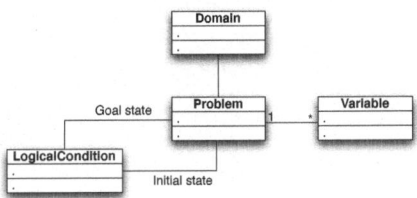

Fig. 7.9 Problem Modeling

Figure 7.9 shows the meta-model of the problem to solve with its link to the domain. It is characterised by the initial state, a logical condition that expresses the state of the world to consider for beginning the solution, and the goal state, a logical condition that identifies the state of the world that must be reached, with a set of actions. It is worth noting that the abstract concept of Action is then connected to the more concrete Activity of the planned process.

7.4.2 Architecture

The proposed meta-model suggested the architecture (see figure 7.10) of a system, called SAWE (Semantic and Autonomic Workflow Engine), able to use the meta-information to automatically perform several activities that, in traditional systems, are executed by users and administrators.

The system comprises three main components: *Configurator*, *Engine* and *Manager*. They interact among each other, and with the external environments that host the services and the resources to use, that virtualise service and goods to handle in the business environment. The Engine is responsible for processes execution. The Manager is in charge of monitoring. The Configurator is in charge of suggesting workflows or sub-workflows to the engine on the basis of inputs derived from users or from the manager (even at runtime).

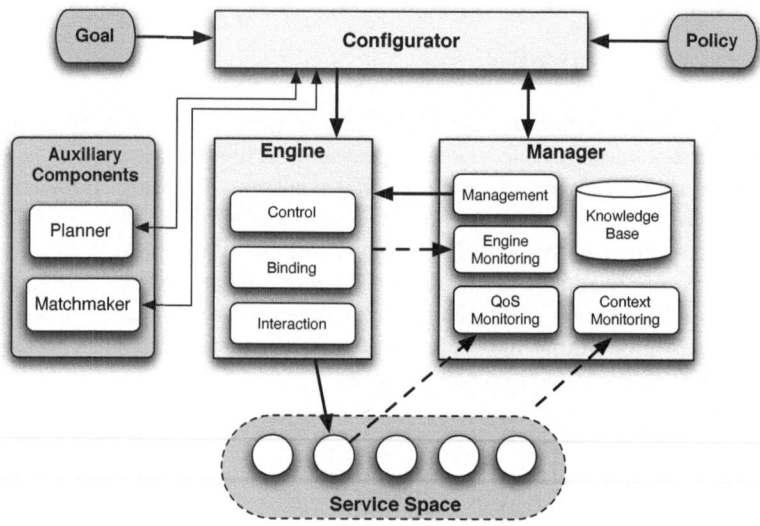

Fig. 7.10 Conceptual architecture

The Engine may execute processes with different levels of details. Goal-based descriptions are a way to state a problem that the system has to solve by defining a plan before executing. Abstract description is used for processes that have only a functional description. Concrete description is for processes that are detailed in every part and contains the resource allocation for the activities.

The Engine is a three layer system. The Control Layer is responsible for receiving the process description, create and navigate the process graph managing the control flow, and evaluate the activities that may be enacted. The Binding Layer is the component performing the selection of the more appropriate resource to execute a function. The Interaction Layer is in charge of contacting the resource and managing the invocation of a functionality. The three layers are responsible for the overall action flow: from the overall process to the function through the resource selection.

In the opposite direction, the Manager handles the reaction flow of anomalies. The primary activity of the Manager is to monitoring the execution. The resource that is executing an activity is monitored for checking compliance with the defined

QoS. The execution environment is monitored to retrieve information from the context. Workflow execution engine levels are monitored themselves to gather information about events and states of the system. All these sources are able to feed the monitoring system with events that are used to trigger rule checking, i.e. condition verification, and, where appropriate, actions may be performed by the manager. Depending on the architectural layer interested by the event, the action may be the generation of another event or a message for the Management component.

The Engine is provided with an API for retrieving information about the execution state. Such events and data reach the internal managers and are collected using a Knowledge Base that hosts ECA rules obtained from the policy defined by the user. Such rules are checked against events and data collected and if it is necessary an action is taken. The action affects specific layers of the Engine with different behaviors: at Interaction Layer, an action can be suspended or an operation can be invoked again; at binding Layer, an activity can be bound to a different service; at control Layer, a dynamic adaptation of the control flow can be performed.

7.4.3 Technologies

A prototype implementation of SAWE has been developed in Java. The system is highly modular and exploits many open source technologies. The Control Layer receives the process description in XPDL (XML Process Definition Language) [9] or WSBPEL, creates and navigates the process graph, and chooses the activities that could be executed according to the activation conditions and the control flow of the executing workflow. To enable the execution of autonomic workflow, programmers need to exploit a variant of XPDL, called XPDL*, and a rule based language, called SPL (SAWE Policy Language). XPDL* integrates specific extended attributes and is used to program the imperative behaviour of an autonomic workflow, whereas SPL exploits ECA rules for defining the reactive behaviour. Their combination is able to stimulate the different components of the SAWE engine in order to execute a workflow towards its completion even in the presence of unexpected events during the normal execution flow.

The Binding Layer performs the binding with the resources to use for the execution. To this end it can interact with a Matchmaker for performing QoS-based discovery and selection of the most appropriate Web Service instance to be assigned to an activity for the execution, when the binding is not specified statically at design time.

The Interaction Layer is able to create connectors between SAWE and the resources by exploiting different technologies: RMI, Web Services, POJO, or HTTP. The interaction with the resources is demanded to a specific adapter for each technology and is managed using a standard Resource Interface that abstracts from the technology the communication with the upper layers.

The current implementation of SAWE exploits JESS (Java Expert Shell System) to host and reason about rules and facts (knowledge base) provided to trigger the reactive behaviour of the system. The engine provides an API enabling some reflective operations: *ADD* an activity - serially, concurrently or iteratively, *DROP* an activity, and *CHANGE* the timing constraints of an activity. Therefore, the monitors get data from the environment and triggers some low-levels rules. This rules generate some facts that in turn trigger rules that can act on the current workflow structure with the aim of changing it.

The external world of services is modeled with OWL-S Profiles and Models and the associated WSDL service interfaces, which are stored in a specific repository. This information is translated in the internal description of the problem domain and used to drive the planner for an initial configuration, by using the PDDL language.

7.5 Rule-Based Workflow Adaptation

The reactive behavior of SAWE is based on monitoring to detect significant changes of the execution environment. Monitoring is intended as a planned or event-driven measurement of one or more properties of a workflow instance. The properties may be directly connected with the executing process (e.g. the actual QoS of the provided services for comparison to the declared one, specified in SLAs), the artefacts managed by the process (e.g. the degree of completion of a document or more generally a measure of a quality parameter of artefacts) or the infrastructure (e.g. a problem to the network infrastructure that produces a temporary unreachability of a service/resource).

The approach proposed to perform these measures is based on the concept of *checkpoint* (see also [28] for further details). This is a point in the control flow that triggers the monitoring system to perform a measurement, evaluate the overall state of the running workflow, and eventually program some actions.

In the proposed approach, checkpoints are defined by the process designer using a semi-automated approach. A checkpoint is only virtually associated to the control flow. It may be linked to a single activity, a set of activities or a process. Therefore the set of checkpoints represents a reactive control flow overlapped to the normal flow of execution.

Checkpoints present several advantages:

- They enable the separation of the workflow definition from the monitoring policy. Workflow enactment and checkpoint execution are two orthogonal and concurrent executing flows.
- Different process instances can be monitored with the same checkpoints and it is possible to change or redefine them for each instance to be enacted.
- It is possible to use the same measurement tools and management actions in several situations, by reusing checkpoints.

The conceptual behaviour of a checkpoint is based on the Event Condition Action paradigm (ECA rules). Rules are expressed according to the following grammar, described with the Extended Backus-Naur form (we omitted angle brackets for readability. See also Fig. 7.11 for a UML representation of the meta-model).

```
EXP ::= EXP_TYPE; | ;
EXP_TYPE ::= CONSTRAINT | RULE
CONSTRAINT ::= ACTIVITY<->ACTIVITY | ACTIVITY->!ACTIVITY
ACTIVITY ::= IDENTIFIER
IDENTIFIER ::= STRING
RULE ::= MNG_RULE | MON_RULE
MNG_RULE ::= on CEVENT (if CCONDITION)?
OP ACTIVITY (WHERE ACTIVITY | WHERE INTEGER)?
MON_RULE ::= on CEVENT (if CCONDITION)?  assert EVENT
CCONDITION ::= (CONDITION) | CONDITION
CONDITION ::= CEVENT (&& CCONDITION)*
CEVENT ::= (EVENT) | EVENT
EVENT ::= IDENTIFIER (COMPAREOP CONSTANT
COMPAREOP IDENTIFIER)?
CONSTANT ::= INTEGER | FLOAT
OP ::= add | drop | replace | move
WHERE ::= after | before | with
COMPAREOP ::= > | < | == | >= | <=
```

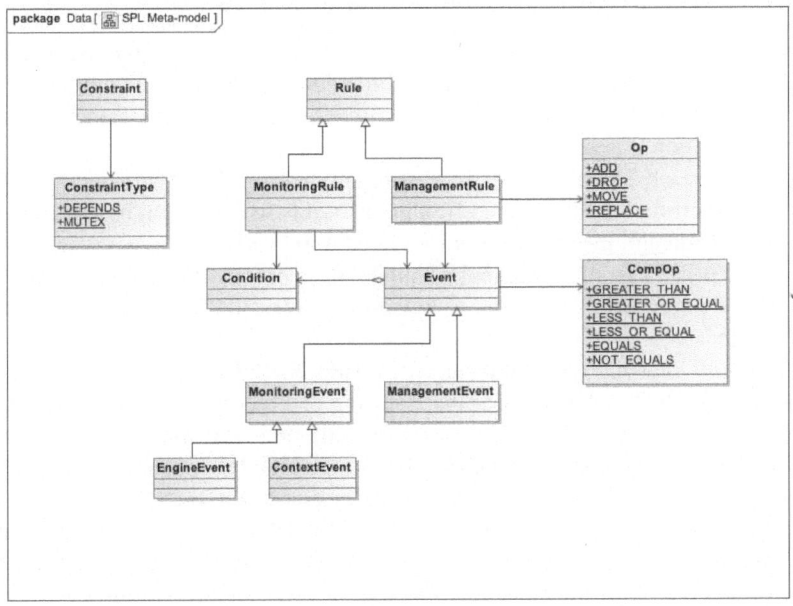

Fig. 7.11 SPL metamodel

EXP is the starting symbol that may be used for defining a rule (RULE) or a constraint (CONSTRAINT).

Constraints can be *mutual exclusion* (ACTIVITY →! ACTIVITY) and *dependence* (ACTIVITY ↔ ACTIVITY) between two entities. Mutual exclusion (MUTEX) means that the two involved activities may be not present in the same workflow. Dependency (DEPEND) means that if one of the involved activity is present in the workflow also the other one have to be present.

Rules can be used for: monitoring (MON_RULE), which generate higher level events by producing facts in the knowledge base; management (MNG_RULE), which are able to manage the current workflow.

Events (EVENT) are identified in the ontology of the application domain and describe the activities (ACTIVITY) to be monitored and managed. Conditions (CONDITION) are tied to possible events that may occur in the system and the related parameters to be measured, such as those related to the QoS or to the execution environment (context). IDENTIFIER can be a literal or a combination of letters and numbers, whereas CONSTANT can be an integer or a float. Constraint are the two restriction that is possible to define in a domain: mutual exclusion and dependency.

OP identifies a possible high-level management action belonging to the set {ADD, DROP, REPLACE, MOVE}.

- ADD inserts an activity to the running process;
- DROP removes an activity from the running process;
- REPLACE replaces an activity in the running process;
- MOVE shifts an activity in the running process.

In particular, actions MOVE and REPLACE are defined as a sequence of DROP and ADD: REPLACE removes an activity and adds another in place of that, instead MOVE removes an activity from the current position and adds the same activity in a different position.

Replacing a task *T1* within process *P* with another task *T2* is equivalent to deleting *T1* and inserting *T2*. Consequently, all constraints which might be violated after applying deletion and insertion operations need to be verified.

Moving a task *T1* from its original position within process *P* to a new position *pos* is equivalent to replacing the same task. Consequently, all constraints are satisfied by definition.

Checkpoints that model monitoring rules (MON_RULE) are not blocking, because any action to be taken regards the assertion of a fact that generate one or more higher level events. The Event triggering the rule concerns the parameters monitored; the Condition tests the value of other state variables (but the one tied to the Event); the Action is the assertion that generates a new fact in the knowledge base.

Checkpoints that model management rules (MNG_RULE) are blocking because, in this case, the running workflow could be changed. The Event is a new fact asserted by one or more monitoring rules; the Condition tests state variables related to the overall process environment to decide whether starting the Action; the Action can be the generation of a new fact or a reflective action on the running workflow devoted to change its structure.

Structural changes of workflows are not light actions. Different aspects of the workflow need to be checked to ensure correctness also after the changes. Three levels of correctness that must be ensured has been identified:

- *structural correctness*: the system has to check if the consistency of the input/output in the resulting process is preserved.
- *local semantic correctness*, the system verifies that after an adaptation the pre- and post-conditions are fulfilled.
- *global semantic correctness*, the system verifies that semantic constraints on the process are respected. Such constraints depends on the particular domain and are derived by the knowledge as elicited by domain experts. Examples of global constraints are: *dependency* and *mutual exclusion*

To express correctness constraints, the following rule schema is used:

```
RULE ID: MNG_RULE
ON EVENT: Event
IF CONDITION: Structural && localSemantic && globalSemantic
ACTION: ...
```

Structural and local semantics are verified, implicitly, by the manager. Global restrictions are derived by the user explicitly defined policies. In the following, examples of templates for expressing constraints for the adaptation meta-operations are presented.

Meta-operation ADD:

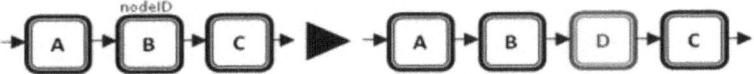

Fig. 7.12 Add operation

```
Structural
(OUTPUT(nodeID)=INPUT(D)) && (OUTPUT(D)=INPUT(nodeID.NEXT))
localSemantic
(POSTCONDITIONS(nodeID).CONTAIN(PRECONDITIONS(D))) &&
     (POSTCONDITIONS(D).CONTAIN(PRECONDITIONS(nodeID.NEXT)))
globalSemantic
(!(D.MUTEX(ACTIVITIES(processID)))) &&
     ((!(D.DEPEND(ACTIVITIES(OTHER)))) ||
     (D.DEPEND(ACTIVITIES(processID)))))
```

where nodeID is the activity that precedes the activity to add, in this case B, and NEXT points to the successor nodeID. The designer must check whether the activity is to be included in mutual exclusion with some other activity in the process, or if a dependency exists with other activities.

Meta-operation DROP:

Fig. 7.13 Drop operation

```
Structural
(OUTPUT(nodeID.PREV)=INPUT(nodeID.NEXT))
localSemantics
(POSTCONDITIONS(nodeID.PREV).CONTAIN
    (PRECONDITIONS(nodeID.NEXT)))
globalSemantics
(!(ACTIVITIES(processID).DEPEND(nodeID)))
```

where nodeID is the task to be deleted, in this case C, NEXT points to the successor of nodeID and PREV points to the predecessor of the same. In this case, the only semantic constraint to check is whether the overall process activities have a dependency with the task to be deleted.

Meta-operation REPLACE:

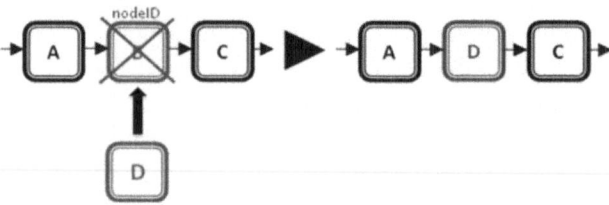

Fig. 7.14 Replace operation

```
Structural
(OUTPUT(nodeID.PREV)=INPUT(D)) &&
(OUTPUT(D)=INPUT(nodeID.NEXT))
localSemantics
(POSTCONDITIONS(nodeID.PREV).CONTAIN(PRECONDITIONS(D))) &&
    (POSTCONDITIONS(D).CONTAIN(PRECONDITIONS(nodeID.NEXT)))
globalSemantics
(!(D.MUTEX(ACTIVITIES(processID)-nodeID))) &&
    ((!(D.DEPEND(ACTIVITIES(OTHER)))) ||
    (D.DEPEND(ACTIVITIES(processID)-nodeID))) &&
    (!(ACTIVITIES(processID).DEPEND(nodeID)))
```

where nodeID is the activity to be replaced with D, in this case B, NEXT points to the successor of nodeID and PREV points to the predecessor of the same. In this case, the semantic constraints to consider regard the ones discussed for DROP and ADD.

Meta-operation MOVE:

Fig. 7.15 move Operation

```
Structural
(OUTPUT(B.PREV)=INPUT(B.NEXT)) && (OUTPUT(nodeID)=INPUT(B)) &&
    (OUTPUT(B)=INPUT(nodeID.NEXT))
localSemantics
(POSTCONDITIONS(B.PREV).CONTAIN(PRECONDITIONS(B.NEXT))) &&
    (POSTCONDITIONS(nodeID).CONTAIN(PRECONDITIONS(B))) &&
    (POSTCONDITIONS(B).CONTAIN(PRECONDITIONS(nodeID.NEXT)))

globalSemantics none
```

where nodeID is the activity that precedes the new position of the moving activity, in this case C, NEXT points to the successor of nodeID and PREV points to the predecessor of the same.

7.6 Business Process Modelling for Autonomic Workflows

The adoption of autonomic workflows in the context of networked enterprises requires extensions to the common approach (business process modelling) adopted for statically defined workflows. This is due for two main reasons: (1) the adoption of the SOA paradigm and the related technologies; (2) the need to specify additional knowledge that guides WfMSs to take proper decisions as much as possible without human interventions.

The SOA context asks for a (partial) bottom up approach: enterprises that wish to belong to a business network must provide services with adequate semantic descriptions. These services are indexed by specific registries used by the enterprises network. The basic technology to adopt for describing services is WSDL. These descriptions are then semantically annotated by using OWL-S. It can be also adopted to add QoS guarantees (by using onQoS) to services to add the semantic ability of reasoning about them.

Designers of autonomic workflows can exploit extensions of commonly used tools. The first step is the definition of the business process in two forms:

- an abstract description, based on the BPMN language, of the flow of the activities designed by the domain expert;

- an abstract, semantic description, based on the OWL language or PDDL, of the problem to solve in terms of initial state and goal. PDDL can be preferred in some cases for its expressiveness and concise syntax.

The former model is useful to exploit domain knowledge owned by experts. The latter is important to provide a support to the run-time environment for taking proper decisions during the execution when anomalies occur. In particular, BPMN modelling can be extended with further information that regards specific activities and the overall workflow. This info is formalized though the policy language (PL) and regards the handling of QoS violations, specific context events, and workflow constraints.

The abstract descriptions provided by domain experts with the previous languages (OWL-S, BPMN, PL) are used by SAWE and additional modelling tools with different purposes. Starting from the BPMN representation of the workflow, a more concrete description, based on XPDL* and SPL, is obtained.

The OWL-S based representation of the problem could be used to drive an automatic generation of the whole workflow on the basis of the formalized domain knowledge provided by the published services. Even if, this is theoretically possible, in practical cases it is difficult to achieve due to the complexity of the whole workflows. Therefore, the formalized description of the problem represents a strategy to satisfy when decisions are taken during the execution. In particular, it is useful to re-plan the workflow after a serious fault that is not possible to handle at lower levels.

7.6.1 Application Example

The example discussed in this section is related to a value chain of companies cooperating with a winery. The overall process (composed of three phases: *cultivation*, *production* and *distribution*) is managed with a workflow and all the partners in the value chain interact by exposing their activities as semantically annotated services.

We will focus on the cultivation phase. During this phase, a grape grower is responsible for managing the activities in the vineyard, handling the cultivation operations, monitoring the status and the quality of the grapes. In a real environment, more than one vineyard may be considered, managed by several grape growers.

The operations to be performed in the vineyard are of two types: *automated* or *human-oriented*. Both of them are handled by the autonomic workflow system, the former with semantically annotated services, the latter using tasks performed by humans through a user interface.

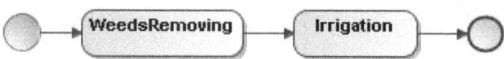

Fig. 7.16 Optimistic daily process

The grape grower handles operation in one or more vineyards. The activities in such environment are planned, manually or using the planning approach described in Section 7.3. A process is started every day for each vineyard, containing the planned activities to be performed. During process execution, the system monitors the environmental parameters, i.e. `temperature`, `humidity`, `wind`, etc. Such parameters are then analysed for identifying emerging problems. The manager, then, is able to plan a reaction, based on the policy rules, and apply it to the running workflow instance. For example, rain is a problem if in the daily process a human activity is planned in the vineyard, as *weed removing*. In this case, the activities must be postponed or removed. If the weather is dry, then an irrigation task may be added to the process. A daily process could be defined also with pure event-driven logic. In this case, no activity is planned, but a listening interval for possible events must be defined. Monitoring the environmental conditions in the vineyard may cause adaptation of the running process, which is involved in the same vineyard.

The process is monitored by exploiting monitoring rules, as reported in the following:

```
on (Pluviometer > 1) assert Rain;
```

The rule checks pluviometer readings, and in the presence of rain, generates a `Rain` event for the management component. In this case, the policy is used for planning reactions to events in order to avoid performing activities when they are not needed or not advisable to perform. Domain experts may assert rules, without needing to write complex code. They are required to express their knowledge that is formalized in high-level rules.

```
on (Rain) drop Irrigation;
on (Rain) drop WeedsRemoving;
```

In this case, the management rules act on the process, dropping the activities that can not be performed. The reaction is asynchronous of the process and depends on the monitored condition of the vineyard. We may consider an analogy with a conditional instruction which depends on a negative condition: if a condition (rain in our case) is not verified, then one or more instructions (vineyard operations) are performed.

The differences are: (1) the "if" control structure is executed only one time while monitoring is asynchronous and may executed in every moment; (2) the process is simpler since we may consider that each "if" control structure introduces two paths in the control flow, whereas each rule removes a path from the control flow. An example of a traditional process that explicitly checks the condition before executing is represented in Fig. 7.17.

The traditional process is more complex than the autonomic one. Moreover it is not able to perform asynchronous checking of the condition.

A different example that we may consider is a pure reactive situation, in which the process is planned without activities and all the actions are executed only when a condition is verified.

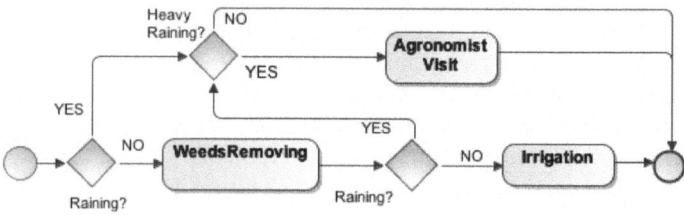

Fig. 7.17 Daily process with a traditional approach

```
on (Temperature > 30)
if ((Humidity > 75) && (Wind > 9))
assert PowderyMildew;
```

In this case, the environmental conditions favorable to the formation of powdery mildew (a vineyard disease) are checked. If they are detected, an appropriate event is passed to the management component, which may react according to the following rule:

```
on (PowderyMildew) add Spraying;
```

The reaction introduces an activity in the process for performing spraying with adequate substances, to prevent the disease. In this case, we may consider an analogy with another control structure: a loop that cycles doing nothing, with a conditional instruction inside (see Fig. 7.18). If the condition is verified, a set of instructions is then executed.

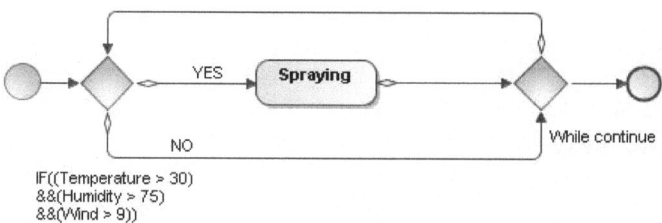

Fig. 7.18 Another daily process with a traditional approach

Both monitoring and management rules expressed in SPL are currently translated in CLIPS rules, for their execution in a Jess engine. In the following, the translation of the management rules for the daily processes previously described is reported:

```
(assert (ADD (event PowderyMildew) (activity Spraying)))
```

Translation is based on templates to factorise the common behaviour of the operation and to perform verification before executing the adaptation. The following fragment of code shows an example for the ADD meta-operation.

```
(deftemplate ADD (slot event) (slot activity))

(defrule ADD-Management
    (ADD (event ?e) (activity ?a))
    (ContextEvent {type == CONTEXT}
      (name ?e) (sourceId ?sId) (processId ?pId) (value ?e)
      (measurementUnit ?mu) (description ?d) (areaId ?aId))
    (not (ConstraintViolation {type == Mutex}
      (activityA ?a) (activityB ?aB) (procId ?pId)))
    (not (ConstraintViolation {type == Dependence}
          (activityA ?a) (activityB ?aB) (procId ?id))))
    =>
    (call ?manager addActivity ?a ?pId)
)
```

The benefits of SAWE are more evident if we consider a larger set of monitoring rules, checking for several kinds of conditions, and a richer set of management rules. This kind of processes should very difficult to manage without the support provided by our approach.

In order to avoid conflicts between rules it is possible to define domain constraints, using our policy language (called SPL - SAWE Policy Language). In the following, two examples of rules that express mutual exclusion between activities are reported:

```
Sprying !-> Irrigation;
WeedsRemoving !-> Sprying;
```

SAWE also supports interprocess adaptation, i.e. monitoring a process in its own environment, which may be adapted, and the adaptation may be used as an additional source of information for the management component.

We consider the case of a daily process monitored with the following rule:

```
on (Pluviometer > 20) assert HeavyRain;
```

The condition for the HeavyRain event includes the condition for the Rain event; thus the two events may be handled together by the manager, reacting to both the situations.

```
on (HeavyRain) add AgronomistVisit;
```

The agronomist must check if the weather is hailing; in this case he must assess the quality of the grapes, and decide if they may be used for making the wine, without degrading the required quality. The agronomist may generate an event for the management component of the higher level process, the one managed by the winery main company.

```
on (GrapesQuality.<ID>==LOW) assert NotHarvesting.<ID>;
```

With the rule above, we are considering the agronomist visit in any one of the vineyard used for growing the grapes. The NotHarvesting.<ID> event asserts that in the vineyard ID it is not possible to harvest grapes. This event will be taken into account by the Grape Grower, which will look for another use of the damaged grapes. At same time, the main company reacts to the event scheduling an activity, in the overall process: *buying the same grapes quality, to be performed instead of the harvesting.*

```
on (NotHarvesting.<ID>) replace Harvesting.<ID> with Buying.<ID>;
```

The vineyard ID is used for identifying the variety of the grapes to be bough when the activities will be actually executed.

In addition to domain-specific rules, also cross-domain rules can be used. QoS violations or service unavailability are two examples of events that can be handled with such rules. In the case of QoS violation, a re-binding mechanism may be used to implement a recover action. However, even this attempt to recover from a QoS failure could fail (for example, when no service is available in the service registry that is able to satisfy the required functionality). In SAWE, this further failure starts a planning process to identify an equivalent subprocess to be used instead of the failed activity. If such a subprocess is found, it is returned to the manager, for being used to satisfy the re-binding. Otherwise a composition failure event is generated.

Such cross-domain rules, expressed with SPL, are the following:

```
on BindingFailure compose <BindFailure>.service;
```

```
on ComposingSuccess invoke <ComposingSuccess>.service;
```

The first monitoring rule detects a problem during binding and then requires an action from the manager. This may try a re-binding or may require a run-time composition. If run-time composition is chosen, the corresponding rule invokes a component able to perform service composition (the planner for example) using as input the current failed activity's semantic description. In the case of successful composition, the third rule is fired. It is responsible for retrieving the process description and performing the invocation, by substituting the failed activity and so allowing to resume the execution.

In the case of a composition failure, the manager may attempt to re-plan the process, defining a planning problem with the actual state of the process as starting situation and the original process goal as planning objective to be reached. If a plan is found that allows for reaching the goal, it is translated into XPDL*, and provided to the manager that takes care of adapting the running instance to the new plan. If a failure occurs also in this case, the manager resorts to a human intervention, sending a message and suspending the process execution.

7.7 Conclusion

This chapter presented the main features of a novel approach for dynamically handling workflows. The approach is based on autonomic computing principles and provides (i) an execution model and a method that aid users to design autonomic workflows, and (ii) a system that is able to run those workflows with the ability of handling execution anomalies with a limited human intervention.

The main objective of the system is to enable workflows management during the entire lifecycle, by collecting and organizing the knowledge from the operative environment through monitoring. This knowledge, in the form of ECA rules, is then used for adapting and improving workflows at run-time.

The chapter presented the motivating scenarios, the dynamic model of autonomic workflows, the meta-model, the overall architecture, the implementation, some details about the language for reactive behaviour and a specific business process modelling.

Future activities are related to the management of domain knowledge to improve the ability of planning reactive actions. Moreover, coupled with ECA rules, predictive techniques will be investigated in order to avoid fast reactions that can create instability of the manager behaviour. To this end, mechanisms based on temporal logics and probabilistic processes will be explored.

References

1. Akkiraju, R., Srivastava, B., Ivan, A.-A., Goodwin, R., Syeda-Mahmood, T.: Semaplan: Combining planning with semantic matching to achieve web service composition. In: ICWS 2006: Proceedings of the IEEE International Conference on Web Services, pp. 37–44. IEEE Computer Society, Washington, DC (2006)
2. Akkiraju, R., Verma, K., Goodwin, R., Doshi, P., Lee, J.: Executing abstract web process flows. In: Proceedings of the ICAPS Workshop on Planning and Scheduling for Web and Grid Services, pp. 9–15 (2004)
3. Baresi, L., Guinea, S.: Towards Dynamic Monitoring of WS-BPEL Processes. In: Benatallah, B., Casati, F., Traverso, P. (eds.) ICSOC 2005. LNCS, vol. 3826, pp. 269–282. Springer, Heidelberg (2005)
4. Bradley, S.P.: Sense and Respond: Capturing Value in the Network Era. Harvard Business School Press, Boston (1998)
5. Brooks Jr., F.P.: The mythical man-month. In: Proceedings of the International Conference on Reliable Software, p. 193. ACM, New York (1975)
6. Brown, A.B., Patterson, D.A.: To err is human. In: Proceedings of the First Workshop on Evaluating and Architecting System dependabilitY, EASY 2001 (2001)
7. Buhler, P.A., Vidal, J.M.: Towards adaptive workflow enactment using multiagent systems. Inf. Technol. and Management 6(1), 61–87 (2005)
8. Canfora, G., Di Penta, M., Esposito, R., Villani, M.L.: A framework for qos-aware binding and re-binding of composite web services. Journal of Systems and Software 81(10), 1754–1769 (2008)
9. Workflow Management Coalition. Xpdl, http://www.wfmc.org/xpdl.html
10. Fecondo, G., Santagata, A., Perrina, F., Zimeo, E.: A platform for collaborative engineering. IT Professional 8, 25–32 (2006)
11. Ganek, A.G., Corbi, T.A.: The dawning of the autonomic computing era. IBM Systems Journal 42(1), 5–18 (2003)
12. Giallonardo, E., Zimeo, E.: More semantics in qos matching. In: SOCA, pp. 163–171. IEEE Computer Society (2007)
13. Han, M., Thiery, T., Song, X.: Managing exceptions in the medical workflow systems. In: ICSE 2006: Proceedings of the 28th International Conference on Software Engineering, pp. 741–750. ACM, New York (2006)

14. Heinis, T., Pautasso, C., Alonso, G.: Design and evaluation of an autonomic workflow engine. In: ICAC, pp. 27–38. IEEE Computer Society (2005)
15. Huebscher, M.C., McCann, J.A.: A survey of autonomic computing—degrees, models, and applications. ACM Comput. Surv. 40(3), 1–28 (2008)
16. Kephart, J.O., Chess, D.M.: The vision of autonomic computing. IEEE Computer 36(1), 41–50 (2003)
17. Kesselman, C., Foster, I.: The Grid: Blueprint for a New Computing Infrastructure. Morgan Kaufmann Publishers (November 1998)
18. Klusch, M., Gerber, A.: Evaluation of service composition planning with owls-xplan. In: WI-IATW 2006: Proceedings of the 2006 IEEE/WIC/ACM International Conference on Web Intelligence and Intelligent Agent Technology, pp. 117–120. IEEE Computer Society, Washington, DC (2006)
19. Lee, K., Sakellariou, R., Paton, N.W., Fernandes, A.A.A.: Workflow adaptation as an autonomic computing problem. In: WORKS 2007: Proceedings of the 2nd Workshop on Workflows in Support of Large-Scale Science, pp. 29–34. ACM, New York (2007)
20. Ly, L.T., Rinderle, S., Dadam, P.: Integration and verification of semantic constraints in adaptive process management systems. Data Knowl. Eng. 64(1), 3–23 (2008)
21. Ter Hofstede, A., Dumas, M., van der Aalst, W.: Process-aware information systems: bridging people and software through process technology. John Wiley and Sons (2005)
22. Müller, R., Greiner, U., Rahm, E.: Agent work: a workflow system supporting rule-based workflow adaptation. Data Knowl. Eng. 51(2), 223–256 (2004)
23. Nau, D.S., Au, T.-C., Ilghami, O., Kuter, U., William Murdock, J., Wu, D., Yaman, F.: Shop2: An htn planning system. J. Artif. Intell. Res (JAIR) 20, 379–404 (2003)
24. Pautasso, C., Heinis, T., Alonso, G.: Autonomic execution of web service compositions. In: ICWS, pp. 435–442. IEEE Computer Society (2005)
25. Di Penta, M., Esposito, R., Villani, M.L., Codato, R., Colombo, M., Di Nitto, E.: ws binder: a framework to enable dynamic binding of composite web services. In: SOSE 2006 Proceedings of the 2006 International Workshop on Service-Oriented Software Engineering, pp. 74–80 (2006)
26. Shishkov, B., Cordeiro, J., Ranchordas, A. (eds.): ICSOFT 2009 - Proceedings of the 4th International Conference on Software and Data Technologies, Sofia, Bulgaria, July 26-29, vol. 2. INSTICC Press (2009)
27. Strohmaier, M., Yu, E.S.K.: Towards autonomic workflow management systems. In: Erdogmus, H., Stroulia, E., Stewart, D.A. (eds.) CASCON, pp. 365–368. IBM (2006)
28. Tretola, G., Zimeo, E.: Monitoring workflows execution using eca rules. In: Shishkov, B., Cordeiro, J., Ranchordas, A. (eds.) ICSOFT (2), pp. 423–428. INSTICC Press (2009)
29. Verma, K., Sheth, A.P.: Autonomic Web Processes. In: Benatallah, B., Casati, F., Traverso, P. (eds.) ICSOC 2005. LNCS, vol. 3826, pp. 1–11. Springer, Heidelberg (2005)
30. Zeng, L., Benatallah, B., Ngu, A.H.H., Dumas, M., Kalagnanam, J., Chang, H.: Qos-aware middleware for web services composition. IEEE Transactions on Software Engineering 30, 311–327 (2004)

Chapter 8
Verification and Analysis of Autonomic Systems for Networked Enterprises

Antonia Bertolino, Guglielmo De Angelis, Felicita Di Giandomenico,
Eda Marchetti, Antonino Sabetta, and Paola Spoletini

Abstract. Autonomic Computing is an innovative research area, that proposes self-management features for dynamic configuration, healing purpose, optimization and protection. Autonomic systems adapt themselves quickly to changes in the environment in which they operate, but, while this feature helps the automatic management of complex systems, it makes the job of validating and verifying them extremely difficult. In this chapter we point out the major challenges in validating such systems and we overview some proposals and methodologies for supporting their validation, analyzing in particular model checking techniques, testing methodologies, model-based dependability analysis and monitoring. Moreover we propose our model checking approach for verifying the autonomic workflow, developed for ArtDeco project and described in Chapter 7.

8.1 Challenges in Validating and Verifying Autonomic Systems

The purpose of validation and verification (V&V) of (software) systems is to show that the system under analysis meets the specified requirements, respecting given constraints and fulfills its intended use when placed in its environment. There are different techniques which may be used for V&V, such as model checking, qualitative analysis, testing, and monitoring. While model checking and qualitative

Antonia Bertolino · Guglielmo De Angelis · Felicita Di Giandomenico ·
Eda Marchetti · Antonino Sabetta
ISTI-CNR, Via Moruzzi 1, Pisa
e-mail: {antonia.bertolino,guglielmo.deangelis,felicita.digiandomenico,
          eda.marchetti,antonino.sabetta}@isti.cnr.it

Paola Spoletini
Universitá dell'Insubria, Via Ravasi 2, Varese
e-mail: paola.spoletini@uninsubria.it

G. Anastasi et al. (Eds.): Networked Enterprises, LNCS 7200, pp. 143–169, 2012.

analysis are often used to check and analyze representations of the software system such as specifications or models at design-time, testing and monitoring are generally used on the implemented system to examine its implementation and its behavior at run-time.

The traditional monolithic or centrally developed systems can be analyzed mainly at design- and at deployment-time, without requiring a continuous monitoring at run-time, but the growth of open, dynamic and decentralized systems arose the need of continuously checking the system while it is running. This highlights the need of complementing the usual verification and testing phases with some other verification techniques. Run-time analysis is needed alongside with offline checking to create a lifelong verification process. Indeed, at development-time, techniques such as model checking analyze the model of the system against properties. This has the obvious limitation that is not the system itself to be verified but its model and it is not possible to guarantee the correspondence between the two objects. V&V at development-time is especially difficult in open environments, where the system is generally a composition of processes, possibly developed by different parties. Within a single component development, the others may not be owned, and consequently known. This arises the need of making assumptions on the interaction with other components while verifying a single component to assess the system behavior. To assure the correctness of the system it is therefore necessary to online check that the made assumptions (or the properties themselves) hold at run-time.

Even more complicated is the scenario when the system to be verified is autonomic. Autonomic computing is an innovative research area that, in the last years, emerged to respond to the growing demand for architectures and systems that can adapt themselves quickly to functional and non-functional requirements and unexpected changes of the external and internal environment. Among the many different context in which autonomic computing is adopted, within the ArtDeco project, the focus has been to apply autonomic systems in the Small and Medium Enterprises (SME) context for reducing the degree of complexity that maintainers have to deal with and to enable new collaboration paradigms among industries sharing common goals. To meet this goal, both an autonomic workflow (see Chapter 7) and the Selflet model (see Chapter 6) have been introduced to implement a sense-and-respond system, able to sense the real-time business environment and react automatically, choosing the most appropriate actions. Within the project, the workflow is manly used for the self-managing of SME, while the Selflet acts as an autonomic interfaces among the different workflow in an enterprises network.

Either looking at the system within a single enterprise or in a network of enterprises, this new way of conceiving systems provides innovative features such as:

- self-configuring, i.e. the system can reconfigure itself "on-the- fly" and within the e-business infrastructure of the enterprise or the network;

- self-optimizing, i.e. the system should maximize the utilization of different resources to meet the requirements of a dynamically managing workload;
- self-healing, i.e. the system must be able to detect autonomously a failed component and recover from it;
- self-protecting, i.e. the system should monitor access to each element on the basis of a number of known technologies.

Figure 8.1 illustrates this control loop, which is referred as MAPE-K (monitor-analyze-plan-execute-knowledge).

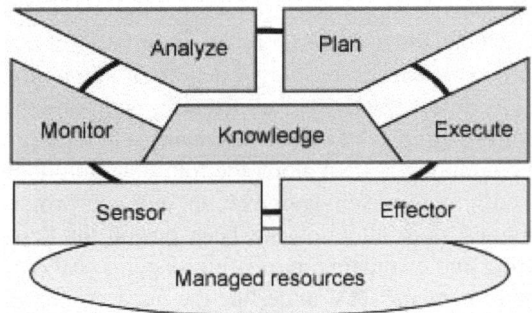

Fig. 8.1 MAPE-K control loop

Obviously this new way of designing systems may improve the performance of enterprises and network of enterprises reducing the cost of management to react to internal and external solicitations, but only if the systems and its self-management are correct and robust. These issues arise the problem of how to develop V&V techniques able to face the new challenges introduced by the features of autonomic systems. It could be even legitimate to doubt that V&V is feasible when the system possesses the ability to adapt and learn, especially in dynamic and not constrained environment. Indeed, together with the classical problems related with open and decentralized systems, autonomic systems evolves autonomously to react to the environment, dynamically changing their structure and their individual and, if networked, global behavior. All these features make the classical approaches of testing and verification not adequate, since the self-management may change the checked model while is running. Moreover, in this context, as shown in Figure 8.1, monitoring becomes part of the control loop and it is used to check the policies that help the system in self-managing and self-adapting, hence, since it, in general, produces some overhead in the system performance, it is not be clear if it can be used also for V&V purposes.

V&V of autonomic systems will be even more challenging for networked systems that require coordinated interactions among several autonomic elements such as the network of enterprises considered in ArtDeco, because it becomes hard to predict the environment behavior. Indeed, it becomes impossible to verify offline a model of the system or to test the system itself capturing the complexity of realistic systems.

Hence, it becomes clear that ad-hoc V&V techniques and methodologies need to be developed to face the challenges of capturing the "self-" features of autonomic systems. Besides the proof of correctness of autonomic components against safety and liveness properties and the monitoring to guarantee the self-management, evaluation of autonomic computing should imply the definition of the properties/measures against which quantitative and qualitative analysis can be performed. An attempt to identify metrics appropriate to measure the effectiveness of autonomic systems is in [43], where the authors focus on quality of service, cost, granularity/flexibility, failure avoidance (robustness), degree of autonomy, adaptivity, time to adapt and reaction time, sensitivity, stabilization, and combination of previous ones. In [14, 57] some examples of benchmark approaches are proposed to quantitatively evaluate a computing system along the four core autonomic dimensions of self-healing, self-configuration, self-optimization, and self-protection. Apart from these specific examples, poor attention has been put on the definition of specific metrics for measuring and evaluating autonomic systems. All these remarks about challenges and open issues in V&V underline the need of overviewing the different qualitative and quantitative approaches proposed for V&V to see how they are adapted to cope with autonomic system features and to propose an approach feasible for ArtDeco architecture.

The rest of the chapter is structured as follows. Section 8.2 overviews the recent achievement in facing the above mentioned challenges. In particular we consider the strategies and the frameworks available for testing an autonomic system as well as the techniques of model checking that can be readapted to this new environment. Moreover, we also analyze the problem of the validation of quality properties, the stochastic model-based dependability analysis and the current proposals for monitoring autonomic systems. All the mentioned techniques were evaluated as a tool to solve problem emerged during the developing of the project. While the previous section deals with validation of autonomic systems, in general, Section 8.3 introduces our proposal to verify both at design-time and at run-time the autonomic workflow developed for the ArtDeco project and Section 8.4 shows how to apply our approach on the project case study. Conclusions together with some new research line in verification of autonomic systems are drawn in Section 8.5.

8.2 An Overview on Verification and Analysis of Autonomic Systems

This section provides an overview of the approaches analyzed to deal with autonomic systems within the project. These approached are categorized through different criteria that can be used to select the convenient verification and analysis technique to assess a specific metric for a certain system under specified conditions.These criteria are the following:

Phase: The stage of a system life-cycle at which a method is applied: specification, design, implementation, deployment, and run-time Some techniques are applicable since the early stages, while others require that at least a prototype of the system already exist;

Quantitative/Qualitative: It indicates whether the method allows for assessing required properties in a quantitative way or in a qualitative way;

Scalability: The ability of the method to scale from relatively small to large, geographically distributed systems;

Accuracy: The ability of a method to provide evaluation results that satisfies specified levels of correctness;

Time: The time necessary to the method to perform the evaluation;

Cost: The cost implied by the application of a method estimated in terms of necessary instrumentation;

Tool Support: Availability of tools or frameworks according to the following scale : Low (manual), Med (semi-automatic), High (automatic);

Target property: The ability of a method to accommodate the evaluation of user-defined properties/metrics, in addition to standard, predefined ones.

One of the target of the verification and validation activity inside ArtDeco was to identify the sets of techniques that better contribute to the improving of the quality and the level of the project products. For this reason, we summarized in Table 8.1 [1] the possible applicable techniques that will be described in the following. However, as evidenced by the Table 8.1, some of the techniques proposed in this chapter were applicable only during the last phases of the product development when the implementation is stable and almost complete, as the testing phase, or even at run time, as monitoring and trust management. Hence, since the development of the ArtDeco case studies were accomplished almost with the end of the project, the above mentioned techniques could not be applied on them for lack of time. For the sake of completeness, however, in the rest of this chapter, we provide an overview of the variety of options that we have investigated during the ArtDeco lifetime (even if we did not directly applied these techniques on the case studies).

Table 8.1 can be consider as a summary of the analysis we have done studying V&V techniques for autonomic systems and as a reference for possible future researches. Notice that, as evidenced by the table, none of the mentioned methods is able to satisfy all the criteria at the highest level. However, when a specific criteria is more relevant, the table suggests that more than a technique can be applied.

---

[1] In the table the possible phases in the system life-cycle are reported as SPE(specification), DES (design), IMP (implementation), DEP (deployment), and RT (runtime).

Table 8.1 An overview of rules to select the most appropriate validation or verification method

	Testing	Monitoring	Model checking	SLA-driven validation approaches	Trust Management	Stochastic State-based Approaches
Phase	IMP, DEP, RT	RT	SPE, DES, RT	DES, IMP, RT	RT	SPE, DES, DEP
Scalability	Low	Med or High	Low or Med	Med		Med
Accuracy	Med	Can be very high, with potentially high overhead	Med	Med		Med
Time	Med		Depending on level of detail			Depending on the context
Cost	Depending on level of detail	Depending on level of detail	Low	Add some computational overhead	Add overhead with respect to both computational, and governance aspects	Low
Tool Support	Low	High	High	Med	Low	High
Target property		Performance, dependability and security			Security	Performability, dependability and security

8.2.1 Model Checking

Model checking, proposed independently by Clarke and Emerson [19] and Queilly and Sifakis [53] in the 80s, allows the automatic analysis of all possible behaviors of a system against any behavioral property, producing a counterexample if the model of the system does not satisfy the property. The systems to be verified can greatly differ from each other, as example we can distinguish between finite and infinite systems, sequential and concurrent systems, deterministic and nondeterministic systems, software and hardware, and synchronous and asynchronous systems. Depending on the characteristic of the system, different techniques of model checking can be adopted. A traditional categorization is based on the class of temporal logic used to formalize the property, that in general can be branching temporal logic, such as CTL (Computation Tree Logic), and linear temporal logic, such as LTL (Linear Temporal Logic). Usually the first uses a symbolic representation of the system, while the latter an automata-based one.

Even though recently many different kinds of model checking approaches were proposed, originally model checking was developed for systems on discrete time without considering metric. Indeed, it was originally developed for hardware, that is also representable as a finite synchronous deterministic model. However, there are further extensions, to cover a wider range of characteristics such as the possibility of verifying real-time[30] or infinite state models [39], [21] offers a brief overview on these techniques.

Independently from the chosen model checking approach, there are some characteristics typical of the technique itself. Indeed, model checking is generally used at specification and design phase, to check a system before it is implemented.

Model checking does not requires particular technologies to be applied and is, in general, not demanding in terms of time. Instead, it is particularly demanding in terms of space and it is affected by the state explosion problem[20], that makes the space needed for the verification growing fast with respect to the size of the model. To overcome this problem, recent tools implements different optimization techniques, such as partial order reduction and abstraction. The mostly used formal tools allowing verification by model-checking are Spin [6], NuSMV [2], Uppaal [5], Emc [18], Tav [37], [3], and XTL [42], but there are many others.

Even if model checking is a widely studied technique, also in terms of application to different context, its application to autonomic environments is still a complex task due to the complexity of state representation, especially because of the ability of the system to self-manage.

In this context one of the emerging proposal is the ASSL (Autonomic System Specification Language) Approach to Autonomic Computing [59], that proposes a frameworks for supporting the specification, verification [60] and the code generation of self-management systems. The current validation approach in ASSL is a form of consistency checking performed against a set of semantic definitions. Currently ASSL has been used to specify autonomic properties and generate prototype models of the NASA ANTS concept mission and NASAs Voyager mission [59, 60].

In particular Vassev et al. in [59] propose an approach where an ASSL specification is translated into a state-transition model, over which model checking is performed to verify whether the ASSL specification satisfies correctness properties. The authors use temporal logic formulae expressed over sets of ASSL constructs and discuss possible solutions to the state-explosion problem in terms of state graph abstraction and probability weights assigned to states.

Furthermore, Bakera et al. propose another approach for mapping ASSL specifications to special Service Logic Graphs [4], which supports reverse model checking and games, and enables intuition of graphical models and expression of constraints in mu-calculus.

To the best of our knowledge these are the only significant attempts to apply model checking to autonomic systems. Both of them are however unable to face some of the challenges highlighted in Section 8.1 such as the possibility of follows the satisfaction of properties verified at design-time when the system changes its configuration. Indeed, these approaches are more a support for developing autonomic systems than for verifying it.

In Section 8.3 we propose an application of the model checking verification for the autonomic workflow developed in the ArtDeco project using both the traditional offline approach and a novel online technique to deal with the system evolution.

8.2.2 Testing

Classical testing approaches need to be revisited and extended to face all the challenges arisen by the "self-" features of autonomic systems. Indeed, since the system capability of evolving, testing can not be longer focused only on defect detection after code completion, but it has to be considered an integrated and significant activity performed during the whole software life cycle. However because of the vastness of the topics involved, in this section we go into depth only on those we considered more interesting from the ArtDeco point of view.

In particular, as evidenced in Table 8.1, the selected proposals are applicable during the implementation, development and runtime phases of the software development and are focused mainly on qualitative aspects. Test results can be evaluated both off-line, mainly applying simulation approaches, or on-line, by analyzing the system during its real execution. However in general the testing activity in the context of autonomic systems still requires further investigations as highlighted in Table 8.1 by the medium/low level of supporting tools, scalability and accuracy.

Among the various topics for the ArtDeco context we focus on defining the metrics and attributes to be observed, providing tools and facilities to execute the testing and collect the results, and assuring that the system works correctly even after recovery actions, modifications or fault corrections.

A common problem in testing is the evaluation of the effectiveness of the test strategies applied. As for the standard systems, the possibility of quantifying the completeness and the maturity of the developed autonomic system is an important task. However while for classical systems several proposals, metrics and approaches for measuring the effectiveness are available, only few ones have been developed for the autonomic systems. Among them, [43] discussed about the evaluation of autonomic systems and proposed a preliminary set of metrics for them, while [43] proposed to quantitatively evaluate a computing system along the four core auto-nomic dimensions of self-healing, self-configuration, self-optimization, and self-protection.

The concept of autonomic self-configuration and the dynamic adaptation of the test suite provide hints for generating test harnesses and testbeds able to simulate the system execution or its behavior by means of platform specific test scripts. In this direction, [50] introduces a self-configuring autonomic test harness for web applications, which is able to: automatically select a testing tool capable of validating any technology-specific features; generate an automated test script, and run it on the web application.

Automat [61] is, instead, an open testbed architecture applicable for analysis and exploration of self-repair and self-configure by automated fault attribution. This lets the simulation of self-management at the hosting center level by the instantiations of virtual data centers comprising subsets of the center's hardware, definition of con-trollers that govern resource arbitration, resources selection, and dynamic migration within their virtual data centers. Another framework for evaluating the self-healing of system is VM-Rejuv [29], a virtual machine based rejuvenation scheme for web-application servers. This framework combines analytical models with runtime fault-injection with the purpose of providing a practical evaluation approach that can be generalized to other self-healing systems.

Different frameworks have been developed for testing autonomic systems at run-time. One of the first proposals is [58], in which the authors introduce the concept of an autonomic container. An autonomic container has been defined as a data structure with self-managing capabilities and the the implicit ability to self-test. The frame-work in particular provides the possibility of automatic runtime testing of autonomic systems after a change has been made and the automatic generation of test results after runtime testing has been performed.

A critical point for autonomic systems is to check that after corrective, adaptive actions or modifications, the system is still able to work as before and no new errors of problems are introduced. In classical systems a standard way to approach this problem is the application of regression testing strategies, i.e. the re-execution of an already run test suite (or an opportunely selected subset) so to verity that no negative effect has been introduced. In autonomic systems the application of standard regres-sion testing methodologies is not always possible because runtime modifications, either of the structure or of behavior of the system, may invalidate the applicability of the test cases or require the generation of new test cases [35].

Thus the development of self-managing features for autonomic systems consist on the validation of changes and the adaptation of the test cases to them. However, the concepts of self-testing is not a novelty introduced for autonomic systems. Sev-eral authors, such as [36, 25, 12, 38, 16], have worked on this topic and proposed different testing approaches. However, considering the autonomic system environ-ment, probably the most closely related work is the one proposed in [38], which introduced the concept of self-testable component. Inspired from that work various frameworks and methodologies have subsequently been defined.

From the current available approaches we mention here the framework proposed in [34], that dynamically validates the changes, discards the test cases no longer applicable and generates, wherever necessary, new test cases to ensure that functional and non functional requirements are still being met after each change. Similarly in [35] the authors proposed approaches for self-testing in the context of a realistic problem, as job scheduling, and the notion of self-protection to provide a more comprehensive application for investigating testing autonomic systems. Indeed, the dynamic nature of autonomic systems is a continuous source of challenges for research on self testing and new and innovative methodologies for test case generation are necessary.

8.2.3 Qualitative Aspects in Autonomic Compositions of Services

Validation of Quality of Service (QoS) properties in the context of traditional enterprise systems is typically done by means of specific models capturing the its major aspects. For example, in the context of performance prediction, performance models aims at capturing the system behavior under load [45]. However, in most of the cases, these techniques traditionally assume that the system is static and that both the available resources and their workload are a priori quantifiable or known. All these assumptions usually do not reflect the flexibility of autonomic applications built from the composition of heterogeneous self-maintained systems interacting in highly distributed and dynamic environments. Furthermore, such self-maintained systems are typically designed, developed and deployed by disjoint administrative domains. Eventually, the openness of the environment characterizing autonomic applications naturally led to the pursuit of mechanisms for specifying the granted levels of QoS of third-party systems/services and establishing an agreement on them.

An interesting solution investigated by both industry and academia, is the use of Service Level Agreements (SLA). SLA can be established and enforced both globally and locally at the services involved in the execution of tasks. Over the last years several proposals on specification languages for SLAs have beeing proposed [31, 56, 27], with the intent to provide a standard layer to build agreement-driven systems. In general, the main ingredients of the languages concern the specification of domain-independent elements of a simple contracting process. Such generic definitions can be augmented with domain-specific concepts.

Even though an SLA is a specification describing the agreements that a system commits to accomplish when processing a request, the information it contains could be also exploited in order to asses and validate the QoS features of an autonomic systems running in an agreement-driven environment. As described in Table 8.1, the SLAs can be used offline during the design and the implementation of autonomic services assessing quantitative/qualitative system properties. For example in [8], SLAs are used in order to empirically evaluate the QoS features of a system under development by generating a test-bed that reproduces realistic run-time scenarios with respect to the agreed non-functional behaviors.

Autonomic systems may appear or disappear also because the nodes can move in the scenario in which they are running. The movement of nodes is one of the relevant factors of context change in ubiquitous systems and a key challenge in the validation of context-aware applications. Specifically, the mobility of the nodes may give rise to different perceptions of QoS properties even though all the systems in the scenario are respecting the contractual agreements. In this direction, some validation techniques considering context-changes in context-aware applications were proposed [9, 63, 28]. In particular, [9] presents a QoS validation approach that takes into account both SLAs, and the mobility models of the nodes in the a network of systems.

Table 8.1 also reports that the SLAs can be used at runtime in order to address the need for performance prediction in dynamic environments. For example, [44] proposed solutions where the performance models can be generated and analyzed on-the-fly reflecting changes in the environment. In this cases, part of the management of the QoS is delegated to the middleware hosting the autonomic services. In particular, the middlewares include resource managers and load-balancing mechanisms. In [48], the authors presented a methodology for designing autonomic QoS-aware resource managers that have the capability to predict on-the-fly the performance of the autonomic services they manage and allocate resources in such a way that SLAs can be fulfilled. This solution makes it possible to deal with QoS aspects that depend on incoming workload but for whose services no workload model is available.

The risks of failures with respect to quality aspects in an autonomic system can be mitigated by means a feedback-based system for managing trust and detecting malicious behavior in autonomically behaving networks. Thus, an effective trust management is usually a crucial qualitative asset for the acceptability of highly pervasive applications and networking [23].

In general, trust management concerns those approaches aiming at making decisions about interacting with something or someone we do not completely know, establishing whether we should proceed with the interaction or not [22]. According to Table 8.1, the frameworks for trust management are usually based on run-time approaches that deal with both technical solutions, and organizational aspects (i.e. the system governance) [10]. Specifically, trust management systems combine distributed access with the control policies, digital credentials, and logical deduction. Since in an autonomic network there are multilateral communications among self-managing and self-preserving partners, there is a pressing need for suitable models/schemes for establishing and maintaining trust relationships between those partners [23]. In other words, a way should be provided to rating participants in any choreography of autonomic nodes, helping "good" nodes to avoid interacting with "malicious" ones. For example, the trust management system proposed in [26] rates autonomic nodes running in a network with respect to three main assets: reputation, quality, and credibility. In particular, the authors contributed to the understanding of how trust information should be aggregated and how much reliance should be attached to reported trust values by other nodes in the network.

Qualitative aspects such as trust, confidentiality, and reputation are strictly related with the concept of identity. However, while in a static scenario digital identity management does not present much of a problem, it emerges as an important issue when autonomic nodes dynamically join different alliances [23]. Many approaches and technologies were studied and developed in order to support distributed identity management. Among the others, some of the more relevant examples are the OASIS Security Assertion Markup Language (SAML) [54], the Liberty Alliance project and WS-Federation [2], and the Microsoft .Net Passport infrastructure [49].

8.2.4 Stochastic Model-Based Dependability Analysis

Stochastic model-based (both analytical and simulative) approaches for dependability evaluation have been proven to be useful and versatile in all the phases of the system life cycle. A model is an abstraction of a system that highlights the important features of its organization and provides ways of quantifying its properties, while neglecting all those details that are relevant for the actual implementation, but that are marginal for the objective of the study. Several types of models are currently used in practice. The most appropriate type of model depends upon the complexity of the system, the specific aspects to be studied, the attributes to be evaluated, the accuracy required, and the resources available for the study. For an already existing system, models allow an "a posteriori" dependability and performance analysis, to understand and learn about specific aspects, to detect possible design weak points or bottlenecks, to perform a late validation of the dependability and performance requirements (this can also be useful in certifying phase) and to suggest sound solutions for future releases or modifications of the systems. The modeling also represents an effective tool to foresee the effects of the system maintenance operations and of possible changes or upgrades of the system configuration. In view of these appealing features, autonomic systems such as those addressed by the ArtDeco project appear to greatly benefit from the application of model-based validation techniques to forecast whether required dependability or QoS levels are satisfied under the planned system reconfiguration.

Of course, the above discussed positive aspects come with some negative counterparts. The most critical problem is complexity: it is not an easy task at all to reflect in a model all the relevant aspects of a complex system. To cope with this problem, a modeling methodology is needed so that only the relevant aspects can be detailed thus allowing numerical results to be effectively computable. In addition, simplifying hypotheses are very often necessary to keep the model manageable. Complexity is also mastered through hierarchical and modular approaches, by structuring the model in different submodels and levels separated by well identified interfaces. From the solution point of view, techniques dealing with "largeness avoidance" and "largeness tolerance" of complex models have been developed [47]. Overall, the scalability of model-based approaches towards increasing size of the system under analysis is rather good. Moreover, models of complex systems usually require many parameters (the meaning thereof is not always intuitive for designers), and require determining the values to assign to them (usually by way of experimental tests), which may be very difficult. To cope with this problem, sensitivity analysis are performed, which allow evaluating a range of possible system scenarios by varying the values of model parameters, so to identify those that are more sensible.

Concerning the various classes of modeling methodologies that have been developed over the last decades to provide dependability engineers the support tools for defining and solving models, a distinction can be made between methodologies that employ combinatorial models (non-state space models) like fault-trees and reliability block diagrams, and those based on state space oriented representations (state space models), such as Markov chains and Petri net models, depending on the nature of their constitutive elements and solution techniques. A short survey of such methods can be found in [47, 13]. Model-based approaches typically target quantitative assessment of dependability properties, although combinatorial methods such as fault-trees can be employed to assess qualitative properties as well.

Dependability evaluation is typically an offline activity and model-based approaches are well consolidated to this purpose. However, when used to support the control loop of autonomic systems, online evaluation becomes necessary. In fact, an important issue to be considered is the dimension of the decision state-space problem. Considering systems which are the result of the integration of a large number of components, as it is more and more the case in modern and future application fields, it could be not feasible to evaluate and store offline all possible decision solutions for each case it may happen. These cases are determined by the combination of external environmental factors and internal status of the system (which are not predictable in advance or too many to be satisfactorily managed [15]) and by the topology of the system architecture which can vary along time. In this scenario, online decision making solutions have to be pursued.

Of course, although appealing, the online solution shows a number of challenging problems requiring substantial investigations. Models of components have to be derived online and combined to get the model of the whole system. In this context, how to deal with model generation and, especially, model solution so as to provide feedback from the analysis in proper time to be profitably used are the big challenges. How to feed such models with accurate values for their input parameters integration with monitoring activities is another additional issue. To this purpose, continuos monitoring, typical in autonomic systems, constitutes a paramount support to online assessment. To help managing the complexity of the generation of the overall system model, a compositional approach is desirable, starting from simple, possibly pre-defined sub-models or customizable ones from a library of predefined models. Thus, definition of appropriate compositional rules and the resulting complexity of the combined model solution appear to be the most critical problems to be properly tackled to promote the applicability of this dynamic approach to reconfiguration, as discussed in [52, 55].

In [46], a hybrid approach combining analytical models with measurement based techniques and monitoring is proposed for autonomic management of system availability in heterogeneous and highly interconnected systems. The overall objective is to automate the whole process of availability management to ensure maximum system availability, thus making the system autonomic. In [24], the validation of large and complex software autonomic systems is addressed. A static approach is proposed to validate autonomic policies, e.g. to detect possible cycles in the application of the rules composing such policies. The validation is performed by extending the DACAR Metamodel, a framework to deploy generic models of autonomous component-based distributed applications, with adequate concepts to deal with validation of the deployment and autonomic concerns of a system. A transformation approach from a design oriented model of adaptive reactive systems to an analysis oriented model suitable for QoS analysis is at the basis of the work in [28, 51]. Specifically, a two-step approach is proposed centered around the construction of a bridge model from the design model to the target analysis model (DSPN, Deterministic and Stochastic Petri Net), expressed in the D-KLAPER intermediate modeling language for dynamically adaptable systems. Resorting to a software dependability growth model based on the self-reconfiguration method applied by autonomic systems and analyzed through Markov Regenerative Stochastic Petri Net (MRSPN) for performance purposes is the approach preliminarily presented in [62].

Although studies are still at a preliminary stage, stochastic model-based has potential to be a powerful evaluation approach to support autonomic systems, both to help selecting the right adaptation actions to match current application requirements and existing operating conditions, and to assess QoS metrics on the overall system to forecast future (non-functional) behavior.

8.2.5 Monitoring

Monitoring has an essential role in the realization of feedback loop that is a key characteristic of the autonomic computing vision, as already illustrated in Figure 6.1. Besides, monitoring is used as a supporting V&V activity together with other verification and analysis approaches, such as those discussed earlier in this chapter. Monitoring is intrinsically a run-time activity, therefore its application belongs to the post-deployment phase, when the monitored system is in operation (see Table 8.1). In general, both quantitative and qualitative properties can be observed by monitoring, and in fact it can be applied to a variety of purposes, including the assessment of performance, reliability, security, availability, and others. Certain monitoring frameworks, such as Glimpse [7], allow custom-defined metrics to be specified by combining a set of primitive event types; in such frameworks, application-specific properties can be easily defined and monitored. When monitoring large distributed systems, scalability issues may arise. In the literature, this problem has been addressed primarily by adopting a layered hierarchical architecture, whereby observations of finer granularity are aggregated at a local level first and then handed over to a global correlation node [41, 40]. Very detailed or precise monitoring comes at the expense of efficiency, and therefore a balance between the two must be pursued; approaches such as the one presented in [11], in the context of Service Level Agreement (SLA) checking, aim to optimize the cost of monitoring by assignign higher priority (and higher-precision observations) to the components of a server that are more likely to violate a SLA.

Despite the key role of monitoring in autonomic computing, very few works in the literature address monitoring from a perspective that is specific to autonomic computing[2].

The very fact that monitoring is key to autonomic computing explains this apparently surprising lack of dedicated literature. In fact, since monitoring is so pervasive in the implementation of any autonomic system, monitoring approaches are described in many works that present autonomic computing approaches but do not address monitoring as their primary concern, taking it as an implied element of the system. Furthermore, many more works address monitoring in the context of self-* systems, without explicit reference to autonomic computing, but with an effective overlap of topics. This makes it harder to take stake of the situation of research on monitoring for autonomic computings.

It is very common for autonomic systems to be constructed out of pre-existing subsystems integrating them with a feedback control loop, implemented by means of a monitoring element coupled with another element that functions as the effector. In this perspective, a very important problem to address is how a feedback control loop can be matched with an existing system, and in particular how it is possible to attach a monitoring infrastructure to it.

In systems that include logging facilities, such as those provided by the Generic Log Adapter and the Log and Trace Analyzer from the IBM Autonomic Toolkit [1], these are used to collect raw information that is then aggregated and interpreted by a separate correlating module.

Adding autonomic features to legacy systems that do not natively support them is the specific target of an implementation of the full MAPE-K loop implemented in Kinesthetics eXtreme [33]. This system addresses the cases where it is not possible to modify the system to monitor, so it adds sensors on top of the existing API and interfaces, so that the monitoring is totally decoupled from the monitored system.

The IBM Agent Building and Learning Environment (ABLE) is a toolkit for constructing agent-based self-managed systems. The toolkit is based on Java technologies and includes a library of JavaBeans implementing intelligent software components, called AbleBeans that can be composed to form AbleAgents. Monitoring in Able is achieved exploiting an event-based paradigm, whereby AbleBeans can exchange AbleEvents.

The monitoring systems mentioned so far and the majority of other frameworks for AC take a generic approach to monitoring, which is considered as a generic element, whose functionality is not really dependent on the application or on the

---

[2] As of this writing, a search in the ACM Guide to Computing Literature (http://portal.acm.org/guide.cfm?coll=ACM&dl=GUIDE, accessed 22 March 2010) for papers published on either journals or conference proceedings from 2000 to 2010 that contain both "monitoring" and "autonomic" in the abstract returns only 144 papers. A similar search but including all publications that have monitoring in the abstract counts 10577 papers. The publications that have the word "autonomic" in the abstract are 1294.

context, and that can be used in multiple situations without the need for specializations.

Lately, this view is being revised, and the monitoring itself is seen as a dynamic element, that is capable of self-adapting to cope with variable operating conditions. In a very recent work, Janik and Zielinski [32] propose an approach to monitoring based on aspect-oriented programming; the approach is called AAOP (Adaptive AOP), since monitoring is implemented through aspects that are activated and deactivated dynamically based on an adaptation strategy.

8.3 Offline and Online Model Checking in ArtDeco

The previous section has overviewed the main techniques for validating and verifying systems, showing the few approaches proposed in literature to deal with the typical features of autonomic systems. However, the overview together with the considerations pointed out in Section 8.1 shows that there is not a standard methodology that proves to be feasible in any context. This is due to the complexity of representing all the features of such systems that need to be analyzed under different points of view.

Indeed, at design-time, they need not only to be verified against suitable properties (as in traditional systems), making a set of hypotheses on the environment they will work in (as for dynamic, decentralized systems), but they also need to be verified considering the way in which they can possibly evolve in. Moreover, at run-time, the rules that manage the "self-" features of autonomic systems may be changed or increased or deleted, making the analysis performed at design-time no more valid.

Although it is a crucial support within any software development lifecycle, V&V has been exploited in a rather limited form in the ArtDeco project. This was mainly due to having the effort primarily devoted to analyze the existing techniques in order to identify the activities needed for autonomic systems and to the fact that concrete applications of ArtDeco concepts have appeared at the rather late stage of the project development. Hence, we concentrated our effort in verifying the correctness of the autonomic workflow described in Chapter 7, applying mainly model checking. Indeed, the adaptation of this technique for autonomic systems was developed in parallel with the case study. Moreover, despite traditional model checking, the proposed approach can be used in the whole life of the workflow, from the developing to the execution.

Before proposing an approach, we need to identify what we mean as workflow correctness and what is involved in checking it. Obviously, besides the typical self-adaptation features of autonomic systems, the workflow as a standard behavior that needs to be verified against desired properties and has to be shown to be deadlock free. Then, since the workflow refers to a Knowledge Base (KB), that contains rules in the form of Event–Condition–Action (ECA), and embeds assertions, in order to manage the adaptation, these features, that cause changes of configuration, need to be checked too. Finally, if some input motivates the need of modifying the KB, the assertions, or both, the proposed technique has to be able to re-verified the system against the same set of desired properties or an enriched/modified one.

To meet all these goals, we decide to adopt a model checking technique, which, as seen in Section 8.2.1, is an automatic technique for formal verification of reactive systems behavior. Model checking has been performed by formally describing the target system in the adopted verification tool formalism, specifying system properties in temporal logic and checking the model against the property using the model checker engine.

To choose the best model checking tool for this purpose, we have evaluated the procedure using two different model checkers: the symbolic model checker NuSMV [2] and the model checker Spin [6]. In the following we will describe both the models. Indeed, even if we developed a tool only for the second approach that was shown to be more efficient in this case, the first approach present interesting features for modeling the system that can help the designer to better analyze it.

In the first approach, the model of the workflow and its adaptation rules have been formalized through Petri nets. Then, once Petri nets are translated in the checker input language, the branching temporal logic CTL has been used for stating the correct behavior that we want to guarantee. In the Spin approach, the process has been modeled using a finite state machine and the external events that occurs with the time evolution has been introduced using one or more dedicate processes (depending on the complexity of the environment). The automaton and the external components are easily mapped in Promela, the input language of Spin and, then, the overall behavior was then checked using Spin. The second solution leads to better results in terms of efficiency and memory usage, hence we have developed a translations, that can be performed fully automatically, to generate Promela code starting from a workflow and its adaptation rules, as represented in Figure 8.2. The translation process consists mainly of four functions, that will be described in detail in Chapter 21. The model together with the assertions and the adaptation rules is checked against a set of properties that are pre-defined by the user, and, if the model satisfies them, the workflow can be executed, otherwise, using the counter example provided by Spin as a guideline, the workflow is modified to meet the requirements.

However this solution only covers the systems behavior and the evolution of the system foreseen by the rules provided at design-time. Moreover, since model checking techniques work only under the hypothesis of close world, some assumption on the environment are made and the performed verification only guarantees the correctness of the model if the environment respect these hypotheses. To overcome this last problem, we add the hypotheses made on the environment as monitoring rule in the workflow life cycle. Instead, to cope with the changes of the KB, every time the KB is modified a new Promela model is generated ad is checked against predefined safety and liveness properties, that once verified can be assumed as guaranteed for the new process. In this way it should be possible to verify, at run-time and automatically, if the alteration of the adaptation rules causes the workflow to be still correct or to become not correct. In such view a verification of correctness have to be performed when the rules are asserted or retracted in the KB.

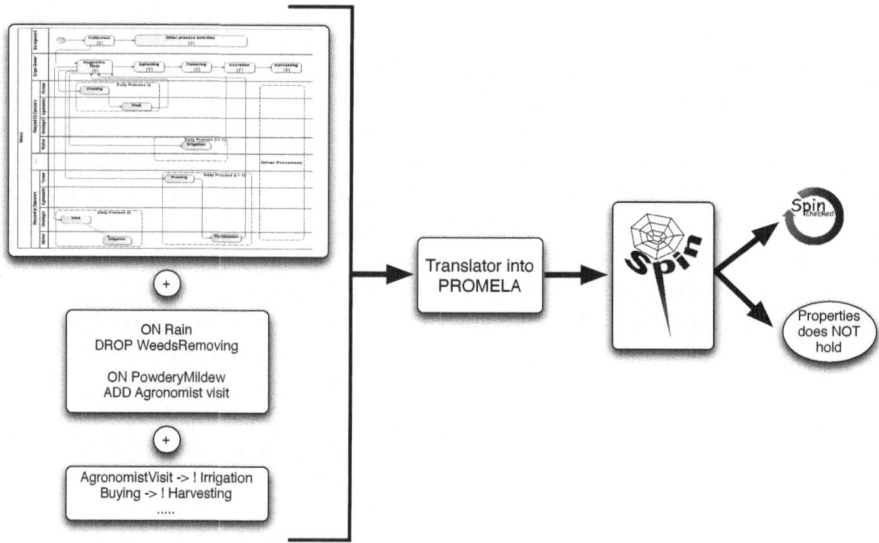

Fig. 8.2 Verification process

When the system is re-checked online and some property is not longer satisfied, this information can be used in a double way: indeed, if the violated property is critical, the workflow execution needs to be interrupted and a designer intervention is required, but, if it is not the case, the property can be monitored online to check if the sequence of events at runtime is a sequence that will violate it.

8.4 Focusing on the Case Study

In this section we show the two model checking approaches mentioned in the previous section, applied on the ArtDeco case study. Here we only show the models and their peculiarities. For further details on the automatic translation technique, refer to Chapter 21. Notice that, even if we chose an approach with respect to the other, we still describe the neglected one because it presents some interesting peculiarity that can be useful for modeling purpose and in other contexts.

Consider the wine production line example, neglecting the implementation details, in order to be able to extract the specification of its control. As presented in Chapter 21 the wine production line is divided in four phases:

- management of the grape growing
- harvesting, fermentation, maturation
- distribution
- selling

Since the first phase needs more than the others to be controlled by an autonomic process, we will focus on it. In particular, we consider a single wine producer responsible of a vineyard. Notice that this is not a limitation of the proposed approach since, as explained shortly, it can be straightforwardly extended to N vineyards. The producer goal is to produce high quality wine. At this stage the quality cannot be measured on the product, but can be estimated using the index Q_{v_t}, that represents the estimated quality of the wine produced by the tth vineyard. Q_{v_t} is computed using a bunch of parameters on the vineyard and in particular: temperature, humidity percentage, light and wind speed and direction.

During the grape growing some critical events may occur, such as freezing cold, heat (and consequently water stress), parasites and hail. These natural disasters affect the grape growing in different ways depending on the stage of the growing, hence different ECA rules are adopted to monitor them. An example of ECA rule is the following:

Growing stage (Oct-Jan) $\Rightarrow$ Occurred calamity: freezing cold $(T < 3^oC) \Rightarrow$ Alarm

For verification purpose, we can neglect the actual parameter values and the critical thresholds. Indeed, we are only interested in:

- the stage of growing, that sets the calamity to be monitored;
- the occurring calamity, that sets the action to be performed.

Moreover, the estimated quality concurs with the calamity to set the action to be performed, but only when it decreases under a critical value. However, in this case, the only possible action to be performed is to buy the grapes from an external vendor and this decision can be made only by the producer and not automatically by the system. Table 8.2 shows the link among growing stage, calamity and actions.

Table 8.2 Relations to define ECA rules

Growing stage	Occurring calamity	Action
Vegetative rest (Oct–Jan)	Freezing cold	Critical alarm
Sprouting (March–May)	Freezing cold	Critical alarm
	Heat	Irrigation
	Parasites	Insecticide
Flowering (May)	Heat	Irrigation
	Parasites	Insecticide
Accretion (June–Sept)	Heat	Irrigation
	Parasites	Insecticide
	Hail	Critical alarm

Starting from the ECA rules, automata based models were built in order to perform verification through model checking [21].

Consider first the symbolic approach based on NuSMV, that, as already noticed, was not chosen as final approach to verify the autonomic workflow, but that represents an alternative approach for modeling autonomic systems that can be more effective in other context. In this case, we represent the overall autonomic process through the Petri Net reported in Figure 8.3. In the model, it is possible to clearly identify the four phases (vegetative rest, sprouting, flowering, and accretion), that are one consequence of the other when no calamity occurs or when calamities are "solved". Observe that more then one calamity can occur in the same stage and the number of manageable calamities depends on the initial marking of the net: the number of tokens in a place bounds the number of calamities.

The Petri net in Figure 8.3 describes a vineyard lifecycle. The resources available to manage the vineyard are represented, as already explained, by the number of tokens in the original marking. If we want to deal with more than a vineyard at the same time, we can describe them independently and then we just constrain the total number of tokens that represents the resources. This means that we just reproduce the same net as many times as the number of considered vineyards and we reduce the number of possible behaviors constraining the resources.

The obtained Petri net is then represented in NuSMV through a set of boolean variables. Each place corresponds to a set of true variables and a transition firing changes this set. Notice that if we have a model with multiple vineyards and each of them has independent resources, we can check each vineyard in isolation.

On this model we have analyzed the following properties:

1. "When a calamity occurs, the system reacts with the adequate action":

$$AG(\text{calamity}=\text{'freezing cold'} \Rightarrow AX(\text{action}=\text{'alarm'}))$$
$$AG(\text{calamity}=\text{'parasite'} \Rightarrow AX(\text{action}=\text{'insecticide'}))$$
$$AG(\text{calamity}=\text{'heat'} \Rightarrow AX(\text{action}=\text{'irrigation'}))$$
$$AG(\text{calamity}=\text{'hail'} \Rightarrow AX(\text{action}=\text{'alarm'}))$$

2. "The grape growing stages must be in the correct order":

$$AG(\text{stage}=\text{'vegetative rest'} \Rightarrow AX(\text{stage}=\text{'vegetative rest'} \vee \text{stage}=\text{'sprouting'}))$$
$$AG(\text{stage}=\text{'sprouting'} \Rightarrow AX(\text{stage}=\text{'sprouting'} \vee \text{stage}=\text{'flowering}))$$
$$AG(\text{stage}=\text{'flowering'} \Rightarrow AX(\text{stage}=\text{'flowering'} \vee \text{stage}=\text{'accretion'}))$$
$$AG(\text{stage}=\text{'accretion'} \Rightarrow AX(\text{stage}=\text{'accretion'} \vee \text{stage}=\text{'vegetative rest'}))$$

3. "The monitored calamity must be relevant for the current grape growing stage':

$$AG!(\text{stage}=\text{'vegetative rest'} \wedge (\text{calamity}=\text{'parasite'} \vee \text{calamity}=\text{'heat'} \vee \text{calamity}=\text{'hail'}))$$
$$AG!(\text{stage}=\text{'sprouting'} \wedge \text{calamity}=\text{'hail'})$$
$$AG!(\text{stage}=\text{'flowering'} \wedge (\text{calamity}=\text{'freezing cold'} \vee \text{calamity}=\text{'hail'}))$$
$$AG!(\text{stage}=\text{'accretion'} \wedge \text{calamity}=\text{'freezing cold'})$$

Not surprisingly, since the size and the peculiarities of the model, all the properties are easily verified by the model checker. Indeed, considering the Petri Net model, the proposed properties may appear meaningless. Instead, all the model checking process, that aims to verify these properties, allows to specify the model of the process in the correct way.

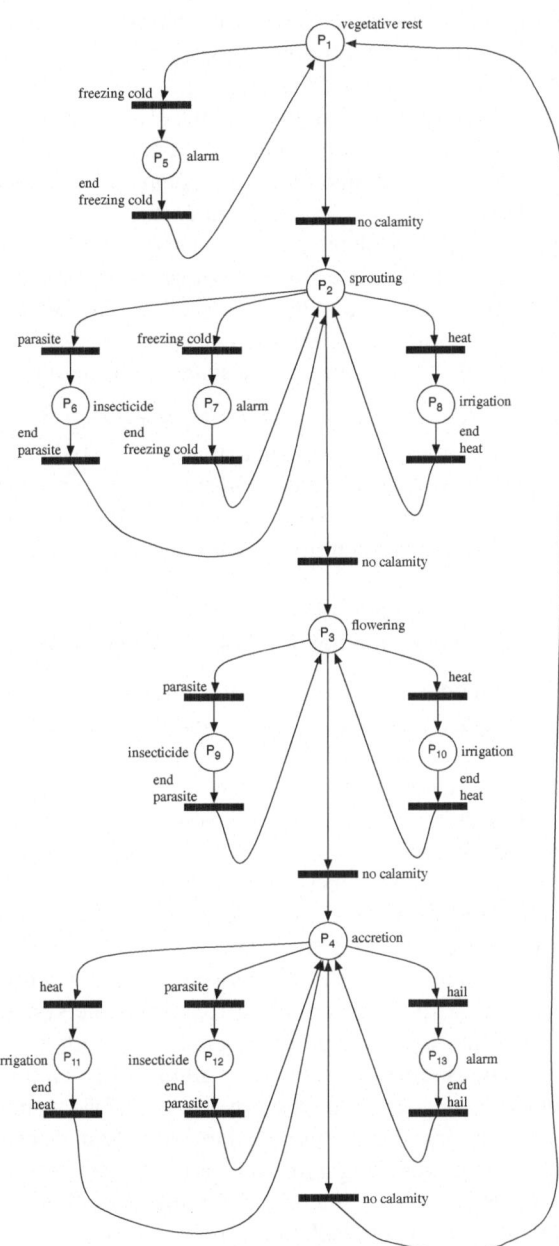

Fig. 8.3 Petri net model of the grape growing control process

The proposed model was also analyzed in an enriched version, that includes timing information. In details, a counter was explicitly added to the model in order to count the months during the evolution of the system. This allows a better understanding of the different stages of the grape growing. Notice that, even if this approach was proven to be less efficient in the grape example than the approach presented below, it is still very interesting. Indeed, Petri nets offer an intuitive model of the system that can help in the design phase to reason about the described system. Moreover, it could be more effective in other contexts.

The second approach leans on the model checker Spin, which requires a different representation of the model, namely a Büchi automaton. This model is less intuitive and popular than Petri net, but it is still very effective. The Büchi automaton that models the grape growing process is represented in Figure 8.4. It is composed by the following elements, that correspond to the states of the automaton:

- Grape growing stages: Vegetative Rest (VR), Sprouting (Sp), Flowering (Fl) and Accretion (Ac); these are represented as white colored in the automaton;
- Calamities, represented as grey linear blend states; for each stage, the calamities that are significant at that stage are represented. Notice that in the figure the states representing the same calamity have the same name, but they are different states;
- Catastrophic states are the grey colored states. These states represent the case in which the system reaches a too low level of estimated quality after a calamity.

In the automata-based model, we have also considered the case in which a calamity cannot be overcome and the system reaches a catastrophic state. Notice that the states that represent a catastrophe are sink states in order to show that once a too low level of quality is reached that system has no more the control of the grape growing. This model can be described using Promela, the input language of Spin, and it results in a Promela process.

The input of this process (and, hence, of the system) represents the nature and is not obviously controlled by any process. As a consequence, in order to verify our autonomic system, we need to add a Promela process that represents the nature. It generates randomly a calamity, as normally happens in the real world. This Promela process is synchronized with the process that represents the system though a rendezvous channel on which signals, representing natural events are sent. Notice that, while the month passing is represented in order, all the other natural events are generated without following any schema. Observe also that if we want to include more than a vineyard in our description, we just need to add a Promela process for each of them in the overall system. Verifying more vineyards at the same time is meaningful only if they are constrained one to the other in terms of used resources. if it is not the case, verifying the vineyards in isolation is more convenient.

On this model we have analyzed the following properties:

1. "Every time the stage is sprouting and there is heat from the nature, the system starts to monitor the heat in order to perform the required action":

$$G(\text{state}=\text{`Sp'} \wedge \text{calamity}=\text{`heat'} \Rightarrow \text{state}=\text{`Sp'} U \text{state}=\text{`Sp:He'})$$

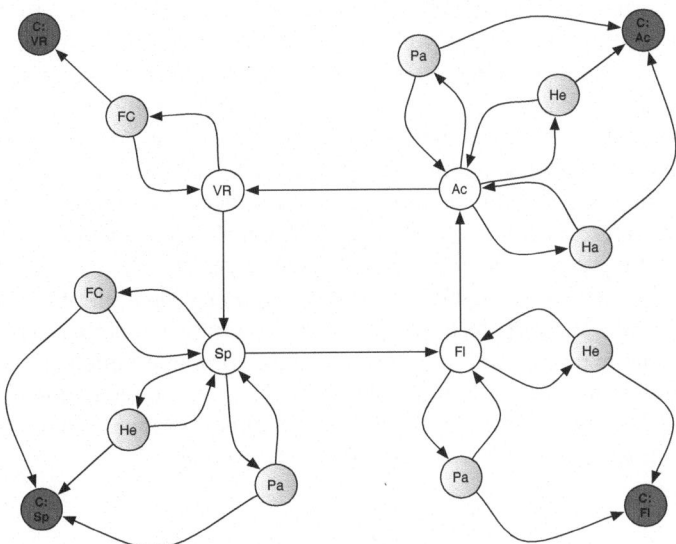

Fig. 8.4 Automata model of the grape growing control process

Analogous properties are checked with respect to all the ECA rules. Notice that in this model we do not care about the action, that is implicitly considered executed in the monitoring state of the system.

2. "In the vegetative rest, the heat is not an admissible calamity".

Notice that we have modeled the nature to generate events randomly. This means that the Promela process representing the nature could generate heat when the system is in vegetative rest. This problem could be overcome by generating calamities conditioned to the season. However, if we do not want to change the model it is only important to assure that the system neglect some calamities in certain stages, as follows:

$$G(\text{state}=\text{'SR'} \wedge \text{calamity}=\text{'heat'} \Rightarrow$$
$$\text{state}=\text{'VR'} \, U(\text{calamity}!=\text{'heat'} \vee \text{state}=\text{'Sp'}))$$

Analogous properties are checked also for other stages, with respect to calamities not considered in that stage.

8.5 Conclusions

In this chapter we provided an overview of some recent proposals and methodologies for verification and analysis of autonomic systems for networked enterprises trying to outline in particular the strategies and methodology most suitable for the ArtDeco context. For this we analyzed the available frameworks and techniques for testing and model checking and we introduced the problem of the validation of quality properties and the model-based dependability analysis. Moreover we also considered the current proposals for monitoring the autonomic systems.

Then we presented an approach developed for verifying the autonomic workflow of ArtDeco and we have validated it through an example from the case study. More precisely we focused verification of the autonomic workflow for the control process of wine production. This system was developed using a "sense and respond" model, i.e., the whole production line is equipped with devices able to observe and collect information on the environment and the model uses this information to reconfigure, changing, if it is the case, the requirements themselves in order to meet some goal.

Despite the various proposals the verification and validation of the autonomic systems have still many open challenges. Current approaches are far to be sufficient for a complete analysis of all the aspects of the self configuration and adaptation. The complexity and the evolving of the automatic system technologies serves as continuous hints and stimulus for the generation of new proposals and facilities that could face the many aspects of this important environment.

References

1. Ibm autonomic computing toolkit
2. The liberty alliance project
3. Arnold, A., Bégay, D., Crubillé, P.: Construction and analysis of transition systems with MEC. World Scientific Pub. Co. Inc. (1994)
4. Bakera, M., Wagner, C., Margaria, T., Vassev, E., Hinchey, M., Steffen, B.: Component-oriented behavior extraction for autonomic system design. In: Methods Symposium, p. 66 (2009)
5. Behrmann, G., David, A., Larsen, K.G.: A Tutorial on Uppaal. In: Bernardo, M., Corradini, F. (eds.) SFM-RT 2004. LNCS, vol. 3185, pp. 200–236. Springer, Heidelberg (2004)
6. Ben-Ari, M.: Principles of the Spin model checker. Springer-Verlag New York Inc. (2008)
7. Bertolino, A., Calabrò, A., Lonetti, F., Sabetta, A.: GLIMPSE: A generic and flexible monitoring infrastructure - techn. rep. 2010-tr-024 (2010)
8. Bertolino, A., De Angelis, G., Frantzen, L., Polini, A.: Model-Based Generation of Testbeds for Web Services. In: Suzuki, K., Higashino, T., Ulrich, A., Hasegawa, T. (eds.) TestCom/FATES 2008. LNCS, vol. 5047, pp. 266–282. Springer, Heidelberg (2008)
9. Bertolino, A., De Angelis, G., Lonetti, F., Sabetta, A.: Let the puppets move! automated testbed generation for service-oriented mobile applications. In: SEAA, pp. 321–328. IEEE (2008)
10. Bertolino, A., De Angelis, G., Polini, A., Sabetta, A.: Trends and research issues in soa validation. In: Cardellini, V., Casalicchio, E., Regina, K., Branco, J.L.C., Estrella, J.C., Monaco, F.J. (eds.) Performance and Dependability in Service Computing: Concepts, Techniques and Research Directions. IGI Global (2011) (to appear, accepted for publication)
11. Bertolino, A., De Angelis, G., Sabetta, A., Elbaum, S.G.: Scaling up SLA monitoring in pervasive environments. In: ESSPE, pp. 65–68 (2007)

12. Blum, M., Luby, M., Rubinfeld, R.: Self-testing/correcting with applications to numerical problems. In: Proceedings of the Twenty-Second Annual ACM Symposium on Theory of Computing, pp. 73–83. ACM, New York (1990)
13. Bondavalli, A., Chiaradonna, S., Di Giandomenico, F.: Model-Based Evaluation as a Support to the Design of Dependable Systems. In: Diab, H.B., Zomaya, A.Y. (eds.) Dependable Computing Systems: Paradigms, Performance Issues, and Applications, pp. 57–86 (2005)
14. Brown, A., Redlin, C.: Measuring the Effectiveness of Self-Healing Autonomic Systems. In: Proceedings of the Second International Conference on Autonomic Computing. IEEE Computer Society (2005)
15. Chohra, A., Di Giandomenico, F., Porcarelli, S., Bondavalli, A.: Towards optimal database maintenance in wireless communication systems. In: Proceedings of the 5th World Multi-Conference on Systemics, Cybernetics and Informatics, ISAS-SCI 2001, pp. 571–576 (2001)
16. Chow, M., Gott, R., Lei, C., Siddigue, N.: Method and System for Autonomic Verification of HDL Models Using Real-Time Statistical Analysis and Layered Feedback Stages (2009)
17. Cimatti, A., Clarke, E., Giunchiglia, E., Giunchiglia, F., Pistore, M., Roveri, M., Sebastiani, R., Tacchella, A.: NuSMV 2: An OpenSource Tool for Symbolic Model Checking. In: Brinksma, E., Larsen, K.G. (eds.) CAV 2002. LNCS, vol. 2404, pp. 359–364. Springer, Heidelberg (2002)
18. Clarke, E., Emerson, A., Sistla, A.: Automatic verification of finite-state concurrent systems using temporal logic specifications. ACM Transactions on Programming Languages and Systems (TOPLAS) 8(2), 263 (1986)
19. Clarke, E.M., Emerson, E.A., Sistla, A.P.: Automatic verification of finite-state concurrent systems using temporal logic specifications. ACM Transactions on Programming Languages and Systems 8, 244–263 (1986)
20. Clarke, E.M., Grumberg, O., Jha, S., Lu, Y., Veith, H.: Progress on the state explosion problem in model checking. In: Informatics - 10 Years Back. 10 Years Ahead, pp. 176–194. Springer, London (2001)
21. Peled, D., Pelliccione, P., Spoletini, P.: Model checking. In: Wiley Encyclopedia of Computer Science and Engineering, John Wiley & Sons, Inc. (2008)
22. den Hartog, J.: Trust Management Architecture Design. The Trusted Architecture for Securely Shared Services (TAS^3) Consortium (June 2009)
23. Dobson, S., Denazis, S.G., Fernández, A., Gaïti, D., Gelenbe, E., Massacci, F., Nixon, P., Saffre, F., Schmidt, N., Zambonelli, F.: A survey of autonomic communications. TAAS 1(2), 223–259 (2006)
24. Dubus, J., Merle, P.: Towards Model-Driven Validation of Autonomic Software Systems in Open Distributed Environments. In: Proceedings of the 1st Workshop on Model-Driven Software Adaptation, pp. 39–48 (2007)
25. Ergun, F., Kumar, S., Sivakumar, D.: Self-testing without the generator bottleneck. SIAM Journal on Computing 29(5), 1630–1651 (2000)
26. Garg, A., Battiti, R., Costanzi, G.: Dynamic Self-management of Autonomic Systems: The Reputation, Quality and Credibility (RQC) Scheme. In: Smirnov, M. (ed.) WAC 2004. LNCS, vol. 3457, pp. 165–178. Springer, Heidelberg (2005)
27. Global Grid Forum. Web Services Agreement Specification (WS–Agreement), version 2005/09 edition (September 2005)
28. Grassi, V., Mirandola, R., Randazzo, E.: Model-Driven Assessment of QoS-Aware Self-Adaptation. In: Cheng, B.H.C., de Lemos, R., Giese, H., Inverardi, P., Magee, J. (eds.) Self-Adaptive Systems. LNCS, vol. 5525, pp. 201–222. Springer, Heidelberg (2009)

29. Griffith, R., Kaiser, G., López, J.: Multi-perspective evaluation of self-healing systems using simple probabilistic models. In: Proceedings of the 6th International Conference on Autonomic Computing, pp. 59–60. ACM (2009)
30. Henzinger, T., Nicollin, X., Sifakis, J., Yovine, S.: Symbolic model checking for real-time systems. Information and Computation 111, 394–406 (1992)
31. IBM. WSLA: Web Service Level Agreements, version: 1.0 revision: wsla-2003/01/28 edition (2003)
32. Janik, A., Zielinski, K.: Aaop-based dynamically reconfigurable monitoring system. Inf. Softw. Technol. 52(4), 380–396 (2010)
33. Kaiser, G., Parekh, J., Gross, P., Valetto, G.: Kinesthetics extreme: an external infrastructure for monitoring distributed legacy systems. In: Autonomic Computing Workshop, 2003, pp. 22–30 (June 2003)
34. King, T., Babich, D., Alava, J., Stevens, R., Clarke, P.: Towards self-testing in autonomic computing systems. In: ISADS, vol. 7, pp. 51–58. Citeseer
35. King, T., Ramirez, A., Cruz, R., Clarke, P.: An integrated self-testing framework for autonomic computing systems. Journal of Computers 2(9), 37–249 (2007)
36. Kumar, S., Sivakumar, D.: Efficient self-testing/self-correction of linear recurrences. In: Proc. 37th Foundations of Computer Science, pp. 602–611 (1996)
37. Larsen, K.: Efficient Local Correctness Checking. In: Probst, D.K., von Bochmann, G. (eds.) CAV 1992. LNCS, vol. 663, pp. 30–43. Springer, Heidelberg (1993)
38. Le Treon, Y., Deveaux, D., Jezequel, J.: Self-testable components: from pragmatic tests todesign-for-testability methodology. In: Proceedings of Technology of Object-Oriented Languages and Systems, 1999, pp. 96–107 (1999)
39. Leuschel, M., Massart, T.: Infinite State Model Checking by Abstract Interpretation and Program Specialisation. In: Bossi, A. (ed.) LOPSTR 1999. LNCS, vol. 1817, pp. 62–81. Springer, Heidelberg (2000)
40. Mansouri-Samani, M., Sloman, M.: GEM: a generalized event monitoring language for distributed systems. Distributed Systems Engineering 4(2), 96–108 (1997)
41. Massie, M.L., Chun, B.N., Culler, D.E.: The Ganglia distributed monitoring system: design, implementation, and experience. Parallel Computing 30(7), 817–840 (2004)
42. Mateescu, R., Garavel, H.: XTL: A meta-language and tool for temporal logic model-checking. In: Software Tools for Technology Transfer STTT 1998, p. 33 (1998)
43. McCann, J.A., Huebscher, M.C.: Evaluation Issues in Autonomic Computing. In: Jin, H., Pan, Y., Xiao, N., Sun, J. (eds.) GCC 2004. LNCS, vol. 3252, pp. 597–608. Springer, Heidelberg (2004)
44. Menascé, D.A., Bennani, M.N., Ruan, H.: On the Use of Online Analytic Performance Models, in Self-Managing and Self-Organizing Computer Systems. In: Babaoğlu, Ö., Jelasity, M., Montresor, A., Fetzer, C., Leonardi, S., van Moorsel, A., van Steen, M. (eds.) SELF-STAR 2004. LNCS, vol. 3460, pp. 128–142. Springer, Heidelberg (2005)
45. Menascé, D.A., Dowdy, L.W., Almeida, V.A.F.: Performance by Design: Computer Capacity Planning By Example. Prentice Hall PTR, Upper Saddle River (2004)
46. Mishra, K., Trivedi, K.S.: Model Based Approach for Autonomic Availability Management. In: Penkler, D., Reitenspiess, M., Tam, F. (eds.) ISAS 2006. LNCS, vol. 4328, pp. 1–16. Springer, Heidelberg (2006)
47. Nicol, D.M., Sanders, W.H., Trivedi, K.S.: Model-Based Evaluation: from Dependability to Security. IEEE Transactions on Dependable and Secure Computing 1, 48–65 (2004)
48. Nou, R., Kounev, S., Julià, F., Torres, J.: Autonomic qoS control in enterprise grid environments using online simulation. Journal of Systems and Software 82(3), 486–502 (2009)

49. Oppliger, R.: Microsoft.net passport and identity management. Information Security Technical Report 9(1), 26–34 (2004)
50. Pava, J., Enoex, C., Hernandez, Y.: A self-configuring test harness for web applications. In: Proceedings of the 47th Annual Southeast Regional Conference, p. 66. ACM (2009)
51. Perez-Palacin, D., Mirandola, J., Merseguer, R., Grassi, V.: QoS-Based Model Driven Assessment of Adaptive Reactive Systems. To appear in Proceeding VIDAS 2010 (2010)
52. Porcarelli, S., Castaldi, M., Di Giandomenico, F., Bondavalli, A., Inverardi, P.: A Framework for Reconfiguration-Based Fault-Tolerance in Distributed Systems. In: de Lemos, R., Gacek, C., Romanovsky, A. (eds.) Architecting Dependable Systems II. LNCS, vol. 3069, pp. 167–190. Springer, Heidelberg (2004)
53. Queille, J.P., Sifakis, J.: Fairness and related properties in transition systems - a temporal logic to deal with fairness. Acta Inf. 19, 195–220 (1983)
54. Ragouzis, N., Hughes, J., Philpott, R., Maler, E., Madsen, P., Scavo, T. (eds.): Security Assertion Markup Language (SAML) – Technical Overview. The OASIS Consortium (March 2008)
55. Simoncini, L., Di Giandomenico, F., Bondavalli, A., Chiaradonna, S.: Architectural challenges for a dependable information society. In: Proceedings of the IFIP World Computer Congress, pp. 282–304. Springer, Boston (2004)
56. Skene, J., Lamanna, D., Emmerich, W.: Precise Service Level Agreements. In: Proc. of ICSE 2004, pp. 179–188. IEEE Computer Society Press (2004)
57. Sterritt, R., Bustard, D.W.: Autonomic computing - A means of achieving dependability? In: Proc. of ECBS, pp. 247–251. IEEE Computer Society (2003)
58. Stevens, R., Parsons, B., King, T.: A self-testing autonomic container. In: Proceedings of the 45th Annual Southeast Regional Conference, pp. 1–6. ACM, New York (2007)
59. Vassev, E., Dublin, I., Hinchey, M., Limerick, I., Quigley, A.: Model checking for autonomic systems specified with ASSL. In: Formal Methods Symposium, p. 16 (2009)
60. Vassev, E., Hinchey, M., Lu, L., Kim, D., Barringer, H., Groce, A., Havelund, K., Smith, M., Cimatti, A., Roveri, M., et al.: Developing Experimental Models for NASA Missions with ASSL. Arxiv preprint arXiv:1003.0396 (2010)
61. Yumerefendi, A., Shivam, P., Irwin, D., Gunda, P., Grit, L., Demberel, A., Chase, J., Babu, S.: Towards an autonomic computing testbed. In: Proceedings of the Workshop on Hot Topics in Autonomic Computing (2007)
62. Zhao, Q., Wang, H., Lv, H., Feng, G.: A Software Dependability Growth Model based on Self-Reconfiguration. In: Proceedings of the 11th Joint Conference on Information Science. Atlantis Press (2008)
63. Wang, Z., Elbaum, S., Rosenblum, D.S.: Automated generation of context-aware tests. In: Proc. of the 29th International Conference on Software Engineering (ICSE), pp. 406–415 (May 2007)

Part III
Knowledge Elicitation and Management

Data and information constitute the main building blocks of an information system, a patrimony that must be conveniently exploited to provide the enterprise with the appropriate knowledge, at each decision level and for each business need. This fact is particularly true when the enterprise itself co-operates with fellow-enterprises to amplify its prospects and opportunities.

This part provides an overview on the collection of methodologies, techniques and tools proposed by the ArtDeco project for the purpose of context-aware knowledge elicitation and exploitation in the networked enterprise. Most of these constitute state-of-the-art research, however some more traditional tools are also mentioned, coupled with the most advanced ones, to produce a fully integrated corpus of equipment for: (i) discovery of the useful information sources, in terms of the enterprise documentation, of structured and unstructured data coming from the networked information systems, of web-available knowledge, of sensor data, of event flows within the business processes; (ii) on-the-fly extraction of synthetic knowledge from these information sources, in terms of a common, semantic-based data model represented as an ontology; (iii) dynamic interpretation and integration of the acquired information; (iv) analysis and dissemination of such knowledge to all decisional levels, appropriately adapting it to the user's function and context.

The chapters that follow discuss how both the explicit and the unexpressed knowledge of the various enterprises can come together to enhance the business capabilities of the network in the various phases of the wine production process. Challenges to be faced are the different formats and nature of the available information, from structured data to natural language to sensor-retrieved information, the dynamicity requirement for the data integration process, and the need to spot, dig-out, and analyze the information that is most appropriate for the current operator in the current context of use.

In the first chapter of this part, entitled *Ontology-based Knowledge Elicitation: An Architecture*, Montedoro et al. give a comprehensive overview of the overall architecture of the ArtDeco Web Portal, which, located at the intermediate level (*Shared Information*) of Figure 1.1, provides the following main functionalities: taxonomy-driven word tagging and tag-based querying; knowledge extraction from natural language sources, and concept-based natural language querying; knowledge extraction from applications and sensor networks; ontology extraction from structured data sources; and collection of data from enterprise processes. In this architecture, research prototypes are mixed up with more consolidated tools, yielding a complete framework for the elicitation and exploitation of knowledge. The following chapters give an account of the knowledge-related research that has been carried out in ArtDeco.

The second chapter, named *Knowledge Extraction from Natural Language Processing*, presents a model for knowledge extraction from documents written in natural language. The model relies on a clear distinction between the conceptual level, where the domain knowledge is represented, and the lexical level, which contains the domain vocabulary. The mapping between these two levels is stored as a stochastic model, which takes in account the context of words. The documentswords are appropriately disambiguated during the indexing phase. The engine supports simple keyword-based queries, as well as *natural language phrase* queries. While extracting and interpreting new terms from the encountered documents, the engine is also able to extend the domain ontology by adding the newly-discovered concepts.

In the chapter *Knowledge Extraction from Events Flows* an analysis of the approaches and methods available for the automated extraction of knowledge from business event flows is presented, focussing on the reconstruction of processes from automatically-generated event logs. Assuming that knowledge can be directly gathered by reconstructing business-process models, various authors in the literature propose that the knowledge acquired be exploited to detect anomalous behaviours in the execution of activities or violations to high level plans. Several of the proposed techniques are briefly examined, and advantages and drawbacks of the different approaches are highlighted, trying to propose a uniform framework for such analysis.

The fourth chapter, entitled *Context-aware Knowledge Querying in a Networked Enterprise*, starts from the idea that the formidable amount of heterogeneous information, accessed by the networked enterprise through all the available channels, makes it difficult for users to find the right information at the right time and at the right level of detail. The research proposes the use of contextual meta-data about the system and the users to reduce this plethora of information, providing high-quality, focussed knowledge to users and applications at all decision-making points. The applicability of the proposed context-aware design methodology and techniques is illustrated within the wine production scenario, where several classes of users access the networked-enterprise data sources, the sensors used for monitoring the productive cycle, and external sources of different nature.

In the ArtDeco scenario data integration of small pieces of heterogeneous information must often be performed on-the-fly, without relying on manual intervention for mapping design; here Semantic Web technologies such as ontologies, which might fail on large-scale data integration, may play an important role. The chapter *On-the-fly and Context-Aware Integration of Heterogeneous Data Sources* describes our ontology-driven framework for dynamic data integration of heterogeneous data sources, where, again, user- and application-queries are dealt with in a context-aware fashion in order to keep the information noise at bay.

Finally, in *Context Support for Analytical Queries* Bolchini et al. propose a solution to analytical query formulation that supports the designer in the OLAP query formulation activity by using the knowledge of the context the decision-maker is currently experiencing. Thus, this last chapter addresses another typical problem of knowledge elicitation, that is, the possibility to provide historical, current, and predictive views of the business operations of the networked enterprise, typical of the Business Intelligence scenario. The major innovative aspect of the proposal is the possibility to connect the relevant contexts of a target application, expressed by using a quite general tree-based dimensional model, with the definition of the analytical queries useful in those contexts; later on, such queries can also be used to select the views that must be transiently or permanently materialized in an data warehouse.

Chapter 9
Ontology-Based Knowledge Elicitation: An Architecture

Marcello Montedoro, Giorgio Orsi, Licia Sbattella, and Roberto Tedesco

Abstract. This chapter overviews the process of collection and automatic analysis of data and documents both inside and outside the Networked Enterprise. We will address the following research problems: discovery of the useful information sources, in terms of the enterprise documentation, of structured and unstructured data provided by existing information systems, of web-available knowledge, of event flow within the business processes; extraction of synthetic knowledge from these information sources, possibly in terms of a common, semantic data model; automatic interpretation and integration of the acquired information; analysis and dissemination of such knowledge to all decisional levels, appropriately adapting it to the user's function and context.

9.1 Introduction

In a networked enterprise, a large amount of information is exchanged among the players. Thus, methodologies and tools are needed which can identify, elicit, and

Licia Sbattella
Dipartimento di Elettronica e Informazione - Politecnico di Milano, via L. da Vinci 32, 20133 Milano, Italy
e-mail: sbattell@elet.polimi.it

Roberto Tedesco
MCPT - Politecnico di Milano, via L. da Vinci 32, 20133 Milano, Italy
e-mail: roberto.tedesco@polimi.it

Marcello Montedoro
IBM Italia S.p.A., Circonvallazione Idroscalo - 20090 Segrate, Italy
e-mail: marcello_montedoro@it.ibm.com

Giorgio Orsi
Department of Computer Science - University of Oxford, OX13QD Oxford, UK
Dipartimento di Elettronica e Informazione - Politecnico di Milano, via L. da Vinci 32, 20133 Milano, Italy
e-mail: giorgio.orsi@cs.ox.ac.uk

G. Anastasi et al. (Eds.): Networked Enterprises, LNCS 7200, pp. 175–191, 2012.
© Springer-Verlag Berlin Heidelberg 2012

represent knowledge (often tacit) where it manifests its relevance for the network partners. Since collective knowledge evolves dynamically, sometimes it may manifest itself upstream in the supply (or design) chain, sometimes downstream (e.g., at the consumers level). The ArtDeco project focuses on the problems of finding, extracting, representing and formalising knowledge, in all various forms in which it may be embedded (in particular, natural language) in order to build a semantic model (via ontologies) of the business domains of networked enterprises. In this scenario we have implemented the ArtDeco Web-Portal, which aims at providing access to an advanced document and data repository, shared by the network partners.

9.2 System Architecture

Figure 9.1 depicts the main architecture of the ArtDeco Web Portal, which provides five main functionalities: taxonomy-driven word tagging and tag-based querying (the OmniFind module); knowledge extraction from natural language sources and concept-based natural language querying (the Knowledge Indexing & Extraction module); data and knowledge extraction from sensor networks and (semi-)structured data sources (the ArtDeco Dynamic Data Integration Systems - AD-DDIS module); and capture & analysis of data collected from enterprise processes (the PROfessional Metrics - PROM module).

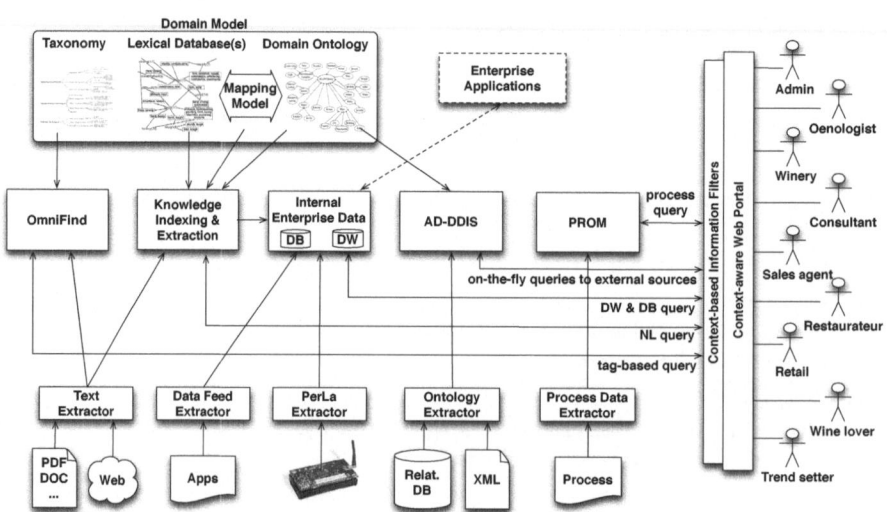

Fig. 9.1 The ArtDeco Context-aware Web Portal

Such modules rely on information stored in the Domain Model. The Domain Model is composed of two sub models: the first one describes the domain by means of a Taxonomy, while the second one relies on a Domain Ontology, a Lexical Database, and a Mapping Model.

The Taxonomy, compiled by a domain expert, provides a hierarchy of tags that associate labels to the words that lexicalise them. This simple model is then used by the OmniFind module to provide taxonomy-driven word tagging and support tag-based queries (see Section 9.3).

The Domain Ontology, a knowledge base defined by means of Description Logics [3], contains the *concepts* relevant for the application domain. As, in our opinion, the model should represent the domain at several levels of abstraction, we made a clear distinction between *conceptual level* and *lexical level*. The Domain Ontology represents the conceptual level model, which contains all the knowledge about the domain we need to represent. This level is language independent, as the Ontology contains the *definition* of concepts [12], without considering any possible linguistic representation. The Domain Ontology is defined by a domain expert, starting from, and extending, the Taxonomy.

The lexical level provides the vocabulary, i.e. terms that the system is able to recognise. In particular, we adopted a Lexical Database [6] as a lexical-level model. The Lexical Databases connect words using specific linguistic relationships, such as synonymity, antinomy, hyponymy, etc. These relationships enable vocabulary navigation in order to discover word similarities and meanings.

Domain Ontology and Lexical Database are connected by means of a stochastic Mapping Model, which permits to *lexicalise* the concepts—it enables the translation of abstract definitions to concrete words.

Domain Ontology, Lexical Database, and Mapping Model are used by the Knowledge Indexing & Extraction module to provide knowledge extraction from natural language sources and support concept-based natural language queries (see Section 9.4).

The Extractors gather data from heterogeneous sources, such as textual documents, web pages, applications, sensor networks, database, XML files, and processes.

The OmniFind and the Knowledge Indexing & Extraction modules analyse and index the content of textual documents. Data coming from application and sensor networks are given to the Internal Enterprise Data module (see Section 9.5), while data captured from processes enter the PROM module (see Section 9.6); finally, ontologies extracted from structured repositories, such as relational databases and XML files, are used by the AD-DDIS module to allow on-the-fly access from external applications (see Section 9.5).

Users interact with the system by means of the Context-aware Web Portal. The portal provides different user profiles with customised views on the data, thanks to the Context-based Information Filtering facility. Within ArtDeco, context-awareness is achieved through the Context-Based Information Filters that correspond to context-aware user views over the database or the data warehouse. Each view corresponds to a different working context of the ArtDeco Web Portal and it is determined on the basis of a context-model and a methodology for Context-Aware View Design (see Chapter 14).

User queries coming from the Web Portal will be answered by means of the views associated to the current context instead of resorting to the whole database or data warehouse schemata that may contain unnecessary information.

Enterprise Applications can exchange information with the Internal Enterprise Data database, in order to take advantage of the data collected by the system, or interact with the portal to gain information from the other data sources, possibly filtered on the basis of the context.

In the following the aforementioned modules will be briefly presented.

9.3 The OmniFind Module

As shown in the picture below, we have used for our project the IBM best in-class technology. Not by chance we used the same technology IBM chose to realize the latest supercomputer called Watson[1].

Actually, the winning component of Watson is UIMA (Unstructured Information Management Architecture)[2]. Unstructured information management (UIM) applications are software systems that analyze unstructured information (text, audio, video, images, and so on) to discover, organize, and deliver relevant knowledge to the user. In analyzing unstructured information, UIM applications make use of a variety of analysis techniques, including statistical and rule-based Natural Language Processing (NLP), Information Retrieval (IR), machine learning and ontologies. IBM's Unstructured Information Management Architecture (UIMA) is an architectural and software framework that supports creation, discovery, composition, and deployment of a broad range of analysis capabilities and the linking of them to structured information services, such as databases or search engines. The UIMA framework provides a run-time environment in which developers can plug in and run their UIMA component implementations, along with other independently-developed

---

[1] Some months ago the supercomputer Watson won the final round in the epic man vs. machine battle in the *Jeopardy!* Game. *Jeopardy!* is a famous American game covering a broad range of topics, such as history, literature, politics, arts and entertainment, and science. *Jeopardy!* poses a grand challenge for a computing system due to its broad range of subject matters, the speed at which contestants must provide accurate responses, and because the clues given to contestants involve analyzing subtle meaning, irony, riddles, and other complexities in which humans excel and computers traditionally do not. Thanks to its technology Watson has the ability to scan and analyze data from far more sources than a human ever could in a short period of time. Watson evaluates the equivalent of roughly 200 million pages of content (about one million books worth), written in natural human language, to find correct responses to complex *Jeopardy!* clues. Watson is an application of advanced Natural Language Processing, Information Retrieval, Knowledge Representation and Reasoning, and Machine Learning technologies to the field of open-domain question answering.
[2] See: incubator.apache.org/uima

components, and with which they can build and deploy UIM applications. By detecting important terms and topics within documents, semantic search engines provide the capability to search for concepts and relationships instead of keywords. IBM's enterprise search solutions such as IBM OmniFind Enterprise Edition, have such semantic search capabilities. They allows UIMA annotators to be plugged into the OmniFind processing flow, enabling semantic search to be performed on the extracted concepts. OmniFind permits to associate words with tags defined in the Taxonomy. The tag-based query mechanism is then provided to the user, as a set of pre-defined forms. Each form is designed to meet a specific user requirement; forms are associated to user profiles, thus each profile has its own view on the data.

Figure 9.2 shows the architecture of OmniFind, based on functionalities provided by IBM OmniFind Enterprise Edition. In the following, the components of OmniFind are introduced.

The Document Crawler performs the function of crawling the various data sources (usually, web sites) at intervals configured by the administrator, and populates a raw data store (file system based) with the contents extracted from the data sources. Users can define one or more *collections* where crawlers can store documents.

The Knowledge Extractor analyses documents collected by the crawler, parses and indexes them. The parser analyses documents' content and meta-data; users can improve the quality and precision of the parsing phase by integrating custom text processing algorithms in the parser. These text processing algorithms are developed using UIMA. The parser performs the following tasks on each document: extracts the text; detects the source language; applies parsing rules specified for the collection, for example it is possible to define *categories* so that users can search documents by means of the categories they belong to. Moreover, the parser associates texts' words with related *labels* defined into the Taxonomy. Finally, the parser stores the results of the analysis in a file temporary system data store.

The indexer may be scheduled at regular intervals in order to periodically read from the temporary store, adding information about new and changed documents to an index that is stored in a file system. In practice, the index is split into (i) a *main index*, which has the global ranking order of all the documents and (ii) the *Delta index*, which corresponds to those documents that have not yet been merged into the main index.

The Search Component relies on OmniFind Search Servers to provide search functionalities to the user. The search application receives input from a browser and invokes the search servers through the Search and Index API (SIAPI). Search Servers execute the query against the index and returns a list of results to the search application; each result includes a "quick link", which displays a preview of the related document.

The Web Application represents the external interface through which users interact with OmniFind. This component is also used to create and administer collections, start and stop other components (such as the crawler, parser, indexer, and search), monitor system activity and log files, configure administrative users, and associate search applications with collections. Moreover, Web Application has been customised to provide the query forms. Such queries depend on the user profile, provided by the ArtDeco Context-aware Web Portal.

9.4 The Knowledge Indexing and Extraction Module: Concepts from Texts

Figure 9.3 depicts the functionalities provided by the Knowledge Indexing & Extraction module, which –after the training phase– is able to extract and index *concepts* from documents written in natural language (e.g., PDF or Word files, web pages, etc.), execute natural language based queries, extend the Ontology, and export concepts to the Internal Enterprise Data module for further elaboration.

For example, an Oenologist designing a new red wine could be interested in gathering information from specialized reviews available on Internet. Using the system Web Portal, she/he inserts the URLs of such web sites; then, the Text Extractor downloads the web pages and extracts the text, while the Knowledge Indexing & Extraction module discovers and indexes the relevant concepts. The advanced query typologies supported by the system provide a powerful tool for both documental and conceptual search, where a large variety of different requests can be expressed.

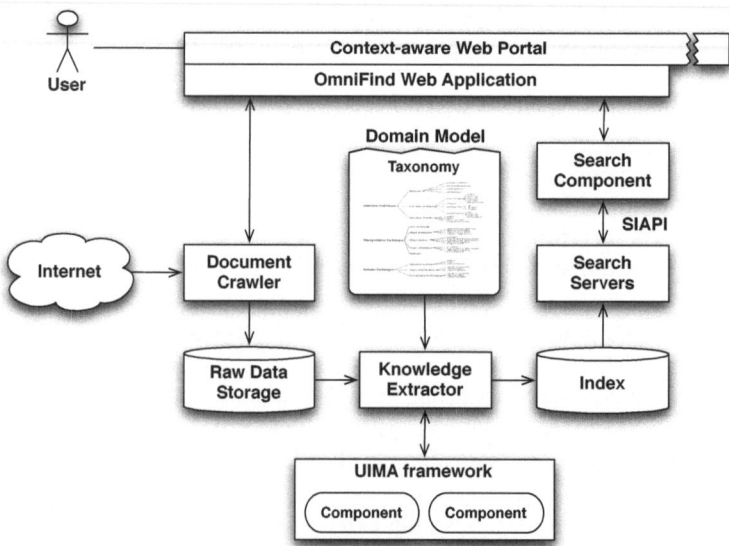

Fig. 9.2 The IBM OmniFind Enterprise Edition High Level Architecture

In the following such functionalities will be described; for an in-depth presentation, as well as experimentation results, see Chapter 10.

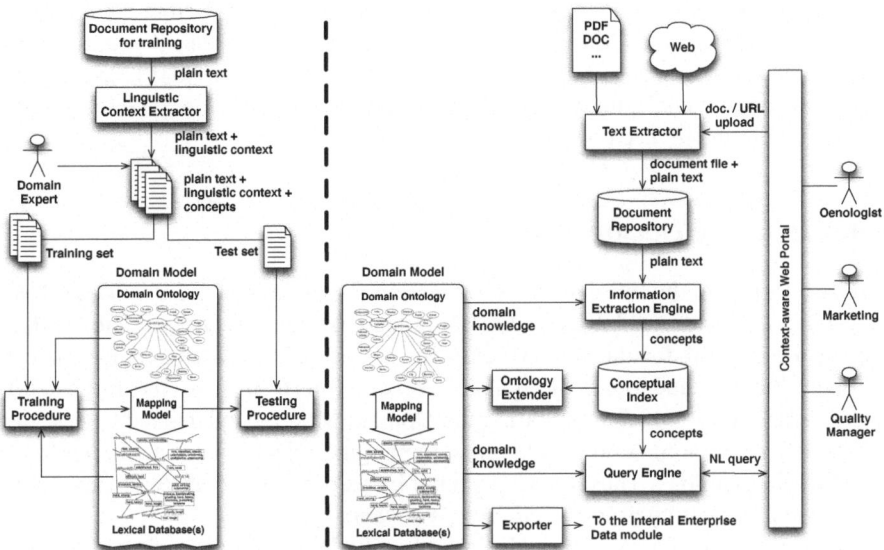

Fig. 9.3 Functionalities provided by the Knowledge Indexing & Extraction module: model training (left); indexing, querying, extending, and exporting the model (right)

9.4.1 Training the Mapping Model

Words are usually polysemic: depending on the *linguistic context*[3], the same word carries different meanings. Thus it is, in general, not possible to map a given word onto one and only one concept. Instead, the relationship between concepts and words is in general many-to-many. We rely on a particular combination of stochastic models to provide such many-to-many mappings: the Mapping Model.

The Mapping Model's training process is divided into several phases. First of all, a specific set of documents in the Document Repository enter the Linguistic Context Extractor that retrieves morphologic and syntactic information (the linguistic "context"), by means of tools such as Freeling [2], the Stanford POS tagger [19], the Stanford parser [14, 5], and the JavaRAP [15] coreference resolutor.

Then, a human expert associates each word with the right concept, creating two sub-sets: the *training set* and the *test set*. The Training Procedure, relying on both the Ontology and the Lexical Database, analyses the training set documents, and generates the statistical model. Finally, the Mapping Model is checked against the test set.

---

[3] The linguistic context represents the information one can extract by reading the text; the linguistic context of a word is composed of the surrounding words, augmented with morphologic and syntactic information.

As an example, consider the sentence "The bottle with red label contains Barolo, a red wine". The human expert associates words and concepts, producing "The bottle/WineBottle with red label/WineLabel contains Barolo/barolo wine, a red/red wine wine/Wine". Notice that the adjective "red" just before the noun "label" is not tagged, as its meaning (the colour of a label) is not considered useful for the domain (i.e., there is no related concept into the Domain Ontology). Instead, just before the noun "wine", the adjective "red" refers to a wine colour, and therefore it is tagged with the related concept.

9.4.2 Indexing Documents

The Text Extractor module gathers text from several document formats (PDF, HTML pages, etc.), provided as URLs or directly uploaded by users. The extracted text, as well as the original document file (if any), are stored into the Document Repository.

Then, the text undergoes the indexing process, performed by the Information Extraction Engine, which –exploiting the Domain Model– updates the Conceptual Index: For each word of the documents, the engine selects the related *concept* of the Domain Ontology. Thus, the Conceptual Index provides an abstract view of the documents, permitting users to *search for concepts*.

The Information Extraction Engine searches the Document Repository for newly added documents, retrieves morphologic and syntactic information (the "context"). Then, a mapping algorithm, relying on the Domain Model, maps documents' words to the related Ontology concepts. Finally, a classic TF-IDF scheme is applied, in order to generate the documents' concept vectors, stored into the Conceptual Index. The engine can also decide that a word does not fit any of the Domain Ontology concepts, as its meaning is outside the domain of interest.

The concepts-words mappings generated during the indexing procedure are then arranged as an index and stored into the Conceptual Index.

Continuing the example, assume the system be trained to recognise concepts in the following set {WineBottle, WineLabel, barolo wine, red wine, Wine}. If the document set is composed of D_1="Barolo is a small village where good wine is produced" and D_2="Barolo is a red wine", the system extracts the following sets of concepts C_1={Wine}[4] and C_2={barolo wine, red wine, Wine}. The Concept Frequency Analyser calculates the weight $w_{c,d}$ associated to each concepts c, for each document d; thus, the following vectors are generated: V_1=[0, 0, 0, 0, w_{Wine,D_1}] and V_2=[0, 0, $w_{\text{barolowine},D_2}$, w_{redwine,D_2}, w_{Wine,D_2}].

9.4.3 Querying Documents

The Query Engine, exploiting both the Conceptual Index and the Domain Model, permits to formulate concept-based queries on the document collection. Searching for a given concept, the system finds every mapped word (as well as its synonyms,

---

[4] Notice that in D_1 "Barolo" refers to a village and should not be considered.

thanks to the Lexical Database), and calculates a ranked list of documents. The Query Engine accepts several query models: from simple lists of keywords, as usually found in traditional search engines, to complex sentences expressed in natural language.

Keyword based queries are composed of a sequence of words, connected by either AND or OR boolean logic operators. The system maps each word to the related concept, and then search the Conceptual Index for these concepts; in the AND case only documents containing all the concepts are returned; in the OR case, documents containing at least on concept are returned.

As an example of AND query, if the user issues Q="Barolo wine", the system indexes the text and produces the following sequence of concepts: $V_Q=[0, 0, w_{\texttt{barolowine},Q}, 0, w_{\texttt{Wine},Q}]$. Then, compares the vector V_Q against V_1 and V_2, and finds the nearest document.

Phrase based queries are written in natural language, as questions or descriptions. Such queries relies on three main steps. First, the phrase issued by the user is indexed as it was a sentence in a document; the result is a sequence of concepts. Second, such concepts, as well as the contextual information, are analysed by a unification based parser, which finds the appropriate structure tree (relying on a feature based grammar, defined by an human expert, which defines the structure of phrases the system should be able to parse.) Third, an SQL query template (also defined by a human expert) is selected and filled with nodes of the parse tree, and the resulting query is executed.

As an example of phrase based query, assume that the Domain Ontology contains information about the facts that wines have colours, and that Barolo's colour is red. Moreover, assume that an human expert prepared the following query template[5]: T=<"what", wine-attribute-name, wine-instance>. If the user issues Q="what is the colour of Barolo?", the system extracts C_Q=[WineColor, barolo wine], finds the phrase template T, fills the placeholders: <"what", WineColor, barolo wine>, and executes a parametric SQL query S_T associated by the same human expert, to the template T. Notice that the template T is actually able to capture several requests, e.g. the colour of wine, the price of wine, etc. In other words, requests with similar structure (in terms of sequence and type of concepts) can be captured by means of the same template.

Notice that, as a further advantage of the concept-based queries with respect to traditional word-based engines, the search is multi-language in nature: the language used to search for documents does not depend on the language used to write them.

9.4.4 Extending the Domain Ontology

The Ontology Extender relies on Drools[6], a production-rule engine. Such rules predicate on information collected while indexing the text, to infer that a given

---

[5] Just to get the idea, the actual definition relies on a feature based grammar, which permits to specify templates in a way that can capture several different superficial forms of the same request.

[6] See: www.jboss.org/drools

word could be a new concept to be added to the Ontology. However, rules cannot guarantee that the newly discovered association text-concept is sound; therefore, the system tags these new concepts as "guessed" to distinguish them from concepts defined by the human expert.

As an example, assume that the Domain Ontology contains information about the fact that wines have production regions. Moreover, assume that the system indexes the sentence"Piemonte produces wines" and associates words to concepts as illustrated in Figure 9.4. A simple rule could state that:

IF subject_of(W_1, "to produce") $\land$ direct_object_of(W_2, "to produce") $\land$
    mapped_on_class(W_2, Wine) $\land$ mapped_on_relationship("to produce", R) $\land$
    connected_by_relationship(R, Wine, C)
THEN
    W_1 is a new individual of C

and the system could infer that W_1="Piemonte" should generate a new individual piemonte belonging to the class C= Region.

Fig. 9.4 Extending the Domain Ontology: an example

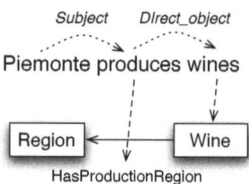

9.5 The Internal Enterprise Data Module

In the ArtDeco scenario the involved actors are not only large enterprises but also small companies or even single users: a dedicated, human-intensive solution is not a reasonable way to integrate such a variety of heterogeneous information, and automated data wrapping and integration are needed. This is even more evident as, in the ArtDeco environment, the data sources are not necessarily stable, and moreover they may vary from structured to semi-structured or even totally unstructured and dynamic, such as web pages. In order to be manageable and meaningful to the users, the input information of the ArtDeco system must be machine processable and entails the need for explicit representation of data semantics.

The information management sub-system of the ArtDeco project is thus devoted to providing a uniform, Ontology-based semantic access to information coming from heterogeneous data sources. In ArtDeco, the choice was taken to keep a centralised repository, in the form of a Relational Database with an associated Data Warehouse, to store and manage all the interesting corporate information [16].

9.5.1 Structuring Web Data

In ArtDeco the decision-making process is largely based on information obtained through the analysis of natural language documents published on the Web. There is no doubt that the entire process may greatly benefit (both in terms of effectiveness and efficiency) from a structured representation of this information.

The need for structured data management is justified by several reasons. In the first place, the Knowledge Indexing and Extraction module produces as output a large number of *facts* about the domain for each document being processed (e.g., "Nebbiolo" is a DOCG lightly-coloured red wine, it is produced in Piemonte and has certain aromas). The size of these facts combined with the number of documents collected from the Web may produce a huge dataset that needs to be efficiently managed. The dataset produced by the extraction process must be eventually handed over a data-mining process [17] in order to extract some useful knowledge from it. Despite data-mining techniques for semi-structured and unstructured information exists in the literature, the ones developed for structured information (e.g., relational databases and data-warehouses [13]) are long established and far more effective, especially for business and market intelligence.

For this reason, we propose a (conceptual) modelling pattern for extending enterprise databases to enable Web-document storage and analysis. With reference to Figure 9.1, the DB component represents the relational (and possibly federated) database giving access to a suitable and integrated view of the enterprise data legacy that could be used to support knowledge extraction from Web documents. The schema of this database is constituted by two sets of tables: the first is a set of views over the enterprise database (*Enterprise Tables*) giving access to relevant information for the extraction process (e.g., a product's description), the second is a set of tables used to structure the information coming from the extractors (*Web-Search Tables*). These tables are domain-independent, and thus can be used as a database modeling-pattern for ArtDeco-style applications.

The reference Relational Schema for the Enterprise Tables is shown in Table 9.1 while the conceptual and relational schemas for the *Web-Search Tables* are shown in Figure 9.5(a) and Figure 9.5(b) respectively.

The core table is FINDING which represent a finding in a web document that is considered "relevant" by the extraction engine. Tables DOCUMENT and SOURCE represent detailed information about the document and the data source (e.g., the website publishing the document). Table EVALUATION stores the associations between adjectives used to give a judgment or an opinion about a source or a finding in a web document (i.e,. "wonderful" or "awful") and a numeric value describing if a particular adjective has to be considered positive or negative.

Example 1. (Structuring Web Data)
Consider, as an example, the following sentence appeared in a blog entry titled "Barbera 2010: Pride in Simplicity?" found in the blog of a wine-expert in the US:

> "[...]so she poured us the unoaked wines, a fantastic Langhe Nebbiolo and a Barbaresco.[...]"

Table 9.1 The ArtDeco Enterprise Tables

> WINE(appellation, category, vinification)
> REF-COMP (appellation, tech-name, min_%, max_%)
> GRAPEVINE (tech-name, name, variety)
> HARVEST (h-id, vineyard-id, note)
> CELLAR (cl-id, material, type, vineyard-id, harvest-id)
> BARREL (bl-id, wood, type, cellar-id)
> COMPOSITION (bottle-id, barrel-id, wine_%)
> BOTTLE (b-id, bott-date, harvest-date, price, appellation)
> LOT (l-id, pkg-date, pkg-type, return-date, sell-date, cr-id, cs-id)
> CUSTOMER (c-id, name, address)
> VINEYARD (v-id, hectares, fraction, municipality, district, region, zone)
> ROW (r-id, vineyard-id, plant-date, tech-name, phenological-phase)
> PHENOMENON (vineyard-id, phenomenon, date, type, notes, emergency-plan)
> DAMAGE (vineyard-id, row-id, name, date, analysis)
> SENSOR-BOARD (sb-id, coordinates, act-date, obj-id)
> SENSOR (s-id, type, model, meas-unit)
> MEASURE-DATA (date, time, sensor-id, value)

From this sentence we can retrieve at least some interesting data such as the fact that "Langhe Nebbiolo" and "Barbaresco" are "unoaked wines" and the fact that the author of the entry considers them "fantastic". Moreover, the title of the document mentions a Wine Festival event in Italy ("Barbera 2010") which is referring to a particular family of wines (i.e., the "Barbera") and the two nouns "pride" and "simplicity" can be used to associate a measure of the consideration that the entry's author has of this event. Each of these findings correspond to tuples in FINDING along with their position in the document. Information about the document and the website address will be stored in the corresponding table, along with the references to the tuples containing positive judgments for the adjectives "pride", "simplicity" and "fantastic" in the EVALUATION table.

As already said, the Web-Search Tables are a modelling-pattern and they can be linked to the enterprise tables if necessary. Suppose, as an example, that one of the two wines mentioned in the entry of Example 1 is a product of the company running the ArtDeco system. In this case, it might be useful to link the findinds to the information in the enterprise databases such as the product's description, the bottle or grapevine's data when they are available. From the modelling viewpoint, this corresponds to the introduction of a set of 1:n relationships between the table FINDING and the interesting tables as shown in Figure 9.6 where the bridge tables FND-G, FND-W, FND-B represents the conceptual connections between a web-search finding and a grapevine, a wine and a bottle respectively.

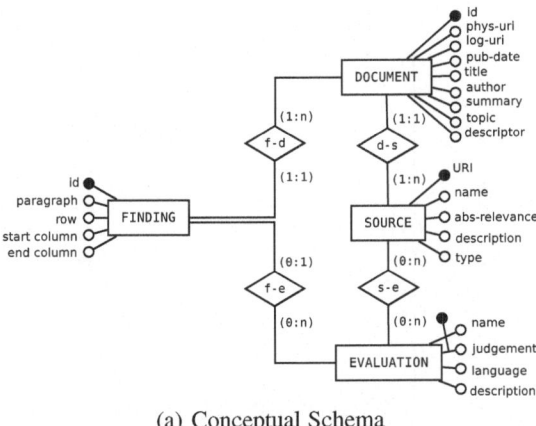

(a) Conceptual Schema

FINDING(id, paragraph, row, start-column, end-column,document, evaluation)
DOCUMENT(id, phys-uri, log-uri, pub-date, title, author, summary, topic, descriptor)
SOURCE(uri, name, abs-relevance, description, type)
EVALUATION(name, judgement, language, description)
S-E(source, evaluation, judgement)

(b) Relational Schema

Fig. 9.5 DB Extension for NL Web Data: Conceptual and Relational Schemata

9.5.2 Dynamic and Heterogeneous Data-Source Integration

While the centralised repository is useful for analytical query processing, mediated on-line access to the original data sources should also be allowed, to provide stakeholders with a query interface for detailed and up-to-date access to the overall network knowledge. In ArtDeco, the dynamic integration of (semi-)structured data sources is carried out by the AD-DDIS component (see Chapter 13 for a detailed description). AD-DDIS first integrates the schemata of the various, heterogeneous data sources that may include XML documents and legacy and current databases from the networked business partners. Sensors data are handled as relational databases since we rely on the Perla Language (see Chapter 17) which provides a relational interface for such devices.

A model of the domain semantics is captured by the Domain Ontology, a shared conceptualisation of the application domain, which is used as Global Schema and later mapped to a set of ontologies, each representing the schema of one data source. Each data source Ontology is an ontological description of the data source schema and it is extracted by means of *domain-aware wrappers* i.e., the extracted description depends from both the target Domain Ontology and the data source schema.

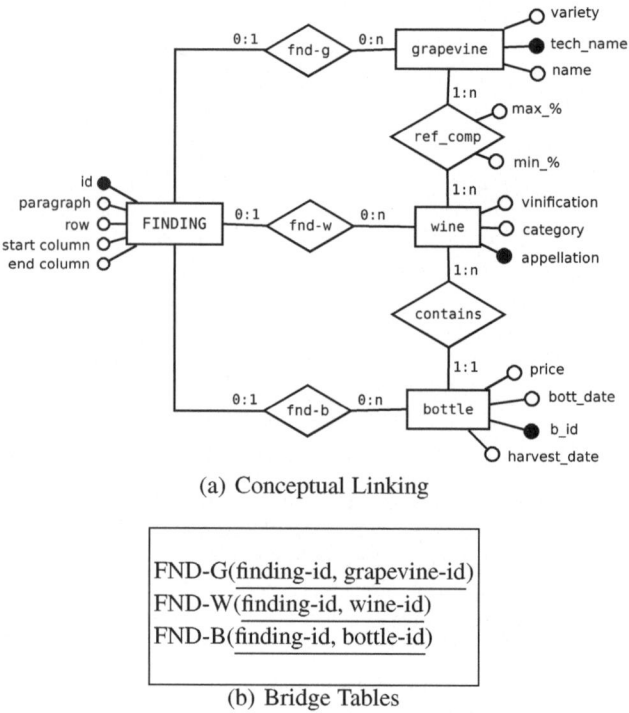

(a) Conceptual Linking

FND-G(finding-id, grapevine-id)
FND-W(finding-id, wine-id)
FND-B(finding-id, bottle-id)

(b) Bridge Tables

Fig. 9.6 DB Extension for NL Web Data: Linkage of Enterprise Views

Data source Ontology are then (semi-)automatically mapped to the Domain Ontology and such information will be later used to enable querying.

9.5.3 Knowledge Extraction

Once the information has been stored and structured into the database, it is possible to exploit standard warehousing and data-mining techniques in order to extract structured knowledge that was not explicitly present into the web pages. This is achieved by building a data-warehouse exploiting the methodology presented in [7]. Context-aware analytical queries are supported by the ArtDeco portal (see Chapter 14). Moreover, the data warehouse will be used to support data mining processes in order to analyse temporal trends [8] (e.g., which products are increasingly mentioned in the web-sphere), derive associations rules [1] (e.g., whose trendsetters participate in certain events), construct clusters of documents with similar features [9] (e.g., all the documents related to successful events) and outliers identification [10] (e.g., products appreciated by trend-setters and not by the market) to the aim of supporting the product design and innovation processes.

Determining such analysis processes is usually responsibility of the product designer and the R&D personnel. Once the various processes have been identified, the data-mining extension of the `ArtDeco` Web Portal will expose the proper graphical user-interfaces for them.

9.6 The PROM Module: Collecting Data from Processes

The PROfessional Metrics (PROM) system[7] consists of a set of tool for automated data collection from enterprise workflows and their subsequent analysis. PROM accesses and collects both code (i.e., interaction traces from running software) and process (i.e., email communication, supply-chain events, etc.) measures. This is important, since a comprehensive approach is the best way to collect data in a more complete way, in order to understand not only what workers produce, but also how they produce it.

The tool focuses on a comprehensive data acquisition and analysis methodology (see Chapter 11) in order to provide elements to improve products [18, 4]. The approach is based on a measurement framework that aims to achieve mainly two goals [11]: (a) to automatically measure the effort spent per artifact, and (b) to provide a framework that integrates existing tools that extract well known metrics with the measurements about the connected effort.

The system provides a non-invasive, automatic and accurate data collection and analysis, which allows to better understand the company-internal workflows. As an example, consider the process of acquisition of grapes in a wine-farm. While some of the producers will be part of an integrated supply-chain and thus supported by a supply-chain management system that can be easily monitored by software probes, some other producers, especially small farmers, will handle their selling process to the wine-farm through paper purchase-orders and receipts. On the wine-farm-side, this corresponds to manual handling of acquisitions with consequent manual insertion of data, email/document archiving etc. In order to understand and improve the process it is thus necessary to monitor as much as possible of these events by collecting data from every application (email clients, spreadsheets, word processors, etc.) used inside the enterprise.

Through an investigation performed over the collected data, it is possible to conduct an analysis of the company processes (with a special focus on its strengths, weaknesses, opportunities and threads). On the base of this analysis is thus possible to improve workflows, to provide decision support for IT purchases, or to increase the quality of developed products by choosing appropriate development methodologies. Using such approach, companies can benefit from a more detailed picture of the state of company-internal workflows and processes, and they can determine the quality of developed products.

---

[7] G. Succi. Professional Metrics (PROM) Home Page: `www.prom.case.unibz.it`

PROM is a client-server system, which is organised on: data collection from plug-ins that are integrated in development tools and other applications, data analysis in a central repository, and knowledge creation in suitable components (see Figure 9.7). The data collection tools store their measurements locally in form of xml files and, when a connection is available to the server, they send them to the central database using web services.

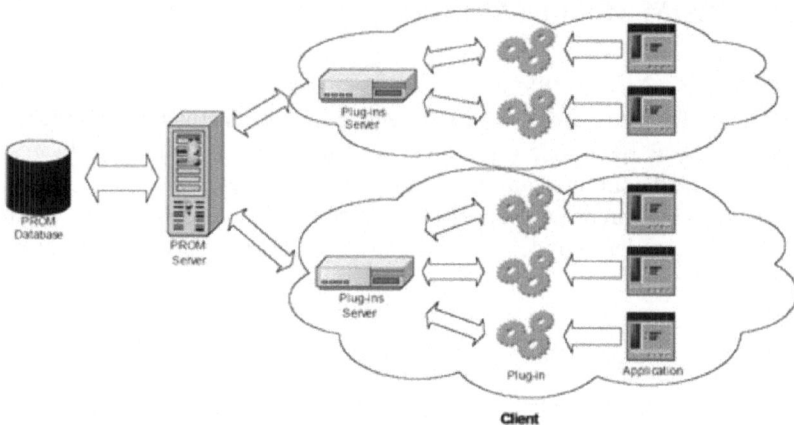

Fig. 9.7 The PROM Architecture

9.7 Conclusions

In this chapter we presented the `ArtDeco` Web Portal, which aims at providing an advanced document and data repository, shared among the network partners.

The portal provides functionalities that permit to discover, extract, and find information from structured, semi-structured, and non-structured data sources.

A Domain Model contains the concepts relevant for the application domain, as well as the related vocabulary.

References

1. Agrawal, R., Imieliński, T., Swami, A.: Mining association rules between sets of items in large databases. ACM SIGMOD Record 22(2), 207–216 (1993)
2. Atserias, J., Casas, B., Comelles, E., González, M., Padró, L., Padró, M.: Freeling 1.3: Syntactic and semantic services in an open-source NLP library. In: Proceedings of the Fifth International Conference on Language Resources and Evaluation (LREC 2006), Genoa, Italy. ELRA (May 2006)
3. Calvanese, D., McGuinness, D., Nardi, D., Patel-Schneider, P.: The Description Logic Handbook: Theory, Implementation and Applications. Cambridge University Press (2003)

4. Coman, I.D., Sillitti, A., Succi, G.: A case-study on using an automated in-process software engineering measurement and analysis system in an industrial environment. In: ICSE 2009: Proceedings of the 31st International Conference on Software Engineering, pp. 89–99. IEEE Computer Society, Washington, DC (2009)
5. de Marneffe, M.C., Manning, C.D.: Stanford typed dependencies manual. The Stanford Natural Language Processing Group, Stanford University, Stanford, California (2008)
6. Fellbaum, C. (ed.): WordNet: An Elettronic Lexical Database. MIT Press (1998)
7. Golfarelli, M., Rizzi, S.: Data Warehouse Design: Modern Principles and Methodologies. McGraw-Hill (2009)
8. Hamilton, J.: Time series analysis. Princeton Univ. Press (1994)
9. Hartigan, J.: Clustering algorithms. Wiley, New York (1975)
10. Hawkins, D.: Identification of outliers. Chapman & Hall (1980)
11. Janes, A., Scotto, M., Sillitti, A., Succi, G.: A perspective on non invasive software management. In: Proceedings of IMTC 2006 – Instrumentation and Measurement Technology Conference, Sorrento, Italy (April 2006)
12. Kashyap, V., Bussler, C., Moran, M.: Ontology Authoring and Management. In: Data-Centric Systems and Applications, ch. 6. Springer, Heidelberg (2008)
13. Kimball, R.: The data warehouse toolkit. Wiley-India (2006)
14. Klein, D., Manning, C.D.: Fast exact inference with a factored model for natural language parsing. In: Advances in Neural Information Processing Systems, vol. 15. MIT Press (2003)
15. Long Qiu, M.-Y.K., Chua, T.-S.: A public reference implementation of the RAP anaphora resolution algorithm. In: Proceedings of the Fourth International Conference on Language Resources and Evaluation (LREC 2004), Lisbon, Portugal, vol. 1, pp. 291–294 (May 2004)
16. Martin, J., Finkelstein, C.: Information Engineering. Savant (1986)
17. Mehmed, K.: Data Mining: Concepts, Models, Methods And Algorithms. John Wiley & Sons (2003)
18. Sillitti, A., Janes, A., Succi, G., Vernazza, T.: Collecting, integrating and analyzing software metrics and personal software process data. In: EUROMICRO 2003: Proceedings of the 29th Conference on EUROMICRO, p. 336. IEEE Computer Society, Washington, DC (2003)
19. Toutanova, K., Klein, D., Manning, C., Singer, Y.: Feature-rich part-of-speech tagging with a cyclic dependency network. In: Proceedings of HLT-NAACL, Edmonton, Canada (2003)

Chapter 10
Knowledge Extraction from Natural Language Processing

Licia Sbattella and Roberto Tedesco

Abstract. This chapter presents a model for knowledge extraction from documents written in natural language. The model relies on a clear distinction between a conceptual level, which models the domain knowledge, and a lexical level, which represents the domain vocabulary. An advanced stochastic model (which mixes, in a novel way, two well-known approaches) stores the mapping between such levels, taking in account the linguistic context of words. Such a stochastic model is then used to disambiguate documents' words, during the indexing phase. The engine supports simple keyword-based queries, as well as natural language-based queries. The system is able to extend the domain knowledge, by means of a production-rules engine. The validation tests indicate that the system is able to extract concepts with good accuracy, even if the train set is small.

10.1 Introduction

Oenologist and Marketing, as a fundamental part of the wine design phase, collect and evaluate advices and tendencies from well-known sources (magazines, web sites, reports, etc.) Once wine has been produced and distributed, Quality Manager is interested in collecting reactions and opinions from customers and wine experts. In both cases, such information is usually extracted from documents written in natural language.

L. Sbattella
Dipartimento di Elettronica e Informazione - Politecnico di Milano, via L. da Vinci 32, 20133 Milano, Italy
e-mail: licia.sbattella@polimi.it

R. Tedesco
MCPT - Politecnico di Milano, via L. da Vinci 32, 20133 Milano, Italy
e-mail: roberto.tedesco@polimi.it

G. Anastasi et al. (Eds.): Networked Enterprises, LNCS 7200, pp. 193–219, 2012.

The `ArtDeco` project aims at addressing such issue by means of tools able to find, extract, represent, and formalise knowledge expressed in natural language. In particular, as a component of the `ArtDeco` Web Portal (introduced in Chapter 9), the Knowledge Indexing & Extraction module permits to define and train a stochastic domain model (see Figure 10.1, on the left), using it to extract and index *concepts* from documents written in natural language, execute natural language based-queries, and extend the Ontology (see Figure 10.1, on the right). A Domain Expert is in charge of defining and training the Domain Model, while Oenologist, Marketing, and Quality Manager interact with the Web Portal user interface, issung queries to the system.

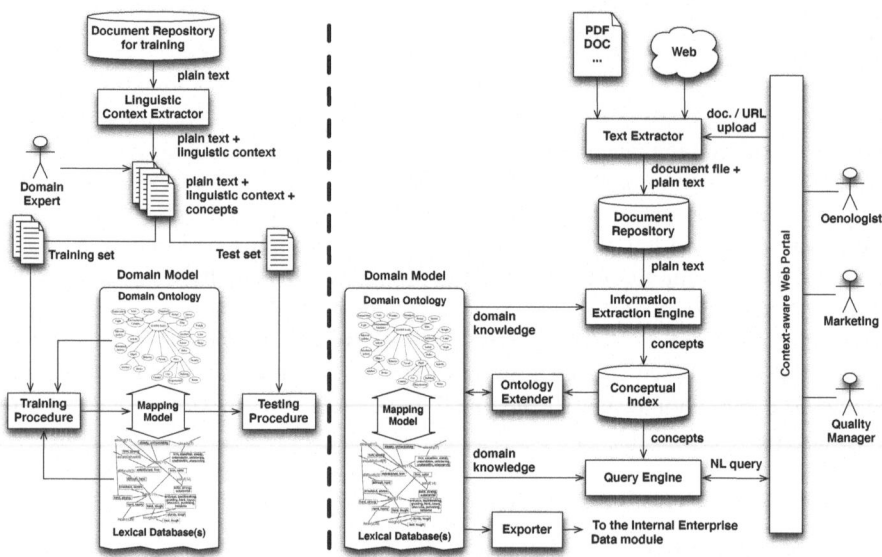

Fig. 10.1 Functionalities provided by the Knowledge Indexing & Extraction module: training (left); indexing, querying, and extending (right)

The Domain Model contains both the domain knowledge (the Domain Ontology), and the domain vocabulary (the Lexical Database); a stochastic model (the Mapping Model) defines how lemmas in the Lexical Database map onto the concepts defined in the Domain Ontology. This mapping is many-to-many, as words are often polysemic and a concept could, in general, be lexicalized by means of several words. Section 10.2 introduces the three components of the Domain Model.

An important element in deciding the right concept to map onto a word is its linguistic context. The linguistic context represents the information one can extract by reading the text; at a minimum, the linguistic context of a word is composed of the surrounding words, but much more can be done by means of Natural Language Processing (NLP) tools, able to extract morphologic and syntactic information. Section 10.3 illustrates how documents' words are associated to their linguistic contexts, what kind of information is stored in the linguistic context, and the tools we rely on to extract such information.

Section 10.4 details our stochastic Mapping Model, shows how to take in account the linguistic context, and explains the training procedure. Figure 10.1, on the left, shows the procedure for training such stochastic model, starting from a manually-tagged corpus.

During the indexing process (see Figure 10.1, on the right), the Information Extraction Engine leverages the NLP tools, analyses documents and transforms each one of them into a sequence of lemma/context information pairs: $< (L_0, I_0), (L_1, I_1), \ldots >$. Relying on this description, a specific algorithm understands whether a given word, taking in account its actual linguistic context, does carry a useful meaning and, if it is the case, selects the right concept to associate; words that do not carry a useful meaning are discarded. At the end of this phase, each retained word is associated to a concept of the Domain Ontology, and documents are represented by a list of lemmas and related concepts $< (L_0, C_0), (L_1, C_1), \ldots >$; such concepts represent the actual information that is stored in the Conceptual Index, following the Term Frequency-Inverse Document Frequency (TF-IDF) schema. This approach permits to enhance the search precision, as a) documents containing irrelevant words are likely to get a low rank; b) the words' meanings are disambiguated and thus the polysemy problem is solved. Section 10.5 describes the process in detail.

The Query Engine, relying on the concept-based index and exploiting the Domain Ontology, permits to automatically extend the query, including synonyms, and other related words. This approach permits to enhance the search recall, as documents containing the searched concept, even if expressed with different words, are captured by the system. Queries can be issued as sequences of words (keyword-based queries), or as question/description written in natural language (natural language based queries). Section 10.6 presents the query engine.

The Ontology Extender is able to discover new concepts, enhancing the Domain Ontology and increasing both precision and recall. Section 10.7 presents the extension mechanism.

Finally, the Exporter permits to send the content of the Domain Model to other modules of the `ArtDeco` Web Portal.

In the rest of the chapter, paragraphs within gray-shaded boxes contain an in-depth description of the model; readers not interested in such details can skip them.

10.2 Defining the Domain Model

Several methodologies exist for knowledge extraction from texts. Among the other approaches, Named Entity Recognition (NER) systems and Ontology-based annotators are often used. NERs, like the Stanford NER tool [8], start from a collection of labels and, usually, a stochastic model that describes how these labels are related to words. At run-time, these systems are able to assign the right labels to the words, resolving ambiguities. As a drawback, the labels are not connected by any structure (i.e. they do not represent an Ontology) and cannot be used for further elaborations such as reasoning or query extension. Ontology-based annotators, like the one described in [29], permits to make inferences on the extracted information, but do not exploit any linguistic information to address the ambiguity of words. Finally, Ontology-based knowledge extractors like Artequakt [1] exploit both an Ontology and a linguistic model. In Artequakt, however, the Ontology and the linguistic model are separated as they are handled by two different modules. Thus, the Ontology can be used for inference but cannot help during the recognition phase.

Our model aims at addressing these issues, providing an unified approach that combines an Ontology and a stochastic model: the domain knowledge stored in the Ontology will help the recognition phase, while the stochastic model will solve linguistic ambiguities. We call this model the Domain Model.

The Domain Model component is the foundation of the `ArtDeco` knowledge management strategy. It represents the specification of the domain being considered (as Figure 10.1 shows, the model needs a training phase, which will be discussed in detail in Section 10.4.4.)

The model should represent the domain at several levels of abstraction, therefore we make a clear distinction between *conceptual level* and *lexical level*.

The conceptual level models the Ontology, which represents all the knowledge about the domain we need to represent. This level is language independent, as the Ontology contains the *definition* of concepts, without considering any possible linguistic representation.

The lexical level provides the vocabulary: words the system is able to recognise. In particular, we adopted a Lexical Database as a lexical-level model. The Lexical Databases connect words using specific linguistic relationships, such as synonymity, antinomy, hyponymy, etc. These relationships permit to navigate the vocabulary, discovering word similarities and meanings. This level permits to *lexicalize* the concepts: translate abstract definitions into concrete words. Changing the language the system supports is a matter of providing the related Lexical Databases.

A mapping level is needed to connect conceptual and lexical levels. Words are often polysemic: depending on the linguistic context, the same word may carry different meanings. Therefore it is, in general, not possible to map a given word onto one and only one concept. Instead, the relationship between concepts and words is in general many-to-many: a word can lexicalize many concepts, and a concept can be lexicalized by many words.

Thus, the mapping level should provide a model to represent such many-to-many mappings. Moreover, when indexing documents, a disambiguation procedure should select the right concept-word association, depending on the linguistic context of the sentence being analysed.

Following the aforementioned ideas, the Domain Model is composed of three parts: The Domain Ontology, the Lexical Database, and the Mapping Model. In the following, these components are explained in detail.

10.2.1 The Domain Ontology

The Ontology model we adopt is based on Description Logics (DL) [3], which subdivides the concepts into two parts: the *terminological component* (TBox) and the *assertion component* (ABox).

Figure 10.2 shows a simple graphical representation of the Ontology; the TBox contains *classes*, which can be thought of as identifiers of sets (e.g., the class Wine is the identifier of the set of wines), and *relationships* among classes (e.g., the relationship HasColour between Wine and WineColour). The ABox contains *individuals*, which can be thought of as identifiers of set elements (e.g., the individual donnafugata nero d'avola identifies the Donnafugata Nero d'Avola wine, an element of the set of wines), and *relationship instances* among individuals (e.g., the relationship HasColour between donnafugata nero d'avola and sicily). Each individual is connected to the related class (the set it belongs to), by means of the InstanceOf relationship. In general, we refer to classes, individuals and relationships as *concepts*.

As an example, the TBox of Figure 10.2 specifies that a wine has a production region (which is part of a country), a colour, and an appellation; moreover, an appellation has a typical colour. The ABox states that the wine named "Donnafugata Nero d'Avola" is produced in Sicily (which is an italian region), is red, and belongs to Nero d'Avola wines, which are red wines.

Authoring ontologies is a challenging task, as methodologies and tools for Ontology definition are still an active research field [15]. In addition, a good ontological representation of the domain is fundamental to the system's performance. Thus, this is probably the hardest step in building the Domain Model.

For our prototype, we developed an Ontology about wine (the one represented in Figure 10.2 is a small subset of the actual Ontology.)

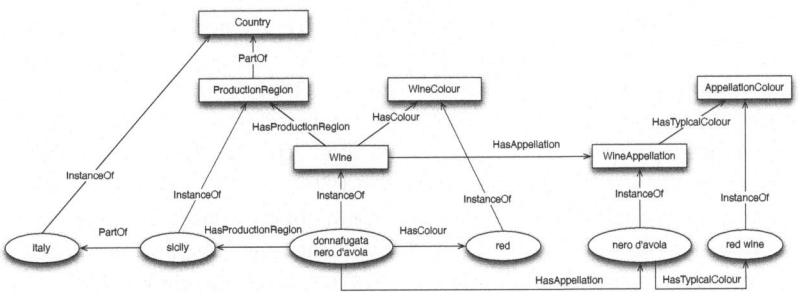

Fig. 10.2 A small Ontology about wine

10.2.2 The Lexical Database

Lexical Databases define the "meanings" of words, in a given language, by means of a graph structure. The most well-known Lexical Database is WordNet [6], the lexical database for the English language. WordNet organizes words into four sections: nouns, adjectives, adverbs, and verbs.

The WordNet's building block is the *synset*. A synset is a set of synonym lemmas (a *lemma* is the base form of a word), which carries a certain "meaning"; if a lemma is polysemic, it will appear in many synsets. Synsets and words are connected to each other by means of particular relationships, such as hypernymy (between a given noun synset and another noun synset carrying a more general meaning; e.g., "red" and "colour"), meronymy (whole/part relationship between two synsets; e.g., "italy" and "sicily"), derivationally related word (between an adjective or a verb, and the noun it derives from; e.g., "sicilian" and "sicily"), etc.

10.2.3 The Mapping Model

So far we defined the domain of interest (authoring the related Ontology), and the language we need to recognise (choosing the appropriate Lexical Database.) The third component we need is a mapping schema between Ontology and Lexical Database. As said before, such schema must take in account that words are, in general, polysemic, that synsets can lexicalize one or more concepts, and that concepts can be lexicalized by one or more synsets.

Figure 10.3 depicts a possible mapping between our exemplified Ontology about wine and the related Lexical Database synsets. Classes and individuals are mapped onto synsets that carry a similar meaning; usually class and individuals map onto nouns, adjectives, and adverbs. Class relationships are often associated with verbs (e.g., the relationship HasProductionRegion between Wine and ProductionRegion can be mapped onto verbs like "to grow"), but also to nouns (the noun "farming"); notice that some relationships cannot be mapped, since the related synset would carry a meaning that is too general (e.g., the relationship HasColour could be mapped onto the verb "to have".) Finally, notice that instance relationships "inherit" the same mappings from the related class relationships (for the sake of simplicity, the figure does not show such mappings.)

Figure 10.3 shows the ambiguities one faces when defining the mapping schema. The first ambiguity issue arises from polysemic lemmas; for example, the word "state" carries the meanings of "country" and "region of a country". The second ambiguity issue is due to the many-to-many nature of the concepts-synsets mapping; for example, both the appellation nero d'avola and the wine donnafugata nero d'avola are lexicalized by the same synset (a many-to-one relationship), while the region sicily is lexicalized by two synsets (a "one to many" relationship). Notice that a full many-to-many relationship is also present: the wine colour red and the typical appellation colour red wine are both mapped onto the same two synsets.

Given such ambiguities, a so-called *word sense disambiguation* procedure is needed, during the indexing phase. The word sense disambiguation procedure,

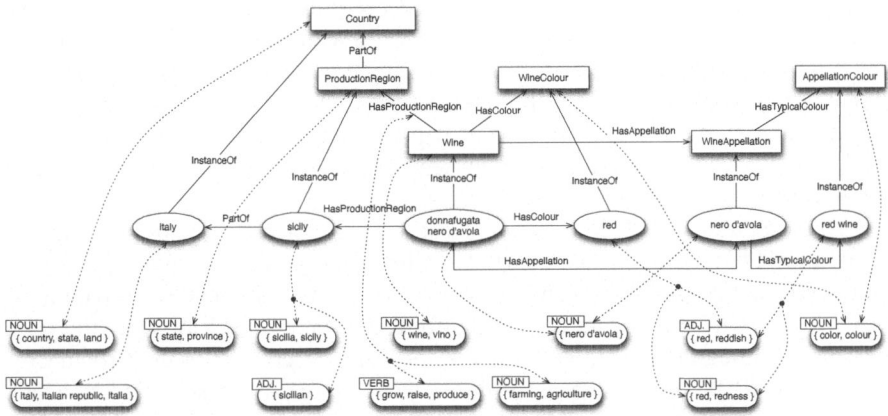

Fig. 10.3 Mappings between Ontology and Lexical Database

performed by the Information Extraction Engine, considers each word and tries to decide whether the word is meaningful, given the linguistic context and the concepts defined by the Ontology; if it is not, the word is not indexed. If the word is meaningful, the engine selects the right concept (class, individual, or relationship) among the possible candidates found in the mapping schema.

The Mapping Model will be presented in detail in Section 10.4.

10.3 Adding Linguistic Context Information to the Domain Model

To enrich the linguistic context, the Linguistic Context Extractor module associates each word with linguistic information extracted by NLP tools. In particular, the text undergoes the following analyses: morphological analysis, POS tagging, dependency parsing, and anaphora resolution (see Figure 10.4).

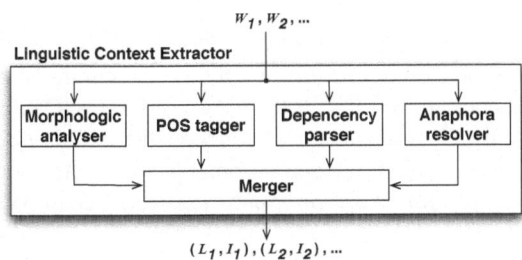

Fig. 10.4 The linguistic context extraction process

Words in texts appear usually in inflected forms (plural forms, verb conjugations, etc.). Morphological analysis associates each word with all the possible base forms, called *lemmas*, with the associated lexical category (or part-of-speech - POS); for example, the word "drunk" can be lemmatised as "drink", if used as a verb, or "drunk", if used an a noun. Then, POS tagging analysis selects the right part of speech (and, as a consequence, the right lemma), examining the linguistic context of the word.

As a morphologic & POS tagger analyser, we combined outputs of Freeling and the Stanford POS tagger. We adopted the Freeling's morphologic analyser [2], which returns all the possible base forms of a given word, with the related probabilities; then, we exploited the output of the Stanford POS tagger [28] to select the right base form, adding the related POS tag (see [9] for a description of the Brown Corpus standard POS's). Inside a proposition, words take part in so-called syntactic relationships; for example, looking at the sentence "They elected John president", word "they" is the subject of the action "elected ", while "John" is the object of the same action. In other words, one can define two relationships, such as subj(they, elected) and obj(John, elected). A dependency parser analyses a sentence and extracts the list of syntactic relationships among words. As a dependency parser, we relied on the Stanford parser [16, 5].

Notice that, in the example above, the subject is a pronoun, which refers to the actual subject (to find in a preceding sentence). This kind of reference (pronominal anaphora) can be addressed by means of specific tools. The anaphora resolution tool links pronouns with the nouns they refer to; our system treats such a link as a special kind of syntactic relationship. JavaRAP [18] is the tool we adopted for this task.

Finally, each word is associated with: the list of synsets it belongs to, the relationships connected to such synsets, the list of concepts it could be mapped onto, and the relationships connected to such concepts.

Thus, the plain text, which is composed of a sequence of words, is transformed into a sequence of lemmas L_t, each of them connected with its *lemma information items* I_t. Such information represents all pieces of knowledge we are able to extract from text, Lexical Database, Ontology, and mapping between them, about lemma L_t. In particular:

$$I_t = (W_t, POS_t, \mathbf{V}_t^{nlp}, \mathbf{S}_t, \mathbf{V}_t^{syn}, \mathbf{C}_t, \mathbf{V}_t^{onto}) \tag{10.1}$$

where W_t is the word whose base-form is the lemma L_t, POS_t is its part of speech, $\mathbf{V}_t^{nlp}$ is a vector of syntactic relationships involving the lemma L_t, $\mathbf{S}_t$ is the vector of synsets the lemma L_t belongs to, $\mathbf{V}_t^{syn}$ is the vector of Lexical Database relationships involving synsets in $\mathbf{S}_t$, $\mathbf{C}_t$ is the vector of concepts the lemma L_t could be mapped onto, $\mathbf{V}_t^{onto}$ is the vector of ontological relationships involving concepts in $\mathbf{C}_t$.

Finally, the *linguistic context information* is the sequence of pairs (L_t, I_t):

$$\mathbf{I} = <(L_0, I_0), (L_1, I_1), \ldots, (L_t, I_t), \ldots, (L_T, I_T)> \tag{10.2}$$

10.4 Defining and Training the Mapping Model

Words (and, after the Linguistic Context Extraction phase, lemmas and lemma information items) represent information the system can directly *observe* by reading the documents. Concepts stored in the Domain Ontology are *not observable*, however, since the information they represent is implicit in the documents.

Concepts do not appear in random order: some sequences of concepts will be more likely to appear than others. So, the outcome of the decision process about a given word depends on the decisions already taken about the preceding words.

Summing up, it seems that a viable model to represent word-concept mappings could be an Hidden Markov Model (HMM) [22], as it permits to deal with observable and non-observable variables; moreover, it permits to implement the disambiguation procedure we need during the indexing phase.

HMM describes the probability distribution of some observable variable, given a hidden variable: the so-called emission probability distribution *p(observable|hidden)*. Moreover, the model describes how the value assumed by the observable variable in the past affects the current hidden variable: the so-called transition probability distribution *p(hidden(now)|hidden(last_step))*.

We actually implemented a slightly more complex HMM-based model, as we had to map concepts onto synsets and synsets onto lemmas (see Figure 10.3); lemmas represent the observable variables, while synsets and concepts are both hidden. Moreover, our model considers two steps in the past (i.e. is a second-order HMM) instead of the usual single step as in regular HMMs; this choice complicates the model but permits to increase the accuracy of the prediction on the current hidden variable, as the model takes in account more information about the past.

An extended HMM *decoding* procedure (see Section 10.5.1) permits, given a sequence of lemmas, to calculate the most probable sequence of concepts and synsets. In other words, such procedure is able to extract concepts from the text.

In the following we define our model, define transition probability distributions (and show how they permit to take in account the linguistic context), define emission probability distributions, and present the training procedure.

10.4.1 The HMM-Based Model

Our model, based on second-order HMMs, is defined by $(\Omega_C, \Omega_S, \Omega_L, A^1, A, B, D, \pi)$. The finite sets Ω define:

$$\Omega_C = \{c_1, \dots, c_{N_c}\}: \text{set of possible hidden concepts}$$
$$\Omega_S = \{s_1, \dots, s_{N_s}\}: \text{set of possible hidden synsets}$$
$$\Omega_L = \{l_1, \dots, l_{N_l}\}: \text{set of possible observed lemmas}$$

Table 10.1 Model assumptions

Second-order *Markov assumption*	Choosing the concept to associate with the *t*-th lemma only depends on the last two concepts
Stationary assumption	The transition probability is independent of *t*; the lemma emission probability and the synset emission probability are independent of *t*, too
Output independence assumption	The value of S_t, the hidden intermediate variable at time *t*, only depends on the value of the hidden state at the same time *t*; the value of L_t, the observable variable at time *t*, only depends on the value of the hidden variables at the same time *t*

while, denoting with random variables C_t the concept at position *i* (hidden state variable), with S_t the synset at position *i* (hidden intermediate variable), and with L_t the lemma observed at position *i* (output variable), the rest of the model defines discrete probability distributions (see Table 10.1 for assumptions the model makes):

$A^1 = \{a^1_{i,j}\}$ s.t. $a^1_{i,j} = p^\diamond(C_1 = c_j | C_0 = c_i, \mathbf{I})$: first transition probabilities

$A = \{a_{k,i,j}\}$ s.t. $a_{k,i,j} = p^\diamond(C_t = c_j | C_{t-1} = c_i, C_{t-2} = c_k, \mathbf{I}), \forall t > 1$: transition probs

$B = \{b_i\}$ s.t. $b_i(k) = p(S_t = s_k | C_t = c_i)$: synset emission probabilities

$D = \{d_{i,j}\}$ s.t. $d_{i,j}(k) = p(L_t = l_k | C_t = c_i, S_t = s_j)$: lemma emission probabilities

$\pi = \{\pi_i\}$ s.t. $\pi_i = p(C_0 = c_i)$: initial (*i* = 0) concept distribution probabilities

Figure 10.5 shows an unrolled view view of the model, as well as the probability distributions we introduced.

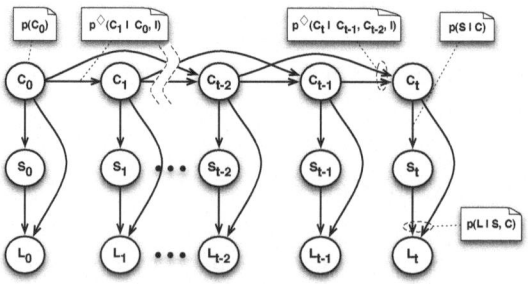

Fig. 10.5 An unrolled view of our HMM-based model

Variables C_t take values from the set of concepts (classes, individuals, relationships) defined by the Domain Ontology. We added to this set the following special concepts: out of context and out of scope. The first one is mapped whenever a lemma, evaluated in its linguistic context, is not meaningful for the domain (e.g., considering the sentence "My car is red", the word "red" should not be mapped neither onto the red concept, nor onto the red wine concept.) The second one is mapped onto words that are not part of the Domain Model—words not associated with any concept of the Domain Ontology (e.g., the word "car".) These two special concepts are referred to as *null concepts*.

Variables S_t take values from the set of synsets associated with at least one concept of the Domain Ontology. We add to this set the following special synsets: out of context and out of scope, with meanings similar to the ones defined for the special concepts. These two special synsets are referred to as *null synsets*.

Variables L_t take values from the set of lemmas present in at least one of the aforementioned synsets.

Transition probability distributions $p^\circ(C_t|C_{t-1},\mathbf{I})$ and $p^\circ(C_t|C_{t-1},C_{t-2},\mathbf{I})$ are actually non-normalised, approximated distributions; moreover, notice that both distributions depend on linguistic context information $\mathbf{I}$. In the following we explain why and how we incorporated linguistic context information into transition probability distributions, and present the model we original adopted to approximate them.

10.4.2 Defining Transition Probability Distributions: The MaxEnt Model

In order to take in account the linguistic context information $\mathbf{I}$, we added it to transition probability distributions. This way, during the disambiguation procedure, the choice of the concept to map onto the current word depends not only on the two last concepts, but also on the information stored in $\mathbf{I}$ (the presence of a particular word, morphologic or syntactic information, etc.)

Conversely, the emission distribution probabilities do not incorporate $\mathbf{I}$; this way, they remain small and easy to calculate.

Adding information to the transition probability distributions could lead to huge memory occupations. So, our model approximates these distributions by means of *functions*; this way, once selected a feasible function model, we only need to store a set of parameters Π. This approach permits to include $\mathbf{I}$ in a simple way, while retaining the useful characteristics of HMMs.

As a function model we chose the Maximum Entropy (MaxEnt) model [12, 23, 19], which is widely used to represent probability distributions, and can easily be adapted to include heterogeneous information. In Tabletab:mx some definitions about our MaxEnt model are summarised.

10.4.3 Defining Emission Probability Distributions: Probabilistic Mapping

The emission probabilities $p(L|S,C)$ and $p(S|C)$ encode the Ontology - Lexical Database mappings. The example of Figure 10.6 shows the probability values

Table 10.2 MaxEnt model definitions

First-transition probability distribution	$a^1_{i,j}(\mathbf{I}) = \frac{\exp(\sum_i^m \gamma_i f_i(C_t,C_{t-1},\mathbf{I}))}{\exp(\sum_i^h \psi_i f_i(C_{t-1},\mathbf{I}))}, t = 1$
Transition probability distribution	$a_{k,i,j}(\mathbf{I}) = \frac{\exp(\sum_i^n \lambda_i f_i(C_t,C_{t-1},C_{t-2},\mathbf{I}))}{\exp(\sum_i^m \gamma_i f_i(C_{t-1},C_{t-2},\mathbf{I}))}, \forall t > 1$
Features	Pattern indicator functions: return 1 whenever a given pattern is found in its arguments, otherwise return 0
$\lambda_i, \gamma_i, \psi_i$	Constants weighting the "importance" of each feature for the calculation of the distribution values
Feature templates	A model that describes a pattern structure. During the training process, the system generates all the patterns that meet the structure; these patterns are called the *feature instances* of the template
The feature selection problem	The feature instances undergo a filtering process that aims at reducing their number, while retaining the most informative ones; the remaining instances are then stored. There are several methods for addressing the feature selection problem; see [24, 4]

associated with a fragment of the mapping depicted in Figure 10.3. Notice that, for example, the probability for synset {color, colour} to emit the word "color" (or, in other words, the probability for the word "color" to carry the meaning intended by the synset {color, colour}) depends on the concept we are considering (`WineColour` or `AppellationColour`).

Notice that the lemma emission probability distribution depends on both synsets and concepts; this is meant to capture the fact that concepts, although lexicalized by the same synset, could not have, in principle, the same lemma distribution.

Figure 10.6 depicts three Ontology-Lexical Database mapping types. The first mapping type is "one to one": concept and synset are univocally connected each other; this is of course the simplest mapping type. The second mapping type is "one to many": a concept is connected to many synsets; this is often used to lexicalize a concept by means of many synset types (e.g., a noun synset and a adjective synset.) The third mapping type is many-to-one: many concepts are connected to the same synset; this is the most interesting type because, starting from a word (e.g., "color"), the model must be able to select the most probable concept; it is easy to recognize that the probability distributions $p(L = \text{"color"}|S = \{\text{color, colour}\}, C)$, in conjunction with prior probability distribution $p(C)$, carry enough information to allow for the selection of the most probable concept.

Now we have all the distributions required by our HMM-based model: the two aforementioned MaxEnt models encodes non-normalised transition distributions $p^\circ(C_t|C_{t-1}, \mathbf{I})$ and $p^\circ(C_t|C_{t-1}, c_{t-2}, \mathbf{I})$, while the mapping model contains both the concept prior probability distribution $p(C)$ and the two emission probability distributions $p(L|S, C)$ and $p(S|C)$.

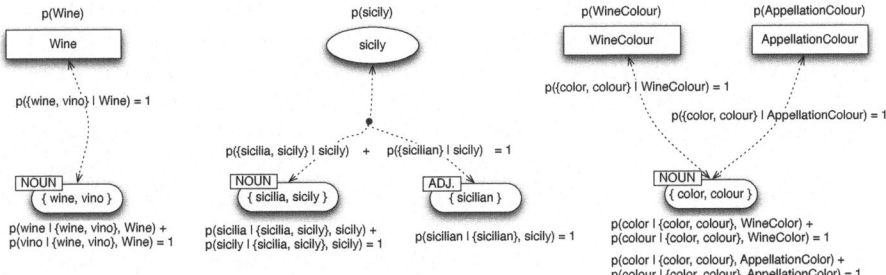

Fig. 10.6 Ontology-Lexical Database mapping types

10.4.4 Training the Model

The training process is divided into several phases (see Figure 10.1). First of all, a specific set of documents in the Document Repository enter the Linguistic Context Extractor that retrieves morphologic and syntactic information. Then, the Domain Expert associates each word with the right concept, creating two sub-sets: the *training set* and the *test set*. If a given lemma, evaluated in its linguistic context, is not meaningful for the domain, the Domain Expert associates it to the out of context concept. Lemmas than are not part of the Domain Model, are associated to the out of scope concept. The Training Procedure, relying on both the Ontology and the Lexical Database, analyses the training set documents, and generates the statistical model. Finally, the Mapping Model is checked against the test set.

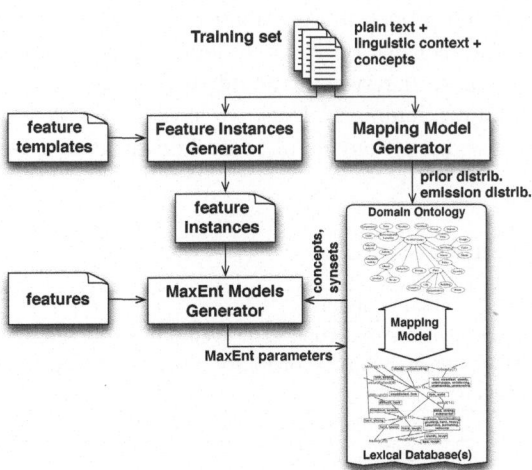

Fig. 10.7 The Training Procedure

As an example, consider the sentence "The bottle with red label contains Barolo, a red wine". The Domain Expert associates words and concepts, producing "The bottle/WineBottle with red label/WineLabel contains Barolo/barolo wine, a red/red wine wine/Wine". Notice that the adjective "red" just before the noun "label" is not tagged, as its meaning (the colour of a label) is not considered useful for the domain (i.e., there is no related concept into the Domain Ontology). Instead, just before the noun "wine", the adjective "red" refers to a wine colour, and therefore it is tagged with the related concept.

Figure 10.7 shows the content of the Training Procedure block. The Mapping Model Generator calculates the prior and emission probability distributions; The Feature Instances Generator reads the set of feature templates and generates the set of feature instances; finally, the MaxEnt Models Generator calculates the MaxEnt models' weights.

Notice that, given the fact that emission probability distributions $p(L|S,C)$ and $p(S|C)$ are not huge (as lemmas are usually mapped onto a few synsets and concepts), and that transition probability distributions $p^\circ(C_t|C_{t-1},\mathbf{I})$ and $p^\circ(C_t|C_{t-1},c_{t-2},\mathbf{I})$ are approximated using a model, we expect that a small training set could still generate an HMM with good performances (see Section 10.8).

The first phase of the Training Procedure calculates prior and emission probability distributions by means of their definitions:

$$p(C) = \frac{c(C)}{\sum_{C'} c(C')}; \quad p(S|C) = \frac{c(S,C)}{c(C)}; \quad p(L|S,C) = \frac{c(L,S,C)}{c(S,C)} \tag{10.3}$$

where the function $c(\ldots)$ counts the number of occurrences.

Notice that, at the end of the calculations, only *known lemmas* L's (i.e., lemmas that appear in the training set) will have a $p(L|S,C) > 0$ associated; *unknown lemmas* will not take part in the probability distribution, even if they belong to the same synset of some L. The fact that only known lemma gets a distribution probability poses a problem, because the system should be able to index and search for all the synonyms of a known lemma L. Thus, we have to complete the model, adding unknown lemmas that are synonyms of some known lemma (see Table 10.3 for details).

Table 10.3 Adding unknown lemmas

$\{L'\}$	The set of unknown lemmas
$\{S_{L'}\}$	The set of synsets that contain a given lemma L' (i.e., the set of synsets whose probability distributions must be changed)
$\{C'\}$	The set of concepts mapped onto synsets in $\{S_{L'}\}$
$p_{L'} = p(L'|S_{L'},C') = \frac{\min_{L,S,C} p(L|S,C)}{|\{L'\}|}, \forall (S_{L'},L',C')$ probability value assigned to each lemma L'	
$p^{new}(L|S_{L'},C') = p^{old}(L|S_{L'},C') \cdot (1 - p_{L'} \cdot |\{L'\}|), \forall (S_{L'},C')$ probability distribution is renormalized, as we added new probability mass	

Table 10.4 The GIS algorithm for unconditioned models $p(C_t, \mathbf{X})$

$(C_t, \mathbf{X})$	Elements of the training set
$f_i(C_t, \mathbf{X})$	Features
$\mathbf{X} = (C_{t-1}, C_{t-2}, \mathbf{I})$	For the first run of the algorithm
$\{\lambda_i\}$	Weights the first run of the algorithm calculates
$\mathbf{X} = (C_{t-1}, \mathbf{I})$	For the second run of the algorithm
$\{\gamma_i\}$	Weights the secont run of the algorithm calculates
$\mathbf{X} = (\mathbf{I})$	For the third run of the algorithm
$\{\psi_i\}$	Weights the third run of the algorithm calculates
$E_{observed}[f_i(C_t, \mathbf{X})] = E_{expected}[f_i(C_t, \mathbf{X})], \forall i$	
Condition to which the GIS converges	

Then, the Training Procedure calculates the transition probability distributions; the Generalized Iterative Scaling (GIS) algorithm for unconditioned models [10, 19] is used to calculate the parameters Π of the MaxEnt models that approximate such distributions (see Table 10.4 for details).

10.4.5 Comparing Our Model with Alternative Stochastic Approaches

The model we defined aims at combining characteristics of HMMs and Max-Ent. Maximum Entropy Markov Models (MEMM) and Conditional Random Fields (CRF) are two popular models that permit to reach the same result. Although powerful, these models are quite complex to train; in contrast, our approach permits to generates models with good performances, even with small training sets.

MEMMs [20] combine Maximum Entropy models and HMM, so that it is possible to decode (by means of a slightly modified Viterbi procedure) sequences of words. MEMMs replace transition and emission probability distributions with a single probability distribution like: $p(C_t|C_{t-1}, C_{t-2}, S_t, \mathbf{I})$, modelled by means of a MaxEnt model. This distribution is bigger and more difficult to approximate than the ones defined in our model.

CRFs [17] combine Maximum Entropy models and HMM and make the constant transition probabilities into arbitrary functions that vary across the positions in the sequence of hidden states, depending on the input sequence. CRFs outperform both MEMMs and HMMs on a number of real-world sequence labeling tasks [30]. Training CRFs is quite difficult, however, as it is not possible to analytically determine the MaxEnt parameter values. Instead, they must be identified using an iterative technique such as iterative scaling or gradient-based methods.

10.5 Indexing Documents: The Disambiguation Procedure

As Figure 10.1 shows, the indexing process exploits the Domain Model and updates the Conceptual Index. In particular, the Information Extraction Engine performs the following steps (see Fig 10.8): the Linguistic Context Extractor generates the linguistic context information; MaxEnt models calculates the transition probability distributions; a disambiguation procedure (the extended Viterbi algorithm) selects the most probable concept-word association; finally, the Concept Frequency Analyser builds up the Conceptual Index.

Continuing the example, assume the system be trained to recognise concepts in the following set {WineBottle, WineLabel, barolo wine, red wine, Wine}. If the document set is composed of D_1="Barolo is a small village where good wine is produced" and D_2="Barolo is a red wine", the system extracts the following sets of concepts C_1={Wine} [1] and C_2={barolo wine, red wine, Wine}. The Concept Frequency Analyser calculates the weight $w_{c,d}$ associated to each concepts c, for each document d; thus, the following vectors are generated: V_1=[0, 0, 0, 0, w_{Wine,D_1}] and V_2=[0, 0, $w_{barolowine,D_2}$, $w_{redwine,D_2}$, w_{Wine,D_2}].

In the following, the indexing process is discussed in detail.

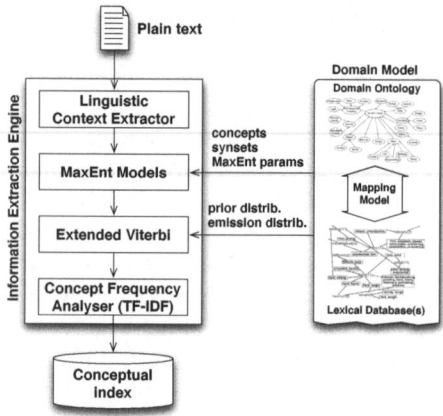

Fig. 10.8 The Information Extraction Engine

10.5.1 Disambiguating Words: Extended Viterbi

The Viterbi algorithm permits to find the most likely sequence of hidden variables that results in a sequence of observed variables (see Table 10.5 for details). Actually, Viterbi only works for first order HMMs but an extended algorithm exists for second

---

[1] Notice that in D_1 "Barolo" refers to a village and should not be considered.

order models [11], which is simple and fast enough for our experiments; temporal complexity is $O(N \cdot |C|^3)$, which could be good enough for several cases. However we are considering to test more sophisticated and efficient algorithms, like the A* decoding [13, 14].

Table 10.5 Details about decoding with the Viterbi algorithm

$p(L|C) = \sum_S p(L|S,C) \cdot p(S|C)$	The emission probability distribution needed by Viterbi
$S_t = \text{argmax}_S \frac{p(L_t|S,C_t) \cdot p(S|C_t)}{p(L_t|C_t)} ; \forall t$	The most probable synset sequence

In Section 10.4.2 we defined non-normalised transition distributions (i.e., scaled by a constant positive factor.) Such distributions do not break the algorithm, as they still lead to calculate the most likely concept sequence. The only shortcoming is that the algorithm is no longer able to calculate the true probability of that sequence, but our model does not make use of it.

Moreover, our transition probability distributions, generated by the MaxEnt models, depend on linguistic context information **I** but, once again, Viterbi still works as **I**, once selected the sentence to decode, is a constant.

10.5.2 Document Representation: Concept Frequency Analyser

The Conceptual Index stores a document $\mathbf{d}_j$ as a *vector of concepts*, in a way similar to the common Vector Space Model (VSM) [25] (actually, the index stores vectors of triples $< (L_0, S_0, C_0), (L_1, S_1, C_1), \ldots, (L_T, S_T, C_T) >$, to preserve the concept/synset/lemma mapping).

Each weight $w_{i,j}$ into the vector refers to the i-th concept of the Domain Ontology; the vector size is given by the amount of concepts into the Domain Ontology.

$$\mathbf{d}_j = (w_{1,j}, w_{2,j}, \ldots) \tag{10.4}$$

For example, consider the following two documents: (1) "(…) red bottle (…) ruby wine (…)"; (2) "(…) red wine (…) ruby liqueur (…)", and assume to map "red" and "ruby" onto `red wine`, while "wine" and "liqueur" are mapped onto `Wine`. Figure 10.9 shows the resulting vector space, after indexing takes place; for the sake of simplicity, for now assume that weights represent concept occurrences. Document (1) is represented by [`WineBottle`, `Wine`, `red wine`]=[1, 1, 1], while document (2) is represented by [`WineBottle`, `Wine`, `red wine`]=[0, 2, 2].

Notice that the word "red" of document (1) is not mapped onto any concept, as it does not mean a colour of wine. Moreover, notice that both words "ruby" and "red" are mapped onto the same concept, and contribute to its occurrence count; this effectively captures the fact that these two words represent, in their linguistic context, the same concept. The same is true for "wine" and "liqueur".

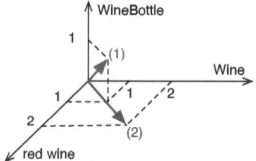

Fig. 10.9 Conceptual space model for two example documents

The Concept Frequency Analyser calculates weight following the TF-IDF schema:

$$w_{i,j} = tf_{i,j} \cdot idf_i; \quad tf_{i,j} = \frac{f_{i,j}}{\sum_i f_{i,j}}; \quad idf_i = \log \frac{N+1}{df_i} \qquad (10.5)$$

where $w_{i,j}$ is the weight of the concept i-th into the document j-th; $tf_{i,j}$ defines the *frequency* of concept i-th into the document j-th; idf_i represents the rarity of concept i-th inside the collection; $f_{i,j}$ is the number of occurrences of concept i-th into the document j-th; N represents the number of documents of the collection; df_i specifies how many documents contain at least one occurrence of concept i-th.

Such weights are then stored in the Conceptual Index. Notice that the index also contains the lemma-synset-concept mappings, the linguistic context information, and the full text of the indexed documents.

10.6 Querying Documents and the Domain Ontology

Queries can be issued in several ways. In particular, we consider the following typologies: Keyword-based and Natural Language-based. The former queries are composed of a sequence of words, eventually connected by boolean logic operators. This kind of query is useful for documental search, where the user specifies the concepts to find. Translation from words to concepts can be done in many ways, as we explain in the following. The latter queries are issued as a question or description, written in natural language. These queries can be used for both documental and conceptual search, where users express a large variety of different requests.

10.6.1 Keyword-Based Queries

Keyword-based queries are composed of a sequence of words; these words can be considered as connected by either AND or OR boolean logic operators. In the first case, only documents containing all the concepts are returned; in the second case, documents containing at least on concept are returned.

The query engine tries to select, for each word W_i in the sequence, the most probable concept C_i to associate (see Table 10.6 for details); the result is the concept sequence $\tilde{C}$.

Table 10.6 Most probable concept

In order to associate concepts to words, the approach we described in Section 10.5 doesn't work, as the given word sequence cannot be treated as a natural language sentence. So, we exploit prior probabilities, stored into the Domain Model, to guess the most probable concept to map. $p(C\|W) = \sum_L \frac{\sum_S p(L\|S,C) \cdot p(S\|C) \cdot p(C)}{\sum_{C'} \sum_S p(L\|S,C') \cdot p(S\|C') \cdot p(C')} \cdot p(L\|W)$; Single word $\tilde{C} = \text{argmax}_C \, p(\mathbf{C}\|\mathbf{W}) = \text{argmax}_C \prod_j p(C_j\|W_j)$; Multiple words

Queries are represented as vectors $\mathbf{q}_k = (w_{1,k}, w_{2,k}, \ldots)$, where concept $\tilde{C}_i$ is represented by weight $w_{i,k}$:

$$w_{i,k} = (0.5 + \frac{0.5 \cdot tf_{i,k}}{\max_l tf_{l,k}}) \cdot idf_i \qquad (10.6)$$

Similarity between document j-th $\mathbf{d}_j$ and query k-th $\mathbf{q}_k$ is given by the following formula, used to calculate and rank the list of documents to return:

$$sim(\mathbf{d}_j, \mathbf{q}_k) = \frac{\mathbf{d}_j \cdot \mathbf{q}_k}{|\mathbf{d}_j| \cdot |\mathbf{q}_k|} = \frac{\sum_i^N w_{i,k} \cdot w_{i,j}}{\sqrt{\sum_i^N w_{i,k}^2} \cdot \sqrt{\sum_i^N w_{i,j}^2}} \qquad (10.7)$$

In our model, however, one could exploit the structure of the Ontology, searching, for example, for sub-classes, part-of classes, and individuals of a given class. In other words, searching for "Italy", one could expect the system to find "Piemonte", as it is a region of Italy.

Starting from a concept, the system adds *derived concepts*, according to the rules:

- Searching for a class → add sub-classes, part-of classes, individuals
- Searching for an individual → add part-of individuals

From the user's point of view, and for the purpose of the query above, the meanings of "Piemonte" and "Italy" are overlapping, since they just represent two ways to lexicalize the idea of "italian territory". Thus, one could think to *merge* the concepts italy and piemonte (and the related axes) into the new concept {italy, piemonte}.

Creating this new concept implies the calculation of the related weight. Given a concept $\tilde{C}$ with weight w_c, and m derived concepts $\{C_d\}$ with weight $\{w_d\}$, the new merged concept $\tilde{C}'$ has weight w'_c, which is calculated as a quadratic mean:

$$w'_c = \sqrt{\frac{w_c^2 + \sum_d^m w_d^2}{m+1}} \qquad (10.8)$$

Thus, the expanded concept $\tilde{C}$ is replaced, into the query vector, by the new concept $\tilde{C}'$ with weight w'_c; expanding each concept of the query generates the new sequence $\tilde{C}'$. Notice that the same process is performed, on the fly, on each document vector that the system consider during the search process.

The Conceptual Index is searched for documents containing all the concepts (for AND), or at least one concept (for OR) in $\tilde{C}'$; the resulting list is ranked using the aforementioned similarity measure, and shown to the user.

10.6.2 Natural Language-Based Queries

The system permits two different kinds of Natural Language-based queries: Disambiguated Word queries and Ontological queries. The first one treats the issued phrase as if it was a small document, and applies the same indexing technique we presented in Section 10.5. The resulting set of concepts is then transformed into a vector q_k as explained in the previous section. For example, issuing Q="red wine", the system could disambiguate the word "red", mapping it onto red wine (it refers to a wine appellation, not to a specific red wine), while in Q="red Donnafugata Nero d'Avola" the word "red", should be mapped onto red (see Ontology in Figure 10.3).

Ontological queries are peculiar, as they permit to search the Domain Ontology for a concept that meet a description or question. The methodology we adopted permits to describe the structure of the requests/questions that the system should recognize, as well as the queries to perform.

Notice that, as this description captures the *structure* of the phrase, not the actual meaning, it is parametric in nature; for example, the sentences "what are the colours of wines?" and "what are the tastes of wines?" share the same structure, while "colors" and "tastes" are parameter instances.

The process is depicted in Figure 10.10 and is divided it two phases. During the first, off-line phase, the Domain Expert prepares the model describing the structure of allowable phrases and the associated query templates. During the second phase, the system recognises the phrase, extracts parameters, prepares and executes the related query.

As an example of Ontological query, assume that the Domain Ontology contains information about the facts that wines have colours, and that Barolo's colour is red. Moreover, assume that the Domain Expert prepared the following phrase template: T=<"what", wine-attribute-name, wine-instance>. If the user issues Q="what is the colour of Barolo?", the system extracts C_Q=[WineColor, barolo wine], finds the phrase template T, fills the placeholders: <"what", WineColor, barolo wine>, and executes a parametric SQL query S_T associated by the Domain Expert, to the template T.

Notice that the phrase template T is actually able to capture several requests, e.g. the colour of wine, the price of wine, etc. In other words, requests with similar structure (in terms of sequence and type of concepts) can be captured by means of the same template. In the following, the definition and the recognition of Ontological queries are described.

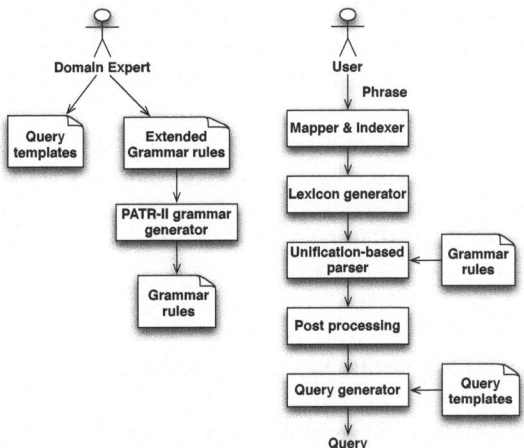

Fig. 10.10 Ontological queries: Phrase description (left) and recognition (right)

10.6.2.1 Ontological Queries: Phrase Description

We adopted the feature-based grammar formalism for the description of phrase templates. In particular, we adopted the PATR-II language [26], implemented by the PC-PATR unification-based [27] syntactic parser.

The feature-based grammars permits to associate attributes with the terminal symbols; the rules can then predicate on these attributes, creating complex conditions that affects the evaluation of the rules. As an example, the following rule:

```
A = b c
    <b x> = <c x>
    <b z> = w
```

only applies when the attribute x of symbol b equals the same attribute of c AND attribute z of b has value w.

Such description is particularly useful in our case, as the attributes can be used to describe and predicate on the morphologic and syntactic information extracted by the linguistic tools, and associated to the words. In particular, we are interested in describing the syntactic relationships among words; for example:

```
A = b c
    <b rel> = <c rel>
    <b rel> = subj
```

but such description poses several issue, the most problematic being the fact that it is not possible to specify the direction of the relationship. Thus, we decided to define an extended syntax ables to provide the semantics we need. The following example:

```
@REQ2 = <<Class>> Class_1#
        @<Class_1 relprep_of> = <Class relprep_of>#
```

describes a phrase composed of two words, both mapped on classes (non mapped words are ignored), where the first one represent the attribute to search for; the condition states that there must be a `relprep_of` relationship from the second word to the first one.

Such extended rules are then translated into the standard PATR-II formalist. The output of the parser will be parsed by a post-processing analyser, in order to enforce the piece of semantics lost during the translation.

Associated with the rules, SQL templates describe the queries to execute whenever a given tree is found.

10.6.2.2 Ontological Queries: Phrase Recognition

The Information Extraction Engine indexes the sentence issued by the user, as it does with sentences in documents; the output is composed by the sequence of concepts and the linguistic context information: $(< C_0, C_1, \ldots, >, \mathbf{I})$. The Lexicon generator transforms each concept in a lexical statements, associating it with information $\mathbf{I}$ (e.g., lemma the concept is mapped onto, concept type, POS tag, relationships connecting its lemma to other lemmas, etc.) Such statements form the PATR-II *grammar lexicon*. Notice that, in contrast with the usual usage of grammars, the lexicon is generated on-the-fly and contains only the lexical statements derived from the phrase being analysed.

The Unification-based parser loads lexicon and rules, analyses the sequence of concepts, and generates a set of parse trees. The Post processing module discards trees that do not satisfy the conditions stated into our extended-syntax rules. The retained tree enters the Query generator, which searches the related SQL template, fills the template placeholders with the content of the tree nodes, and executes the query. The result is the set of concepts that answer the user's request.

10.7 Extending the Domain Ontology

The `ArtDeco` Web Portal provides a way to extend the Domain Ontology, adding new concepts to the Domain Model. In particular, the system is able to add new individuals and relationship instances to the A-Box.

The Ontology Extender relies on Drools, a production-rules engine. Such rules predicates on information $\mathbf{I}$ collected while indexing the text, to infer that a given word could be a new individual of some class, or a new instance of some relationship. Rules cannot guarantee that the newly discovered concept is correct, however. Therefore, the system tags such new concepts as "guessed" to distinguish them from the concepts defined by the Domain Expert.

Figure 10.11 depicts a simple example that could be captured by means of a Drools' rule. The sentence is "Piemonte produces wines"; the dependency parser found that "Piemonte" is the sentence's subject and "wines" is the direct object; moreover, assuming that the Domain Ontology contains information about the fact that wines have production regions, the Information Extraction Engine mapped "wines" onto Wine and "produces" onto HasProductionRegion. A simple rule could state that:

```
IF subject_of(W₁, "to produce") ∧ direct_object_of(W₂, "to produce") ∧
   mapped_on_class(W₂, Wine) ∧ mapped_on_relationship("to produce", R) ∧
   connected_by_relationship(R, Wine, C)
THEN
   W₁ is a new individual of  C
```

and the system could infer that W_1="Piemonte" should generate a new individual piemonte belonging to the class $C=$ Region. See table 10.7 for details.

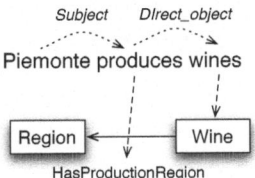

Fig. 10.11 Extending Ontology: an example

Table 10.7 Updating the distributions

As the system discovers new concepts, the probability distributions and the Conceptual Index must be updated	
$\{C\}$	The old set of concepts
$\{C_n\}$	The set of newly discovered concepts
$p(C_n) = \frac{1}{|\{C\}|+|\{C_n\}|}, \forall C_n \in \{C_n\}$ New prior probability values	
$p^{new}(C) = p^{old}(C) \cdot \frac{|\{C\}|}{|\{C\}|+|\{C_n\}|}, \forall C \in \{C\}$ Updated prior probabilities of old concepts	
L_n	the lemma mapped onto C_n
$\{S''\}$	Set of synsets containing L_n
$p(S''|C_n) = 1/|\{S''\}|; \text{if } \{S''\} \neq \emptyset$ $p(S_n|C_n) = 1; \text{if } \{S''\} = \emptyset$, a new synset S_n is added	
$p(L_n|S'', C_n) = 1/|\{S''\}|; \text{if } \{S''\} \neq \emptyset$ $p(L_n|S_n, C_n) = 1; \text{if } \{S''\} = \emptyset$	

10.8 Results

We conducted preliminary tests on our model, using a Domain Ontology about "wine" composed of 70 classes, 45 relationships, and 449 individuals. The Ontology concepts was mapped onto 270 WordNet synsets.

The Document Repository was populated with several HTML pages, gathered from RSS feeds. The topic of some of these RSS's was "wine", and thus they provided in-topic documents; other RSS's were intentionally chosen to provide out-of-topic documents, so that we could check the ability of the system to discard words whenever they carry meanings that do not match the Domain Ontology. We collected 124 HTML documents, gathering 70504 words. Table 10.8 shows how these documents are distributed between training set and test set.

The training set contained 346 concepts (17 classes and 329 individuals), while the test set contained 218 concepts (16 classes and 202 individuals).

Notice that the training set is quite small: only the 4.54% of the entire word collection is devoted to the training procedure. This choice is unusual, as stochastic models typically require huge training sets to give good results. However, as mentioned in Section 10.4.4, one of our goal was to investigate the ability of the model to provide good performances even with small training sets.

Table 10.8 Test set & training set

Training set		Test set	
# In-topic docs	# Out-of-topic docs	# In-topic docs	# Out-of-topic docs
1 (2074 words)	2 (1129 words)	73 (61527 words)	48 (5774 words)

We defined the *degree of ambiguity* $d(W)$ of a word as the number of concepts mappable on the word (notice that if W is not part of the domain, it is mapped only to the out of scope concept, and therefore $d(W) = 1$; for all other words, $d(W) \geq 2$, as they are mapped onto at least one useful concept, and also onto the out of context concept.)

Table 10.9 shows that, in our training set, the majority of words had $d(W)$ between 1 and 2. However, if we ignore words that are not ambiguous, words with $d(W) = 2$ were 88% of the training set. Table 10.9 also shows that the vast majority of the test set was composed of words mapped onto out of scope; this is due to the fact that the test set contained several out-of-topic documents, and, in general, contained "real life" documents, while the training document was edited in order to eliminate useless sentences. Ignoring unambiguous words, about 88% ot the test set contained words whose $d(W)$ is 2 or 3. Finally, notice that the test set contained a subset (about 63%) of the concepts learned by the system using the training set.

As a performance index, we defined the accuracy AC (the probability for a given word to be mapped onto the right concept):

$$AC = \frac{\sum_i^{N_c} a_i}{N_w} \qquad (10.9)$$

where a_i represents the number of words of the test set correctly mapped onto the concept i-th, while N_w in the number of words in the test set s.t. $d(W) > 1$, and N_c is the number of concepts of the test set (not including out of scope). Notice that we

Table 10.9 Degree of ambiguity for words in training and test sets

Training set									
d	1	2	3	4	5	6	7	8	tot
# unique words	531	451	32	8	9	4	3	3	1041
% unique words	51.0	43.3	3.1	0.7	0.8	0.4	0.3	0.3	100
% unique words with d>1		88.0	6.0	1.5	1.7	0.8	0.6	0.6	100
Test set									
d	1	2	3	4	5	6	7	8	tot
# unique words	7178	126	65	16	5	3	2	-	7359
% unique words	97.5	1.7	0.8	0.2	0.0	0.0	0.0	-	100
% unique words with d>1		58.0	29.9	7.3	2.3	1.3	0.9	-	100

chose not to include words with $d(W) = 1$ in our calculations, as we were interested in evaluating the true ability of the model to disambiguate word meanings.

Training the model with the training set, and then indexing the test set, we obtained $AC = 0.79$. To understand whether this accuracy can be considered good, we compared it against the naive classifier. The naive classifier, given a word W of the test set, chooses at random a concept C_i, among the $d(W)$ concepts that in the test set are mapped on W; the probability for the naive classifier to select the right concept for the word W is then $p(C_{sel} = C_i | C_{right} = C_i, W) = 1/d(W)$. Accuracy is defined as $AC_n = p(C_{sel} = C_{right})$; see Table 10.10 for details.

Table 10.10 Naive Classifier

W_{C_i}	The set of words that, in the test set, are mapped onto C_i
$p(W)$	The prior probability of the word W mapped onto C_i
$d(W)$	The number of concepts that, in the test set, are mapped onto the word W
C	The set of concepts that appear in the test set
$p(C_{sel} = C_i \| C_{right} = C_i) = \sum_{W \in W_{C_i}} p(C_{sel} = C_i \| C_{right} = C_i, W) p(W) = \sum_{W \in W_{C_i}} \frac{p(W)}{d(W)}$ The probability for the naive classifier to select the right concept C_i	
$AC_n = \sum_i^{\|C\|} p(C_{sel} = C_i \| C_{right} = C_i) p(C_i) = \sum_i^{\|C\|} \sum_{W \in W_{C_i}} \frac{p(W)}{d(W)} p(C_i)$ The accuracy of the naive classifier	

Relying on the test set (and, as for AC, ignoring words with $d(W) = 1$), we obtain $AC_n = 0.37$. Our model, with $AC = 0.79$, clearly outperforms the naive classifier, even if trained with a small set.

10.9 Conclusions

We presented a model for knowledge extraction from documents written in natural language. In our model, a Domain Ontology models the domain knowledge, while a Lexical Database represents the domain vocabulary; An HMM-inspired, MaxEnt-enhanced stochastic model stores the mappings between such two levels. A second order Viterbi decoder disambiguates documents' words during the indexing phase. Both keyword-based and natural language-based queries are supported by the system, which is able to search both the document collection and the Domain Ontology. Finally, a rule-based component permits to discover new individuals in the document collection, adding them to the Domain Ontology.

As a future work we plan to refine the model, applying smoothing techniques to the MaxEnt models [28]. Moreover, new feature and feature templates, as well as new feature filter algorithms will be tested. Finally, the adoption of standard frameworks, like UIMA (see Section 9.3) will be investigated.

References

1. Alani, H., Kim, S., Millard, D.E., Hall, M.J.W.W., Weal, M.J., Hall, W., Lewis, P.H., Shadbolt, N.R.: Automatic ontology-based knowledge extraction and tailored biography generation from the web. IEEE Intelligent Systems 18, 14–21 (2002)
2. Atserias, J., Casas, B., Comelles, E., González, M., Padró, L., Padró, M.: Freeling 1.3: Syntactic and semantic services in an open-source NLP library. In: Proceedings of the Fifth International Conference on Language Resources and Evaluation (LREC 2006), Genoa, Italy. ELRA (May 2006)
3. Calvanese, D., McGuinness, D., Nardi, D., Patel-Schneider, P.: The Description Logic Handbook: Theory, Implementation and Applications. Cambridge University Press (2003)
4. Charniak, E., Johnson, M.: Coarse-to-fine n-best parsing and MaxEnt discriminative reranking. In: Proceedings of the 43rd Annual Meeting of the ACM, New York, pp. 173–180 (June 2005)
5. de Marneffe, M.C., Manning, C.D.: Stanford typed dependencies manual. The Stanford Natural Language Processing Group, Stanford University, Stanford, California (September 2008)
6. Fellbaum, C. (ed.): WordNet: An Elettronic Lexical Database. MIT Press (1998)
7. Ferrucci, D., et al.: Towards an interoperability standard for text and multi-modal analytics. Research Report RC24122 (W0611-188), IBM (2006)
8. Finkel, J.R., Grenager, T., Manning, C.: Incorporating non-local information into information extraction systems by gibbs sampling. In: Proceedings of the 43nd Annual Meeting of the Association for Computational Linguistics (2005)
9. Francis, W.N., Kucera, H.: Brown Corpus Manual. Department of Linguistics. Brown University, Providence (1979)
10. Goodman, J.: Sequential conditional generalized iterative scaling. In: Proceedings of the 40th Annual Meeting of the Association for Conputational Linguistics (ACL), Philadelphia, USA, pp. 9–16 (July 2002)
11. He, Y.: Extended viterbi algorithm for second order Hidden Markov Process. In: Proceedings of 9th International Conference on Pattern Recognition, Rome, Italy, pp. 718–720 (January 1988)
12. Jaynes, E.T.: Prior probabilities. IEEE Transactions On Systems Science and Cybernetics sec-4(3), 227–241 (1968)

13. Jelinek, F.: A fast sequential decoding algorithm using a stack. IBM Journal or Research and Development 13, 675–685 (1969)
14. Jurafsky, D., Martin, J.H.: Speech and Language Processing, pp. 254–259. Prentice-Hall (2000)
15. Kashyap, V., Bussler, C., Moran, M.: Ontology Authoring and Management. In: Data-Centric Systems and Applications, ch. 6. Springer, Heidelberg (2008)
16. Klein, D., Manning, C.: Fast exact inference with a factored model for natural language parsing. In: NIPS (2003)
17. Lafferty, J., McCallum, A., Pereira, F.: Conditional random fields: Probabilistic models for segmenting and labeling sequence data. In: Proceedings of 8th International Conference on Machine Learning (ICML 2001), Williamstown, MA, USA (June 2001)
18. Qiu, L., Kan, M.-Y., Chua, T.-S.: A public reference implementation of the RAP anaphora resolution algorithm. In: Proceedings of the Fourth International Conference on Language Resources and Evaluation (LREC 2004), Lisbon, Portugal, vol. 1, pp. 291–294 (May 2004)
19. Manning, C., Schütze, H.: Foundations of Statistical Natural Language Processing. MIT Press, Cambridge (1999)
20. McCallum, A., Freitag, D., Pereira, F.: Maximum entropy markov models for information extraction and segmentation. In: Proceedings of 7th International Conference on Machine Learning (ICML 2000), Stanford, USA, pp. 591–598 (June 2000)
21. Quillian, M.: Semantic Memory. Semantic Information Processing, pp. 227–270. MIT Press, Cambridge (1968)
22. Rabiner, L.R., Juang, B.H.: An introduction to hidden markov models. IEEE ASSP Magazine (January 1986)
23. Ratnaparkhi, A.: A simple introduction to maximum entropy models for natural language processing. Technical report, Institute for Research in Cognitive Science, University of Pennsylvania (1997)
24. Ratnaparkhi, A., Roukos, S., Ward, R.T.: A maximum entropy model for parsing. In: Proceedings of the International Conference on Spoken Language Processing, pp. 803–806 (1994)
25. Salton, G., Wong, A., Yang, C.S.: A vector space model for automatic indexing. Commun. ACM 18(11), 613–620 (1975)
26. Shieber, S.M.: The design of a computer language for linguistic information. In: Proceedings of Coling84, 10th International Conference on Computational Linguistics, pp. 362–366. Stanford University, Stanford (1984)
27. Shieber, S.M.: An Introduction to Unification Based Approaches to Grammar. CSLI Lecture Notes Series, vol. 4. Center for the Study of Language and Information, Stanford University (1986)
28. Toutanova, K., Klein, D., Manning, C., Singer, Y.: Feature-rich part-of-speech tagging with a cyclic dependency network. In: Proceedings of HLT-NAACL, Edmonton, Canada (2003)
29. Vargas-Vera, M., Motta, E., Domingue, J., Shum, S.B., Lanzoni, M.: Knowledge extraction by using an ontology-based annotation tool. In: Proceedings of the K-CAP 2001 Workshop on Knowledge Markup and Semantic Annotation, pp. 5–12 (2001)
30. Wallach, H.M.: Conditional random fields: An introduction. Technical Report MS-CIS-04-21, University of Pennsylvania CIS (February 2004)

Chapter 11
Knowledge Extraction from Events Flows

Alireza Rezaei Mahdiraji, Bruno Rossi, Alberto Sillitti, and Giancarlo Succi

Abstract. In this chapter, we propose an analysis of the approaches and methods available for the automated extraction of knowledge from event flows. We specifically focus on the reconstruction of processes from automatically generated events logs. In this context, we consider that knowledge can be directly gathered by means of the reconstruction of business process models. In the ArtDECO project, we frame such approaches inside delta analysis, that is the detection of differences of the executed processes from the planned models. To this end, we provide an overview of the different techniques available for process reconstruction, and propose an approach for the detection of deviations. To show its effectiveness, we instantiate the usage to the ArtDECO case study.

11.1 Introduction

Event logs are typically available inside an organisation, and they encode relevant information about the execution of high level business processes. Even if such information is usually available, very often knowledge about processes can be difficult to reconstruct or processes can deviate from the planned behaviour. Thus, retrieving knowledge from such event flows is part of the so-called process mining approach, that is the reconstruction of business processes from event log traces [30].

The reasons for using process mining are usually multiple and multi-faceted. Two are the most relevant. 1) the information can be used to see whether the actions inside an organisation are aligned with the designed business processes (so-called *delta analysis*). 2) process mining can be used to derive existing patterns in users' activities that can then be used for process improvement. In both cases, the

Alireza Rezaei Mahdiraji · Bruno Rossi · Alberto Sillitti · Giancarlo Succi
Center for Applied Software Engineering (CASE) - Free University of Bozen-Bolzano,
Piazza Domenicani 3, 39100 Bolzano, Italy
e-mail: {alireza.rezaei,bruno.rossi,alberto.sillitti,gsucci}@unibz.it

G. Anastasi et al. (Eds.): Networked Enterprises, LNCS 7200, pp. 221–236, 2012.

knowledge that can be gathered allows a more efficient usage of resources inside an organisation [28].

In this chapter, we provide an overview of several techniques and approaches that can be used for process mining from event flows. We analyse such approaches, and we describe how such approaches have been used in the context of the ArtDECO project.

11.2 Knowledge Flow Extraction in the ArtDECO Project

In the context of the ArtDECO project, we have interconnected networked enterprises that do not only use their own internal business processes but need to orchestrate higher level processes involving several companies. Three layers have been identified: the *business process*, the *application*, and the *logical layer*. The *business process* level deals with the interrelations among enterprises at the level of business process models. The *application* level is the implementation of the business processes, running either intra- or inter-enterprises. The *logical layer* is an abstraction over the physical resources of networked enterprises. The contextualisation to the GialloRosso winery case study is discussed in Chapter 2, Figure 1.1.

The vertical view shows all the different abstract processes, while the horizontal line evidences the stakeholders interested in the different information flows. Enterprises are typically integrated horizontally, as such we need to consider the integration of the different business processes. As the case study has been instantiated to the wine domain, we have typically three different stakeholders to consider: the winery, the delivery company and the retailer. Each actor has its own business process that is integrated in the overall view of the general interconnected processes.

Extracting knowledge from low level event flows, poses thus more issues that normal business process reconstruction as we have very fine-grained information as the source of the whole reconstruction process [18, 6].

In this context, a possible approach for process improvement is the following:

1. *Collection of data with client-side plug-ins.* This is a step that can be performed by using systems for automated data collection enabling the organisation to harvest information during the execution of the business processes. Such information can then be used to reconstruct knowledge from the low-level events generated;
2. *Collection of data from sensor networks.* Another source of data comes from sensor networks. They provide low level data that can be used to evaluate inconsistencies between the actual situation and the expected one. Such low-level events can be used for subsequent phases of the process;
3. *Integration of the data sources into a* DBMS. All low-level events need to be integrated in a common centralised repository or *DBMS*. Such point of integration can be queried for the reconstruction of the process and the evaluation of triggers when the expected behaviour is not conformant to the planned business processes;

4. *Reconstruction of the business processes.* To reconstruct the business processes in an organisation, we need to extract knowledge from low-level events data. To this end, we need some algorithms that - given a set of events - are able of reconstructing the ongoing process;
5. *Delta analysis of the deviations from expected behaviour.* Once the business processes have been reconstructed, the difference with the planned models can be detected. If necessary, triggers are set-up to notify stakeholders about unexpected behaviours in the context of the planned high-level processes;

Such approach can be supported by a tool as PROM [27] that allows the collection of low-level events and the subsequent analysis of the processes [3, 4, 9].

In this chapter, we focus on techniques for the extraction of knowledge from event flows and process reconstruction (*Step 4*).

11.3 Overview of Process Mining Techniques

A business process is a collection of related activities that produces a specific service or product for a customer. The success of an organisation is directly related to the quality and efficiency of its business processes. Designing these processes is a time consuming and error prone task, because the knowledge about a process is usually distributed between among employees and managers that execute them. Moreover, such knowledge is not only distributed in a single organisation but, frequently, it is distributed across several organisations (cross-organisational processes) that belong to the same supply-chain. For these reasons, business processes experts that are in charge of the formal definition of processes face an hard task. Additionally, the designed model needs to adapt to changes that the market imposes to organisation to be competitive.

To spend less time and effort and obtain models based on what really happens in an organisation, we can adopt a bottom-up approach, i.e., extract the structure of processes from recorded event-logs [26, 2, 5]. Usually, information systems in an organisation have logging mechanisms to record most of the events. These logs contain information about the actual execution of the processes such as the list of executed tasks, their order and process instances. The method of extracting the structure of process (*a.k.a. process model*) from the event-logs is known as process mining or process discovery or workflow mining [28]. The extracted model can be used to analyse and improve current business, e.g., it can be used to detect the deviations from normal process executions [19].

Each event-log consists of several traces. Each trace corresponds to the execution of an instance process, also known as a *case*. Each case is obtained by executing activities (tasks) in a specific order. A process model is designed to handle similar cases, it specifies tasks to be executed and their temporal order of execution.

11.3.1 A General Process Mining Algorithm

Figure 11.1 shows the steps of a typical process mining algorithm. The first step deals with reading the content of the log. Most of the algorithms assume that the log contains at least the cases identifiers, tasks identifiers, and execution orders of the tasks for each case. This is the minimal information that is needed for process mining. However, in reality, logs contain further information that can be used to extract more dependency information, e.g. if a log contains information about start and completion time for each task (non-atomic tasks), parallelism detection can be done by examining just one trace, otherwise at least two traces are needed [35].

The second step deals with the extraction of the dependency relations (also known as follows relations) among tasks. These relations can be inferred using temporal relationships between tasks in event-log. Task B is dependent on task A iff in every trace of the log, task B follows task A. This definition for real-world logs is unrealistic because logs always contain noise. Noisy data in the logs are either because of a problem in the logging mechanism or exceptional behaviours. A process mining algorithm has to extract the most common behaviour even in presence of noisy data that makes the task of the mining algorithm more difficult and can result in overfitting (i.e., the generated model is too specific and allows only the exact behaviours seen in the log). To cut-off the noise, we need a way (e.g., a threshold) to discard less frequent information and reconstruct a sound business process. Hence, we need to consider the frequency of dependency relation. We can modify the definition of dependency relation as follows: the task B is dependent on task A iff in most traces task B follows task A.

The third step deals with the induction of the structure of the process model . The problem is to find a process model that satisfies three conditions: (i) generates all traces in the log (in case of noise-free logs), (ii) only covers few traces that are not in the log (extra-behaviur), and (iii) has the minimal number of nodes (i.e., steps of the process). For simple processes, it is easy to discover a model that recreates

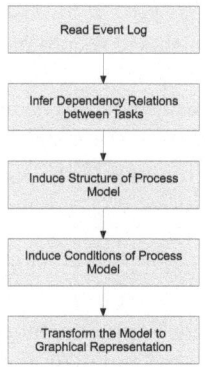

Fig. 11.1 Steps in a General Process Mining Algorithm

the log, but for larger processes it is an hard task, e.g., the log does not contains all combinations of selection and parallel routings or some paths may have low probability and remain undetected or log may contain too noisy data.

The forth step deals with the identification of routing paths among the nodes. The resulting models may include of four basic routing constructs as follows: (i) Sequential: the execution of one task is followed by another task, (ii) Parallel: tasks A and B are in parallel if they can be executed in any order or at the same time, (iii) Conditional (Selection or choice): between tasks A and B, either task A or task B is executed, and (iv) Iteration (Loop): when a task or a set of tasksis executed multiple times.

The induction of conditions phase corresponds to induction of conditions for non-deterministic transitions (i.e., selection) based on process attributes. A process attribute is a specific piece of information used as a control variable for the routing of a case, e.g., the attributes of the documents which are passed between actors. Approaches such as decision rule mining from machine learning can be used to induce these conditions on a set of attributes [16].

The last step converts the result model (e.g., a Petri net) to a graphical representation that is easy to understand for the users.

11.3.2 Challenges of Process Mining Algorithms

In [28], several of the most challenging problems of the process mining research has been introduced. Some of those challenges are now partially solved, such as mining hidden tasks, mining duplicate tasks, mining problematic constructs (e.g., non-free-choice net) and using time information, but some of them still need more research. The current most important challenges of process mining are:

- **Mining different perspectives.** Most of the research in process mining is devoted to mining control flow (How?), i.e., ordering of tasks. There are few works that aim at mining perspective such as organisational perspectives (Who?), e.g., mining social networks of organisational relationships [31].
- **Handling noise.** The noise in real world event logs is unavoidable and each process mining system should deal with noisy data. As mentioned before, noisy logs cause overfitting. The main idea is that the algorithms can be able to detect exceptions from real behaviours. Usually, exceptions are infrequent observations in logs, e.g., infrequent casual relations. Most algorithms provide a threshold to distinguish between noise and correct behaviours. There are several heuristics in literature to handle noise.
- **Underfitting.** It means that the model overgeneralises behaviours seen in the logs and that is because of unsupervised setting, i.e., lack of negative traces in the logs [7]. The logs only contain successful process instances, so to generate negative instance process we need to generate them from the current log, e.g., first we find outliers in the current log and label them as negative traces and then apply a binary classification algorithm.

- **Conformance Testing.** Comparing a prior model of the process with the extracted model from process mining is known as conformance testing or delta analysis. The comparing procedure aims at finding the differences and commonalities between the two models [24].
- **Dependency between Cases.** Most of current approaches assume that there are no dependencies among cases, i.e. the routing of one case does not depend on the routing of other cases or in another words, the events of different cases are independent. Real world data may violate this assumption. For example, there may be a competition among different cases for some resources. These so-called history-dependent behaviours are stronger predictor for the process model [12].
- **Non-Unique Model.** It means that different non-equivalent process models can be mined from one log. Some of them are too specific and some others are too general, but all of them generate the log. By finding a balance between specificity and generality, we can find an appropriate process model.
- **Low-Level Event-Logs.** Most of current techniques on process mining are only applicable for process-aware information systems' logs, i.e., logs that are in task level abstraction. This is not the case in many real-life logs. They usually lack the concept of task and instead they contain many low-level events. Although groups of low-level events together represents tasks, it is not easy to infer those groups. The information systems' logs must be first ported to task level and then a process mining technique can extract the process model [33, 13, 15].
- **Un-Structured Event-Logs.** In real life systems, even if there is a prior process model, it is not enforced and there are lots of behaviours in the logs that are instances of deviation from the model. This flexibility results in unstructured process models, i.e., models with lots of nodes and relations. Most of techniques in literature generate unstructured models in such environments. The resulting models are not incorrect and usually capture all deviations in the log. Two approaches in literature dealing with this issue use clustering and fuzzy techniques [33, 14].

11.3.3 Review of the Approaches

There are several approaches for mining process models from log data [28]. Based on the strategy that they use to search for appropriate process model, we can divide them as either local or global approaches. Local strategies rely only on local information to build the model step by step, while global strategies search in the space of potential models to find the model. The local strategies usually have problems with noise and discovering more complex constructs. Most of the current approaches of process mining are in the first category and only few of them are in the global category. Figure 11.1 depicts the general steps of local algorithms. In the following, we concisely introduce some of these approaches.

In [16, 17], three algorithms are developed, namely, Merge Sequential, Split Sequential, and Split Parallel. These algorithms are the best known for dealing with duplicate tasks. The first and second algorithms are suitable for sequential process models and the third one for parallel processes. They extract process models as a Hidden Markov Model (HMM). HMM is basically a Finite State Machine (FSM), but each transition has a probability associated and each state has a finite set of output symbols. The Merge Sequential and Split Sequential algorithms use the generalization and specialization approaches, respectively. The Merge Sequential is a bottom-up algorithm, i.e., it starts with the most specific process model with one separate path for each trace in the log, then it iteratively generalizes the model by merging states that having same output symbols, till the satisfaction of a termination criterion. To reduce the size of the initial model, prefix HMM can be used, i.e., all states that have a common prefix are mapped to a single state. The problem with specialization operator is the number of merging operations, i.e., even with few states with the same symbols, the number of merging operations is usually very large.

The Split Sequential algorithm aims to overcome the complexity of Merge Sequential. It starts with the most general process model (i.e., without duplicate tasks and able to generate all the behaviours in the log) and then iteratively splits states with more than one incoming transition into two states. The Split Parallel algorithm is an extension of the Split Sequential for concurrent processes. The reason for this extension is that unlike sequential processes, when the split of a node in concurrent processes changes the dependency graph, in this case it may have global side effects. So, instead of applying the split operator on the model, the split operations are done at level of the process instance. It is also top-down and works as follows: suppose that activity A is split. The split operator is able to make distinction among certain occurrences of A, e.g., A1 and A2. Then, it induces a general model based on current instances which contains two nodes A1 and A2 instead of only A. After termination of specialization, re-labelled nodes are changed to their original labels. This approach also uses a decision rule induction algorithm to induce conditions for non-deterministic transitions.

In [25], a new approach based on block-oriented representation to extract minimal complete models has been introduced. In block-oriented representation, each process model consists of a set of nested building blocks. Each building block consists of one or more activities or building blocks that are connected by operators. There are four main operators, namely, sequence, parallel, alternative, and loop. Operators define the control-flow of the model.

The works in [28, 29, 30] are some of the more extensive works in process mining. Authors started by developing an alpha algorithm and over time they extended it with many modifications to tackle different challenges. Authors proved that alpha algorithm is suitable for a specific class of models. In the first version of their alpha algorithm, they assumed that logs are noise-free and complete. The alpha algorithm is unable to find short loops and implicit places, and works based on binary relations in the logs. There are four relations: follows, causal, parallel, and unrelated. Two tasks A and B have a follows relation if they appear next to each other in the log. This relation is the basic relation from which the other relations are extracted. Two tasks A and B have a causal relation if A follows B, but B does not follow A. If B also follows A, then the tasks have a parallel relation. When A and B are not involved in a follows relation, they are said to be unrelated. All the dependency relations are inferred based on local information in the log. Additionally, because the algorithm works based on sets, it cannot mine models with duplicate tasks. The alpha algorithm works only based on follows relation without considering frequency, therefore it cannot handle noise. In [34], the alpha algorithm was enhanced with several heuristics to consider frequencies to handle noise. The main idea behind the heuristics is as follows: the more often task A follows task B and the less often B follows A, it is more likely that A is a cause for B. Because the algorithm mainly works based on binary relations, the non-free-choice constructs cannot be captured. In [21], they first extended the three relational metrics in [34] and added two new metrics that are more suitable in distinction between choice and parallel situations. Then, based on a training set that contains information about five metrics and actual relations between pairs of activities (causal, choice, and parallel), they applied a classification rule learner to induce rules that distinguish among different relations based on the values of five relational metrics.

Clustering-Based Techniques. Most of the approaches in process mining only produce one process model. This single model is supposed to represent every single detail in the log. So, the result model is intricate and hard to understand. Many real-world logs contain such unstructured models, usually because they allow very flexible execution of the process model. Flexible environments generate heterogeneous logs, i.e., they contain information about very different cases. Another reason for unstructured logs is that some logs record information about different processes that belong to the same domain. In both cases, only one process model is unable to describe this kind of log and the result model is very specific (when an exact process model is the goal) or over-general. The basic idea of using clustering techniques in the process mining domain is to divide original logs into several smaller logs where each one of them contains only homogeneous cases. Then, for each partition, a separate process model is extracted.

In the process mining literature, there are several studies using clustering in different ways in process mining context. For example in [1], each trace is represented with a vector of features extracted from different traces, i.e., control-flow, organisation, data, etc. In [10, 11, 23], a hierarchy of process model is generated, i.e., each log partition will be furthered partitioned if it is not expressive enough.

These are are all local algorithms, but there are three main global approaches in literature. Namely, genetic process mining, approach based on first order logic, and fuzzy process mining.

Genetic Process Mining. Since genetic algorithms are global search algorithms, genetic process mining is also a global algorithm that can handle noise and can capture problematic structure such as non-free-choice constructs [22, 32]. To apply genetic algorithms to process mining, a new representation for process models was proposed in [22, 32], known as casual matrix. Each individual or potential process model is represented by a causal matrix. Casual matrix contains information about casual relations between the activities and input/output of each activity. Causal matrix can be mapped to Petri Nets [22, 32], where they used three genetic operators, namely, elitism, crossover, and mutation. Elitism operator selects a percentage of best process models for the next generation. Crossover recombines causality relationships in the current population and mutation operator inserts new casual relationships, and adds/removes activities from input or output of each activity. One of the main drawback of genetic process mining is the computational complexity of the approach.

First-Order-Logic Approach. Another approach based on global search uses a first-order-logic learning approach to process mining [20]. In contrast to other approaches seen so far, this approach generates declarative process models instead of imperative (procedural) models. Imperative approaches define exact execution order for a set of tasks and declarative approaches only focus on what should be done. In [20], the process mining problem is defined as a binary classification problem, i.e., the log contains positive and negative traces.

Advantages of using first-order learning are as follows: (i) it can discover structural patterns, i.e., search for patterns of relations between rows in the event log, (ii) by using declarative representation, it generates more flexible process models, and (iii) it can use prior knowledge in learning.

Fuzzy Process Mining. The third global strategy is a fuzzy approach. The idea of fuzzy process mining is to generate different views of the same process model based on configurable parameters. Based on what is interesting, configuration parameters are used to keep only those parts of the process that are relevant and remove others. To achieve this objective, it considers both global and local information from different aspects of the process such as control-flow and organisational structure to produce a process model. This is why the fuzzy approach is also known as multi-perspective approach [14].

11.4 Application to the GialloRosso Case Study

In this section, we start by defining the process of delta analysis, then we delve into an application to the GialloRosso case study. Knowledge extraction in the ArtDECO Project has been contextualised to the analysis of deviations from

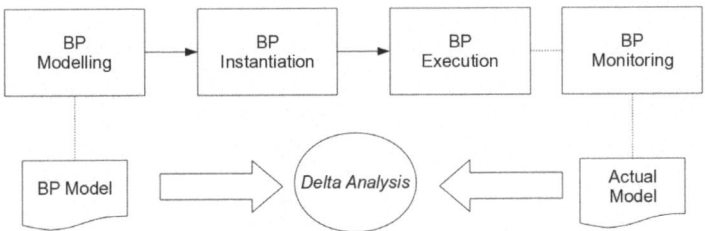

Fig. 11.2 Phases of the lifecycle of Business Processes for delta analysis

expected business models. In the specific, we used a local algorithm based on happens-before relations, and supported by a rule-based system for tasks reconstruction from event logs. Figure 11.2 shows different phases of a business process, from modelling, instantiation, to execution. We start generally with a business process model that represents all the steps as they have been conceived by a business process analyst. Such model is then instantiated and executed: at the same time, several executions of the same process model can be running. Each process execution needs to be monitored to derive indications from the real running processes. Then, this information can be used to reconstruct the real model. Afterwards, delta analysis will be used to compare the planned and the actual models to derive the deviations. For delta analysis, the monitoring phase is critical for the derivation of low-level events that can be analysed for knowledge extraction. Once the actual execution model has been reconstructed by means of algorithms for process mining, delta analysis is used to derive the deviations from the execution traces.

In our case, the process of delta analysis is done by means of the following steps:

(a) determination of the happened-before relations among tasks from the original planned model. For each task, if task A precedes B, then A -> B;
(b) process monitoring for the collection of events for different instances of process execution;
(c) tasks reconstruction from events by means of a rule-based system, as in Figure 11.3;
(d) determination of the violation of happened-before relations from execution traces, when an event in task B is detected while no event from task A happening before. Such violations are annotated;

In particular, to reconstruct the process phases from low-level events, events need to be mapped to higher-level constructs (Figure 11.3). For this, we use a rule-based system that maps all the events according to domain information to the higher levels,

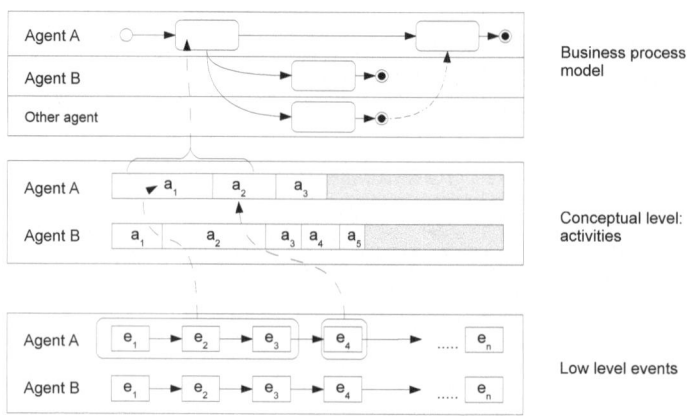

Fig. 11.3 Phases of the lifecycle of Business Processes for delta analysis

so that it is possible to associate each event to a phase of a process. A rule is defined by means of a source application that generated the event, the text of the event, and a discriminator for a particular process instance (e.g. an item the instance of the process is related to). Each rule specifically maps to a task in the original process model. The actor/agent that generated the event is already part of the event metadata. After rules are applied, events that do not comply to a rule are discarded and not considered for delta analysis.

If we consider the case study of the winery (the GialloRosso case, see Chapter 2), we can instantiate the approach to part of the business processes defined to show how the approach has been applied.

The planned behaviour in the case study foresees that when the handling of the wine starts, a message must be sent from the *distributor* of the winery to the *carrier*, responsible for the transportation (Figure 11.4). In parallel, the distributor starts the quality monitoring process for the delivery so that to gather objective data about the performance of the delivery process. Such information is then used to alert the distributors and to update the final wine quality. At this point, the whole process ends.

Figure 11.4 already contains information about different execution traces and their deviation from the planned behaviour. The detail is shown with the process number (e.g. *P1* or *P2*) and an indication of a *(d)eviation* or an *(u)nauthorized* action during the step. We will explain in the following how these activities are detected and the software implementation that has been used for the analysis.

With the scope of explaining the approach undertaken, we just focus on the initial data exchange among the distributor and the carrier (top left part of Figure 11.4). Actions can deviate from the original plans under some circumstances, as the process can start without being triggered by a message in the information system. For example, the process can be started due to personal communication among the two actors of the process. Even in this trivial case, this activity can be detrimental for

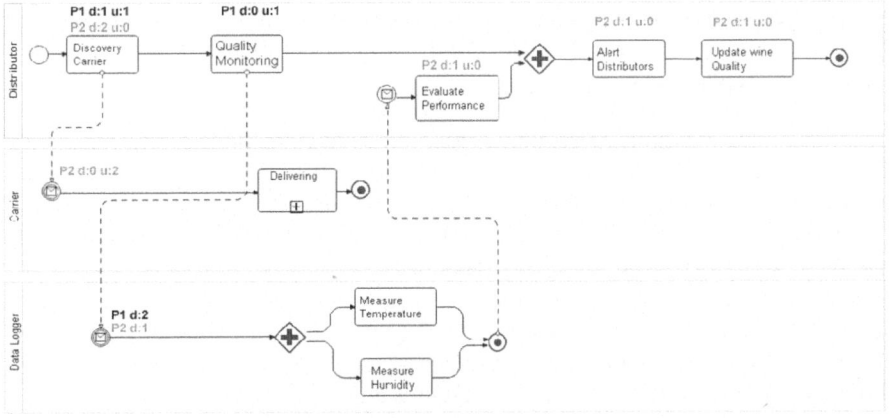

Fig. 11.4 The 'discover carrier' subprocess results from delta analysis: the original workflow is tagged with deviations of two process instances P1 and P2

process improvement analyses, as there is no information about how much time was required to pass from the decision taken at the management level to the control center level. Also, there is no tracking of the original communication, and evaluation of errors in the definition and/or execution of the directives that were assigned. Furthermore, the execution violates the original planned sequence of actions.

This can be enforced by the low-level data process mining. The following preconditions and post-conditions (happened-before relations), can be inferred from the original model. They can be derived from the execution of several execution traces or derived from the original process. If we consider *Handling_Start (H_S)*, *MailSend (MS)*, and *Handling_Execution (H_E)* actions:

```
H_S = pre (MS(Distributor, Carrier))
H_S = post (MS(Carrier, Distributor))
H_S = post (H_E)
```

We derive that the *Handling_start* phase has a precondition that a message must be exchanged between two actors of the business process. Post conditions are another exchange of messages among the actors, and the execution of the successive phase in process. In other terms, these are the conditions for the part of the process that we consider, if we use happened-before notation:

```
MS(Distributor, Carrier) -> H_S;
H_S -> MS(Carrier, Distributor);
H_S -> H_E;
```

Once we have this information for the planned model, we need to focus on the actual instances of the process. If we run several instances of the process monitoring the actors, we collect execution traces that need to be mapped to higher constructs. This is done by a domain-dependent rule-based system that maps events to tasks and process phases. An alternative is to use machine learning approaches that need to be trained by several execution traces. The rule-based system needs also to discriminate the specific instance of the process, so we need to have a way to divide flows of events across different processes. The following is an example of a rule:

```
APP{"any"} AND EVENT{"*warehouse*"} AND DISC{"item *"} -> Task A
```

In this case, we are defining a rule to process events generated by any application that are related to documents that have warehouse in the title and should be divided according to a tagged item. Therefore, events with different items will be mapped to different process flows. All the events that comply to this rule will be mapped to task A, by keeping the timestamp information.

Once events have been associated to the planned tasks, the detection of violations of the original model is done by means of the evaluation of temporal execution of the events associated to a task: if a task is executed before the actual execution of another temporal-related task, such violation is annotated.

In Figure 11.4 there is an example of the original process annotated with violations from two running instances *P1* and *P2*. Each violation means the possibility that the process has been executed without following the original temporal relations among phases. In the case study, this can mean that the communication flows among stakeholders followed different paths rather than those planned. We can see the number of violations per task and we can further focus on each task to inspect the causes of such deviations by looking at the single low level events. We can see two different types of information. The *(d)eviations*, actions undertaken without respecting temporal relations, for example sending the delivery without prior communication by the distributor, or *(u)nauthorized* actions, that is actions that were not allowed at a specified step, like recording information in a warehouse's registry. The latter kind of actions can be specified by the user to be evaluated at runtime against the real execution traces.

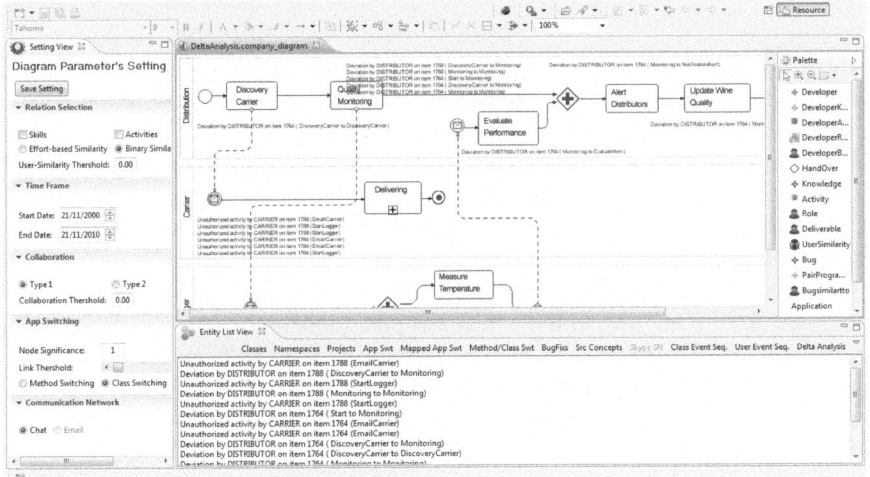

Fig. 11.5 PROM plug-in for the execution of delta analysis

For the execution of process mining and delta analysis, we implemented a plugin for the PROM software [27]. In particular, we took the opportunity to use the low-level events collected by PROM to perform the analysis. The plug-in has been integrated into the *Eclipse*[1] application with the support of the *Graphical Modeling Framework (GMF)*.

Figure 11.5 gives an overview of the software prototype used for the analysis. We see the same business case that has been followed by explaining the approach, loaded into the application's workspace. The current prototype takes three different types of inputs: a) the definition of a workflow, b) the definition of the rules for each node in the workflow, c) a set of events generated from several execution traces.

Given the inputs, the application parses all the events and annotates the business process with relevant information. The relevant quadrants of the application in Figure 11.5 are the top and bottom ones. In the top one, the loaded process is visualized. In the bottom one there is the output of all the deviations detected, and the user can select the different view to activate in the workspace.

---

[1] www.eclipse.org

In the case used to exemplify the approach, two different execution traces have been instantiated by generating events and set as the input for the application. After the analysis has been performed, each node of the loaded process is marked with deviation information derived from delta analysis. This is a particolar scenario, in which the data is analysed ex-post. As noted in section 11.2, the usefulness of the proposed approach is to analyse data in real-time, with execution traces collected and analysed as actions are performed by the actors of the process.

11.5 Conclusions

In this chapter, we proposed an analysis of methods for an automated extraction of knowledge from event flows. We focused specifically on process mining, that is reconstructing business processes from event log traces.

Process mining can be important for organisations that want to reconstruct knowledge hidden in the event logs. Typically any organisation has the opportunity to collect this kind of information. The advantages are multi-faceted, mostly we referred to two specific areas.

On one side, such knowledge can be used to evaluate whether the high level business processes are aligned with the business plan models. As such, process mining can be used to see whether the actual behaviour is deviating from the expected behaviour. On the other side, the knowledge can be used to detect hidden behaviours - i.e. not encoded in high level business processes - inside the organisation. Such behaviours can then be the focus of further analyses to see whether they are really required, resources are wasted, or even process improvement/restructuring opportunities can derive from them.

We proposed an approach based on delta analysis to derive information from low level event flows and reconstruct the original processes. We showed in the context of the case study of the ArtDECO project, the GialloRosso winery, how event flows are used to reconstruct the original processes and detect deviations from the planned model.

References

1. Aires da Silva, G., Ferreira, D.R.: Applying Hidden Markov Models to Process Mining. In: Rocha, A., Restivo, F., Reis, L.P., Torrao, S. (eds.) Sistemas e Tecnologias de Informacao: Actas da 4a Conferencia Iberica de Sistemas e Tecnologias de Informacao, pp. 207–210. AISTI/FEUP/UPF (2009)
2. Coman, I., Sillitti, A.: An Empirical Exploratory Study on Inferring Developers' Activities from Low-Level Data. In: 19th International Conference on Software Engineering and Knowledge Engineering (SEKE 2007), Boston, MA, USA, July 9-11 (2007)
3. Coman, I., Sillitti, A.: Automated Identification of Tasks in Development Sessions. In: 16th IEEE International Conference on Program Comprehension (ICPC 2008), Amsterdam, The Netherlands, June 10-13 (2008)

4. Coman, I., Sillitti, A., Succi, G.: Investigating the Usefulness of Pair-Programming in a Mature Agile Team. In: 9th International Conference on eXtreme Programming and Agile Processes in Software Engineering (XP 2008), Limerick, Ireland, June 10-14 (2008)

5. Coman, I., Sillitti, A.: Automated Segmentation of Development Sessions into Task-related Subsections. International Journal of Computers and Applications 31(3) (2009)

6. Coman, I., Sillitti, A., Succi, G.: A Case-study on Using an Automated In-process Software Engineering Measurement and Analysis System in an Industrial Environment. In: 31st International Conference on Software Engineering (ICSE 2009), Vancouver, BC, Canada, May 16-24 (2009)

7. Cook, J.E., Wolf, A.L.: Discovering Models of Software Processes from Event-Based Data. ACM Transactions on Software Engineering and Methodology 7(3), 215–249 (1998)

8. Ferreira, D., Zacarias, M., Malheiros, M., Ferreira, P.: Approaching Process Mining with Sequence Clustering: Experiments and Findings. In: Alonso, G., Dadam, P., Rosemann, M. (eds.) BPM 2007. LNCS, vol. 4714, pp. 360–374. Springer, Heidelberg (2007)

9. Fronza, I., Sillitti, A., Succi, G.: Modeling Spontaneous Pair Programming when New Developers Join a Team. In: 3rd International Symposium on Empirical Software Engineering and Measurement (ESEM 2009), Lake Buena Vista, FL, USA, October 15-16 (2009)

10. Greco, G., Guzzo, A., Pontieri, L., Sacca', D.: Discovering expressive process models by clustering log traces. IEEE Trans. Knowl. Data Eng. 18(8), 1010–1027 (2006)

11. Greco, G., Guzzo, A., Pontieri, L.: Mining taxonomies of process models. Data & Knowledge Engineering 67(1), 74–102 (2008)

12. Goedertier, S., Martens, D., Baesens, B., Haesen, R., Vanthienen, J.: Process Mining as First-Order Classification Learning on Logs with Negative Events. In: ter Hofstede, A.H.M., Benatallah, B., Paik, H.-Y. (eds.) BPM Workshops 2007. LNCS, vol. 4928, pp. 42–53. Springer, Heidelberg (2008)

13. Guenther, C.W., Van der Aalst, W.M.P.: Mining Activity Clusters from Low-Level Event Logs. BETA Working Paper Series, WP 165. Eindhoven University of Technology, Eindhoven (2006)

14. Günther, C.W., van der Aalst, W.M.P.: Fuzzy Mining – Adaptive Process Simplification Based on Multi-perspective Metrics. In: Alonso, G., Dadam, P., Rosemann, M. (eds.) BPM 2007. LNCS, vol. 4714, pp. 328–343. Springer, Heidelberg (2007)

15. Günther, C.W., Rozinat, A., van der Aalst, W.M.P.: Activity Mining by Global Trace Segmentation. In: Rinderle-Ma, S., Sadiq, S., Leymann, F. (eds.) BPM 2009. LNBIP, vol. 43, pp. 128–139. Springer, Heidelberg (2010)

16. Herbst, J.: Dealing with concurrency in workflow induction. In: Proceedings of the 7th European Concurrent Engineering Conference, Society for Computer Simulation (SCS), pp. 169-174 (2000)

17. Herbst, J., Karagiannis, D.: Integrating Machine Learning and Workflow Management to Support Acquisition and Adaptation of Workflow Models. International Journal of Intelligent Systems in Accounting, Finance and Management 9, 67–92 (2000)

18. Janes, A., Scotto, M., Sillitti, A., Succi, G.: A perspective on non-invasive software management. In: 2006 IEEE Instrumentation and Measurement Technology Conference (IMTC 2006), Sorrento, Italy, April 24-27 (2006)

19. Janes, A., Sillitti, A., Succi, G.: Non-invasive software process data collection for expert identification. In: 20th International Conference on Software Engineering and Knowledge Engineering (SEKE 2008), San Francisco, CA, USA, July 1-3 (2008)

20. Lamma, E., Mello, P., Riguzzi, F., Storari, S.: Applying Inductive Logic Programming to Process Mining. In: Blockeel, H., Ramon, J., Shavlik, J., Tadepalli, P. (eds.) ILP 2007. LNCS (LNAI), vol. 4894, pp. 132–146. Springer, Heidelberg (2008)

21. Maruster, L., Weijters, A.J.M.M., Van der Aalst, W.M.P., Van den Bosch, A.: A rule-based approach for process discovery: Dealing with noise and imbalance in process logs. Data Mining and Knowledge Discovery 13(1), 67–87 (2006)

22. de Medeiros, A.K.A., Weijters, A.J.M.M., van der Aalst, W.M.P.: Genetic Process Mining: A Basic Approach and Its Challenges. In: Bussler, C.J., Haller, A. (eds.) BPM 2005. LNCS, vol. 3812, pp. 203–215. Springer, Heidelberg (2006)

23. de Medeiros, A.K.A., Guzzo, A., Greco, G., van der Aalst, W.M.P., Weijters, A.J.M.M., van Dongen, B.F., Saccà, D.: Process Mining Based on Clustering: A Quest for Precision. In: ter Hofstede, A.H.M., Benatallah, B., Paik, H.-Y. (eds.) BPM Workshops 2007. LNCS, vol. 4928, pp. 17–29. Springer, Heidelberg (2008)

24. Rozinat, A., van der Aalst, W.M.P.: Conformance Testing: Measuring the Fit and Appropriateness of Event Logs and Process Models. In: Bussler, C.J., Haller, A. (eds.) BPM 2005. LNCS, vol. 3812, pp. 163–176. Springer, Heidelberg (2006)

25. Schimm, G.: Mining exact models of concurrent workflows. Comput. Ind. 53, 265–281 (2004)

26. Scotto, M., Sillitti, A., Succi, G., Vernazza, T.: Dealing with Software Metrics Collection and Analysis: a Relational Approach. Studia Informatica Universalis, Suger 3(3), 343–366 (2004)

27. Sillitti, A., Janes, A., Succi, G., Vernazza, T.: Collecting, Integrating and Analyzing Software Metrics and Personal Software Process Data. In: Proceedings of the 29th EUROMICRO Conference (2003)

28. Van der Aalst, W.M.P., Weijters, A.: Process mining: a research agenda. Comput. Ind. 53, 231–244 (2002)

29. Van der Aalst, W.M.P., Van Dongen, B.F., Herbst, J., Maruster, L., Schimm, G., Weijters, A.J.M.M.: Workflow mining: A survey of issues and approaches. Data & Knowledge Engineering 47(2), 237–267 (2003)

30. Van der Aalst, W.M.P., Weijters, A.J.M.M., Maruster, L.: Workflow Mining: Discovering Process Models from Event Logs. IEEE Transactions on Knowledge and Data Engineering 16(9), 1128–1142 (2004)

31. Van der Aalst, W.M.P., Reijers, H., Song, M.: Discovering Social Networks from Event Logs. Computer Supported Cooperative work 14(6), 549–593 (2005)

32. van der Aalst, W.M.P., de Medeiros, A.K.A., Weijters, A.J.M.M.: Genetic Process Mining. In: Ciardo, G., Darondeau, P. (eds.) ICATPN 2005. LNCS, vol. 3536, pp. 48–69. Springer, Heidelberg (2005)

33. Song, M., Günther, C.W., van der Aalst, W.M.P.: Trace Clustering in Process Mining. In: Ardagna, D., Mecella, M., Yang, J. (eds.) BPM 2008 Workshops. LNBIP, vol. 17, pp. 109–120. Springer, Heidelberg (2009)

34. Weijters, A.J.M.M., Van der Aalst, W.M.P.: Rediscovering Workflow Models from Event-Based Data using Little Thumb. Integrated Computer-Aided Engineering 10(2), 151–162 (2003)

35. Wen, L., Wang, J., Van der Aalst, W.M.P., Wang, Z., Sun, J.: A Novel Approach for Process Mining Based on Event Types. BETA Working Paper Series, WP 118. Eindhoven University of Technology, Eindhoven (2004)

Chapter 12
Context-Aware Knowledge Querying in a Networked Enterprise

Cristiana Bolchini, Elisa Quintarelli,
Fabio A. Schreiber, and Maria Teresa Baldassarre

Abstract. In today's knowledge-driven society, the increasing amount of heterogeneous information, available through a variety of information channels, has made it difficult for users to find the right information at the right time and at the right level of detail. This chapter presents the exploitation of context-awareness in order to reduce the plethora of information that,within the networked enterprise scenario, may confuse and overwhelm the users in need of high-quality information coming from the several available and heterogeneous information sources.

12.1 Introduction

A networked enterprise is a scenario rich of information and data, to be shared among different users, each one characterised by a specific role and interests. To solve the information overload problem, researchers have developed systems that adapt their behaviour to the goals, tasks, interests, and other characteristics of their users. The notion of context, emerged in the past years in various fields of research, has more recently received a lot of attention also in the computer science field, because it can help distinguishing useful information from noise, i.e., from all the information not relevant to the specific application; indeed, the same piece of information can be considered differently, even by the same user, in different situations, or places – in a single word, in a different context.

In this chapter we present the methodology and system architecture adopted in ART-DECO to support all the phases of contextual view design: (a) the specification

Cristiana Bolchini · Elisa Quintarelli · Fabio A. Schreiber
Politecnico di Milano, Dipartimento di Elettronica e Informazione, P.zza L. da Vinci, 32, Milano - I20133 Italy
e-mail: {bolchini,quintare,schreiber}@elet.polimi.it

Maria Teresa Baldassarre
Dipartimento di Informatica - Università degli Studi di Bari "'Aldo Moro'"
e-mail: baldassarre@di.uniba.it

G. Anastasi et al. (Eds.): Networked Enterprises, LNCS 7200, pp. 237–258, 2012.

of a context model appropriate for the specific application; (b) the association of each previously defined context with the portion of data relevant to that specific context; (c) the access to contextualised data. We show its applicability in the wine production scenario, where several classes of users access a variety of information, made available from the networked enterprise data sources, from sensors used for monitoring the productive cycle and from external sources. The access to such information is offered through a Context-Aware Web Portal, where the users identify themselves by logging into the system and defining their context by answering a short sequence of pre-defined questions. This initial identification phase aims at semi-automatically determining the active context of the user, in order to customise the portion of information made available for browsing and querying. These context-related behaviours (context specification and context-aware tailoring of data) are the outputs of a design phase where, taking into account the application scenario, the possible contexts are modeled and their relationship with the available data is designed. We focus the attention on the elements of context-based access to data, presenting the various facets of the approach. In particular, in Section 12.2 we present the context related aspects of the architecture, followed in Section 12.3 by the description of the model we use for defining the possible application contexts. Sections 12.4 and 12.5 are devoted to how the context-aware subsystem supports data tailoring and access to the various kinds of data sources, respectively. Section 12.6 shows an application of the theoretical concepts, illustrated in the previous sections, to a real context consisting in a wine production scenario through a Context-Aware Web Portal.

12.2 System Architecture

This section presents the context-aware subsystem of the overall system architecture (shown in Figure 12.1), focusing the attention on the **Context-Aware Web Portal**, the **Design-Time Context Manager** subsystem, the **Context-Aware Application Server**, enabling the context-aware technology, to offer the users the possibility to filter the data (available in the **Data Access Server**) s/he will access. The architecture is the one presented in Chapter 21, analyzed from the point of view of the context-aware data management and access.

The upper layer of this facet of the architecture consists of the Context-Aware Web Portal, which allows the user to access the information requested based on her/his role and on the context. The portal represents the access point for potential users to the data, in that it provides a specific view on a subset of data, based on the role each user has within the particular context, avoiding the need to dodge through not relevant information. Furthermore, all of the elaborations, queries and data retrieval operations necessary to present the information in the web portal are transparent to the user. Indeed, the communication between portal and the system occurs through the web which filters the requests and breaks them down with respect to the single architectural components involved. This occurs independently of the type of information source, either internal, external or coming directly from sensors.

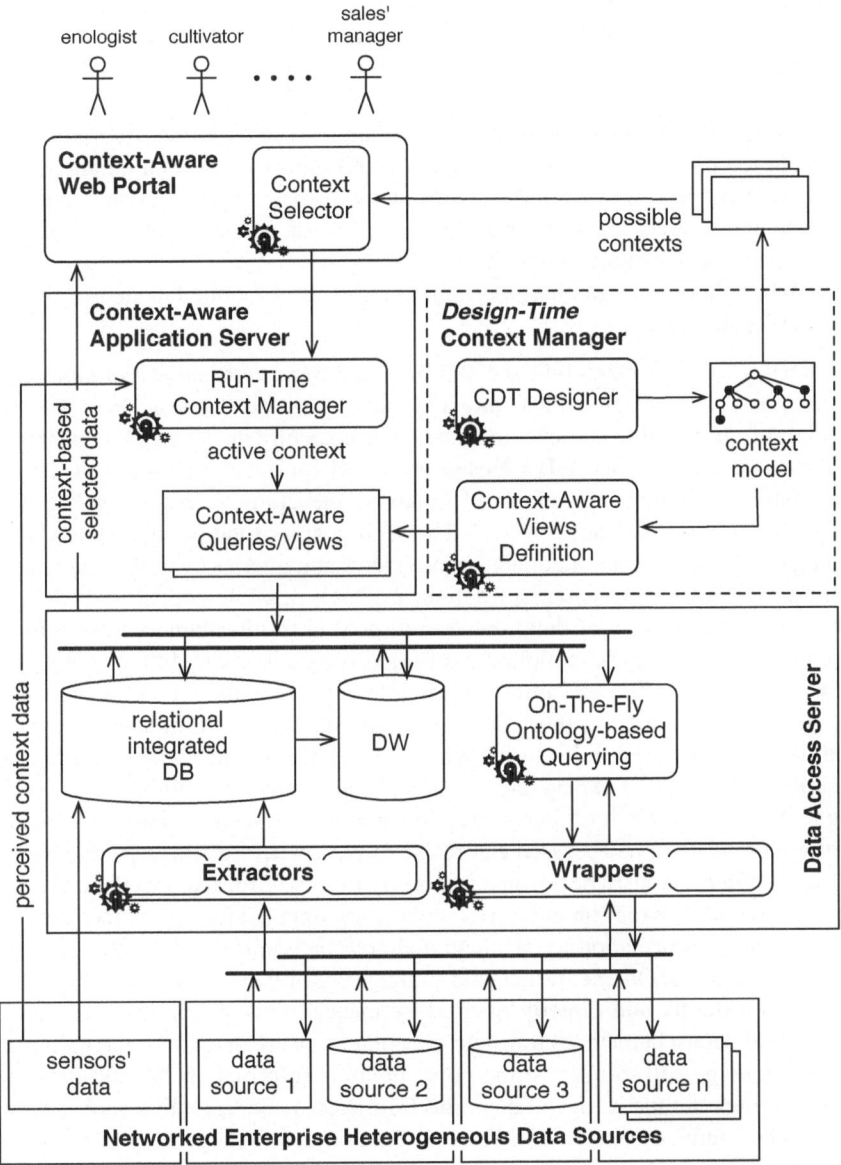

Fig. 12.1 The system architecture of the networked enterprise, from the context-aware data access point of view

The rest of the architecture consists of the above mentioned components, devoted to the following tasks:

Design-Time Context Manager: In charge of supporting, at design time, the modelling of the context of the specific application scenario, and the formulation of the context-aware views and queries over the available data sources, associated with the various possible contexts.

Context-Aware Application Server: In charge of determining the *active context* based on the explicit user's selection and on the contextual parameters autonomously perceived from the environment, dispatching the queries to the different data sources and delivering to the web portal and to the other applications the context-driven retrieved data.

Data Access Server: In charge of maintaining all available data and of the data access technologies.

Indeed, context is taken into account both at *design time* and at *run time*. In the former phase, the Design-Time Context Manager supports the designer in the specification of the possible contexts the users will be acting in. This task is performed by means of the so-called **CDT Designer**, a tool for the specification of the context model according to the adopted formalism, the *Context Dimension Tree*, CDT [2], briefly presented in Section 12.3. The second aim of the Design-Time Context Manager is the definition of context-aware views and queries over the data sources. More precisely, it is necessary to define, for each possible context, which is the context-relevant portion of data; the overall goal is to filter unnecessary information, that would make the exploitation of interesting data more difficult. Therefore, the **Context-Aware Views Definition** module automatically computes such views and queries, starting from the CDT and additional information, according to the methodology presented in [5, 6] for XML and relational data sources. More information on how context affects data querying is presented in Section 12.5, by referring to the different kinds of data sources the application scenario may be accessing. Furthermore, data to be accessed can reside in the enterprise relational database, as the result of a previously performed import by means of format-specific extractors; these components wrap the enterprise-interesting data sources to extract the (possibly external) information to be integrated and materialised within the enterprise relational database. Other external data sources, possibly not known in advance, are accessed on-the-fly and directly queried by means of wrappers, used to interpret the external data schemata, to translate the context-aware query into the native data source language and to retrieve the corresponding information. Should it be necessary to integrate the retrieved data, the Data Access Server offers such a feature to provide a unified answer to the user's data request. Altogether these modules constitute the Data Access layer and the approaches to implement them are further discussed in Section 12.5.

The information on the supported contexts, the context-aware views and queries is exploited at run time, by means of the **Context-Aware Application Server**; two are the main tasks this component carries out: a) it determines the active context, and, on its basis, b) it applies the context-aware views and queries, collecting the data from the sources and providing it to the upper application layer. In particular, the **Run-Time Context Manager** is in charge of receiving both user's explicit characterization (e.g, his/her role, specific interest category) and the perceived context data (for instance, from a GPS or other sensors) to determine the active user's context, used to select the interesting portion of data, by applying the appropriate **Context-Aware Queries/Views**.

The application layer, the **Context-Aware Web Portal** in our scenario, supports the selection of the user's active context, to allow a customised access to data. In particular, some of the aspects characterizing the active context can be automatically derived (such as the location and time of the user, possibly her/his role within the application scenario, the kind of device used to access the information), while others may require an explicit specification by the user (for instance, her/his interests). Therefore, the output of the Design-Time Context Manager component is the list of possible contexts, together with the different aspects characterising each of them. In particular, the context model is such that it is quite immediate to derive an application related **Context Selector**, taking as input the possible contexts and allowing the user to specify her/his active one.

In the next sections, we shall detail the several aspects contributing to the overall picture by starting from the context model, which is the pivot element of the entire approach.

12.3 Context Model

Our context model, called Context Dimension Tree, plays a fundamental role in tailoring the available information according to the user's needs. In particular, the CDT models all possible applicable contexts with respect to the given application scenario, and is then used as the starting point to define which portion of data should be made available in each possible context. It is worth noting that in different working scenarios it is possible to associate with a context not just the relevant portion of data, but, for instance, a set of rules or actions that should be carried out. In this perspective, the proposed context model is a quite general means for context-aware knowledge management.

Figure 12.2 shows an example – discussed below – which models the possible contexts of the wine cultivation scenario used in in Chapter 21. Black nodes represent dimensions and sub-dimensions, white nodes represent dimension values; nodes with a double border are parameters.

In the CDT, the root's children are the context main dimensions, which capture the different characteristics of the users and of their context; by specifying a value for each dimension, a point in the multidimensional space representing the possible

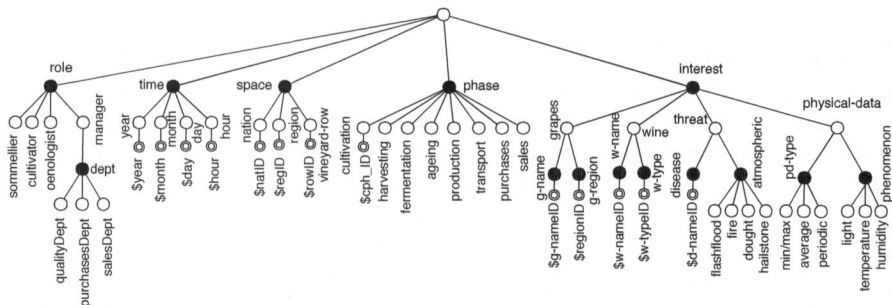

Fig. 12.2 A context model – CDT– for the wine case study

contexts is identified. As an alternative, rather than explicitly listing all possible values, in order to specify the exact value for the (sub-)dimension, it is possible to specify a parameter to be given an actual value at run time. Moreover, a dimension value can be further analysed and specialised, becoming the root of another sub-tree.

Independently of the application scenario, one of the key dimensions is the role, for the specification of the possible classes of users of the application; it is quite common, in fact, that different classes of users are interested in different portions of data, possibly due to access permission limitations. In the present example, seven are the foreseen roles of users accessing the available data, namely the wine *cultivator*, the *oenologist*, the *sommelier* and the *managers*, further classified on the basis of their specific department. Different, possibly partially overlapped, subsets of data will be made available to users of the various classes; when the user identifies him/herself as belonging to one of these context classes, a dedicated tailoring of the information will be provided.

time and space dimensions are often included in the context model, used to tailor data according to the time the user accesses the information or the time-window data belongs to, or to the user's location. In the wine scenario, as far as the temporal aspects are concerned, different granularities are introduced, based on the fact that certain information is analysed on a per-*year* or per-*month* scale, as for instance sales or production in general, whereas, when accessing data related to the productive process, a more frequent access is necessary (*daily* or *hourly*). Similarly, also the location can be used to filter information on wine, for instance with reference to a specific production *nation* or *region*.

The phase dimension is used to model the different application situations the user may be in; the values for this dimension include standard commercial phases such as *purchases* (e.g., of raw components such as sugars and yeast) and *sales* (related to the amount of distributed and sold wine), along with domain-specific phases like *cultivation* (further identified by its specific cph_ID) and *harvesting* of grapes, *fermentation* of grape must, and *ageing*, and *transportation* of wine. While the commercial phases are, in general, used to tailor data stored in the commercial

database, the latter dimensions are mostly used for the analysis of the plants growing process and the environmental phenomena that may be collected from sensors.
Notice that the selection of a particular cultivation phase is important since certain
values perceived by sensors may be considered as "alarms" or not on the basis of
the cultivation phase of the grapevine.

Finally, the interest dimension constitutes the other key dimension, being the
one that mainly drives the tailoring process, since it coarsely partitions data into
chunks, further refined by the other dimension values. Given the coarse granularity, a hierarchy of levels is introduced, to refine the selection criteria, for a more
precise tailoring of data. In the present example, four main interest areas have been
envisioned, namely *grapes*, *wine*, *threat* and *data-from-sensors*. In order to allow
a more refined selection of information, *grapes* can be selected according to their
name (*g-name*) and region (*g-region*), *wine* by name (*w-name*) and (*w-type*). As far
as *physical-data* is concerned, it is selected based on the physical *phenomenon* they
monitor and on the kind of monitoring (*pd-type*) the user is interested in. Moreover,
the *threat* value is selected whenever the analyst is concerned with those phenomena
that might threaten the quality of the final product. Threats include plant *disease* and
strong *atmospheric* phenomena like *flashfloods*, prolonged *drought*, *hailstone* and
fire, provoked either by natural causes (e.g., hot temperatures) or arsons.

A context derives from the specification of one or more of its characterizing dimensions, whereas given a dimension, only one of its values can be selected to
define a single context. For example, given the CDT of Figure 12.2, the following is
a valid context:

$$
\begin{aligned}
C \equiv \quad & (\texttt{role} = \texttt{oenologist}) \\
& \wedge\,(\texttt{time} = \texttt{year}(\$\texttt{yearID} = 2010)) \\
& \wedge\,(\texttt{space} = \texttt{region}(\$\texttt{regID} = \text{``Sicily''})) \\
& \wedge\,(\texttt{phase} = \texttt{production}) \\
& \wedge\,(\texttt{interest} = \texttt{grapes}(\$\texttt{regionID} = \text{``Sicily''}))
\end{aligned}
$$

Quite intuitively, the context refers to an oenologist interested in production data of
grapes in year 2010 and grown in Sicily. White nodes and black leaf nodes are also
called *context elements*, since they constitute the the characterizing elements of a
context.

The model allows the expression of *constraints* or *preferences* among the values
of a context definition to avoid the generation of meaningless ones (more details on
constraints and their use can be found in [6]). As an example, a constraint might
indicate that, in a context for a user with a sale manager role (*dept=salesDept*),
information referring to data located in a certain vineyard row (*vineyard-row*) is not
interesting. This is stated as follows (*useless-context* constraint, uc):

$$
\begin{aligned}
\texttt{uc}((\texttt{dept} = \texttt{salesDept}),(\texttt{space} = \texttt{vineyard--row})) \equiv \\
\neg((\texttt{dept} = \texttt{salesDept}) \wedge (\texttt{space} = \texttt{vineyard--row}))
\end{aligned}
$$

Here we have introduced a set of constraints to separate the two main areas of the application domain, which are related to the wine production process, and to its output as distributed and sold.

A *context*, stemming from this model, is obtained by composing elements of the CDT, and determines a portion of the entire data set, i.e., a *view*, specified at design-time. At run-time, when a context becomes active, the corresponding view is selected and all parameter values involved in the specification of that context are appropriately instantiated.

12.4 Methodology for Context-Aware Data Filtering

The ultimate goal of context-awareness in the ART-DECO project is to filter knowledge and information based on the user's active context. More precisely, given the possible and reasonable contexts modelled by the CDT, it is then necessary to associate with each of them the portion of data to be made available to the user. We have proposed a methodology for semi-automatic computation of the portion of data associated with each possible context, starting from the context model and the schema of the application data. Such a methodology consists of an approach of general validity, specialised for the different nature of the available data sources. The data and application designer is in charge of defining such a relation, at design time, so that, at run-time, once the user's active context is known, s/he receives the selected portion of information.

Two strategies have been proposed to support the designer in his/her activity.

The first strategy works globally on a *per-context* basis, it is precise but burdensome, and it is quite independent of the nature of the data source, since it requires the designer to specify, for each possible significant context, the portion of the data to be considered relevant. In such a perspective, the definition of the so-called *relevant area* is carried out independently for each context, by specifying on the data source what elements are to be discarded and filtered. The activity is though quite demanding, especially because the number of possible contexts derived even from small CDTs is of the order of hundreds. As a result, although many contexts might have similar associated relevant areas, the required designer's effort is particularly high. On the other hand, a great advantage of such an approach is the possibility to precisely tailor the interesting data, with a fine granularity.

The second strategy, working on a *per-context-element* basis (the node of the CDT), has been introduced to cope with the complexity of the former one, to reduce the amount of user intervention, still achieving a relevant filtering power and precision. This approach requires the designer to specify, for each dimension value (white nodes) or dimension node with parameter (black nodes with parameter), the relevant area that should be made available when a context contains such an element. Subsequently, the relevant area corresponding to an entire context is automatically computed by combining the areas of the included context elements, by means of ad-hoc designed operators (see [6] for the relational scenario).

For example, the contextual view (relevant area) for the context C introduced above is obtained by combining the relevant areas of its components, that are

$(\texttt{role} = \texttt{oenologist}), (\texttt{time} = \texttt{year}(\$\texttt{yearID} = \texttt{2010})),$
$(\texttt{space} = \texttt{region}(\$\texttt{regID} = \texttt{"Sicily"})), (\texttt{interest} = \texttt{grapes}).$

In this perspective, the methodology is strictly related to the specific nature of the data source, since operators for combining *partial* relevant areas cannot be general-purpose because they operate on the source schema and instances. Indeed, when dealing with relational data sources, the partial relevant areas are expressed by means of sets of relational views, thus, the intersection operator (a set-intersection operator, dubbed *double intersection*) is applied to sets of virtual tables (views) both at the schema and the instance levels, between pairs of relational views defined on the same relation, thus producing a tailored portion of the database containing only the pertinent common information.

In this chapter we focus on the relational database scenario, used to tailor the enterprise internal database. For dynamically querying external data sources an ontological approach is proposed (see Chapter 11).

12.5 Data Querying

In the ART-DECO scenario we face a situation characterised by three main categories of information sources: i) *internal*, resulting from the extraction of data from different heterogeneous sources, integrated and materialised in a unified relational database, ii) *external*, unknown in-advance and possibly transient data sources, to be queried on the fly, without materializing their information in a permanent data storage, and iii) data coming *from sensors*, used to perceive part of the user's context.

The internal data source is the result of an extraction and integration process carried out on the information coming from the available and stable data sources of the networked enterprise. A set of *extractors* has been designed (Figure 12.1), to bundle the different sources according to their native format, and to extract their information, to be saved in a general, unique relational database. The proposed context-aware data filtering approach accesses this relational data, by selecting a portion of the entire information, according to the user's context. The filtering process, discussed in Section 12.5.1, adopts the per-context-element strategy discussed in Section 12.4, and exploits the use of specific operators to combine the relevant areas of information to determine the portion of the entire database corresponding to each possible context.

When data sources are not stable and possibly not available at all times, a more dynamic access to their data is proposed. In this situation, a set of *wrappers* has been designed and developed, to cope with the different source formats; these wrappers offer the opportunity to interpret the source data, to deliver the query to such sources, and to collect the retrieved information. Section 12.5.2 offers an overview of the adopted approach.

Finally, context awareness has a two-fold role with respect to sensors' data, as discussed in Section 12.5.3: part of the collected information is stored in the integrated internal database, to constitute part of the information to be made available through the application and is affected by context as in the general internal case, part of it is directly used to determine some of the contextual parameters, such as the time of the day or the location (e.g., from GPSes), that is it becomes part of the context itself. In the latter situation, the information is directly exploited by the *Run-Time Context Manager* to actualise the correct context.

12.5.1 Internal Data Sources

Data extracted from the several available sources are integrated and materialised into a unique relational database for a context-based filtered access, as discussed in the following. Moreover, for an effective and efficient analysis of trends, a data warehouse has been developed. Even though the nature of the data is common to both situations, context-awareness can be exploited with different flavours.

Once the context model for the specific application scenario has been defined, the designer must then define the association between contexts and their relevant portion of data. Here we only discuss the strategy requiring the definition of a relevant area for each one of the context elements, to be combined to derive the relevant area for the whole context. In particular, two steps constitute the adopted methodology: i) *partial view* definition and ii) *partial view* combination.

During the former activity, the designer defines a mapping between each *context element* of the CDT and a portion of the relational schema, a *partial view* or *relevant area*. More precisely, the partial view associated with a context element has a tailored schema with respect to the global one, where some relations do not appear and those that appear may have a reduced number of attributes or tuples. In fact, in this operation the designer is defining which subset of the information is of interest any time a context includes such a context element.

As an example, the designer must specify the relevant area, $\mathcal{R}el(\texttt{oenologist})$, associated with the context element *oenologist*, by determining which relations contain data interesting for an oenologist, and with respect to such relations, which attributes, and, eventually, which subset of tuples. The same activity must be carried out for each one of the white nodes in the tree and all other black nodes with a parameter, since they can be part of a context specification.

Once all the partial views have been specified, the second step automatically combines the partial views of the nodes constituting a given context, to obtain a *final view*, which is a tailored schema and dataset over the original one. The computation of the combined view is carried out by means of purpose-defined operators, introduced preliminarily in [6], working on sets of relations. For example, the contextual view (relevant area) for the context C introduced above, is obtained by combining by means of the so-called *double-intersection* operator, $\cap$, the partial relevant areas of its components, that is:

$$\mathcal{R}el(C) = \mathcal{R}el(\text{role} = \text{oenologist}) \cap \mathcal{R}el(\text{time} = \text{year}(\$\text{yearID} = 2010)) \cap$$
$$\cap \mathcal{R}el(\text{space} = \text{region}(\$\text{regID} = \text{``Sicily''})) \cap$$
$$\cap \mathcal{R}el(\text{phase} = \text{production}) \cap \mathcal{R}el(\text{interest} = \text{grapes})$$

Consider now the following partial view defined by the designer, associated with the grapes *interest*, containing information about vineyards, their composition w.r.t. grapevines, and the harvest data:

$$\mathcal{R}el(\text{interest} = \text{grapes}) = \{\text{VINEYARD}, \text{HARVEST}, \text{CELLAR}, \text{ROW}, \text{GRAPEVINE}\}$$

The partial view, associated with the region(\$regID = "Sicily") *space*, includes information about wines, vineyards, and so on, filtered on the base of the specific region:

$$\mathcal{R}el(\text{space} = \text{region}(\$\text{regID} = \text{``Sicily''})) = \{\sigma_{region=\text{``S icily''}} \text{VINEYARD},$$
$$\text{HARVEST} \bowtie_{vineyard\_id=v\_id} \sigma_{region=\text{``S icily''}} \text{VINEYARD},$$
$$\text{CELLAR} \bowtie_{vineyard\_id=v\_id} \sigma_{region=\text{``S icily''}} \text{VINEYARD},$$
$$\text{BARREL} \bowtie_{cellar\_id=cl\_id} \text{CELLAR} \bowtie_{vineyard\_id=v\_id} \sigma_{region=\text{``S icily''}} \text{VINEYARD}, \dots\}$$

The $\cap$ operator applied on partial views gives as result a set of tables obtained by applying the classical intersection operator $\cap$ to all pairs of expressions having the same schemata. On the above partial views we obtain the following set:

$$\mathcal{R}el(\text{interest} = \text{grapes}) \cap \mathcal{R}el(\text{space} = \text{region}(\$\text{regID} = \text{``Sicily''})) =$$
$$\{\sigma_{region=\text{``S icily''}} \text{VINEYARD}, \text{HARVEST} \bowtie_{vineyard\_id=v\_id} \sigma_{region=\text{``S icily''}} \text{VINEYARD},$$
$$\text{CELLAR} \bowtie_{vineyard\_id=v\_id} \sigma_{region=\text{``S icily''}} \text{VINEYARD}\}$$

As another example, the contextual view for a *sommelier*, interested in the *average temperature* of the canteen for the current day, while the wine is *ageing* is computed as the intersection of the relevant areas of the context elements involved in the definition, that is:

$$\mathcal{R}el(C) = \mathcal{R}el(\text{role} = \text{sommelier}) \cap \mathcal{R}el(\text{time} = \text{day}(\$\text{dayID} = \text{today}())) \cap$$
$$\cap \mathcal{R}el(\text{phase} = \text{ageing}) \cap \mathcal{R}el(\text{phenomenon} = \text{temperature})$$
$$\cap \mathcal{R}el(\text{sd}-\text{type} = \text{temperature})$$

It is immediate to see that the number of context elements for which the designer has to define a relevant area is at least an order of magnitude smaller than the number of possible contexts, therefore the per-context-element strategy is particularly efficient. In general, the designer can start by the top context elements in the tree (the ones closer to the root) and define the relevant area for each one of them. Nodes belonging to sub-trees rooted in such upper nodes will have a relevant area that is included in their parent node's area, thus limiting the overall necessary effort. As previously stated, since the combining procedure is automatic starting from the definition of the relevant areas, the resulting contextual view might be slightly imprecise, however the designer can finally refine it to satisfy his/her needs.

12.5.2 External Data Sources

In the ART-DECO scenario, Semantic Web technologies such as RDF [8] and OWL [17], which might fail on large-scale data integration [11], play an important role as tools for on-the-fly integration of small pieces of information belonging to external, independent and heterogeneous data sources. In open environments, *external data sources* may be Relational DBMSs, XML and RDFS files, web-pages, natural language documents, sensor data, etc. Sensor data are taken care of by the PerLa query language and by appropriate middleware layers (see Chapter 18), while web-pages and natural language documents are extracted according to technologies described in Chapters 10 and 9. ART-DECO provides a framework to automatically extract ontological representations from the schemas of Relational databases, XML and RDFS files, and (semi)automatically map them to the application-domain ontology (see Chapter 9), which is used as a uniform representation of the available information and can be queried using SPARQL [13]. Moreover, context-based information space reduction is performed at runtime by selecting fragments of the application-domain ontology which, through the mappings, induce corresponding fragments on the data-source ontologies, and thus a subset of the information space. A complete description of our approach to on-the-fly query distribution, with a focus on relational datasources, is provided by Chapter 13, while here we give a quick overview.

We consider ontologies expressed using CA-$\mathcal{DL}$, a fragment of OWL-DL [17] which (a) can uniformly represent the data and the user context(s) in any application domain and (b) supports context-aware SPARQL query-answering and distribution to (possibly heterogeneous) information sources. The SPARQL query is translated into a context-aware query over the (context-aware) portion of the domain ontology, and thus reduced. Then, the distribution of the reduced query takes place, as described in Chapter 13. All the data-source queries are translated into the native language of each target source by means of automatically generated wrappers, which also provide a CA-$\mathcal{DL}$-compatible representation of the answers. A final step hands the results over to the user after a (possible) further contextual refinement.

12.5.3 Data from Sensors

Thanks to the advent of pervasive systems, sensors became an important source of data; sensors are nearly always organised in networks and often they are tiny devices which communicate over wireless links which constitute a *wireless sensors network* (WSN). From a datacentric point of view a WSN can be seen as a special form of *distributed database* [16]. However, differently from traditional (relational) distributed databases, in which data are permanently stored on stable storage and data sources are mostly known and durable, in a WSN *sources availability greatly vary over location and time* due to one of the most outstanding properties of wireless sensors: their life is bounded to that of their power source. Other reasons for sensors data unavailability lie in the weakness of the communication links and on the physical integrity of the sensor itself, which is often deployed in harsh environments.

Another important feature that distinguish WSNs is a somehow *reversed usage paradigm* with respect to traditional databases: there is no more a passive database and users who actively put queries on it, but queries are embedded into the sensors and actively send, possibly for long periods, *data streams* to a base station, where the application programs consume them. Moreover, sometimes queries can also specify actuation actions on the environment or on the measurement behaviour of the sensor; therefore each device can be, at the same time, a producer and a consumer of data.

Two phases must be considered in the life of a query in a WSN:

- the *query dissemination* phase, where queries are downloaded to the sensor's local memory. As to this phase, we must notice that the same sensor can be used for very different types of applications, with different usage modes. On the other hand, the syntactical structure of queries is fairly standard but for the presence or absence of some constructs and options [16]; *it is the context that makes the difference*. Therefore, starting from a standard skeleton, context can drive *query tailoring* in order to fulfill the application needs both as to the query text and to possible execution parameters, such as setting the sampling frequency or some alarm thresholds.
- the *result collection* phase, where the measurement data are fed back to applications. This phase is heavily affected by power management issues. As it was observed, once the battery is exhausted, the sensor dies since it is difficult or not worth to replace it; therefore energy sparing is one of the main optimization goals in query processing in WSNs. Since data transmission is more energy eager than local data processing, data preprocessing on a single sensor or within a sensors cluster is often needed and complex optimization techniques involving MAC, routing protocols and local processing have been proposed. For these reasons often applications are happy with approximate results, the bounds of which again depend on the *working context* of the application and are embedded in the downloaded query.

Last, but not least, sensors not only feed data directly to the application, but, as shown in Figure 12.1, they also provide numeric values for evaluating the context dimensions. Let us suppose that, during the ageing phase, the wine can be in three different *context situations*: initial, middle, mature; these situations are defined over three physical variables – temperature, acidity, activity – which are respectively sensed by thermometers, Ph-meters, and acoustic meters, and are subject to the following rules:

situation = initial :=
{(*temperature* < *xx C*) AND (*acidity* < *Ph y*) AND (*noise* < *z db*)};
situation = middle :=
{(*temperature* > *xx C*) AND (*acidity* = *Ph y*) AND (*noise* > *z db*)};
situation = mature :=
{(*temperature* < *xx C*) AND (*acidity* > *Ph y*) AND (*noise* < *z db*)};

Therefore, data provided by sensors might directly affect the actual context through a feedback path, possibly inducing a context change. Such systems must be carefully designed and the sensor data must be cleaned and filtered in order to prevent the arising of possible instability situations.

In Chapter 18, a data management language for sensor data developed within the ART-DECO project, is presented, which deals with many of the mentioned functionalities.

12.6 Selecting Context, Query and Query Answering: A Web Portal

The context-aware knowledge querying concepts illustrated in the previous sections have been exploited through a case study in a wine production scenario.

Before going into the details of the specific study case, it is important to provide some general information on the Context-Aware Web Portal. First of all, it should not be seen as a mere container of information, rather it is an access point for creating knowledge by sharing and collecting information and experiences through a combination of contents, processes, technological and human resources. It is a framework for integrating information, people and processes across boundaries, and provides a secure unified access point in the form of a web-based user interface, designed to aggregate and personalize information through application-specific portlets. One feature of the web portal is the de-centralised content contribution and content management, which keeps the information always updated. Fundamental features of the web portal generated for the case study are:

- *Single Sign-On* capabilities between the users and various other systems. As so, the user authenticates her/himself only once;
- *Integration*, the connection of functions and data from multiple systems into new components/portlets/web parts with an integrated navigation between these components;
- *Federation*, the integration of data coming from heterogeneous sources is transparent to the user;
- *Customization*, users can customize the look and feel of their environment. This aspect also refers to the ability to prioritize most appropriate content based on attributes of the user and metadata of the available content;
- *Personalization*, consists in matching content with the user. Based on a user profile, personalization uses rules to match the "services", or content, to the specific user;
- *Access Control*, it is possible to limit specific types of content and services users have access to, based on their role. These access rights may be provided by a portal administrator or by a provisioning process. Access control lists manage the mapping between portal content and services over the portal user base. This assures that each user has a narrower view over the large amount of information contained in the system, i.e. only the portion of information considered relevant

for that user, according to the context, can be accessed, reducing the risk of disorientation and avoiding having to dodge through quantities of unnecessary data, as the information that appears to the user is retrieved dynamically from the system's data sources.

The web portal communicates with the system architecture through web services, which filter the user's requests and break them down with respect to the single architectural components involved. This occurs independently of the type of information source, internal, external or coming directly from sensors used for monitoring the production processes as shown in Figure 12.1.

As previously mentioned, our case study application refers to a wine production context. Here, the contextual views are designed accordingly to the CDT, identifying different classes of users. Any user accessing the system for the first time is requested to register and during this process s/he is guided through a set of selections, conforming to the CDT dimensions, contributing to the determination of the "static" aspects of his/her active context (Figure 12.3).

Once the profile has been defined, each time s/he logs into the system, only the subset of information, relevant to the context, are shown (Figure 12.4). In this

Fig. 12.3 Access to the portal and context determination

Fig. 12.4 User authentication and context-aware interface generated for role=sales dept.manager

scenario the user can carry out a set of actions leading to different results according to the context.

The interface contains various parts. The "Main Menu" section includes, among others, a generic web-link feature common to all roles. It suggests a set of possible links that may be of interest for the specific user. So, although the feature is common to all profiles, the contents differ from one another. Also, each user may upload a document or a URL through a dedicated interface. The documents and links are made available to other users with the same context characteristics and therefore likely to have the same interests (Figure 12.5). A survey has also been structured to acquire the users' viewpoint on usability aspects. Any user is free to answer the questionnaire; only the first question of the survey is shown here, but once the user answers, the entire survey is displayed in a new page.

The specific parts of the portal relate to the "Context Aware Menu" and the "User Profile" in the lower left part. The first contains features for customizing and personalizing one's profile. Here a user can upload a Web Link which is classified as relevant for that certain context. The latter is specific of the role represented. Indeed, the header of the box is one's own role. We will comment the most relevant features of the specific parts related to the wine production scenario. A first functionality is the information retrieval. This is exploited through a semi-structured approach where s/he selects the information in a dedicated interface. A query is processed,

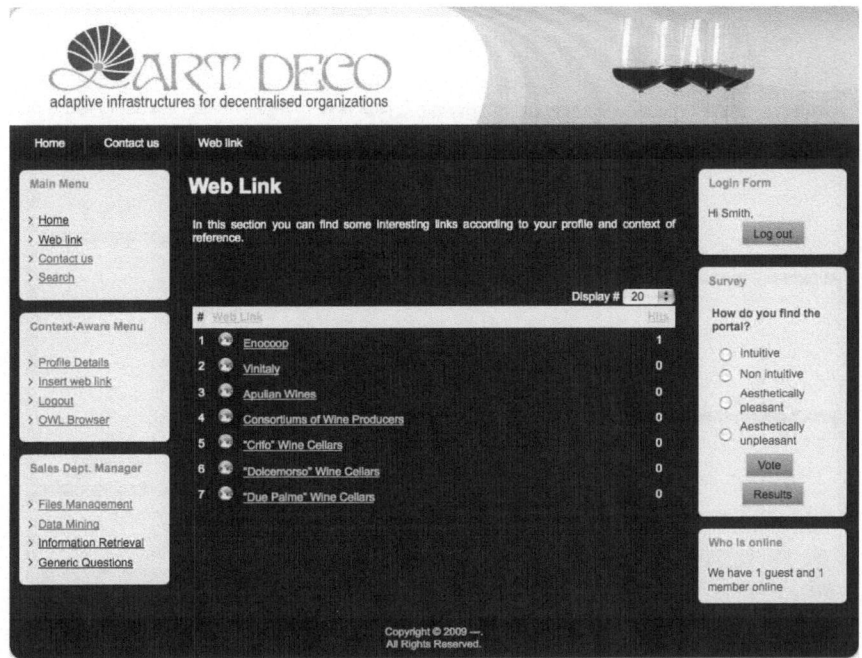

Fig. 12.5 Web-Links of interest for the `sales dept.manager` role

and data is extracted from the data sources. For example, suppose that a sales department manager wants to know the wines that have been top sellers in a specific year (i.e., dept = SalesDept, time = year($yearID =2010), and interest = wine), the result is shown as a list of wines that fulfill the interrogation (Figure 12.6).

The underlying Context-Aware Application Server provides, as the answer, only the portion of data corresponding to the user's active context. So, if on the other hand an oenologist were to ask the same information (time = year($yearID =2010) and interest = wine) but with a different role (role = oenologist), data on the cultivation and harvesting, collected from the sensors, would also appear, because the role is different, and therefore the view on the data changes. To summarize, in accordance to the previous sections, the web portal shows different portions of data to different users according to the combination of context dimensions, given a specific role.

As final application of the case study, it is worth mentioning that a user can also ask generic questions guided through a dedicated interface where s/he selects the parameters and the query is dynamically structured and sent to the system architecture components. Some examples of queries, based on the CDT, may be:

- Average time needed to provide an invoice [supplierID];
- Average temperature in cellar areas during a [day]/[month]/[year]
- Which are the most trendy wine origins/wine-food pairings/glasses/colours during the period [start]-[end]

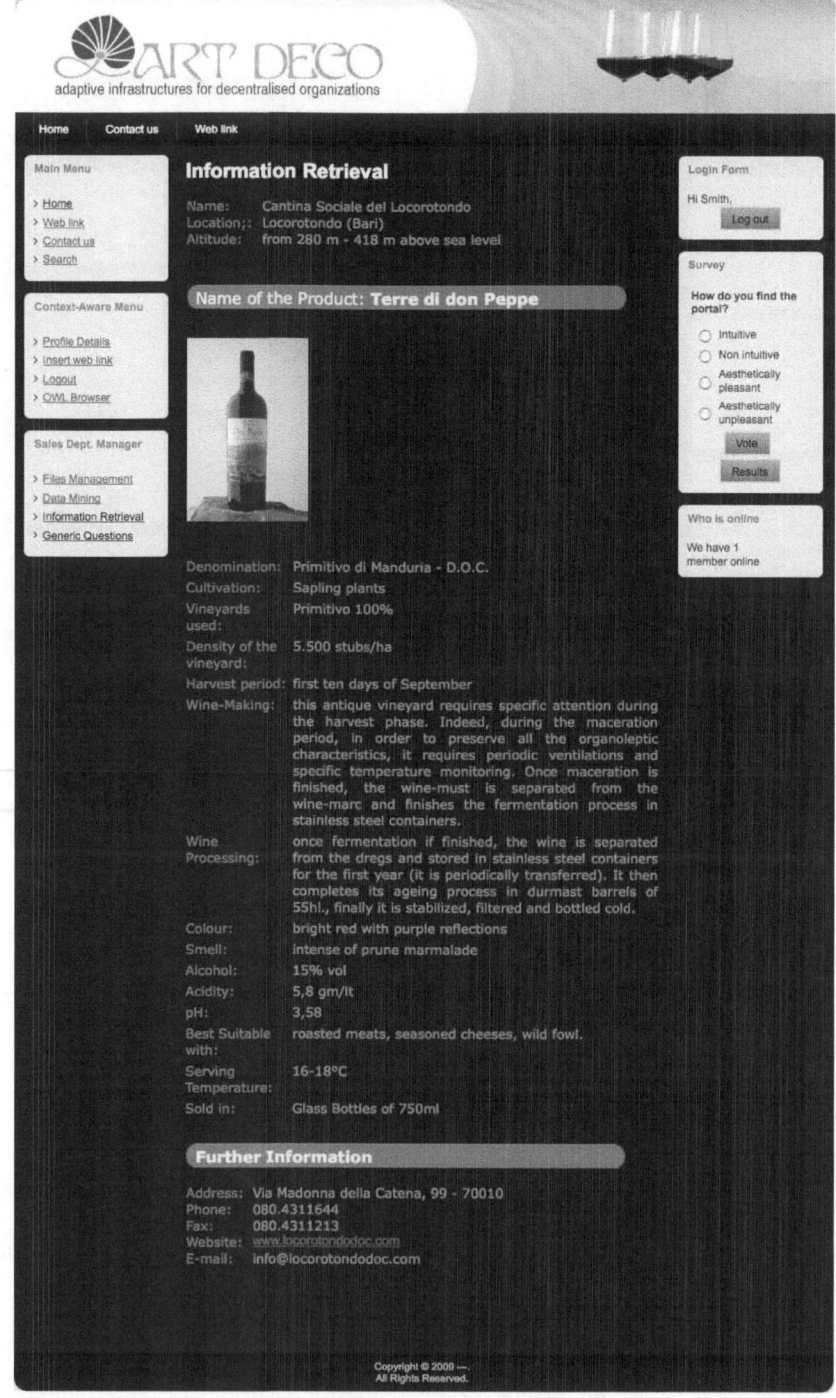

Fig. 12.6 Context-Aware Information Retrieval

- Which wines were drank by the various trend setters during [year]
- Which are the most trendy gift package ideas for DOC red wines produced in Sicily in years [start]-[end]?

These queries are generated from an interface which varies according to the context. Moreover, Figure 12.7 shows an example of the interface for the last generic question of the list.

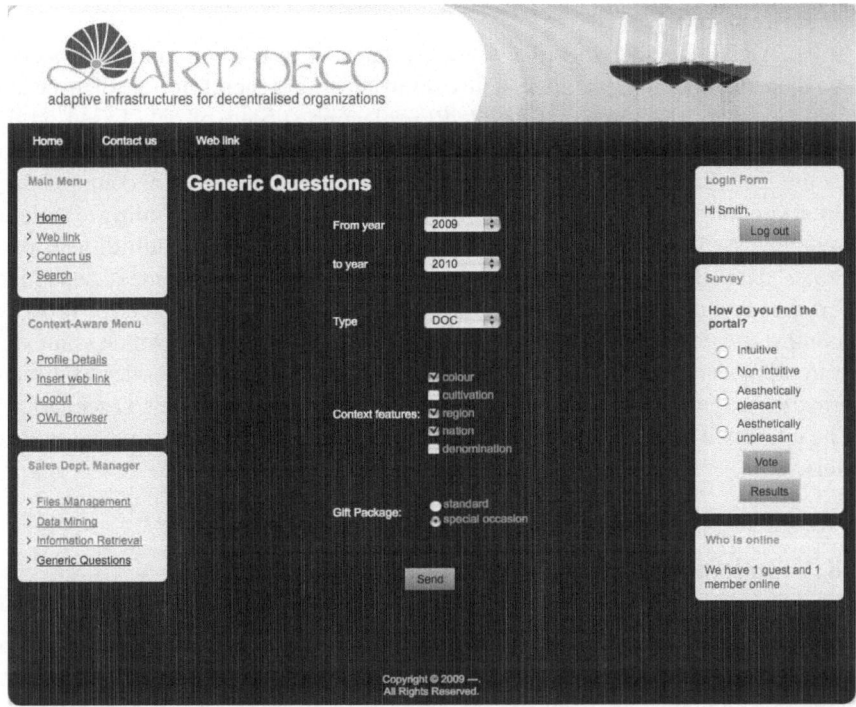

Fig. 12.7 Example of Generic Question

As it can be seen, the entire process of query building, information search and retrieval, and results is completely transparent to the user, who has no perception of which system components the data comes from, nor how his/her view differs from or is similar to other ones. In this way the web portal offers a systematic manner for dynamically extracting a specific context-aware view over the entire system.

12.7 Related Work

Context modelling refers to the representation of a user and his/her context at the level of a system profile. Adequate modelling of the user and his/her context is the basis for every form of personalization support. There are different lines along which a context model should be analysed in order to be adopted: (a) the modelled aspects; the set of context dimensions managed by the model (e.g., time, location, situation, ...); (b) the representation formalism (graph-based, logical, etc.); (c) the language (e.g., XML). These choices heavily influence context manageability and dynamicity.

Sophisticated and general context models have been proposed, to support context aware applications which use them to (a) adapt interfaces [9], (b) tailor the set of application-relevant data [3], (c) increase the precision of information retrieval [18], (d) discover services [14], (e) make the user interaction implicit [12], or (f) build smart environments [10].

In the ART-DECO perspective, we are interested to data-oriented approaches, rather than methodologies and systems referring to a general notion of user and context. Among the approaches devoted to selecting subsets of data based on the notion of context, we count the Context-Relational model (CR) [15], which proposes the use of relational tables having an additional dimension, accommodating different versions of the same information, to be made available according to the current context. Analogously, in [19], context is modelled as a set of multidimensional attributes, data cubes are used to store the dependencies between context-dependent preferences and database relations, and OLAP techniques are adopted for processing context-aware queries. In [7], [21] an approach is presented in which context is used to tailor data, coming from multiple information sources, in order to discard information not relevant to the current situation, like in the ART-DECO scenario.

The interested reader can refer to [20, 1, 4] for comprehensive surveys on context models.

12.8 Conclusions

In this chapter, we showed how the scenario of a a networked enterprise is rich of information and data to be shared among different users, each one characterized by a specific role and interests. Furthermore, other distinguishing aspects may affect the portion of such broad information the user is interested in. As a consequence, the adoption of a contextual approach to data access is particularly interesting, to limit information overhead and noise. In this chapter we have presented a flexible contextual model supporting data tailoring in a rich, heterogeneous data environment, where both internal and external sources are available. Such a model is used as the starting point for the application of a methodology for the association of a portion of data deemed interesting for each possible context the user may be in. This activity is performed at design time, when the application scenario is envisioned, establishing the relation between the aspects characterizing the context, the possible significant contexts and relevant information that should be associated with each of them. At run time, the user's active context is exploited to effectively perform a dynamic tailoring of the available data, to retrieve only the significant information.

Within the ART-DECO project, the methodology has been developed and applied to the wine production case study, here presented, where the application providing the context-aware access to data is a web portal, developed to interact with a context-aware application server to support the specification of the user's active context and its exploitation to access filtered data.

References

1. Baldauf, M., Dustdar, S., Rosenberg, F.: A survey on context-aware systems. Int. J. Ad Hoc Ubiquitous Computing 2(4), 263–277 (2007)
2. Bolchini, C., Curino, C., Quintarelli, E., Schreiber, F.A., Tanca, L.: Context information for knowledge reshaping. Intl. Journal of Web Engineering and Technology 5(1), 88–103 (2009)
3. Bolchini, C., Curino, C., Schreiber, F.A., Tanca, L.: Context integration for mobile data tailoring. In: Proc. 7th IEEE/ACM Int. Conf. on Mobile Data Management, p. 5 (2006)
4. Bolchini, C., Curino, C.A., Quintarelli, E., Schreiber, F.A., Tanca, L.: A data-oriented survey of context models. SIGMOD Record 36(4), 19–26 (2007)
5. Bolchini, C., Quintarelli, E.: Context-Driven Data Filtering: A Methodology. In: Meersman, R., Tari, Z., Herrero, P. (eds.) OTM 2006 Workshops, Part II. LNCS, vol. 4278, pp. 1986–1995. Springer, Heidelberg (2006)
6. Bolchini, C., Quintarelli, E., Rossato, R.: Relational Data Tailoring Through View Composition. In: Parent, C., Schewe, K.-D., Storey, V.C., Thalheim, B. (eds.) ER 2007. LNCS, vol. 4801, pp. 149–164. Springer, Heidelberg (2007)
7. Bolchini, C., Schreiber, F.A., Tanca, L.: A methodology for very small database design. Information Systems 32(1), 61–82 (2007)
8. Carroll, J.J., Klyne, G.: Resource description framework (RDF): Concepts and abstract syntax, W3C recommendation. Technical report, W3C (2004)
9. De Virgilio, R., Torlone, R., Houben, G.-J.: A rule-based approach to content delivery adaptation in web information systems. In: Proc. 7th IEEE/ACM Int. Conf. on Mobile Data Management, p. 21 (2006)
10. Dey, A.K., Sohn, T., Streng, S., Kodama, J.: iCAP: Interactive Prototyping of Context-Aware Applications. In: Fishkin, K.P., Schiele, B., Nixon, P., Quigley, A. (eds.) PERVASIVE 2006. LNCS, vol. 3968, pp. 254–271. Springer, Heidelberg (2006)
11. Franconi, E.: Conceptual Schemas and Ontologies for Database Access: Myths and Challenges. In: Parent, C., Schewe, K.-D., Storey, V.C., Thalheim, B. (eds.) ER 2007. LNCS, vol. 4801, p. 22. Springer, Heidelberg (2007)
12. Petrelli, D., Not, E., Strapparava, C., Stock, O., Zancanaro, M.: Modeling context is like taking pictures. In: Proc. of the Workshop "The What, Who, Where, When, Why and How of Context-Awareness" in CHI 2000 (2000)
13. Prud'hommeaux, E., Seaborne, A.: SPARQL query language for RDF. Technical report, W3C (2007)
14. Raverdy, P.-G., Riva, O., de La Chapelle, A., Chibout, R., Issarny, V.: Efficient context-aware service discovery in multi-protocol pervasive environments. In: Proc. Intl Conf. on Mobile Data Management, p. 3 (2006)
15. Roussos, Y., Stavrakas, Y., Pavlaki, V.: Towards a context-aware relational model. In: Proc. Context Representation and Reasoning - CRR 2005, pp. 7.1–7.12 (2005)
16. Schreiber, F.A.: Automatic generation of sensor queries in a wsn for environmental monitoring. In: Proc. 4th Int. ISCRAM Conference, Delft (NL), pp. 245–254 (2007)

17. Schreiber, G., Dean, M.: OWL web ontology language reference, W3C recommendation. Technical report, W3C (2004)
18. Shen, X., Tan, B., Zhai, C.: Context-sensitive information retrieval using implicit feedback. In: Proc. Int. Conf. on Research and Development in Information Retrieval, pp. 43–50 (2005)
19. Stefanidis, K., Pitoura, E., Vassiliadis, P.: A context-aware preference database system. Journal of Pervasive Computing & Comm. 3(4), 439–460 (2007)
20. Strang, T., Linnhoff-Popien, C.: A context modeling survey. In: Workshop on Advanced Context Modelling, Reasoning and Management at UbiComp (2004)
21. Tanca, L.: Context-based data tailoring for mobile users. In: Proc. BTW 2007, pp. 282–295 (2007)

Chapter 13
On-the-Fly and Context-Aware Integration of Heterogeneous Data Sources

Giorgio Orsi and Letizia Tanca

Abstract. Users and applications of today's networked enterprises are often interested in *high-quality*, integrated information coming from external data sources, whose worth is often flawed by the *information noise* produced by *unfocused or partial information*. In this chapter we describe an ontology-driven framework for dynamic data integration of heterogeneous data sources, where user and application queries are dealt with in a context-aware fashion, i.e., by taking into account contextual meta-data about the system and the users in order to keep the information noise at bay.

13.1 Introduction

In modern enterprises, the proliferation of data-intensive web-applications, the presence of legacy databases and the dynamicity of pervasive technologies generate a need for on-the-fly access to, and thus integration of, external and possibly heterogeneous data sources. On the other hand, the availability of large information masses does not necessarily yield better decision support. Often, information is *noisy*, and should be appropriately shaped on the current needs of the users. One of the possible means to make the answers precisely fit the current user's needs is represented by *context-aware information reduction* [5], which uses *context meta-data* to reduce the

Letizia Tanca
Dipartimento di Elettronica e Informazione - Politecnico di Milano, via L. da Vinci 32, 20133 Milano, Italy
e-mail: tanca@elet.polimi.it

Giorgio Orsi
Department of Computer Science - University of Oxford, OX13QD Oxford (UK)
Dipartimento di Elettronica e Informazione - Politecnico di Milano, via L. da Vinci 32, 20133 Milano, Italy
e-mail: giorgio.orsi@cs.ox.ac.uk

G. Anastasi et al. (Eds.): Networked Enterprises, LNCS 7200, pp. 259–276, 2012.
© Springer-Verlag Berlin Heidelberg 2012

information space. User queries will then be answered using just the right amount of available information, properly focused using context information. A very important side effect of this reduction is that, by concentrating the query only on the context-relevant part of the information space, in general only a portion of the available data sources will be accessed, thus optimizing the query evaluation process.

In such a scenario, Semantic Web technologies such as OWL [15], which might fail on large-scale data integration [13], play an important role as tools for on-the-fly integration of small pieces of information belonging to various (independent and heterogeneous) data sources. In this chapter we describe the ontology-driven framework for dynamic and context-aware data integration of heterogeneous data sources proposed in ART-DECO, named *ART-DECO Dynamic Data Integration System (AD-DDIS)*.In particular, ART-DECO provides a framework, complete with tools and algorithms, to interact, at runtime, with interesting data sources which are external w.r.t. the enterprise: ontological representations are automatically extracted from their schemas and mapped to the enterprise application-domain ontology (henceforth called *ADO*), which is used as a uniform representation of the available information and can be queried using the SPARQL [27] language; note that the ART-DECO portal supports the user for easy query formulation.

The possible contexts envisaged in the application, and used to reduce the information space, are specified at design time by means of the general context model [5] presented in Chapter 12, represented here in ontological form. The information space is reduced at runtime by selecting the context-relevant fragments of the application-domain ontology which, through the mappings, induce corresponding fragments on the data-source ontologies and thus a subset of the available information space. The query issued by the user is translated into a context-aware query over the (context-aware) portion of the domain ontology, and thus reduced. Then, the distribution of the reduced query takes place.

The peculiar feature of an ontology-based Data Integration System (DIS) is the exploitation of a knowledge representation language (such as Description Logics [2]) to represent both the data (i.e., we use the DL as a data model) and the meta-data needed for the integration [16, 25, 6]. The result is, in general, a better description of the semantics of the integration system and, moreover, it is possible to perform inference over the mediated schema, to get richer answers to users queries.

13.2 Preliminary Discussion and Related Work

The problem of data integration has been investigated for over 20 years [18] and despite many advances in this field, the human necessity for effectively and efficiently store information produced many different formalism and several (possibly incompatible) semantics. All this evidence makes the data integration problem always new and challenging. The framework presented in this chapter heavily relies on ontologies: in this section we survey some of the general techniques presented in the literature.

We can reasonably group the various techniques into *mediated-schema approaches* [22], which rely on a global schema and specify mappings between each data source and the global schema, and *Multiple-schema or hybrid approaches* [14], where we usually have a data-source schema for each data source, a global schema and a mapping between the sources and the mediated schema. Similar approaches involve multiple schemas where data source schemas are mapped to one another in a P2P fashion. The advantage of multiple schema approaches is that we are free to specify the mapping between the data source and the mediated-schema without considering other data sources, thus making this method better suited for dynamic, (possibly) heterogeneous information integration tasks.

Orthogonally, it is possible to classify the various approaches also according to the type of mapping between the sources and the global schema (in our case represented by the domain ontology). In *GAV (Global-As-View)* we have a global schema specified as a view over the sources. In *LAV (Local-As-View)* [19, 8], each data source is specified as a view over the global schema. The process of query answering requires *view-based query rewriting*.

Another interesting approach in the direction of semantic integration is that of D2R [3], a mapping language which maps relational schemas to RDF triples. Through the D2R-Server it is possible to query via SPARQL or even to browse with semantic or common browsers relational schemas as if they were RDF databases.

More steps toward more dynamic DISs are represented by [17] and [29], performing query answering over distributed ontologies. The approach virtually integrates distributed and autonomous ontologies using ontology meta-data contained in distributed registries.

The role of *context*, formerly emerged in various fields of research like psychology and philosophy [7], has been recently studied also by the Knowledge Representation and Database communities in order to extend the data models with context-aware features. However, most of the research on context-aware data management systems [4] does not present a clear separation between the (meta)knowledge related to context and the knowledge which is of interest for the user. [28] makes such an attempt; here, context is elevated to the role of first class citizen, and operations on context-aware relations are defined, by extending the (traditional) relational algebra. More on context-related literature has been presented in Chapter 12.

13.3 The AD-DDIS Internal Language and Architecture

In this section we briefly introduce CA-$\mathcal{DL}$, a language which allows us to represent the context model, the domain and data source ontologies, the mappings among them and the conjunctive queries in a homogeneous way. Based on this, we then describe the AD-DDIS architecture and its components.

13.3.1 CA-$\mathcal{DL}$: A Data Language for AD-DDIS

The CA-$\mathcal{DL}$ language, syntactically expressible using a FO-rewritable fragment of OWL2-DL [15], allows us to represent a restricted form of *GAV* and *LAV* mappings, the context model, the domain and data source ontologies, and the conjunctive queries rewritable in the SPARQL syntax. It is worth noting that this language will be confined inside the system: users and applications will be only concerned with SPARQL queries and answers.

An OWL2-DL ontology can be defined as a 4-tuple $O = (N_C, N_R, N_I, A)$ where N_C is the set of class names, N_R is the set of role (i.e., binary relations) names, N_I is the set of individuals and A is a set of axioms of the conceptualization. The full OWL2-DL language provides several operators (called constructors) to build complex class expressions such as union and intersection ($\sqcup, \sqcap$), qualified existential and universal quantification ($\exists R.C, \forall R.C$), quantified existential and universal quantification ($\leq_n R.C, \geq_n R.C$) negation ($\neg C$) and role chaining ($R \circ S$) whose combined expressive power is too high for a query to be rewritable as a *first-order theory*. Concepts and roles are then related by axioms of the form $C \sqsubseteq D$ (concept inclusion axioms - *GIAs*) or $R \sqsubseteq S$ (role inclusion axioms - *RIAs*).

CA-$\mathcal{DL}$ allows us to represent the domain and data-source ontologies, simple forms of *GAV* and *LAV* mappings, the context model and the context-relevant areas of the ontology. It is worth noting that this language will be confined inside the system: users and applications will only be concerned with SPARQL queries and answers.

As usual when defining a DL language, we first describe the constructors that are allowed in the definitions of concepts and roles and how they can be possibly combined. We denote by means of the symbols A, B atomic, named (i.e., not anonymous) concepts and by the symbols R, P the atomic roles. Complex (i.e., constructed) concepts and roles are denoted respectively using the symbols C, D and S, T. Individuals are denoted by the symbols a, b.

Definition 1 (CA-$\mathcal{DL}$ ABox Assertions)
A CA-$\mathcal{DL}$ ABox contains assertions of the following form:

- $C(a)$ (type assertion)
- $S(a,b)$ (role assertion)

nothing else is a CA-$\mathcal{DL}$ ABox assertion.

While we do not impose any restriction on the form of ABox Assertions, in the following we restrict the form of CA-$\mathcal{DL}$ TBox formulas in order to retain decidability and tractability of query answering.

Definition 2 (CA-$\mathcal{DL}$ TBox Formulas)
A CA-$\mathcal{DL}$ TBox contains formulas of the following form:

- $\bigsqcap_{i=1}^{n} C_i \sqsubseteq A$ (concept subsumption)
- $\bigsqcap_{i=1}^{n} S_i \sqsubseteq R$ (role subsumption)

- Dom(S) $\sqsubseteq$ C (domain restriction)
- Ran(S) $\sqsubseteq$ C (range restriction)
- $\sqcap V_C \sqsubseteq \bot$ (concept disjunction)
- $\sqcap V_R \sqsubseteq \bot$ (role disjunction)
- C $\sqsubseteq \exists$R.D (qualified existential quantification)
- C $\sqsubseteq \exists$R.Self (hasSelf restriction)
- C $\sqsubseteq \exists$R.{a} (hasValue restriction)
- funct(S) (functional roles)
- key(S_1, ..., S_n), where none of the S_i is used as role in a hasValue, self restriction or as a right-hand-side of a role subsumption (key constraints).

nothing else is a CA-$\mathcal{DL}$ TBox Formula.

Another important aspect is the class of queries that we consider and that can be answered by taking into account CA-$\mathcal{DL}$ TBoxes. We concentrate on unions of conjunctive queries (UCQs) and represent them using DL syntax.

Based on the definition given in [16], we define a *context-aware mapping system* for AD-DDIS as a triple (*Rel(ADO)*, *DS O*, $\mathcal{M}$), where *Rel(ADO)* is the portion of the application domain ontology, which is relevant to the current context, *DS O* is a data source ontology and $\mathcal{M}$ is the mapping between *Rel(ADO)* and the *DS O*. $\mathcal{M}$ is composed by a set of assertions in the form $q_{ADO} \leadsto q_{DSO}$, where q_{ADO} and q_{DSO} are conjunctive queries over the *Rel(ADO)* and the *DS O* respectively, and $\leadsto$ $\in \{\sqsubseteq, \sqsupseteq\}$.

Note that the nature of the ontologies used within our system is that of terminological ontologies (TBoxes) only related by mapping assertions and employed as a means to rewrite and distribute the queries, while the real data are stored in the data sources, represented in their native models. Since the instances are not physically imported into the domain ontology, but remain in the data sources, in order to retrieve them the system takes a query Q expressed in terms of an application domain ontology, rewrites it in terms of the ontologies of the data sources containing the objects involved in the (contextualized) query , decomposes and forwards it to each data source where it will be finally rewritten in terms of the data-source query language.

13.3.2 The AD-DDIS Architecture

Figure 13.1 presents the three layers of our general architecture for Context-Aware Semantic Query Distribution in AD-DDIS, already quickly introduced in Chapter 12: (A) the *context-aware knowledge tailoring layer*, (B) the *query rewriting layer* and (C) the *query distribution layer*.

The *application domain ontology ADO* is used as a specification of the concepts of the specific enterprise domain along with the relationships among them, and plays the role of the global schema in traditional data integration systems. Figure 13.2) describes a portion of the *ADO* for the domain of wine production, written using

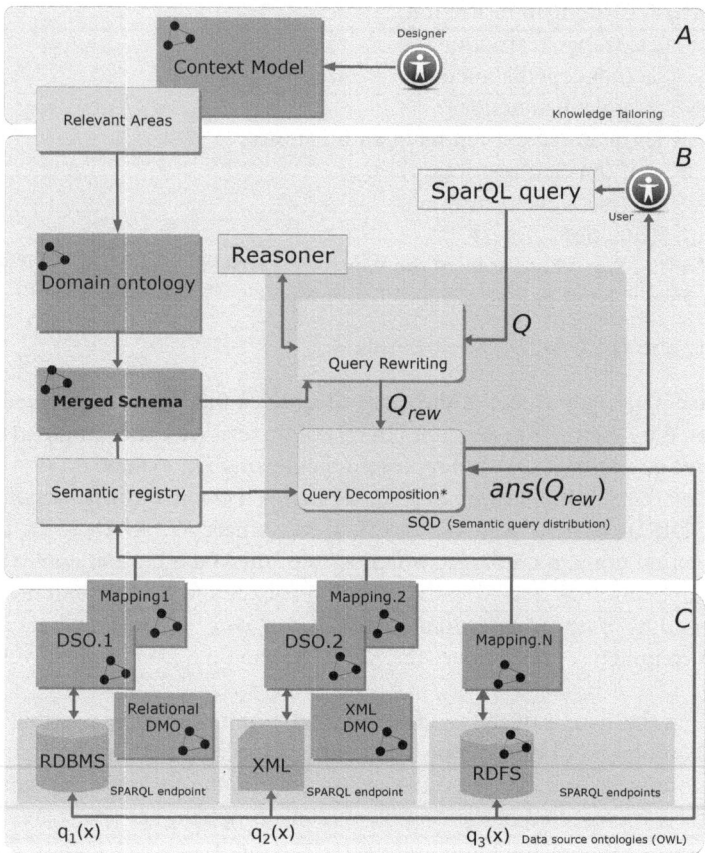

Fig. 13.1 AD-DDIS Architecture

Description Logics Formulae, containing domain concepts representing wines, customers, wine-farms, etc.

The *context model* (from now on *CM*) is used, within the knowledge-tailoring layer, to specify the possible situations in which the system might be operating. Each context determines a portion of the *ADO*, called *relevant area*, that contains the elements of the *ADO* that are relevant in that particular context.

The context model allows the application designer to specify all the possible (meaningful) contexts related to the application situations, and also the association between each context and its relevant area. Once the relevant fragment of the *ADO* has been determined, the *ADO* is matched against a set of *data source ontologies* (henceforth *DSOs*). A *DSO* is a semantic description of a given data source which might be provided by the data source itself or extracted from the data source by means of wrappers. In this case, given a user context, a *domain-aware wrapper* [10] dynamically extracts an ontological representation of the data source containing only the concepts which have a counterpart in the ADO (domain awareness) and are relevant

$Concepts = \{Customer, Man, Woman, Wine, TopClass, MidRange, Discount, WineFarm, Farm, Grower\}$
$Roles = \{purchases, produces, supplies\}$
$Attributes = \{customerName, region, designation, farmName, address\}$

$Customer \sqsubseteq \exists customerName \sqcap \exists purchases.Wine$ $Dom(purchases) \sqsubseteq Customer$
$Ran(purchases) \sqsubseteq Wine$
$Dom(customerName) \sqsubseteq Customer$
$Ran(customerName) \sqsubseteq \texttt{xsd:string}$

$WineFarm \sqsubseteq Farm \sqcap \exists region \sqcap \exists produces.Wine$ $Dom(produces) \sqsubseteq WineFarm$
$Ran(produces) \sqsubseteq Wine$
$Dom(region) \sqsubseteq WineFarm$
$Ran(region) \sqsubseteq \texttt{xsd:string}$

$Grower \sqsubseteq Farm \sqcap \exists supplies.WineFarm$ $Dom(supplies) \sqsubseteq Farm$
$Ran(supplies) \sqsubseteq Farm$
$Dom(address) \sqsubseteq Grower$
$Ran(address) \sqsubseteq \texttt{xsd:string}$

$Wine \sqsubseteq \exists designation$ $Dom(designation) \sqsubseteq Wine$
$Ran(designation) \sqsubseteq \texttt{xsd:enumeration}$

$Farm \sqsubseteq \exists farmName$ $Dom(farmName) \sqsubseteq Farm$
$Ran(farmName) \sqsubseteq \texttt{xsd:string}$

$TopClass \sqsubseteq Wine$
$MidRange \sqsubseteq Wine$
$Discount \sqsubseteq Wine$

Fig. 13.2 Running Example: The Application Domain Ontology (*ADO*)

in the given context (context-awareness) [10]. The run-time translation of (context-aware) queries expressed over the data source ontologies into queries expressed in the native format of the data source, is also the responsibility of the wrapper.

In the following example, oenologists, sommeliers, farmers etc. are interested in data coming from the external data sources DS_1 and DS_2. DS_1 (Figure 13.3) is the relational database supporting a blog of women wine lovers[1], while DS_2 (Figure 13.4) contains data about wine farms producing top-class wines.

```
User(username, email)
Purchase(user, wine)
Wine(designation, rank, producer)
WineFarm(farmName, region)
```

Fig. 13.3 The Relational Datasource DS1

```
WineFarm(name, region)
Wine(designation, producedBy)
Supply(grower, wineFarm)
Grower(company, address)
```

Fig. 13.4 The Relational Datasource DS2

---

[1] http://www.womenandwine.blogs.com/

The corresponding Data Source Ontologies (DSO), representing the concepts and roles expressed by the relational data schema, are shown in Figure 13.5.

DSO_1	DSO_2
$Concepts = \{User, Wine, WineFarm\}$	$Concepts = \{WineFarm, Wine, Grower\}$
$Roles = \{purchases, produces\}$	$Roles = \{isProducedBy, supplies\}$
$Attributes = \{username, rank, region\}$	$Attributes = \{region, company, address\}$
$User \sqsubseteq \exists username \sqcap \exists email$	$WineFarm \sqsubseteq \exists name \sqcap \exists region$
$Dom(username) \sqsubseteq User$	$Dom(name) \sqsubseteq WineFarm$
$Ran(username) \sqsubseteq \texttt{xsd:string}$	$Ran(name) \sqsubseteq \texttt{xsd:string}$
$Dom(email) \sqsubseteq User$	$Dom(region) \sqsubseteq WineFarm$
$Ran(email) \sqsubseteq \texttt{xsd:string}$	$Ran(region) \sqsubseteq \texttt{xsd:string}$
$Wine \sqsubseteq \exists designation \sqcap \exists rank \sqcap \exists producer.WineFarm$	$Wine \sqsubseteq \exists designation \sqcap \exists producedBy.WineFarm$
$Dom(designation) \sqsubseteq Wine$	$Dom(designation) \sqsubseteq Wine$
$Ran(designation) \sqsubseteq \texttt{xsd:string}$	$Ran(designation) \sqsubseteq \texttt{xsd:string}$
$Dom(rank) \sqsubseteq Wine$	$Dom(producedBy) \sqsubseteq Wine$
$Ran(rank) \sqsubseteq \texttt{xsd:string}$	$Ran(producedBy) \sqsubseteq WineFarm$
$Dom(producer) \sqsubseteq Wine$	
$Ran(producer) \sqsubseteq WineFarm$	
$WineFarm \sqsubseteq \exists name \sqcap \exists region$	$Grower \sqsubseteq \exists company \sqcap \exists address$
$Dom(name) \sqsubseteq WineFarm$	$Dom(company) \sqsubseteq Grower$
$Ran(name) \sqsubseteq \texttt{xsd:string}$	$Ran(company) \sqsubseteq \texttt{xsd:string}$
$Dom(region) \sqsubseteq WineFarm$	$Dom(address) \sqsubseteq Grower$
$Ran(region) \sqsubseteq \texttt{xsd:string}$	$Ran(address) \sqsubseteq \texttt{xsd:string}$
$Dom(purchases) \sqsubseteq User$	$Dom(supply) \sqsubseteq Grower$
$Ran(purchases) \sqsubseteq Wine$	$Ran(supply) \sqsubseteq WineFarm$

Fig. 13.5 Running Example: Data Source Ontologies

A *data model ontology* (i.e., *DMO*) is a formal description of the *data model* of a data source (relational, object oriented, etc.) and is used to guide the extraction of *DSO*s from the data sources from a structural point of view (see Section 13.4). All the wrapped resources expose a *SPARQL endpoint* that enables querying of the corresponding *DSO*.

The communication among the different layers is supported by the *Semantic Registry* that provides a service to communicate the availability and the features of new sources, and their mappings w.r.t the ADO. When an unknown data source joins the system, the mappings will be provided either manually (as usual), or in a semi-automatic way, using ontology matching tools [12, 11]. It is important to notice that, in our framework, the mappings are computed w.r.t. the portion of the *ADO* that is relevant for the current context, thus determining, on-the-fly, the context-aware correspondence between the *ADO* and the *DSO*s. The semantic registry contains also the associations between each context and the corresponding relevant area of the *ADO*.

We now sketch how query answering in AD-DDIS can be achieved within the above architecture: the process begins when a SPARQL query is received by the system. The query is parsed by the query rewriting module which translates the SPARQL query into the corresponding definition in description logics formulae (described in Section 13.3.1); the elements composing the query (i.e., the query sub-goals) are matched with concepts and relationships of the relevant areas of the *ADO* for the current context. In a subsequent step, the query rewriting module invokes the SAT service of a reasoner to check the merged schema for satisfiability (i.e., we check if the merged schema has at least one possible model in that context) and then classifies it using the related reasoning service.

Since the rewritings are brought, in the end, to the SPARQL endpoints of the data sources, we need to translate the rewritten conjunctive queries expressed in description logics back into SPARQL syntax. In particular, the algorithm rewrites the mappings by using a unifying function which identifies a query sub-goal with the left-hand-side (resp. the right-hand-side) of a GAV (resp. LAV) mapping definition.

The final rewriting q_{rew} is then decomposed by the query distribution module in order to build a query distribution plan ($q_{rew} := q_{rew\_1} \cup q_{rew\_2} \cup \ldots \cup q_{rew\_n}$), where $q_{rew\_1} \ldots q_{rew\_n}$ are queries over the data source ontologies. Each endpoint receives a query q_{rew_i} and hands a result set $ans(q_{rew\_i})$ over the query distribution module which has finally to aggregate results, evaluating distributed joins on local results. The last step communicates the results in terms of the application domain ontology that are then interpreted by the application.

The three architectural layers are described in the following, starting from the lowest one. The next section provides the description of Layer C, in the case of relational data sources.

13.4 Ontology Extraction: The Relational Case

AD-DDIS avails itself of a generic framework for the automatic generation of ontological descriptions from relational data source schemas [10]. Such an enriched description provides an infrastructure to access and query the content of the relational data source by means of a suitable query language for ontologies such as SPARQL.

The access infrastructure to a data source consists of three ontologies which are used to describe different aspects of its structure:

- *Data Model Ontology* (DMO): it represents the structure of the data model in use. The TBox of this ontology does not change as the data source schema changes, since it strictly represents the features of the data model such as the logical organization of entities and attributes. For the relational model, we adopt the Relational.OWL ontology [21] whose structure is shown in Table 13.1. Since the current version of Relational.OWL does not distinguish between one, *composite* foreign key (that references more than one attribute at the same time) and multiple foreign keys, we extended the Relational.OWL with explicit foreign keys. Our extensions to the Relational.OWL ontology are italicised in Table 13.1.
- *Data Source Ontology* (DSO): it represents the intensional knowledge described by the data source schema. This ontology captures a possible conceptualization of the data source under analysis. The *DSO* does not contain individual

Table 13.1 The Relational.OWL ontology (DMO for the relational case)

Relational.OWL Classes			
rdf:ID	**rdfs:subClassOf**		**rdfs:comment**
dbs:Database	rdf:Bag		Represents the relational schemas
dbs:Table	rdf:Seq		Represents the database tables
dbs:Column	rdfs:Resource		Represents the columns of a table
dbs:PrimaryKey	rdf:Bag		Represents the primary key of a table
dbs:ForeignKey	*rdf:Bag*		*Represents the foreign key of a table*
Relational.OWL Properties			
rdf:ID	**rdfs:domain**	**rdfs:range**	**rdfs:comment**
dbs:has	owl:Thing	owl:Thing	General composition relationship
dbs:hasTable	dbs:Database	dbs:Table	Relates a database to a set of tables
dbs:hasColumn	dbs:Table	dbs:Column	Relates a tables, primary and foreign keys to a set of columns
	dbs:PrimaryKey		
	dbs:ForeignKey		
dbs:isIdentifiedBy	dbs:Table	dbs:PrimaryKey	Relates a table to its primary key
dbs:hasForeignKey	*dbs:Table*	*dbs:ForeignKey*	*Relates a table to its foreign keys*
dbs:references	dbs:Column	dbs:Column	Represents a foreign-key relationship between two columns
dbs:length	dbs:Column	xsd:nonNegativeInteger	maximum length for the domain of a column
dbs:scale	dbs:Column	xsd:nonNegativeInteger	scale ratio for the domain of a column

names (instances), which are stored in the data source and accessed natively on-demand.

- *Schema Design Ontology (SDO)*: this ontology maps the *DSO* to the *DMO* and describes how concepts and roles of the *DSO* are rendered in the particular data model represented through the *DMO*. This ontology enables the separation of the schema's meta-data (by means of the *SDO*) and schema's semantics (described by the *DSO*). We remark that, in general, the *SDO* can be extremely useful during schema evolution, because it describes how the changes in the relational schema are going to affect the semantics of the schema itself by detecting changes in the conceptual model.

The extraction procedure first generates the *DSO* by applying a set of reverse-engineering rules. Concepts and roles of the *DSO* are then connected through mappings of the *SDO* to the corresponding concepts of the *DMO*.

Table 13.2 contains the extraction rules for the relational case. We use the notation C_r to refer to a concept of the DSO obtained from the translation of the relation schema $r \in \mathcal{R}$, R_a to refer to a role obtained from the translation of an attribute $a \in att(r)$ and denote the domain and the range of a role R by $Dom(R)$ and $Ran(R)$ respectively. Moreover, we use the notation $r(a)$ to denote the relational projection $(\pi_a r)$ of an attribute a of a relation r.

13.5 Query Answering in AD-DDIS

In the data integration literature, and also in the present work, the interesting questions that can be asked to a data integration system are often represented as (unions of) conjunctive queries (UCQs) [1]. A Conjunctive Query is a formula of the form

Table 13.2 Relational to ontology translation rules

Rule	Preconditions	Effects
R1	$\exists r \in \mathcal{R}$ such that: $\mid pkey(r) \mid = 1$ or $\mid pkey(r) \mid \geq 1$ and $\exists a \in pkey(r) \mid a \notin fkey(r)$	a concept C_r a role R_a $\forall a \in pkey(r)$ and $a \notin fkey(r)$ an axiom $C_r \equiv \exists R_a.dom(a)$ $\forall a \in pkey(r)$ and $a \notin fkey(r)$
R2	$\exists r_i \in \mathcal{R}$ such that: $\mid att(r_i) \mid > 2$, $\forall a \in pkey(r_i), a \notin fkey(r_i)$, $\forall a \in fkey(r_i) \exists r_j \mid r_i(a) \subseteq r_j(b) \wedge b \in pkey(r_b)$ for some $b \in att(r_j)$	a concept C_{r_i} a role R_a $\forall a \in pkey(r_i)$ an axiom $C_{r_i} \equiv \exists R_a.C_{r_j}$ $\forall a \in pkey(r)$ an axiom $Dom(R_a) \sqsubseteq C_{r_i}$ $\forall a \in pkey(r)$ an axiom $Ran(R_a) \sqsubseteq C_{r_j}$ $\forall a \in pkey(r)$
R3	$\exists r_i \in \mathcal{R}$ such that: $\mid att(r_i) \mid = 2$, $att(r_i) = pkey(r_i)$, $fkey(r_i) = pkey(r_i)$, $\exists r_j, r_k \in \mathcal{R}, \exists a_1, a_2 \in att(r_i) \mid r_i(a_1) \subseteq r_j(b) \wedge r_i(a_2) \subseteq r_k(c)$ for some $b \in att(r_j)$ and $c \in att(r_k)$	a role R_{r_i} an axiom $Dom(R_{r_i}) \sqsubseteq C_{r_j}$ an axiom $Ran(R_{r_i}) \sqsubseteq C_{r_k}$
R4	$\exists r_i \in \mathcal{R}$ such that: $\mid pkey(r_i) \mid \geq 1$ and $\exists a \in att(r_i) \mid a \in pkey(r_i) \wedge a \in fkey(r_i)$, $\exists r_j \in \mathcal{R} \mid r_i(a) \subseteq r_j(b)$ for some $b \in att(r_j)$	a role R_a an axiom $C_r \sqsubseteq \exists R_a.C_{r_i}$ $\forall a$ declared as not null an axiom $Dom(R_a) \sqsubseteq C_{r_i}$ an axiom $Ran(R_a) \sqsubseteq C_{r_j}$
R5	$\exists r_i \in \mathcal{R}$ such that: $\exists a \in att(r_i) \mid a \notin pkey(r_i) \wedge a \in$ $fkey(r_i)$, $\exists r_j \in \mathcal{R} \mid r_i(a) \subseteq r_j(b)$ for some $b \in pkey(r_j)$	a role R_a such that $Dom(R_a) \sqsubseteq C_{r_i}$ and $Ran(R_a) \sqsubseteq C_{r_j}$. an axiom $C_r \sqsubseteq \exists R_a.C_{r_j}$ $\forall a$ declared as not null
R6	$\exists r \in \mathcal{R}$ such that: $\exists a \in att(r) \wedge a \notin fkey(r)$	an attribute T_a with $Dom(R_a) \sqsubseteq C_r$ and $Ran(R_a) \sqsubseteq dom(a)$, an axiom $C_r \sqsubseteq \exists R_a.C_{r_j}$ $\forall a$ declared as not null

Constructor	Syntax	SPARQL translation
Concept	dso:A(x)	?x rdf:type dso:A
Role name	dso:R(x,y)	?x dso:R ?y
Intersection	dso:C(x) $\cap$ dso:D(x)	?x rdf:type dso:C . ?x rdf:type dso:D
Union	dso:C(x) $\cup$ dso:D(x)	{ ?x rdf:type dso:C } UNION { ?x rdf:type dso:D }
Qualified Existential restriction	$\exists$ dso:R(x,y). dso:C	?x dso:R ?y . ?y rdf:type dso: C
Existential restriction	$\exists$ dso:R(x,y)	?x dso:R ?y

Fig. 13.6 Translation rules from CA-$\mathcal{DL}$ to SPARQL

$$q(\mathbf{X}) \leftarrow \phi(\mathbf{X}, \mathbf{Y})$$

Where $\phi(\mathbf{X}, \mathbf{Y})$ is a conjunction of atoms called *subgoals*. In practice, the left-hand-side (i.e., the *head*) of the query corresponds to the SELECT section of a SPARQL query, and the right-hand-side (i.e., the *body*) to the WHERE section. The clauses appearing in the WHERE section are expressed in SPARQL by combinations of operators similar to the ones of relational algebra [9] such as OPTIONAL (i.e., left join), UNION and FILTER (i.e., conditions on constants). We refer to the semantics given in [24] for these operations.

Existentially quantified variables are explicitly represented in SPARQL through blank nodes (usually in the form : $b0$) and cannot appear in the SELECT clause. As seen before, CA-$\mathcal{DL}$ is a data language tailored in order to be rewritable into SPARQL syntax. The translation problem arises especially when GAV mappings are involved, because we substitute an atom belonging to the *ADO* with a set of atoms belonging to the *DS Os* (while in the case of LAV mappings, the rewriting corresponds to a single atom). In Figure 13.6 we list the translation rules used in our system.

The SPARQL query posed to AD-DDIS is expressed in terms of the *ADO*. The first step is to ask a reasoner for classification of the *ADO*, and then to use the inferred model (i.e., that now contains all the containment relationships among classes) to expand the query. After this expansion our algorithm handles GAV and LAV mappings.

Our aim is to rewrite the original query q into another query q' which takes into consideration also the dependencies defined by the TBox, that in our language, are just inclusion dependencies. Basically, this step iterates over each sub-goal g_k of the query, and adds to each sub-goal a new OR child whenever it finds elements x_i of the application domain ontology linked to the query sub-goal by ISA relationships. In practice, we adopt a lazy evaluation policy which considers the query as-is, and retrieves the subclasses (sub-properties) from the inferred model (thus traversing a sub-graph of the application domain ontology) only when we try to unify query sub-goals correspond, through mappings, to concepts in the data sources.

GAV mappings rewrite one atom of the query expressed in terms of *ADO* into a set of sub-goals expressed in terms of *DSOs* through *unfolding*, i.e., by simply expanding the head of the mapping by means of the body of the mapping. This is ensured by the absence, in the *ADO*, of constraints that it is not possible to express with CA-$\mathcal{DL}$ statements.

By contrast, since LAV mappings specify atoms belonging to *DSOs* as queries over the *ADO*, the algorithm should infer the inverse views to rewrite a query q, originally expressed in terms of the *ADO*, in terms of data source ontologies. This problem is well-known in data integration and query optimization for relational databases and has been extensively studied [20]. According to [20], given a query q and a set of view definitions $\mathcal{V}_1,\ldots,\mathcal{V}_m$, a rewriting of q using the views is a query expression q' whose sub-goals are either view relations $\mathcal{V}_1,\ldots,\mathcal{V}_m$ or comparison predicates. Roughly speaking, the problem is to translate a query q into a query q' that contains terms belonging to views. In order to limit the number of all the possible legal rewritings of the original query q, we adapt an existing, general rewriting algorithm (i.e., MiniCon [26]) to our setting.

The next section introduces the top layer of the AD-DDIS architecture.

13.6 The Context Model

In a common-sense interpretation, the application context is perceived as a set of variables whose values may be of interest for an agent (human or artificial) because they influence its actions. Within AD-DDIS we represent in CA-$\mathcal{DL}$ an extended, ontological version of the *Context Dimension Tree (CDT)* defined in Chapter 12, along with the association between each context and the *ADO relevant areas*, that will be used to answer the queries (also called, elsewhere, *context-aware chunks*).

In Figure 13.7 a reduced version of the CDT for the wine domain is shown. In this example, context is analysed with respect to the dimensions (black nodes) which are common to most applications: the `role`, representing the user's role (e.g., `sales` or

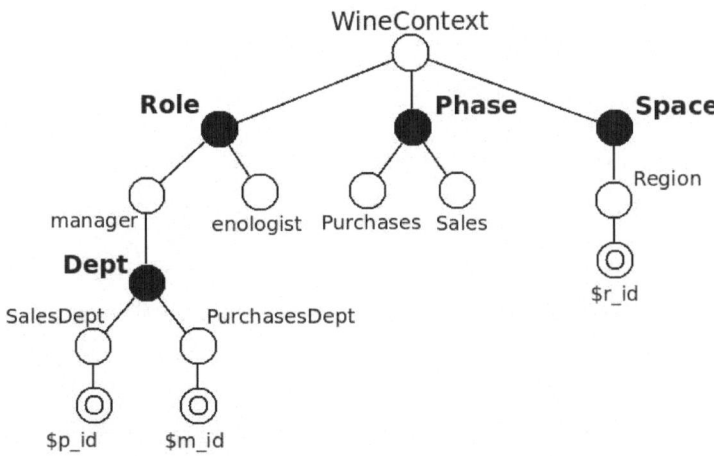

Fig. 13.7 The Context Dimension Tree (CDT) for the wine Domain

purchases departments), the region and the phase of the production of interest for the user. We represent the context model in ontological terms introducing the following structures:

- the *context-vocabulary* ontology defines the vocabulary (i.e., the meta-model) used to build the context models. This vocabulary is application-independent.
- the *context-model* ontology is an instantiation of the context-vocabulary meta-model and specifies the (possibly hierarchical) context dimensions for the specific application, along with their possible values. The vocabulary used in this ontology is application-dependent.
- the *context-configurations* ontology is an instantiation of the context-model ontology and enumerates all the valid (i.e., consistent with the context-model) contexts for a particular application.
- the *relevant-areas* ontology represents the associations between each valid context and the corresponding fragment of the Application Domain Ontology.

Once the context model for the specific application has been produced, the set of valid contexts is represented by a concept called *Context* whose instances are all and only the valid contexts for that given application. The context model for our running example represented in CA-$\mathcal{DL}$ is shown in Figure 13.8. The dimensions are subclasses of the concept *Dimension* and, in our example, the *Space*, the *Phase* and the *Role* of the user are represented. Moreover, the *Role* dimension is further specified into the concept *Dept* (i.e, the Department) that is valid only when the user role is *manager*. Constraints of the form $\exists hasValue.\{a\} \sqcap \exists hasValue.\{b\} \sqsubseteq \bot$ are used to prevent certain dimensions to assume certain values in the same context such as the fact that the values of the same dimension are always mutually disjoint.

Each distinct individual satisfying the concept *Context* represents a context, which is connected by one or more *hasValue* relationships to one or more

Meta-model: Values:

$Context \sqsubseteq \exists hasValue.Value$ $Value \sqsubseteq \top$
$Dimension \sqsubseteq \exists hasValue.Value$ $SpaceValue \sqsubseteq Value$
$Dom(hasValue) \sqsubseteq Dimension$ $PhaseValue \sqsubseteq Value$
$Ran(hasValue) \sqsubseteq Value$ $RoleValue \sqsubseteq Value$
                                                      $DeptValue \sqsubseteq RoleValue$
Model: $SpaceValue(Region)$
                                                      $PhaseValue(purchases)$
$Space \sqsubseteq Dimension \sqcap \exists hasValue.RegionValue$ $PhaseValue(sales)$
$Phase \sqsubseteq Dimension \sqcap \exists hasValue.PhaseValue$ $RoleValue(manager)$
$Role \sqsubseteq Dimension \sqcap \exists hasValue.RoleValue$ $DeptValue(saleDept)$
$Position \sqsubseteq Role \sqcap \exists hasValue.\{marketing\}$ $DeptValue(purchaseDept)$

Constraints: Instantiation:

$\exists hasValue.\{saleDept\} \sqcap \exists hasValue.\{purchaseDept\} \sqsubseteq \bot$ $Context(c_1)$
$\exists hasValue.\{sales\} \sqcap \exists hasValue.\{purchaseDept\} \sqsubseteq \bot$ $hasValue(c_1, manager)$
$\exists hasValue.\{purchases\} \sqcap \exists hasValue.\{saleDept\} \sqsubseteq \bot$ $hasValue(c_1, sale)$
                                                      $hasValue(c_1, region)$
                                                      $hasValue(c_1, sales)$

Fig. 13.8 The Context-model

dimensions with the relative values. A valid context (i.e., instantiation) is represented by an ABox fragment consistent with the context model.

Once all the valid contexts have been generated, it is possible to associate to each of them its *relevant area* of the *ADO*. This is a design-time activity which is carried out by the designers who are well familiar with the possible application context as well as with the application domain.

To define a *relevant area* we associate with each context a set of named concepts, a set of roles and a set of axioms involving them. Note that no other concept or role of the *ADO*, outside the relevant area, will be used during query answering. The relevant area for c_1 is shown in Figure 13.9. Such associations between the *ADO* and the contexts are given in terms of GAV mappings of the form $N_{ADO} \sqsubseteq \exists relevantArea.\{v_1\}$ where v_1 is a value for a given dimension and N_{ADO} represents a named resource. The relevant area of the context is then defined by a set of context-mapping axioms of the form $\exists relevantArea.\{c_1\} \sqsubseteq \exists relevantArea.\{v_1\} \sqcap \exists relevantArea.\{v_2\} \sqcap \ldots \sqcap relevantArea.\{v_n\}$. The concept Rel_c_1, describing the final relevant area, is then defined as $\exists relevantArea.\{c_1\} \sqsubseteq Rel\_c_1$. As an example consider the following SPARQL query expressed over the ADO asking to count all the recorder purchases of Nebbiolo D.O.P.:

```
select count(?cust) as ?count
where {
?cust ado:purchases ?wine.
?wine rdf:type ado:Wine.
?wine ado:designation ?wineName.
FILTER (?wineName = 'Nebbiolo D.O.P.')
}
```

$Concepts_{Rel(ADO)} = \{Customer, WineFarm, Wine, Farm, TopClass, MidRange, Discount\}$
$Roles_{Rel(ADO)} = \{purchases, produces\}$
$Attributes_{Rel(ADO)} = \{customerName, region, designation, farmName\}$

$Customer \sqsubseteq \exists customerName \sqcap \exists purchases.Wine$	$Dom(purchases) \sqsubseteq Customer$
	$Ran(purchases) \sqsubseteq Wine$
	$Dom(customerName) \sqsubseteq Customer$
	$Ran(customerName) \sqsubseteq xsd:string$
$WineFarm \sqsubseteq Farm \sqcap \exists region \sqcap \exists produces.Wine$	$Dom(produces) \sqsubseteq WineFarm$
	$Ran(produces) \sqsubseteq Wine$
	$Dom(region) \sqsubseteq WineFarm$
	$Ran(region) \sqsubseteq xsd:string$
$Wine \sqsubseteq \exists designation$	$Dom(designation) \sqsubseteq Wine$
	$Ran(designation) \sqsubseteq xsd:enumeration$
$Farm \sqsubseteq \exists farmName$	$Dom(farmName) \sqsubseteq Farm$
	$Ran(farmName) \sqsubseteq xsd:string$
$TopClass \sqsubseteq Wine$	
$MidRange \sqsubseteq Wine$	
$Discount \sqsubseteq Wine$	

Fig. 13.9 Running Example: The Relevant Area for c1

Once cleared from the procedural statement count and the SPARQL syntactic sugar, the query above is equivalent to the following query expressed in Datalog-style syntax.

$$q(X) \leftarrow purchases(X,Y) \wedge Wine(Y) \wedge designation(Y, \text{'Nebbiolo} \\ \text{D.O.P.'})$$

and equivalent to the following DL definition

$$q \sqsupseteq \exists purchases.(Wine \sqcap \exists designation.\{\text{'Nebbiolo D.O.P.'}\})$$

Assuming that the context c1 is active, we will actually execute the query over the corresponding relevant area avoiding the burden of taking into account the entire ADO during the rewriting process. We then rewrite the query over the relevant area by using the mappings between the ADO and the data-source ontologies DSO_1 and DSO_2. Since the information about purchases is only contained in the first data source, the rewriting process will produce only queries over DSO_1 as the following:

```
select count(?usr) as ?count
where {
?usr dso1:purchases ?wine.
?wine rdf:type dso1:Wine.
?wine dso1:designation ?wineName.
FILTER (?wineName = 'Nebbiolo D.O.P.')
}
```

The above query will be then translated in SQL through the mappings generated by means of the Relational.OWL ontology as follows:

```
SELECT COUNT(username) AS count
FROM Purchase P join Wine W
where W.designation = 'Nebbiolo D.O.P.';
```

13.7 Conclusions and Future Work

This chapter described a framework for *Context-aware Semantic Query distribution* in a *dynamic* data integration system(AD-DDIS). Context-awareness is the fundamental basis of our proposal for overcoming information overload, with the two-fold purpose of overcaming information noise and optimizing query distribution. Further steps in this direction are oriented towards the introduction of context-aware preference mechanisms [23], where user preferences are used to focus the answer by means of appropriate ranking of the query results.

The immediate work that awaits us now is the experimentation of the system on massive examples, which will give us feedbacks to act towards effective optimizations. The process of unification between query sub-goals and mappings could be very expensive (in terms of time) whenever we consider a large application domain ontology. This is surely compensated by the context-based reduction of the search space; however, we are also devising further, appropriate optimization techniques.

From a formal point of view, we also plan to lift the expressiveness of the mapping language and of the application domain ontology, treating also the problem of expressing constraints over the global schema. From the context-awareness viewpoint, another interesting research issue on our agenda is the autonomous management of run-time-discovered new possible contexts, not envisaged at design time.

References

1. Abiteboul, S., Hull, R., Vianu, V.: Foundations of Databases. Addison-Wesley (1995)
2. Baader, F., Calvanese, D., McGuinness, D.L., Nardi, D., Patel-Schneider, P.F.: The Description Logic Handbook: Theory, Implementation, and Applications. Cambridge University Press (2003)
3. Bizer, C.: D2rq - treating non-rdf databases as virtual rdf graphs. In: Proc. of the 3rd Int. Semantic Web Conf. (ISWC) (poster session), Hiroshima, Japan (2004)
4. Bolchini, C., Curino, C.A., Quintarelli, E., Schreiber, F.A., Tanca, L.: A data-oriented survey of context models. SIGMOD Rec. 36(4), 19–26 (2007)
5. Bolchini, C., Schreiber, F.A., Tanca, L.: A methodology for very small database design. Information Systems 32(1), 61–82 (2007)

6. Calvanese, D., de Giacomo, G., Lembo, D., Lenzerini, M., Rosati, R.: Tailoring owl for data intensive ontologies. In: Proc. of OWL: Experiences and Directions Workshop (OWLED), Galway, Ireland (2005)
7. Chalmers, M.: A historical view of context. Computer Supported Cooperative Work 13(3), 223–247 (2004)
8. Chen, H., Wu, Z., Wang, H., Mao, Y.: Rdf/rdfs-based relational database integration. In: Proc. of the 22nd Int. Conf. on Data Engineering (ICDE), p. 94 (2006)
9. Codd, E.F.: The relational model for database management: version 2. Addison-Wesley Longman Publishing Co., Inc., Boston (1990)
10. Curino, C., Orsi, G., Panigati, E., Tanca, L.: Accessing and Documenting Relational Databases through OWL Ontologies. In: Andreasen, T., Yager, R.R., Bulskov, H., Christiansen, H., Larsen, H.L. (eds.) FQAS 2009. LNCS, vol. 5822, pp. 431–442. Springer, Heidelberg (2009)
11. Curino, C., Orsi, G., Tanca, L.: X-som: A flexible ontology mapper. In: Proc. of the 1st Intl Work. on Semantic Web Architectures for Enterprises (SWAE), pp. 424–428 (2007)
12. Euzenat, J., Shvaiko, P.: Ontology matching. Springer, Heidelberg (2007)
13. Franconi, E.: Conceptual Schemas and Ontologies for Database Access: Myths and Challenges. In: Parent, C., Schewe, K.-D., Storey, V.C., Thalheim, B. (eds.) ER 2007. LNCS, vol. 4801, p. 22. Springer, Heidelberg (2007)
14. de Giacomo, G., Lembo, D., Lenzerini, M., Rosati, R.: On reconciling data exchange, data integration, and peer data management. In: Proc. of the 26th ACM SIGMOD-SIGACT-SIGART Symposium on Principles of Database Systems (PODS), pp. 133–142 (2007)
15. Group, W.O.W.: OWL 2 web ontology language. W3C recommendation, W3C (2009), http://www.w3.org/TR/owl2-overview/
16. Haase, P., Motik, B.: A mapping system for the integration of owl-dl ontologies. In: Proc. of the 1st Int. Work. on Interoperability of Heterogeneous Information Systems (IHIS), pp. 9–16 (2005)
17. Haase, P., Wang, Y.: A decentralized infrastructure for query answering over distributed ontologies. In: Proc. of the ACM Symp. on Applied Computing (SAC), pp. 1351–1356 (2007)
18. Halevy, A., Rajaraman, A., Ordille, J.: Data integration: the teenage years. In: Proc. of the 32nd Intl Conf. on Very Large Data Bases (VLDB), pp. 9–16 (2006)
19. Halevy, A.Y.: The information manifold approach to data integration. IEEE Intelligent Systems 13, 12–16 (1998)
20. Halevy, A.Y.: Answering queries using views: A survey. Int. Journal on Very Large Data Bases (The VLDB Journal) 10(4), 270–294 (2001)
21. de Laborda, C.P., Conrad, S.: Relational.owl: a data and schema representation format based on owl. In: Proc. of the 2nd Asia-Pacific Conf. on Conceptual Modelling APCM 2005, vol. 43, pp. 89–96 (2005)
22. Lenzerini, M.: Data integration: a theoretical perspective. In: Proc. of the 21th ACM SIGMOD-SIGACT-SIGART Symposium on Principles of Database Systems (PODS), pp. 233–246 (2002)
23. Miele, A., Quintarelli, E., Tanca, L.: A methodology for preference-based personalization of contextual data. In: Proc. of 12th Intl Conf. on Extending Database Technology (EDBT), pp. 287–298 (2009)
24. Pérez, J., Arenas, M., Gutierrez, C.: Semantics and Complexity of SPARQL. In: Cruz, I., Decker, S., Allemang, D., Preist, C., Schwabe, D., Mika, P., Uschold, M., Aroyo, L.M. (eds.) ISWC 2006. LNCS, vol. 4273, pp. 30–43. Springer, Heidelberg (2006)

25. Poggi, A.: Structured and semi-structured data integration. Ph.D. thesis, Universita' Degli Studi di Roma "La Sapienza" (2006)
26. Pottinger, R., Halevy, A.: Minicon: A scalable algorithm for answering queries using views. Int. Journal on Very Large Data Bases (The VLDB Journal) 10(2-3), 182–198 (2001)
27. Prud'hommeaux, E., Seaborne, A.: Sparql query language for rdf (working draft). Tech. rep., W3C (2007)
28. Roussos, Y., Stavrakas, Y., Pavlaki, V.: Towards a context-aware relational model. In: Proc. Context Representation and Reasoning (CRR), pp. 7.1–7.12 (2005)
29. Xing, W., Corcho, O., Goble, C., Dikaiakos, M.D.: Active ontology: An information integration approach for dynamic information sources. In: Proc. of the 4th European Semantic Web Conference (ESWC) - Poster Session (2007)

Chapter 14
Context Support for Designing Analytical Queries

Cristiana Bolchini, Elisa Quintarelli, and Letizia Tanca

Abstract. Data repositories of complex organisations are often very large, and understanding which analytical queries are interesting for different kinds of decision makers, in the various possible situations, may be difficult. We propose to support the formulation of OLAP queries by using the knowledge of the context the decision maker is currently experiencing.

14.1 Introduction

Business processes of complex organisations require to efficiently manage a large amount of possibly distributed information. Effective collection and analysis of data coming from the business environment may influence the success of the organisation itself and the possibility to discover in advance potential causes of future concerns. In such integrated environments decision support queries have a practical impact and often require the analysis of aggregate data, stored in huge historical repositories, grouped on the basis of different, *dimensional* attributes. Data Warehouses have represented an effective solution to implement such aggregate queries; they are defined as time-variant, integrated, non-volatile, and subject-oriented data storage to support reporting, analysis, and decision-making [1, 2]. On-Line Analytical Processing (OLAP) technologies have been designed to improve the querying of large amounts of data stored in data warehouses by using multidimensional structures [3].

As observed in [4], analytical queries that can be applied to data warehouses may be influenced by the context , since interests normally vary according to the context the user is currently in. Consider as a practical application field the wine production running example of ART-DECO described in Chapter 21, where the wine quality depends on all the phases of the production process and consequently continued

Cristiana Bolchini · Elisa Quintarelli · Letizia Tanca
Politecnico di Milano, Dipartimento di Elettronica e Informazione, P.zza L. da Vinci, 32,
Milano - I20133 Italy
e-mail: {bolchini,quintare,tanca}@elet.polimi.it

G. Anastasi et al. (Eds.): Networked Enterprises, LNCS 7200, pp. 277–289, 2012.
© Springer-Verlag Berlin Heidelberg 2012

observation and awareness is a fundamental part of a complex process: here, collecting the most appropriate information for each process phase assumes strategic relevance. The different roles (e.g. oenologist, sales' manager and quality manager) of the actors correspond to different portions of data they need to analyse for their activities and OLAP queries may be of precious use. For example, the sales' manager may be interested in *total sales in all regions, for all wines, for the current period and the previous one*, whereas the cellar manager may be interested in *total sales for all wines (or some specific wines) produced in his/her own cellar*. The oenologist may be interested, after the annual maturation of wine, in *productivity per vineyard*. The marketing manager may be interested also in analysing the trend of publications about the produced wines, as an example *the total number of documents published on a specific wine in each year*. From these simple examples, we can note that the typical analysis dimensions of OLAP queries can be, at least partially, deduced on the basis of the current context of the analyst: e.g. the actor, the fact of interest, and the phase of the production process – in other words, the relevant queries are connected to the notion of *context*. Indeed, context has often a significant impact on the way humans act, and on how they interpret things, therefore a change in context causes a change in the relative importance of information to a user.

In Chapter 12 we have described how the current user (or application) context can be used to distill relevant information out of a large amount of data; in this chapter we show how the contextualisation operated by the ART-DECO portal also supports the users in the formulation of OLAP queries. We do not propose yet another methodology for data-warehouse design; instead, we discuss the personalisation of analytical queries on the basis of the current context of the decision maker.

The structure of the chapter is as follows. Section 14.2 describes the literature about OLAP query formulation and personalisation. In Section 14.3 we introduce some innovative features of the Context Dimension Tree (CDT) context model which allow the specification of contextual aggregate queries, Section 14.4 shows some examples of contextual queries in the application scenario of wine production. Section 14.5 concludes the chapter.

14.2 Related Work

We investigate a solution to analytical query formulation that allows the designer to specify the contextual aggregate queries that can be used to deliver specific information satisfying particular needs of each decision maker. The major innovative aspect of the proposal is the possibility to connect the relevant contexts of a target application, expressed by using a quite general tree-based context model, with the definition of the analytical queries useful in those contexts; later on, such queries can also be used to select the views that must be transiently or permanently materialised in a data warehouse. Therefore, two are the literature fields related to this topic: personalisation of OLAP databases and queries, and view materialisation in data warehouses.

Among the recent approaches to personalise multidimensional databases, [5] introduces the notion of user preferences in the data warehouse field; the authors define a framework called MDX that allows the user to express complex queries on multidimensional data and to specify how results will be presented on the screen. Subsequently, [6] proposes an enriched framework to personalise the visualisation of the result of OLAP queries: the user specifies both his (her) preferences and a visualisation constraint (e.g., the limitations imposed by the device used to display the answer to a query) and then, for each query, the framework computes the part of the answer that respects both the user's preferences and the visualisation constraint. The work that is closest to our main aim is [4], which proposes a framework based on active rules to personalise elements of a data warehouse by considering the personalisation of both instances and structures, and by introducing a rich user model describing user needs, preferences, and contexts. The approach defines OLAP personalisation at a conceptual level, that will produce only at run-time the effect of the specified personalisation actions.

Analytical query processing speed-up is the main data management issue in data warehouses; the materialisation of appropriate views, which constitute precomputed – possibly partial – queries, produces a trade-off between the needed storage space and the gain in terms of efficiency. The problem of determining which views have to be materialised has received a lot of attention in the past years, see for example [7, 8, 9, 10]; [7] formalises a methodology to select the materialised views to be maintained at the warehouse in order to minimise total query response time and the cost of maintaining the selected views. [8] proposes a technique and the algorithms to materialise only the queries that are really needed by the users instead of materialising all the elements of the data cube. The problem is to find the set of views to materialise that maximises the performance of a multidimensional DBMS in answering the represented queries. Agrawal et al. in [9] present some heuristics for materialised view maintenance. More recently, [10] shows an integrated approach to materialised view selection and maintenance: the authors propose a technique to find an efficient plan for the maintenance of a set of materialised views, by using common sub-expressions between different view expressions.

The tree-based representation proposed in this work may well be the basis for the elicitation of a hierarchy of more and less abstract OLAP queries, among which the views to be materialised are chosen.

14.3 Context Modelling Revisited

The context model described in Chapter 12 – and previously introduced in [11] – supports different roles (e.g. oenologist, sales' manager, etc.) in their different activities and in particular in the query formulation process, based on the assumption that the data that are crucial to support the decision process are influenced by the user's current context. Albeit the hierarchical representation provided by the Context Dimension Tree (CDT) seems most appropriate for modelling different points of view for analysing data, some slight modifications are in order, so that its dimensions and concepts can more appropriately support the formulation of complex aggregate queries, with the ultimate purpose of answering business questions.

The main innovations to the CDT, which becomes the CDT^O (OLAP Context Dimension Tree) are now described.

Dimensions and Concepts. In the CDT, dimensions model the different perspectives used to tailor the data related to a specific scenario, whereas the concepts represent dimension values, with different levels of granularity. In a data warehouse perspective, the dimensions are the points of view used to aggregate data. In addition to **classical** context dimensions , such as the user role or situation, useful to identify the classes or current circumstances of people interested in an analytical query, we explicitly introduce the **warehouse** dimensions, which account for the aggregation perspectives and for the other important elements encompassed by OLAP queries .

Figure 14.1 shows the CDT^O of our running example. Some classical dimensions often used to tailor the relevant portion of data, such as situation/phase and location, are omitted to concentrate on the new warehouse dimensions.

A warehouse dimension labeled fact lists as values the concepts that are relevant for the decisional process; typically, they model the phenomenon we want to analyse. In our running example, sales, web_docs, production, and storage are the values for the fact dimension (see Figure 14.1).

Another warehouse dimension is labeled olap-dimension and lists the fact properties (i.e. analysis perspectives), each of them defined w.r.t. a finite domain. For example, time-hier, space-hier, grape-hier, wine-hier are different perspectives for analysing the sales, web_docs, production, and storage facts (see Figure 14.1). In a given context, more than one warehouse dimension might be chosen to analyse a given fact; thus, all the values for the olap-dimensions node are collected as sibling sub-dimensions (i.e. sibling black nodes). The concept (white) node all_values is used when the query related to a context does not specify grouping criteria.

The numerical properties of the considered facts are grouped as concept children of the warehouse dimension measure. In our running example, the measure values are: income, qty_bottles, qty_liters, stock_level, and number.

Aggregation requires to specify an operator to combine values, thus, the last warehouse dimension is operator that lists all the possible aggregate operators that can be used, in the target scenario, to compose OLAP queries. Since the nature of a measure determines the set of operators that can be used to aggregate its values [1], most likely some constraints will have to be introduced to prevent inappropriate use of the operators. For example, not all measures are *additive*: a measure is additive w.r.t. a given dimension if and only if the SUM operator is applicable to that measure along that dimension; for example, income is additive w.r.t. the time-hier OLAP dimension, while the measure stock-level is not.

Hierarchies in OLAP dimensions. Typically, in data warehouses, dimensions are hierarchical, and hierarchies are determined by the business need to group and summarise data into usable information. Similarly, our OLAP-dimension may assume a hierarchy of values when the analysis of contextual data must be performed at different levels of aggregation. For example, the `time-hier` dimension can be further specialised in a 5-years window or in a single-year window. In our model, we recognise a hierarchical dimension from the simple chain of alternating concepts (with an attribute) and dimension nodes starting from the dimension itself; for instance the `time-hier` and `space-hier` dimensions are hierarchical with multiple levels of aggregation. Note that the attribute connected to each concept node of a hierarchical dimension allows us to specify in a context the generic concept (e.g., `5yrs_window`) used to aggregate data, without instantiating the parameter, or a concept with a parameter that will be instantiated at run-time (e.g., `5yrs_window(fromYear=$year)`).

As an example, imagine a sales' manager interested in the total income for all wines during the last 5 years; his/her current contexts will contain the context element `time-hier=5yrs_window(fromYear=$year)`. Then, at run-time the sales' manager specifies that the temporal window starts from the year 2007.

The roll-up and drill-down operators allow to aggregate or dis-aggregate the data of a measure by climbing up and down the values of an OLAP dimension. We will show that context-aware aggregate queries can follow the same paradigm.

Master dimensions.   A flexibility aspect of the original CDT is that a context can lack some dimension value(s): this means that those dimensions are not taken into account to tailor data, i.e., the view corresponding to that context does not filter the data for these dimensions, and consequently, contains the data relevant for all the values of the non-instantiated dimensions. In CDT^{O} there are some *master dimensions* that are not optional and must be always instantiated when specifying the current context. The rationale of this feature is that contexts without such a personalising dimension would actually be meaningless. In our running example, the master dimensions are are `fact`, `measure`, and `operator`, which must be present because they determine the focus of the decision process.

The additional aspects of the CDT^{O} w.r.t. the CDT context model are summarised in Table 14.1.

Table 14.1 The specializing aspects of the CDT context model for analytical queries

Characterizing Element in CDT^{O}	Value	In our example
Warehouse dimensions	*facts*	`fact`
	olap dimensions	`olap-dimension`
	measures	`measure`
	operators	`operator`
OLAP dimensions	*time*	`time-hier`
	space	`space-hier`
	product	`grape-hier`
		`grape-hier`
Master dimensions	*facts*	`fact`
	measures	`measure`
	operators	`operator`

14.3.1 Constraints on the CDT^O

Once the CDT^O has been designed, all possible contexts can be combinatorially generated as conjunctions of context elements, where each context element describes a dimension and its value. For example, the context describing the situation of a sales' manager who is interested in total sales in Italy, for all wines, for the current period is specified as:

$$
\begin{aligned}
C_1 \equiv \quad & (\texttt{role} = \texttt{sales\_mngr}) \wedge (\texttt{fact} = \texttt{sales}) \\
& \wedge\, (\texttt{space\_hier} = \texttt{nation(nationID} = \texttt{\$nat)}) \\
& \wedge\, (\texttt{measure} = \texttt{income}) \wedge (\texttt{operator} = \texttt{sum}) \\
& \wedge\, (\texttt{year\_window} = \texttt{year(yearID} = \texttt{\$y)})
\end{aligned}
$$

C_1 contains the specification of the role dimension (i.e., $\texttt{sales\_mngr}$), and the values for the master dimensions: the fact of interest is $\texttt{sales}$, which is evaluated w.r.t. to the total income by using the sum operator. Two parameters need to be actualised at run time, the specific nation ($\texttt{\$nat}$) and year ($\texttt{\$y}$) of interest for the analysis. As an example, when interested in the relevant sales in Italy in the current year, we will have, at run-time, $\texttt{nationID}$ = "Italy" and $\texttt{yearID}$ = 2010, respectively.

However, not necessarily all combinatorially generated contexts make sense and do originate aggregate queries for a given scenario: consider the example above of non-additive measures that cannot be combined with the SUM operator; thus many "meaningless" contexts should be discarded by introducing constraints.

Master dimensions do not admit the ALL value, because the corresponding context would be too generic to hint a tailored portion of data to be analysed. In our

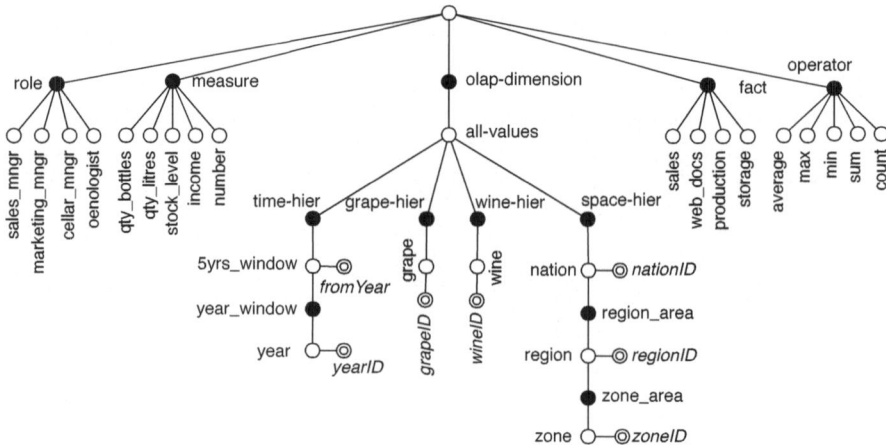

Fig. 14.1 The OLAP Context Dimension Tree for the wineries case study

running example we have imposed such constraint to the fact, measure, and operator dimensions. The (*md*) constraints are specified as:

- $md(\texttt{fact}) \equiv \neg(\texttt{fact} = \texttt{ALL})$
- $md(\texttt{measure}) \equiv \neg(\texttt{measure} = \texttt{ALL})$
- $md(\texttt{operator}) \equiv \neg(\texttt{operator} = \texttt{ALL})$

Useless constraints allow the context designer to specify context element combinations that are not significant, i.e. that represent semantically meaningless context situations or are irrelevant for the application. For example, there is no operational context where someone with the oenologist role is interested in the sales fact; this is formally specified as

$$\texttt{uc}((\texttt{role} = \texttt{oenologist}), (\texttt{fact} = \texttt{sales})) \equiv$$
$$\neg((\texttt{role} = \texttt{oenologist}) \wedge (\texttt{fact} = \texttt{sales}))$$

Another useless constraint is the one stating that the production fact cannot be combined with the income measure. The formal specification is:

$$\texttt{uc}((\texttt{measure} = \texttt{income}), (\texttt{fact} = \texttt{production})) \equiv$$
$$\neg((\texttt{measure} = \texttt{income}) \wedge (\texttt{fact} = \texttt{production}))$$

14.4 Query Formulation

Given the possible and meaningful contexts, the subsequent work of the designer consists in associating them with the appropriate OLAP queries to gather the important information used for decision making. The association between contexts and OLAP queries will be used by the Context-Aware Web Portal described in Chapter 12, to distill the appropriate query on the basis of the user active context.

In the ART-DECO scenario, the networked enterprise data is stored in a relational database, used also as a source for data warehousing. Therefore, we refer to such data model and in particular we express queries in SQL .

Figure 14.2 shows the design process: the data stored in the relational database are re-modeled to be (virtually or physically) stored as *data marts* according to the data cube metaphor, where OLAP queries can be subsequently applied. In our proposal, the decision maker's context is used to automatically formulate the queries over the data mart. The advantage of such an approach is in the reduced amount of data to be managed and manipulated in order to summarise the required information, thus limiting also all costs related to view materialisation (e.g., storage, maintenance, etc.). The starting point is the analysis of interesting contexts the user will be possibly in, and the output is the corresponding analytical query that retrieves all and only the interesting data.

Consider the ART-DECO relational database of Figure 14.3, including the tables about documents describing the production process (as described in Chapter 21) and

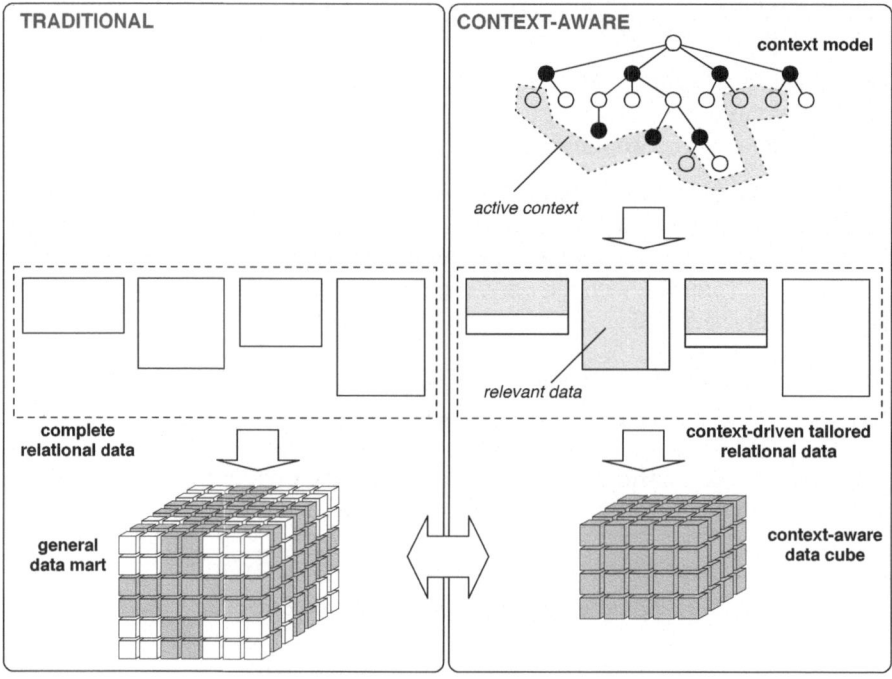

Fig. 14.2 Associating contexts with aggregate queries

the CDT^O reported in Figure 14.1 (Section 14.3). The following is the context of *an oenologist who is interested in information related to the total production per year for a specific wine*:

$$C_1 \equiv \quad (\text{role} = \text{oenologist})$$
$$\wedge \, (\text{fact} = \text{production})$$
$$\wedge \, (\text{measure} = \text{qty\_bottles})$$
$$\wedge \, (\text{operator} = \text{sum})$$
$$\wedge \, (\text{wine\_hier} = (\text{wine}(\text{wineID} = \$\text{wine})))$$
$$\wedge \, (\text{year\_window} = \text{year})$$

Note that in C_1 the fact of interest is the production evaluated with respect to the number_of_bottles. The wine_hier OLAP dimension explicitly indicates the parameter that will be instantiated at run-time with the specific wine appellation (e.g., "Barolo"). On the contrary, year_window assumes as value a concept node without the parameter, thus, this measure will determine the use of a GROUP BY operator (w.r.t. the production year) in the contextual SQL query.

VINEYARD(v_id, region, district, municipality, fraction, zone, hectares)
HARVEST(h_id, vineyard_id, note)
CELLAR(cl_id, materia, type, harvest_id,vineyard_id)
BARREL(bl_id, type, wood, cellar_id)
BOTTLE(b_id, harvest_date, bottle_date, price, wine_appellation, lot_id)
COMPOSITION(bottle_id, barrel_id, wine_%)
LOT(l_id, pkg_date, pkg_type, sell_date, return_date, customer)
CUSTOMER(c_id, address, name)
WINE(appellation, category, vinification)
REF_COMP(wine_appellation, grapevine_name, min_%, max_%)
GRAPEVINE(tech_name, variety, name)
ROW(r_id, vineyard_id, plant_date, phenological_phase, grapevine_name)
FINDING(id, paragraph, row, start_column, end-column)
DOCUMENT(id, phys-uri, log-uri, pub-date, title, author, summary, topic, description,d-id,e-id)
SOURCE(uri, name, abs-relevance, description, type)
EVALUATION(name, judgment, language, description)
S-E(source, evaluation, judgment)
FND-G(finding-id, grapevine-id)
FND-W(finding-id, wine-id)
FND-B(finding-id, wine-id)

Fig. 14.3 Database schema of the running example

A possible SQL aggregate query associated with context C_1 is thus the following:

SELECT   DATEPART(yyyy,bottle_date), COUNT(*)
FROM   BOTTLE
WHERE   wine_appellation=$wine
GROUP BY DATEPART(yyyy,bottle_date)

In particular, the designer deems significant for context C_1 the number of the specific wine bottles produced every year. The wine of interest is filtered in the **WHERE** clause by using its appellation, that is specified at run-time by the user who assigns a value to the $wine parameter.

As another example, let us consider the context of *a sales' manager who is interested in information related to the total income per wine for the current period*:

$$C_2 \equiv \quad (\text{role} = \text{sales\_mngr})$$
$$\wedge\,(\text{fact} = \text{sales})$$
$$\wedge\,(\text{measure} = \text{income})$$
$$\wedge\,(\text{operator} = \text{sum})$$
$$\wedge\,(\text{wine\_hier} = \text{wine})$$
$$\wedge\,(\text{year\_window} = \text{year}(\text{yearID} = \$\text{yid}))$$

Note that in C_2 the total income is computed by using the SUM operator; furthermore we consider all the produced wines, i.e., we do not filter w.r.t. a specific wine appellation, but evaluate the total income only for the current year (the year concept has the parameter).

Assume the designer associates it with the following SQL aggregate query:

SELECT	wine_appellation, SUM(price)
FROM	BOTTLE, LOT
WHERE	LOT.return_date IS NULL AND BOTTLE.lot_id=LOT.l_id AND
	DATEPART(yyyy,sell_date)=$yid
GROUP BY wine_appellation	

In particular, the designer deems significant for the specific context, the total income obtained by selling wine bottles in the current year, grouped on the basis of the wine appellation. The current year is specified at run-time by actualising parameter value $yid. The sold bottles are those that have a NULL value in the attribute *LOT.return_date*.

If we only change in context C_2 the operator used to evaluate the sales fact, and consider the context of *a sales' manager who is interested to information related to the average price of bottles sold in each year of a 5-years temporal window*:

$$C_3 \equiv \quad (\texttt{role} = \texttt{sales\_mngr})$$
$$\wedge\,(\texttt{fact} = \texttt{sales})$$
$$\wedge\,(\texttt{measure} = \texttt{income})$$
$$\wedge\,(\texttt{operator} = \texttt{avg})$$
$$\wedge\,(\texttt{time\_hier} = \texttt{5years\_window(fromYear} = \texttt{\$fy}))$$

The aggregate query becomes the following, which groups the information on the basis of the sale year (on a 5-years temporal window) and, for each group, computes the average price of the sold bottles, that have not been returned by the customers.

SELECT	DATEPART(yyyy,sell_date), AVG(price)
FROM	BOTTLE, LOT
WHERE	LOT.return_date IS NULL AND BOTTLE.lot_id=LOT.l_id AND
	DATEPART(yyyy,sell_date) BETWEEN $f AND $f+5
GROUP BY DATEPART(yyyy,sell_date)	

Let us now exploit the possibility to use the CUBE operator in order to show all the possible tuple aggregations of the wine sales. Consider the context of *a sales manager who is interested to information related to the total income per wine and per year (in a 5 years window), in a certain region – and s/he wants to analyse all the possible aggregations*:

$$C_4 \equiv \quad (\text{role} = \text{sales\_mngr})$$
$$\wedge\, (\text{fact} = \text{sales})$$
$$\wedge\, (\text{measure} = \text{income})$$
$$\wedge\, (\text{operator} = \text{sum})$$
$$\wedge\, (\text{time\_hier} = \text{5yrs\_window}(\text{fromYear} = \$\text{fy}))$$
$$\wedge\, (\text{wine\_hier} = \text{wine})$$
$$\wedge\, (\text{region\_area} = \text{region}(\text{regionID} = \$\text{rID}))$$

The SQL aggregate query using the CUBE operator is:

SELECT	wine_appellation, region, DATEPART(yyyy,sale_date), SUM(price) AS SALES
FROM	BOTTLE, LOT, COMPOSITION,BARREL,CELLAR, HARVEST,VINEYARD
WHERE	LOT.return_date IS NULL AND BOTTLE.lot_id=LOT.l_id AND BOTTLE.b_id=COMPOSITION.bottle_id AND COMPOSITION.barrel_id=BARREL.bl_id AND BARREL.cellar_id=CELLAR.cl_id AND CELLAR.vineyard_id=VINEYARD.v_id AND VINEYARD.region=$rID AND DATEPART(yyyy,sale_date) BETWEEN $fy AND $fy + 5
GROUP BY	(wine_appellation, region,DATEPART(yyyy,sale_date)
WITH CUBE	

The SQL query gives as result all the possible tuple aggregations of sales of all the produced wines sold in the 5-years window specified at run-time (fy parameter), coming from vineyards of the region specified by the user (rID parameter). Let us suppose that the sales' manager be interested in the total income in the years starting from 2005, produced with grapes coming from the "Valpolicella" location; a possible result is the following table:

wine_appellation	region	year	SALES
Valpolicella	Valpolicella	2008	2000
Amarone	Valpolicella	2008	2500
Recioto	Valpolicella	2008	2800
Valpolicella	Valpolicella	2009	3000
Amarone	Valpolicella	2009	3200
Recioto	Valpolicella	2009	2700
...	...	...	...

As a last example, consider the scenario described in Chapter 21, where the decision-making process is based on information obtained from Web documents, and in particular the context of *a marketing manager who is interested to the total number of Web sites publishing information about a specific wine per year*:

$$C_5 \equiv \quad (\texttt{role} = \texttt{marketing\_mngr})$$
$$\wedge\, (\texttt{fact} = \texttt{web\_docs})$$
$$\wedge\, (\texttt{measure} = \texttt{number})$$
$$\wedge\, (\texttt{operator} = \texttt{count})$$
$$\wedge\, (\texttt{year\_window} = \texttt{year})$$
$$\wedge\, (\texttt{wine\_hier} = \texttt{wine}(\texttt{wineID} = \$\texttt{wID}))$$

The SQL aggregate query associated with the context C_5 is:

SELECT	DATEPART(yyyy,pub-date), COUNT(DISTINC(d-id))
FROM	DOCUMENT, FINDING, FND-W, WINE
WHERE	FINDING.d-id=DOCUMENT.id
	AND FINDING.id=FND-W.finding-id,
	AND FND-W.wine-id=WINE.appellation
	AND WINE.appellation=wID
GROUP BY	DATEPART(yyyy,pub-date)

The SQL query gives as result, for each year, the number of distinct Web documents publishing information about the wine specified at run-time (wID parameter).

14.5 Conclusions

In general, context information can be used to personalise and tailor the information to be provided to the user; here we aim at going one step further, and exploit contextual information also to automatise the formulation of context-specific analytical queries over general data marts. The proposal consists in the $\text{CDT}^{\circlearrowleft}$, a variant of the context model of Chapter 12, which accommodates elements characterising OLAP data processing. Analytical queries can be semi-automatically built by associating SQL statement and clauses with the OLAP-related dimensions and values of the $\text{CDT}^{\circlearrowleft}$, in such a way that, given a context, it is possible to derive the corresponding query. We have presented some examples related to the winery running example, showing the potential of the approach. Steps forward in the formulation and formalisation of the methodology are left as future work.

References

1. Berson, A., Smith, J.: Data Warehousing, Data Mining & OLAP. McGraw-Hill (1997)
2. Hammer, J., Garcia-Molina, H., Widom, J., Labio, W., Zhuge, Y.: The stanford data warehousing project. IEEE Data Eng. Bull. 18(2), 41–48 (1995)

3. Choong, Y.W., Laurent, D., Marcel, P.: Computing appropriate representations for multidimensional data. Data Knowl. Eng. 45(2), 181–203 (2003)
4. Garrigós, I., Pardillo, J., Mazón, J.-N., Trujillo, J.: A Conceptual Modeling Approach for OLAP Personalization. In: Laender, A.H.F., Castano, S., Dayal, U., Casati, F., de Oliveira, J.P.M. (eds.) ER 2009. LNCS, vol. 5829, pp. 401–414. Springer, Heidelberg (2009)
5. Mouloudi, H., Bellatreche, L., Giacometti, A., Marcel, P.: Personalization of MDX queries. In: BDA (2006)
6. Bellatreche, L., Giacometti, A., Marcel, P., Mouloudi, H., Laurent, D.: A personalization framework for OLAP queries. In: ACM 8th International Workshop on Data Warehousing and OLAP, DOLAP 2005, pp. 9–18. ACM (2005)
7. Gupta, H., Mumick, I.S.: Selection of views to materialize in a data warehouse. IEEE Trans. Knowl. Data Eng. 17(1), 24–43 (2005)
8. Baralis, E., Paraboschi, S., Teniente, E.: Materialized views selection in a multidimensional database. In: Proceedings of 23rd International Conference on Very Large Data Bases VLDB 1997, pp. 156–165. Morgan Kaufmann (1997)
9. Agrawal, S., Chaudhuri, S., Narasayya, V.R.: Automated selection of materialized views and indexes in SQL databases. In: El Abbadi, A., Brodie, M.L., Chakravarthy, S., Dayal, U., Kamel, N., Schlageter, G., Whang, K.-Y. (eds.) Proceedings of 26th International Conference on Very Large Data Bases VLDB 2000, pp. 496–505. Morgan Kaufmann (2000)
10. Mistry, H., Roy, P., Sudarshan, S., Ramamritham, K.: Materialized view selection and maintenance using multi-query optimization. In: SIGMOD Conference, pp. 307–318 (2001)
11. Bolchini, C., Curino, C., Quintarelli, E., Schreiber, F.A., Tanca, L.: Context information for knowledge reshaping. Intl. Journal of Web Engineering and Technology 5(1), 88–103 (2009)

Part IV
Management of Peripheral Devices

Many enterprise processes are affected by the external physical environment, e.g., grapes cultivation, wine production, wine distribution and commercialization, just to name a few in the wine business domain. Data acquisition from the external environment is thus an important task. This part of the book discusses how data can be acquired from the external environment through specialized peripheral devices and made available to the networked enterprises information system for appropriate processing and decision. To allow a continuous, automated, and punctual monitoring of external environmental conditions, data acquisition can be performed through tiny and pervasive devices such as RFID (Radio Frequency IDentification) tags and wireless sensor nodes. In particular, sensor nodes are able to self-organize autonomously to form wireless sensor networks (WSNs) that can be used for continuous and fine-grained monitoring of large geographic areas, e.g., a vineyard or a wine cellar. However, as sensor nodes have a limited energy budget, power must be managed in a very efficient way to ensure an appropriate network lifetime. Efficient power management is required also when energy can be scavenged from the external environment, e.g., through solar cells. The following chapters discusses how peripheral devices, especially WSNs, can help in different phases of the wine business process. Challenges to be faced for their effective utilization in this specific application domain are addressed, and some novel solutions are proposed. Special emphasis is devoted to the way data are extracted from peripheral devices and made available to the information system of the networked enterprise. In the first chapter of this part (Chapter 15), entitled Wireless Sensor Networks for Monitoring Vineyards, Alippi et al. design a single-hop WSN for continuous and real-time monitoring of physical quantities that affect grapes production in a vineyard. Data acquired by sensors are transferred to a local Gateway (through short-range wireless communication) and, then, to the Data Server (through long-range wireless communication). An adaptive energy harvesting system for powering the Gateway node and, possibly, all sensor nodes as well is also proposed. Finally, energy conservation at sensor nodes is achieved through a polling-based communication scheme that allows sensor nodes to transition to sleep mode when communication is not required. If the geographic area to be monitored is very large, a single-hop sensor network may not be an appropriate solution due to the limited transmission range of sensor nodes. A more suitable option for such a scenario is a multi-hop sensor network. In Chapter 16 entitled Design, Implementation, and Field Experimentation of a Long-lived Multi-hop Sensor Network, Anastasi et. al design and implement an adaptive scheme to address the problem of energy conservation in a multi-hop sensor network for vineyard monitoring. The proposed approach adapts the duty cycle of sensor nodes depending on the (time-varying) operating conditions, thus minimizing their energy consumption. When dealing with WSNs, two important issues are related with (i) how to extract data from sensors, and (ii) how to manage and control the behaviour of each single sensor and the overall network. To solve these problems two alternative solutions have been investigated in the project. In Chapter 17, entitled Extracting Data from WSNs: a Service Oriented Approach, Bini et al. propose a service-oriented approach based on contracts. According to this approach, an application process can establish a service agreement with the underling

WSN. For the specific wine business domain three service contracts have been defined, i.e., *Periodic Measurement* (to report data periodically), *Event Monitoring* (to notify the occurrence of an event) and *Network Management* (to control the WSN behaviour). The alternative approach is described in Chapter 18, entitled Extracting Data from WSNs: a Data Oriented Approach. Schreiber et al. propose a middleware layer and a specific language to install queries and extract data from peripheral pervasive devices (e.g., RFID tags, sensors, WSNs). Specifically, three different types of queries are defined, i.e., *Low Level Queries* (to access data produced by sensor nodes), *High Level Queries* (to perform data manipulation operations), and *Actuation Queries* (to set parameters at sensor nodes). In chapter 19, entitled Optimal Design of Wireless Sensor Networks, Mura and al. illustrate how the issues of a WSN design can be tackled starting from the first phases of the design cycle. A tool supporting the design phase to perform architectural choices for sensor nodes and network topology, taking target performance goals and estimated costs into account, is described. Then, a methodology that allows analysing and optimising the power performances in a hierarchical fashion, encompassing various abstraction levels, is presented. Finally, in Chapter 20, entitled Enabling Traceability in the Wine Supply Chain, Cimino et al. address the problem of traceability in the wine supply chain. They propose a system that is able to systematically store information about products and processes throughout the entire supply chain, from the grapes grower to the retailer. The proposed system also manages quality information, thus enabling an efficient analysis of the supply chain processes.

Chapter 15
Wireless Sensor Networks for Monitoring Vineyards

Cesare Alippi, Giacomo Boracchi, Romolo Camplani, and Manuel Roveri

Abstract. Vineyards are complex microclimate-affected cultivations that require the continuous and careful presence and intervention of farmers to guarantee the quality of the final product. Distributed monitoring systems may support the farmers' work by providing real-time and spatially dense measurements of climatic and plant-related parameters that influence the grape production. We propose a WSN-based monitoring system for grape production that, differently from what proposed in the literature, is characterized by adaptation to topological changes of the network due to permanent or transient faults that affect the sensor nodes, energy availability, and plug&play insertion or removal of units. Sensed data are remotely transmitted to a control room for storage, visualization and interpretation.

15.1 Introduction and Motivation

Vitis vinifera is a plant able to grow on all continents. Vineyards can be found in Europe (from Sicily to the Scandinavian Peninsula), in Africa (the Mediterranean area and South Africa), in America (from Canada to Argentina), China, Australia and New Zealand. Despite the high adaptability to several climates (the plant supports a reasonable stress induced by heat, drought and frost) the grape production (and hence the quality of the vine) requires a careful presence of the farmers. For instance, human intervention is requested to remove unnecessary leaves retaining local humidity, to apply pesticide to combat oidium and downy mildew, to identify the right moment for the vintage. Microclimate, in terms of humidity at the ground and leaves level, and temperature play a main role here, account for the reproduction of parasites and ultimately interfere with the wine quality. Unfortunately, traditional weather forecasting provides meteorological information at a coarse spatial

Cesare Alippi · Romolo Camplani · Giacomo Boracchi · Manuel Roveri
Politecnico di Milano, Italy
e-mail: {alippi,boracchi,camplani,roveri}@elet.polimi.it

G. Anastasi et al. (Eds.): Networked Enterprises, LNCS 7200, pp. 295–310, 2012.
© Springer-Verlag Berlin Heidelberg 2012

resolution and, consequently, cannot describe the complex micro-climate ingredients, which so heavily influence the grape growth [1]. Nowcast solutions, possibly integrated with specific leaves humidity sensors, would help but their cost is very high [2]. Differently, distributed monitoring systems are effective, satisfy the constraints and can provide the due information at a reasonable cost. Such systems can be realized as wireless sensor networks (WSNs) with sensor nodes cooperating to wirelessly transmit the acquired data to a remote control room; in some cases, hybrid wired-wireless architectures can be considered, with the wireless unit becoming the node of the WSN.

In the last decade, precision agriculture [3] and, in particular, vineyard monitoring, have drawn the attention of the Wireless Sensor Networks (WSNs) community and several distributed monitoring systems have been deployed [1][4][5][6][7][8][9]. More specifically, a distributed system for monitoring the conditions influencing the grape growth has been suggested in [1]. There, nodes are deployed in a regular grid of 20 meters and organized hierarchically into a two-layer multi-hop network. Unfortunately, the information acquired by the sensor nodes restricts to temperature measurements, units are battery-powered and energy harvesting mechanisms or power-aware solutions have not been taken into account. A system for vineyard monitoring is also presented in [4] where a 30-acre vineyard has been covered by sensor nodes measuring environmental information (e.g., temperature, humidity). The proposed system relies on the MeshScape Wireless Sensor Networking System that guarantees adaptability to topological changes in the network (due to immediate insertion of new nodes) without introducing modifications to the rest of the system. Unfortunately, the units are battery powered and no energy harvesting techniques have been considered. [5] proposes a WSN for real-time monitoring of various environmental properties such as soil pH, temperature and moisture and electrical conductivity; sensing nodes are organized into a mesh network. The routing algorithm proposed by the authors assumes no changes in the network topology, an hypothesis that limits the applicability of the system. Neither power-aware solutions nor energy harvesting mechanisms have been reported. A large-scale WSN for vineyard monitoring is presented in [6] where 64 sensing nodes, deployed over a squared grid, acquire moisture and light measurements: a reactive routing protocol guarantees a reliable data transfer, but neither energy harvesting mechanisms nor power-aware solutions have been considered. Differently, [7] faces the application aspects of a WSN-based micro-environmental monitoring system, rather than the technological ones. This system focuses on the processing, storage and visualization of acquired data together with a web-based surveillance subsystem for remote control and monitoring. Distributed measurement systems have also been suggested for wine-cellar monitoring [8], the vinification process and the wine storage [9].

Here, we present a monitoring pervasive system, which integrates with the technologies presented in the rest of the book, aiming at effectively and timely providing the information acquired from the environment to the higher levels of the information processing (e.g., Chapter 6, Chapter 7 and Chapter 17). In particular, we propose a novel WSN-based distributed monitoring system that is adaptive, robust and power-aware . More in detail, the proposed system adapts to topological changes of the network due to insertion/removal of units, is robust to permanent or transient faults that can affect the sensing nodes and includes power-aware strategies for an energy-efficient management both at the network and unit levels. Moreover, units are endowed with solar panels and energy harvesting mechanisms to prolong the WSN lifetime. The paper is organized as follows. Section 15.2 presents the architecture of the proposed monitoring system and the robust and adaptive transmission protocol is detailed in Section 15.3. Section 15.4 briefly presents the real-time data storage and the data visualization system.

15.2 Designing the System: Hardware Aspects

15.2.1 The Architecture of the Monitoring System

The proposed WSN-base monitoring system is structured as a two-layer hierarchical architecture (see Figure 15.1). The network is composed of independent clusters spread out over the field, a cluster for each microclimatic niche. Sensor nodes composing a cluster are coordinated by a gateway connected with a single hop solution to the remote control room. The network topology is extremely flexible and general since each unit can be seen as the terminal communication point of a multi hop sub-network.

The local (sensor nodes-gateway) and remote (gateways-control room) communication may adopt the same protocol as the one suggested in Section 15.3. This protocol inherits the simplicity and energy-efficiency of a traditional time division multiple access (TDMA) and handles topological changes in the network associated with an online node insertion/deletion. Nevertheless, for the gateway level, other transmission protocols (e.g., see Section 15.3) could be considered as well.

15.2.2 The Units: Sensor Nodes and Gateway

Sensor units are composed of four main modules (see Figure 15.2.b): the data processing and control module, the signal acquisition and conditioning module, the module for local transmission and the energy harvesting module. A gateway (see Figure 15.2.a) is a straight evolution of a sensor node and differentiates for the additional presence of a long range communication module.

The *control and data processing module* is composed of a CrossBow MPR2400 (MicaZ) [10] unit comprising a low-power Atmel ATmega128L microcontroller (128 Kb self-programming Flash Program Memory, 4 KB SRAM). We opted for the MicaZ platform because of its large diffusion in the WSN community, the reduced power consumption, and the possibility to rely on the open-source TinyOS operating system.

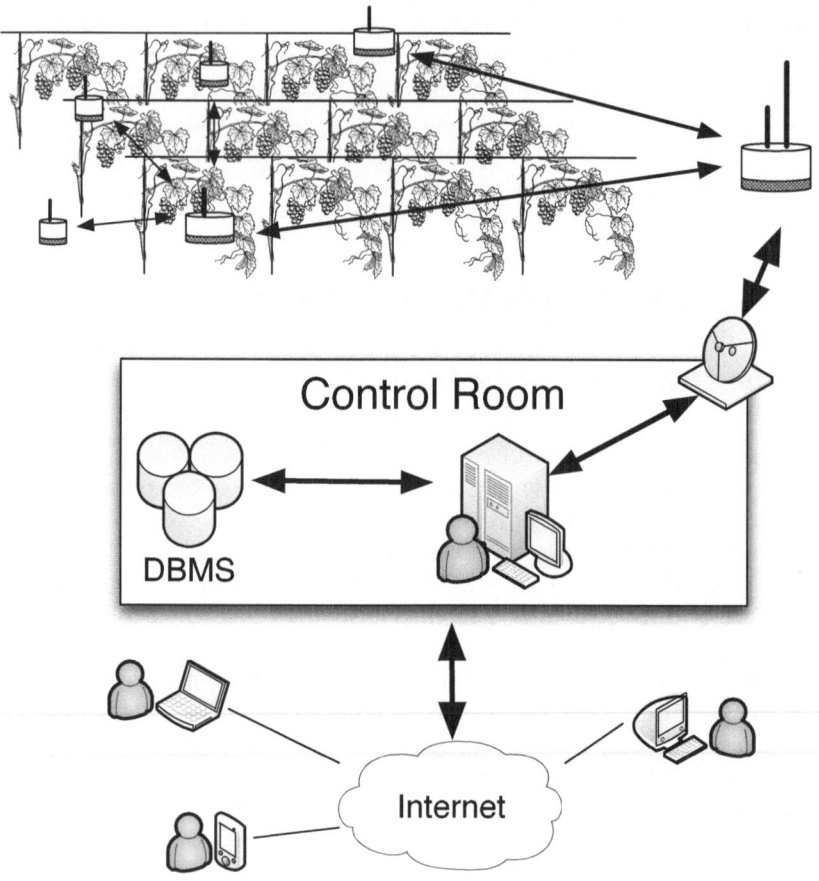

Fig. 15.1 The architecture of the proposed WSN-based monitoring system

The *local transmission radio* is controlled by the Chipcon CC2420 low range transceiver (single-chip 2.4 GHz 802.15.4 compliant RF transceiver with a 250kbps data rate). These transceivers are particularly suited for local transmission as they allow the unit to easily cover an area of 2800 m^2 (30m radius) with a reduced power consumption (35mW in transmission and 38mW in reception). However, different radio modules such as the Jennic Wireless Microcontroller [11] (that fully support the ZigBee protocol) could be integrated with minor modifications at the electronic level of the units.

The *remote transmission module* of the gateway relies on the MaxStream 2.4GHz X Stream Radio Link Modem [12]. The gateway uses the CC2420 transceiver for the communication with the sensor nodes of the cluster (local transmission) and the 2.4GHz radio link for the remote transmission to the control room. Unfortunately,

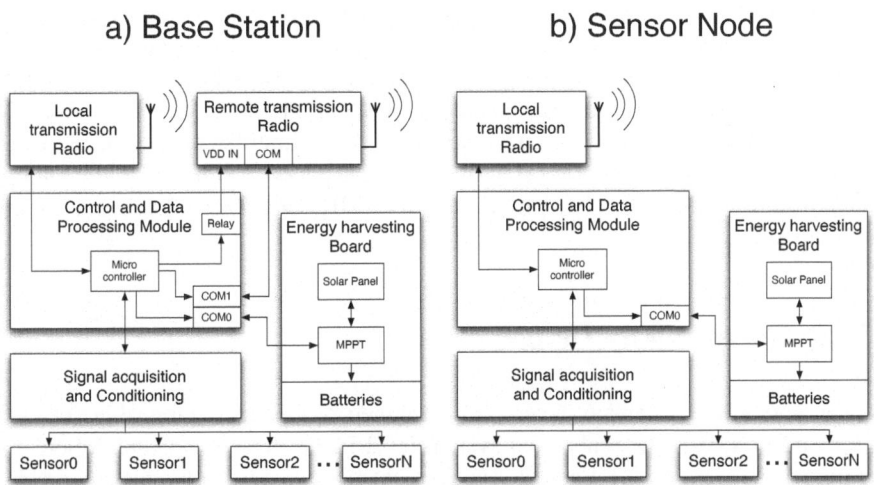

Fig. 15.2 a) Gateway architecture; b) Sensor node architecture

the radio link is power eager (750 mW in transmission) and an effective duty cycle mechanism must be adopted to keep the radio off as much as possible. The radio is switched on by the control and data processing module only when a remote transmission is required and switched off immediately afterwards.

The *signal acquisition and conditioning module* acquires analog data from the sensors and provides digital output streams. The analog signal provided by the sensor requires a conditioning stadium starting with a low pass filter (to avoid aliasing effects) followed by a gain amplification stadium. The analog-to-digital converter finally provides a digital outcome on 12 bits. The list of sensors considered in this application is given in Table 15.1.

Table 15.1 The sensors. Legend: nd - not declared

Sensor	Producer	Model	Range	Accuracy	Power Consumption
Temperature	Texas Instr.	TMP275	$-20°C / +100°C$	$±0.5°C$	0.27mW
Humidity	Sensirion	SHT11	0 to 100% RH	pm3.5% RH	3mW
Rain gauge	AANDERAA	3864	max 200mm/sampl. int.	± 2%	0.27mW
Leaf Wetness	Decagon Dev.	LW05-rc	nd	nd	5mW

The *energy harvesting module* is composed of a dedicated low-power 32kHz 8-bit microcontroller that control a step-up DC/DC converter based on the Maximum Power Point Tracker (MPPT) circuit [13]. Since the photovoltaic panel is essentially a diode, the relation between the voltage imposed on the panel and the provided power is non-linear and dependent on the radiating power. The MPPT automatically sets the voltage of the solar panel to maximize the power transferred to the energy storage mean. Since the radiating solar energy changes over time, the MPPT tracks the solar dynamic to maximize energy extraction. Units are endowed with two 0.5W polycristalline solar cells for each sensor node and eight cells for the gateway. In both cases, the same MPPT circuit guarantees an optimal harvesting of the solar energy by adapting the working point of the solar panels to maximize the energy transferred to the rechargeable batteries (four NiMH cells battery packs with 4.8V nominal voltage).

15.2.3 The Hardware Design

Sensor nodes and gateways have been designed to be interchangeable (i.e, the unique difference at the embedded system level is the presence of the remote radio board on the gateway). Each unit is composed of two vertically-stacked circular electronic boards, linked together by means of specific connectors. The upper board includes the control and data processing module, the signal acquisition and conditioning module and the transmission radio module, while the lower one contains the energy harvesting module. The developed sensor node boards prototype are presented in Figure 15.3.

Since the remote units might be lost in the environment, we used Restricted of Hazardous Substance (RoHS) compliant electronic components in order to significantly reduce the amount of hazardous materials (e.g., lead, mercury, cadmium).

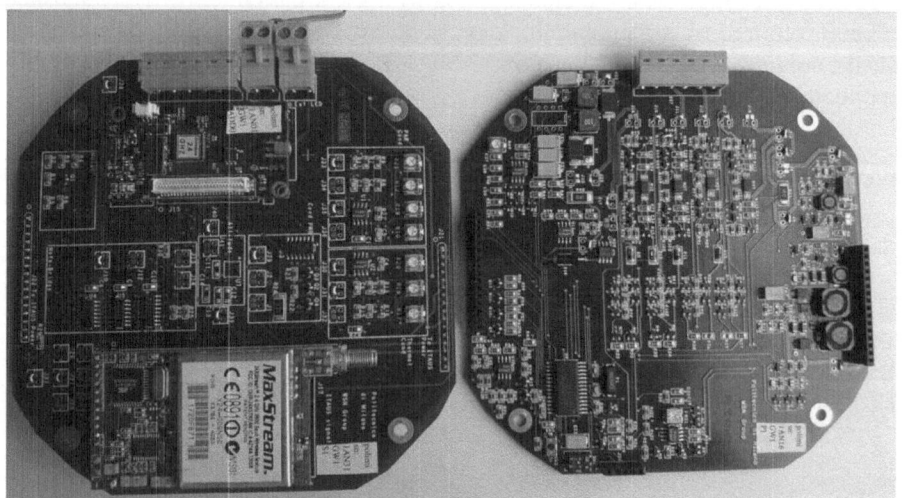

Fig. 15.3 The boards composing the sensor nodes

In addition to the application-specific sensors presented above, we included some utility sensors (e.g., temperature and humidity) to detect the occurrence of critical situations within the units, such as water infiltration or condensation or an excessive temperature (industrial electronics operates in the -20 C, 80 C range).

15.3 Designing the System: Software Aspects

Measurements acquired by the sensing nodes are collected by the gateway, which in turn acquires the data from its own sensors. After a possible processing at the cluster level data are forwarded to the remote control room for storage, visualization and interpretation.

Unfortunately, local communication could be corrupted by disturbances and the network topology may change during the operational life. For instance, the former phenomenon is associated with the lack/availability of energy or transient/persistent faults in the sensing nodes, while the latter arises when some nodes are switched on/off according to energy availability or when new nodes are introduced in the deployment area. It is clear that adaptation is requested at the routing level to guarantee flexibility but, due to energy constraints only simple and power-aware routing algorithms can be considered.

Several power-aware routing algorithms have been presented in the literature. The Self Organization Medium Access Control (SMAC) [15] protocol allows sensor nodes selecting a transmission frequency to communicate with adjacent nodes. Unfortunately, most of the common off-the-shelf sensor units (e.g., Xbow MICA units [10], Jennic Wireless Microcontroller [11], T-Mote Sky Moteiv [14]) can receive only from a single frequency channel (no frequency division multiple access, or FDMA). The Eavesdrop-And-Register (EAR) [15] protocol allows the management of both fixed and mobile sensing units, by introducing energy-efficient routing mechanisms. However, since our nodes are static and the network topology does not require sophisticated routing mechanisms, the EAR protocol is over-dimensioned for the proposed solution. The TDMA [16] protocol is an efficient solution as the nodes transmit data to the gateway only at predefined time slots. This protocol allows the units to undergo an energy-efficient management as radios are turned on only during the assigned time slots. Unfortunately, TDMA has not been tailored to WSNs and does not provide adaptation mechanisms to manage network topology changes. Moreover, TDMA is not power-aware for gateways (that remain always active). The Hybrid TDMA-FDMA [16] protocol relies on a combined TDMA/FDMA approach to transmit more data in the allocated time slots with a specific frequency modulation but, again, common sensing units cannot receive simultaneously data on more than one frequency. The Carrier Sense Multiple Access (CSMA) protocol prevents message collision by listening to the channel before enabling the transmission. This approach is not power-aware as multiple access to the channel might be required for a single transmission. The ZigBee protocol [17] could be a realistic solution although it does not allow the unit for a fine-tuning of the radio module (that is autonomously managed by the ZigBee protocol) and it requires the gateway to be active (i.e., does not allow duty-cycling).

We propose a modified TDMA protocol that supports power-awareness and dynamic adaptation to network topology changes. In particular, the proposed protocol provides plug-and-play mechanisms for the insertion/wake up and removal of units and, at the same time, an energy-efficient management of the radio module both for sensor nodes and gateways. A high-level description of the suggested adaptive power-aware TDMA protocol is shown in Figure 15.4. Once woken up by a timing interrupt, the generic sensor node acquires data and transmits data to the gateway that, in turn, aggregates and transmits them to the control room through the radio link.

The suggested protocol, which was designed and developed to guarantee both the adaptability to topological changes in the network and the robustness to unit and communication faults, revealed to be particularly effective in the considered deployment. In principle, other communication protocols could be considered as well. Nevertheless, we emphasize to consider communication protocols which provide a trade-off between robustness/adaptability and energy consumption.

The suggested adaptive power-aware TDMA for the unit (Figure 15.5) and the gateway (Figure 15.6) acts as follows.

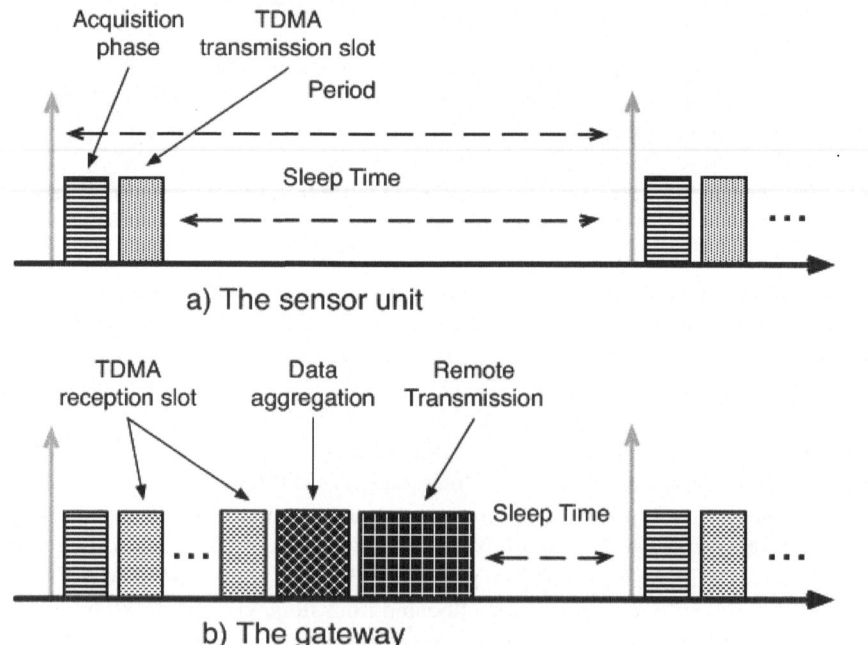

Fig. 15.4 The software phases of the suggested adaptive power-aware TDMA protocol

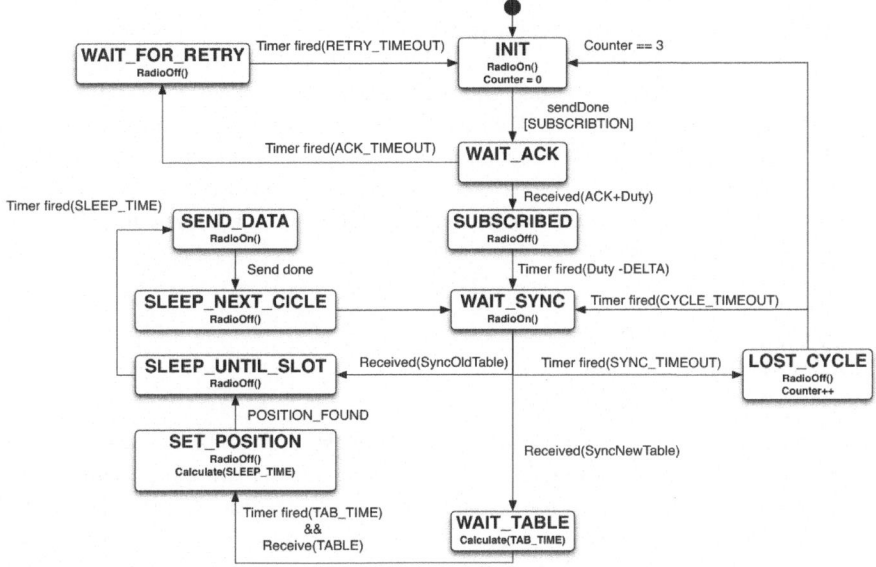

Fig. 15.5 FSM of the sensor node protocol

15.3.1 Sensor Node

In the initial INIT state, the sensor node does not know about the possible presence of the gateway and signals its presence by turning on the radio in transmission modality (TX mode) and sends a SUBSCRIBE message. After the transmission, the radio module moves into a reception modality (RX mode) and waits for an ACK message from the gateway. If the ACK message does not arrive within ACK_TIMEOUT seconds, the sensor node switches off the radio module, sleeps for RETRY_TIMEOUT seconds and then restarts from the INIT state.

When a sensor node receives an ACK message it *has been* successfully registered by the gateway, which has modified the TDMA table to account for its presence. The gateway, also specifies within the ACK message the sleeping time (DUTY_DELTA) the sensor node has to set. In particular, after DUTY_DELTA seconds, the sensor node wakes up in the WAIT_SYNC state, turns on the radio in RX mode and waits for the SYNC message from the gateway. If the sensor node does not receive the SYNC message within SYNC_TIMEOUT seconds, it switches off the radio module, sleeps for CYCLE_TIMEOUT seconds (LOST_CYCLE state) and starts again from the WAIT_SYNC state. If the SYNC message is not received for three consecutive times (this parameter can be set by the WSN designer), the unit moves to the INIT state. Conversely, when the SYNC message is received from the gateway, the sensor node moves to the SYNCHRONIZED state. The SYNC message includes a field specifying whether the network topology has changed (due to the insertion or removal of a node) or not, and thus whether an updated TDMA table needs to be sent to the satellite units or not. If the sensor node does not need to wait for a new TDMA table it moves to the SLEEP_UNTIL_SLOT state and sleeps until the next transmission slot. If a new TDMA table must be sent, the sensor node changes state into WAIT_TAB state and waits for the TAB_MESSAGE. If the message does not arrive within TAB_TIMEOUT seconds, the sensor node disconnects itself from the network and moves back into the INIT state. On the contrary, when the TAB message is received, the sensor node is able to identify its own time slot and computes the amount of time it can sleep (SLEEP_TIME) and moves to the SLEEP_UNTIL_SLOT state.

15.3.2 Gateway

At the INIT state, the gateway turns on the radio in RX mode and waits for the SUBSCRIBE messages coming from the sensor nodes. This approach minimizes the power consumed by the sensor nodes during the registration phase but, at the same time, forces the gateway to wait for messages (with a possibly relevant power consumption). To mitigate this effect the gateway is provided with a 4W solar panel for energy harvesting.

As soon as a SUBSCRIBE message is received, the gateway activates a timer (for PERIOD seconds) and then moves to the REGISTER_NODE state to register the new sensor node, updates the TDMA table, and sends back the ACK message to the registered node. Afterwards, the gateway moves to the WAIT_FOR_SUBSCRIPTION state, waiting for SUBSCRIBE messages to be sent from other sensor nodes. If another SUBSCRIBE message is received, the gateway moves again to the REGISTER_NODE state to register the node and so on. When the timer generates the interrupt (i.e., PERIOD seconds have elapsed from the reception of the first SUBSCRIBE message), the gateway moves to the SEND_SYNC state, broadcasts the SYNC message and activates the TAB_TIMEOUT timer. As stated in the Section 15.3.1, the SYNC message also provides the information about the need to send a new TDMA table to sensor nodes. In case of a TDMA table transmission the gateway moves to the SEND_TABLE state, sets the radio in TX mode, broadcasts the new TDMA table and, finally, reaches the RADIO_SLEEP state. On the contrary, if the TDMA table does not need to be changed, the gateway moves directly to the RADIO_SLEEP state.

The gateway wakes up activated by the TAB_TIMEOUT timer and moves into the WAIT_FOR_DATA state. In this state the gateway sets the radio to the RX mode and waits for data coming from the registered sensor nodes according to the slot allocation specified in the TDMA table. For each received message the gateway reaches the REGISTER_DATA state, stores the data in a specific memory location and moves back to the WAIT_FOR_DATA state. When a sensor node does not transmit its own message, the TAB_TIMEOUT timer sends the interrupt, the gateway remains in the WAIT_FOR_DATA state and passes to the next time slots. When the time slots for all the registered nodes have elapsed, the gateway is ready (DATA_READY state) to remotely transmit data collected from the sensor nodes in the current TDMA cycle.

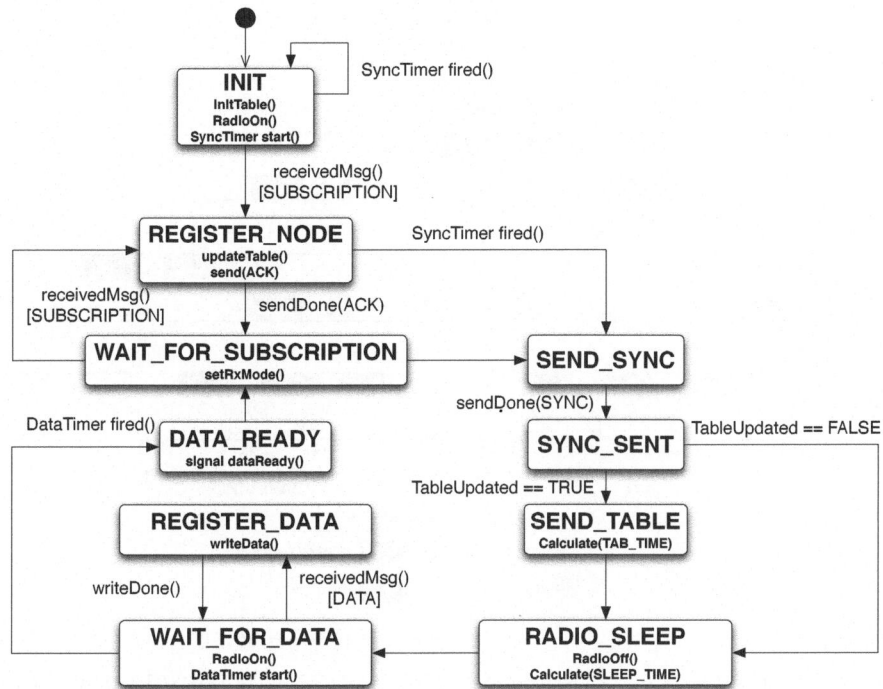

Fig. 15.6 FSM of the gateway protocol

Afterwards, the gateway returns to the WAIT_FOR_SUBSCRIPTION state to possibly register other nodes.

15.3.3 Robustness of the Protocol

Robustness to transmission errors and unit faults is a key issue that has been taken into account in the design of our adaptive TDMA protocol. In more detail, and referring to Figure 15.7, the suggested protocol is robust w.r.t. the miss of the SUBSCRIBE message (Figure 15.7.a), the SYNC message (Figure 15.7.b), the TDMA table (Figure 15.7.c) for the sensor nodes and the DATA message for the gateway (Figure 15.7.d).

As presented in Figure 15.7.a, if the sensor node sends its SUBSCRIBE message but does not receive the ACK within ACK_TIMEOUT seconds, a retry-mechanism forces the node to sleep for RETRY_TIMEOUT seconds and starts again in the INIT state. This procedure is repeated until the node is registered by the gateway (i.e., the ACK is received). In case of a miss of the SYNC message (Figure 15.7.b), the sensor node returns to the WAIT_SYNC state and sleeps for the remaining TDMA cycles. If

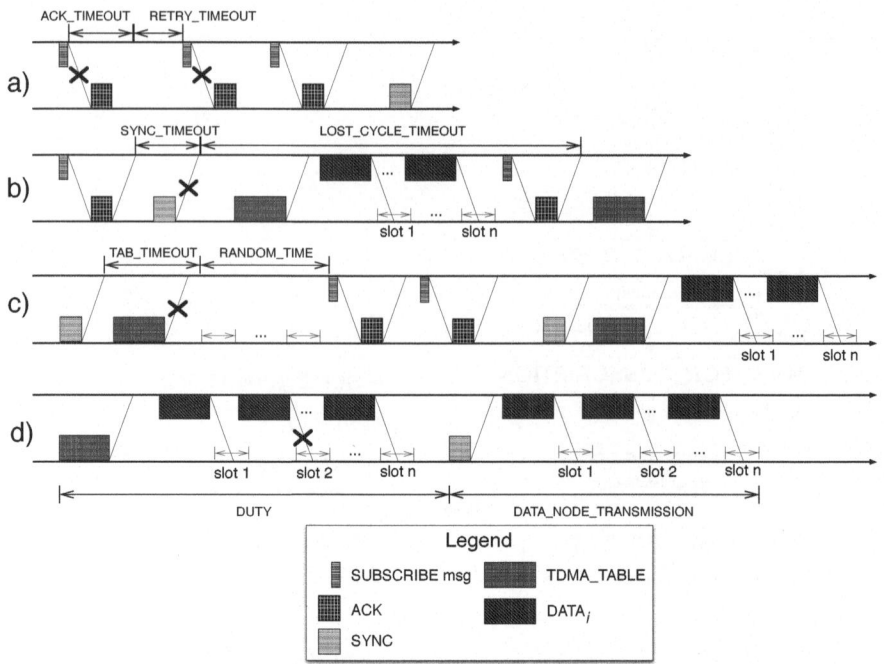

Fig. 15.7 Robustness of the suggested protocol to the loss of: a) the SUBSCRIBE message; b) the SYNC message; c) the TDMA table for the sensor nodes; d) the DATA message for the gateway

the sensor node does not receive the SYNC message for RETRY consecutive cycles (in our case RETRY = 3), it returns to the INIT state. The miss of the TDMA table from a sensor node is particularly critical since it could corrupt the TDMA transmission time slot of other nodes. For this reason, it disconnects itself from the network by returning to the INIT state (Figure 15.7.c). Instead, in the case of DATA message loss the gateway moves to the next transmission time slot: this allows not preventing messages reception from sensor nodes even if the DATA transmission is corrupted. In addition to the transmission errors presented above, the suggested protocol faces those situations where the gateway activates after the sensor nodes (Figure 15.8.a) or switches off before sending the SYNC message (Figure 15.8.b) or the TDMA table (Figure 15.8.c) after the registration and the synchronization of the nodes. In particular, the protocol does not require the gateway to be activated before the sensor nodes (Figure 15.8.a). In fact, sensor nodes keep sending the SUBSCRIBE message until an ACK from the gateway is received. If the gateway switches off before sending the SYNC message, the sensor nodes wait for RETRY cycles and then move to the INIT state (Figure 15.8.b). On the contrary, if the sensor nodes are waiting for the TDMA table and the gateway switches off, the nodes move immediately to the INIT state (Figure 15.8.c) so that when the gateway is back the sensor units are ready to recreate the network. (Figure 15.8.a).

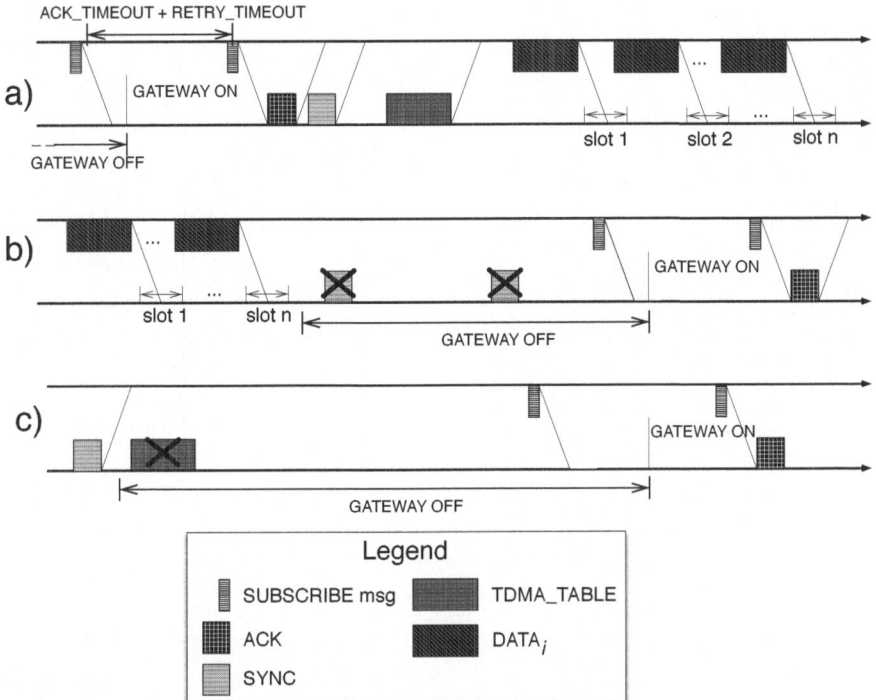

Fig. 15.8 Robustness of the suggested protocol to the gateway faults: a) the gateway activates after the sensor nodes; b) the gateway switches off before sending the SYNC message; c) the gateway switches off before sending the TDMA table

15.4 Designing the System: Data Storage and Presentation

Measurements (together with timestamps) are provided to the data storage system (DSS) that stores them in a MySQL Server 5.0 DBMS running on a Linux OS (See Figure 15.1). For each sensing node of the network the DSS also stores the status (i.e., registered, not-connected, connected), the voltages of the batteries, the current provided by the solar panels and the temperature and the humidity inside the case.

The information stored in the DBMS can then be processed by opening views of the data according to the needs of the operators (e.g., monitoring historical data, detecting potentially dangerous situation for the vineyard by defining alert-thresholds on the acquired measurements, etc.). A proprietary visualization SW, SensorWeb, which can be used at the control station or remotely through the Internet, provides pre-defined views and allows users to define their own views. In more detail, SensorWeb, which is a Tomcat-based application, allows to visualize the status and the measurements of each sensing node of the network by simply specifying the node ID and the time horizon of interest. Moreover, SensorWeb provides graphical tools to compare the

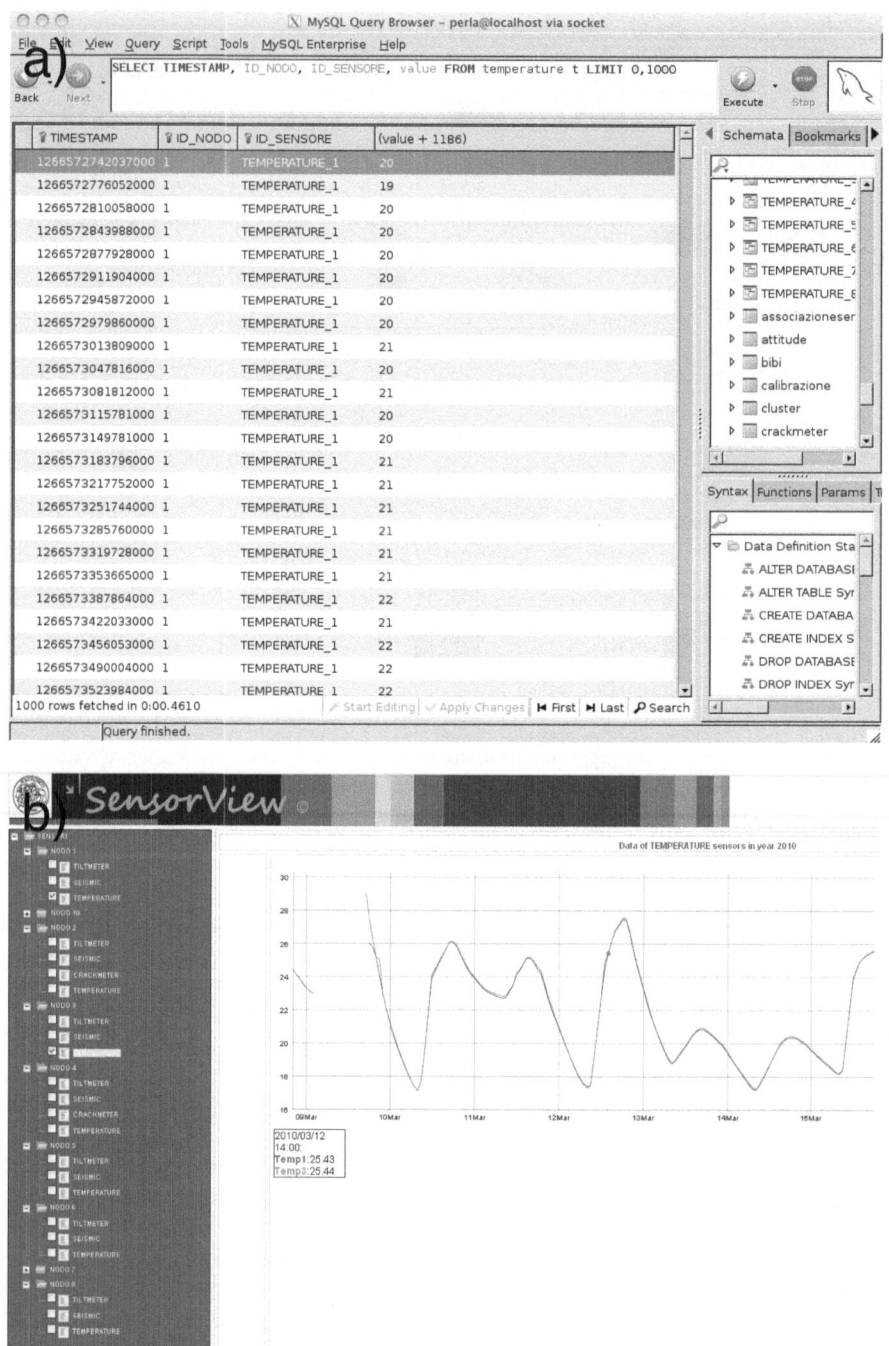

Fig. 15.9 Database and visualization software

measurements of different sensing nodes over time. An example of a predefined view aiming at presenting the status of the sensing node of the network and the temperature (in air) measurements in the last 24 hours is presented in Figure 15.9.

15.5 Conclusions

We presented a WSN-based distributed monitoring system for vineyard monitoring. Adaptability, robustness and energy-efficiency have been envisaged in the by considering plug-and-play mechanisms for the insertion/removal of nodes, robust communication protocols (with respect to faults in the sensor nodes and the gateway) and energy harvesting mechanisms to prolong the lifetime of the WSN. All the aspects of the presented monitoring system such as signal and conditioning board, processing unit, energy harvesting circuits, local and remote transmission protocols, data storage and visualization software have been designed and implemented to guarantee the credibility of the system both from the application and the energy point of view.

References

1. Beckwith, R., Teibel, D., Bowen, P.: Unwired wine: Sensor networks in vineyards. Proceedings of IEEE Sensors (2004)
2. Pasero, E., Moniaci, W.: Artificial neural networks for meteorological nowcast. In: IEEE International Conference on Computational Intelligence for Measurement Systems and Applications, CIMSA 2004, pp. 36–39 (2004)
3. Baggio, A.: Wireless sensor networks in precision agriculture. In: ACM Workshop on Real-World Wireless Sensor Networks (REALWSN 2005), Stockholm, Sweden (2005)
4. Accenture Remote Sensor Network, http://accenture.net/NR/rdonlyres/ 16F0544A-F2E0-4A38-B9BA-6E2F175CA7EE/0/pickberry.pdf
5. Anurag, D., Siuli, R., Somprakash, B.: AGRO-SENSE: Precision Agriculture Using Sensor-based Wireless mesh Networks. In: Proc. of Innovations in NGN: Future Network and Services, K-INGN 2008. First ITU-T Kaleidoscope Academic Conference, Geneva, Switzerland (May 2008)
6. Jardak, C., Rerkrai, K., Kovacevic, A., Riihijarvi, J., Mahonen, P.: Email from the vineyard, 5th International Conference on Testbeds and Research Infrastructures for the Development of Networks and Communities and Workshops (2009)
7. Cao, X., Chen, J., Zhang, Y., Sun, Y.: Development of an integrated wireless sensor network microenvironmental monitoring system. ISA Transactions 47(3), 247–255 (2008)
8. Costa, N., Ferreira, F., Santos, N., Pereira, A.: WSNet–WineCellar An Evolutionary Wireless Sensor Network to Monitor Wine-Cellars. In: Proceedings of the Second International Conference on Systems and Networks Communications, p. 81. IEEE Computer Society (2007)
9. Connolly, M., O'Reilly, F.: Sensor networks and the food industry. In: Workshop on Real-World Wireless Sensor Networks (2005)
10. XBOW (2010), http://www.xbow.com
11. JENNIC Wireless Microcontroller (2010), http://www.jennic.com

12. MAXSTREAM (2010), http://www.maxstream.com
13. Alippi, C., Galperti, G.: An adaptive system for optimal solar energy harvesting in wireless sensor network nodes. IEEE Transactions on Circuits and Systems I 55(6), 1742–1750 (2008)
14. SENTILLA (2010), http://www.sentilla.com/
15. Sohrabi, K., Gao, J., Ailawadhi, V., Pottie, G.J.: Protocols for self organization of a wireless sensor network. IEEE Personal Communications 7(5), 16–27 (2000)
16. Goldsmith, A.: Wireless Communications, Cambridge (2005)
17. ZIGBEE (2010), http://www.zigbee.org/

Chapter 16
Design, Implementation, and Field Experimentation of a Long-Lived Multi-hop Sensor Network for Vineyard Monitoring

Giuseppe Anastasi, Marco Conti, Mario Di Francesco, and Ilaria Giannetti

Abstract. Precision agriculture can particularly benefit from wireless sensor networks, as they allow continuous and fine-grained monitoring of environmental data, which can thus be used to reduce management costs and improve crop quality. Such applications typically require long-term and unattended monitoring of large geographical areas. Therefore, sensor nodes must be able to self-organize and use their limited energy budget very efficiently, so as to prolong the network lifetime to many months or, even, years. In this chapter we present ASLEEP, an adaptive strategy for efficient power management in multi-hop WSNs targeted to periodic data collection. The proposed strategy dynamically adjusts the active periods of nodes to match the network demands with the minimum energy expenditure. In this chapter we focus on the implementation and the experimental evaluation of ASLEEP on a real testbed deployed in a vineyard, according to the case study considered in the project. We show that our adaptive approach actually reduces the energy consumption of sensor nodes, thus increasing the network lifetime up to several years.

16.1 Introduction

A wireless sensor network (WSN) consists of a number of tiny sensor nodes deployed over a geographical area. Each node is a low-power device which is able to sense physical information from the environment (e.g., temperature), process the acquired

Giuseppe Anastasi · Ilaria Giannetti
Dept. of Information Eng., Univ. of Pisa, Italy
e-mail: {g.anastasi,ilaria.giannetti}@iet.unipi.it

Marco Conti
Institute of Informatics and Telematics, CNR, Italy
e-mail: marco.conti@iit.cnr.it

Mario Di Francesco
Dept. of Computer Science and Engineering, Aalto University,
and CReWMaN, University of Texas at Arlington
e-mail: mario.di.francesco@aalto.fi

G. Anastasi et al. (Eds.): Networked Enterprises, LNCS 7200, pp. 311–327, 2012.

data locally, and send them to one or more collection points, referred to as *sinks*, *base stations* [1], or *gateways* (Chapter 15). The number of potential WSN applications is extremely large. However, precision agriculture is one application that can particularly benefit from WSNs. In fact, instrumenting crops with wireless sensor nodes enables continuous and fine-grained monitoring of environmental data — such as temperature, humidity, soil moisture — which can be used by farmers to optimise their strategies, reduce management costs, improve quality and, ultimately, increase profits ([31, 5], Chapter 15). Typically, such applications require monitoring large geographical areas for very long times (e.g., several months, or even years). The key issue is that sensor nodes are generally powered by batteries with a limited capacity and, often, cannot be replaced nor recharged, due to environmental or cost constraints. Therefore, efficient power management strategies should be devised at sensor nodes to extend the network lifetime as much as possible, and avoid frequent battery replacements. Efficient power management is required also when energy can be harvested from the external environment (e.g., through solar cells) (Chapter 15), since a battery has to be used as a buffer anyways.

Since in these applications the radio component accounts for the major energy consumption of a sensor node, even when it is idle [22], the most effective approach to energy conservation is *duty-cycling*, which consists in putting the radio in the (low-power) sleep mode during idle periods. Thus, sensor nodes should coordinate their sleep and wakeup periods, and agree on a network-wide sleep schedule in order to make communication feasible and efficient [3]. In the previous chapter a TDMA (Time Division Multiple Access) based duty-cycling scheme was proposed for a star network where all nodes are directly connected to the gateway (Chapter 15). This scheme is very effective for single-hop networks. However, it exhibits limited scalability and flexibility properties in multi-hop sensor networks. In this chapter we focus on multi-hop WSNs, which are required when the monitoring area is large. Most duty-cycling schemes for multi-hop WSNs use fixed parameters i.e., the sleep/wakeup periods of different sensor nodes are defined before the deployment and cannot be changed during the operational phase. Fixed duty-cycling schemes require simple coordination mechanisms but, typically, have non-optimal performance. In this chapter we present an *Adaptive Staggered sLEEp Protocol (ASLEEP)* for multi-hop WSNs, which automatically adjusts the activity of sensor nodes, achieving both low power consumption and low message latency. With respect to other similar approaches, our scheme has two major strengths. First, it is able to quickly adapt the sleep/wakeup periods of each single node to the actual operating conditions (e.g., traffic demand, link quality, node density etc.), resulting in a better utilization of the energy resources and, hence, in a longer network lifetime. In addition, it is not tied to any particular MAC (Medium Access Control) protocol and, thus, it can be used with any sensor platform.

A detailed simulation analysis of ASLEEP has been carried out in [4]. The obtained results have shown that ASLEEP largely outperforms commonly used fixed duty-cycling schemes for multi-hop WSNs in terms of energy efficiency, message latency, and delivery ratio. In this chapter we focus on the implementation of the proposed protocol in actual sensor nodes and its experimental evaluation in a real environment. To this end, we deployed a testbed in a vineyard, according to the case study considered in the project, and performed several experiments, in both stationary and dynamic conditions. The experimental measurements confirm previous simulation results, and show that ASLEEP increases significantly the network lifetime, thus making the deployment of a long-lived WSNs really possible.

The remainder of this chapter is organized as follows. Section 16.2 surveys the related work. Section 16.3 introduces the reference system model and outlines the main design principles. Section 16.4 describes the ASLEEP protocol. Section 16.5 introduces the testbed used for the experimental analysis, while Section 16.6 presents the obtained results. Finally, Section 16.7 concludes the chapter.

16.2 Related Work

Power management can be implemented either at the MAC layer — by integrating a duty-cycling scheme within the MAC protocol — or as an independent sleep/wakeup protocol on top of the MAC layer (e.g., at the network or application layer). Since the solution proposed in this chapter belongs to the latter class, below we will focus on independent sleep/wakeup schemes only.

General sleep/wakeup schemes can be broadly classified into three main categories: *on-demand*, *asynchronous* and *scheduled rendezvous* schemes. *On-demand* schemes assume that destination nodes can be awakened somehow just before receiving data. To this end, two different radio transceivers are typically used [24, 30]. The first radio (*wakeup radio*) — typically a very low-power radio — is used to wake up a target node when needed, while the second radio (*data radio*) is used for the regular data exchange. These schemes can achieve a high energy efficiency and a very low latency. However, they cannot be always used in practice because commonly available sensor platforms only have one radio.

A different option is using an *asynchronous* scheme [11, 20, 30, 32]. In this case a node can just wakeup whenever it wants and still be able to communicate with its neighbours. Although being robust and easy to implement, asynchronous schemes generally present high latency in message forwarding and are not very suitable to manage broadcast traffic.

The last class of general sleep/wakeup protocols is represented by *scheduled rendezvous* schemes, which require that sensor nodes are synchronized and nodes that need to communicate wake up at the same time. Our ASLEEP protocol belongs to this category. A possible approach to scheduled wakeups consists in establishing a coarse-grained TDMA schedule defined at the application layer, and exploiting an underlying MAC protocol for actual data transfer. This approach is used by *Flexible Power Scheduling* (FPS) [8, 9], which includes an on-demand reservation mechanism capable to dynamically adapt to traffic demands. Since slots are relatively large, a strict synchronization among nodes is not required. However, FPS borrows some drawbacks [23] from TDMA schemes, i.e., it has limited scalability and flexibility in adapting to traffic and topology changes in a multi-hop network.

Most solutions following a scheduled rendezvous approach use a simple duty-cycle-based scheme. For instance, the well-known TinyDB query processing system [27] includes a sleep/wakeup scheme based on a fixed duty-cycle. All sensor nodes in the network wake up at the same instant and remain active for a fixed time interval. An improvement over this simple approach is the staggered scheme included in TAG (Tiny AGgregation) [16], which relies on a routing tree rooted at the sink node. In this scheme, the active times of sensor nodes are staggered according to their position in the routing tree. Nodes located at different levels of the routing tree wake up at different, progressive times, like in a pipeline. Due to its nice properties, this scheme has been considered and/or analysed in many subsequent papers ([6, 13, 14, 15, 19, 12] among the others).

Although providing a basic form of adaptation (wakeup times are staggered to the network topology), this scheme is not able to react to varying operating conditions as active times are fixed and equal for all nodes in the networks. This constraint simplifies the coordination among nodes, but results in low energy efficiency and high message latency. Like TAG, our proposal leverages a staggered approach. However, in our proposal nodes' active periods are dynamically adapted to the observed network conditions and can be tailored to the actual needs. By minimizing the active period of each single node, our adaptive protocol significantly increases network lifetime and reduces message latency.

16.3 Network Model and Design Principles

We address the problem of data collection in dense WSNs, where nodes are assumed to be static. We also refer to the common *convergecast* scenario where data typically flow from sensor nodes to the sink, while data in the opposite direction are much less frequent. We assume that nodes are organized to form a logical *routing tree* (or *data gathering* tree), rooted at the sink, for data forwarding[1]. The routing tree may change over time due to link/node failures or nodes running out of energy. Also, it might be recomputed periodically to better share energy consumption among nodes. However, as nodes are static, we assume that the routing tree — once established — remains stable for a reasonable amount of time. We also assume that sensor readings are periodically reported to the sink node, e.g., every minute. Specifically, nodes share a common notion of time and communicate each other in *communication periods* that repeat periodically. Each communication period includes an *active interval* during which nodes are awake and can communicate with their neighbours (by using the underlying MAC protocol), and a *silence interval* during which nodes turn their radio off to save energy. The active interval can be made of one or more *talk intervals*, defined as the time shared between a given node and its children. It is wortwhile noting that the active interval is made of only a single talk interval for both the sink (which

---

[1] Many routing protocols for WSNs rely on a routing tree, e.g., [10, 14, 16, 18, 26, 27].

has only its own talk interval with its children) and leaf nodes (which have only the talk interval with their parent).

We referred to the following design principles to effectively design our power management strategy.

- *Low-latency and Energy-efficient Operations.* Although the duty-cycle mechanism helps to reduce the energy consumtpion, the power management strategy has to be carefully defined, in order to keep overheads (e.g., radio state switching) as low as possible. In addition, the latency of data collection should not be significantly affected by the duty-cycle scheme.
- *Adaptive Duty-Cycle.* Since the actual time required for transmitting/receiving all data depends on the time-varying operating conditions, the active interval should be adjusted over time. In addition, as sensor nodes at different locations experience different traffic and network conditions, nodes should set the length of the active interval on an individual basis.
- *Distributed and Local Computation.* The algorithm used by nodes for the active interval calculation should be local and simple. A global algorithm would require the exchange of information among nodes, thus consuming additional energy. In addition, a complex algorithm might not be suitable for devices with limited computational capacity.
- *Coordinated and Robust Network-wide Sleep Schedule.* A variation in the active interval of a single sensor node should not compromise the correctness and energy efficiency of the global schedule. Hence, a robust cooperation mechanism is required for nodes to manage the network-wide sleep schedule.

In order to address all those principles, we defined the following power management strategy. First of all, each parent node is responsible for choosing its own talk interval with the children. As a consequence, parent nodes can define the most appropriate talk interval duration, depending on their specific needs. The decision is not centralized, hence parent nodes can choose a specific talk interval duration based on local measurements of the current network activity. In addition, to achieve a low latency, the active intervals are *staggered* according to the position of nodes along the routing tree. Specifically, sensor nodes at a given level in the routing tree wake up earlier than their ancestors, and the active intervals of intermediate nodes span over two adjacent talk intervals (Fig. 16.1). In a staggered scheme, latency of messages flowing from source nodes to the sink is bounded by the sum of the talk intervals of traversed nodes, and is independent from the duration of the communication period [12]. In addition, latency can be further reduced if sensor nodes remain active for the minimum time needed for message exchange, i.e., sending and/or receiving all messages addressed to the sink. In order to approximate this behaviour, the power management strategy includes an estimation algorithm such that parent nodes can estimate the actual talk interval they need. Finally, the power management strategy provides a sleep coordination algorithm to organize and update the newtork-wide sleep schedule according to the talk intervals dynamically selected by individual parent nodes.

16.4 Protocol Description

In this section we present the ASLEEP protocol. We will describe first the *talk interval prediction algorithm* used by each node for estimating its forthcoming talk interval, then the *sleep coordination algorithm* used to propagate the new sleep schedule throughout the network.

Throughout we will refer to the talk interval shared by a generic node j and (all) its children during the m-th communication period γ^m as τ_j^m (Fig. 16.1). For convenience, the duration of the talk interval is defined as an integer number of slots, whose duration is set to q. In addition, part of the talk interval is reserved to the transmission of control messages. This part is called *beacon period*, and its duration is denoted as β.

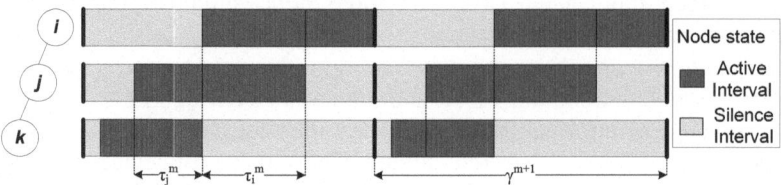

Fig. 16.1 Sleep scheduling parameters

16.4.1 Talk Interval Prediction Algorithm

The duration of the talk interval to be shared with the children in the next communication period is dynamically estimated by each parent node. In principle, any algorithm can be used to obtain such an estimate. In order to address the requirements outlined in Section 16.3 we used the algorithm discussed below.

Each parent node measures and stores two values: the *message inter-reception time* (Δ_i), i.e., the time elapsed between two consecutive messages which are correctly received in a specific communication period; and the total *number of messages* (n) correctly received in the same communication period. The time needed to receive all messages sent by children in the next communication period is then estimated as $\bar{\Delta} \cdot N$, where $\bar{\Delta}$ and N are the average inter-reception time and the maximum number of received messages over the observation window made of the last L communication periods, respectively. Using N is a conservative choice to cope with possible message losses. The former time interval is increased so as to accommodate the beacon period. Finally, the expected talk interval is discretized into an integer number of time slots. Hence, the expected talk interval for the $(m + 1)$-th communication period can be expressed as $\hat{\tau}^{m+1} = \lceil (\bar{\Delta} \cdot N + \beta)/q \rceil \cdot q$. The duration q of the talk interval slot should be chosen as a trade-off between efficiency and stability. A low value of q allows a fine granularity for the talk interval duration, but may introduce frequent changes in the sleep schedule. Conversely, a large value of q makes the schedule more stable, but may lead to talk intervals much larger than necessary, thus wasting energy[2].

Advertising $\hat{\tau}^{m+1}$ to children as the next talk interval might lead to frequent variations in the schedule parameters of sensor nodes. Hence, the talk interval for the next communication period, τ^{m+1} is determined as follows. If $\hat{\tau}^{m+1} - \tau^m > 0$, then $\hat{\tau}^{m+1}$ is chosen as the estimated value and advertised to children (i.e., $\tau^{m+1} = \hat{\tau}^{m+1}$). If the predicted talk interval is below the current value and the difference is greater than, or equal to a guard threshold g (i.e., $\tau^m - \hat{\tau}^{m+1} \geq g$, with $g \geq 2q$), then the talk interval is decreased by just one time slot ($\tau^{m+1} = \tau^m - q$). Finally, if $0 < \tau^m - \hat{\tau}^{m+1} < g$, the talk interval is not immediately decreased. However, if the same condition persists for a number l of consecutive communication periods, then the talk interval is reduced anyway. According to the rules discussed above, an increase in the talk interval is managed less conservatively than a decrease. This is because a more aggressive increase tends to minimize the probability that a node can miss messages from its children.

16.4.2 Sleep Coordination Algorithm

Although parent nodes can independently set their talk intervals, a collective effort is needed for the schedule of the whole network to remain consistent and energy efficient. Hence, as a result of a change in the talk interval of a single parent node, the network-wide schedule needs to be re-arranged. This is accomplished by appropriately shifting the active intervals of nodes to ensure that: *i*) the active intervals of all nodes are properly staggered; and *ii*) the two talk intervals of parent nodes are contiguous. Two special messages, *direct beacons* and *reverse beacons*, are used for propagating schedule parameters to downstream and upstream nodes, respectively.

Direct beacons are broadcast at *each* communication period by *every* parent node at the end of the talk interval, during the beacon period. They include the schedule parameters for the next communication period. Specifically, the direct beacon sent by a node j in the m-th communication period contains: *i*) the length of the next communication period γ^{m+1}; *ii*) its next wakeup time $t^{m+1}_{parent,j}$; and *iii*) the length of the next talk interval to be shared with its children (τ_j^{m+1}).

Conversely, reverse beacons are sent by child nodes. They might be sent at any time during the talk interval, and only include the amount of time the talk interval of the parent node has to be shifted. As schedules are local, nodes only have to coordinate with their parent, i.e., they have to know the wakeup time of their parent, and use it as a basis for establishing schedules with their children.

As beacon messages are critical for correctness, ASLEEP also includes mechanisms to: *i*) increase the probability of successful transmission of direct beacons; and

---

[2] It is worthwhile noting that the expected talk interval cannot be lower than one slot. This guarantees that any child has always a chance to send messages to its parent, even after a phase during which it had no traffic to send.

ii) enforce a correct (even if non-optimal) behaviour of nodes in case they miss a direct
beacon. The interested reader can find the description of these mechanisms in [2].

Algorithm 1. Actions performed by an intermediate node j as a parent

---

1 **upon** *wakeup as a parent*
2 $\quad s_j^{m+1} = 0$
3 $\quad$ schedule direct beacon transmission
4 $\quad$ **do**
5 $\quad\quad$ wait for messages from a generic child node k
6 $\quad\quad$ **if** *message = reverse beacon(r_k^{m+1})* **then**
7 $\quad\quad\quad s_j^{m+1} = \max(s_j^{m+1}, r_k^{m+1})$
8 $\quad\quad$ **else if** *message = data* **then**
9 $\quad\quad\quad$ update statistics
10 $\quad\quad\quad$ store message in the local queue
11 $\quad$ **until** *direct beacon transmission time*
12 $\quad$ obtain τ_j^{m+1} from statistics
13 $\quad$ schedule next talk interval at
$\quad\quad t_{parent,j}^{m+1} = t_{parent,j}^m + \gamma^{m+1} + s_j^{m+1} + \max(0, \tau_j^m - \tau_j^{m+1})$
14 $\quad$ send direct beacon($\gamma^{m+1}, \tau_j^{m+1}, t_{parent,j}^{m+1}$)

Obviously, the specific actions performed by each single node depend on its position on the routing tree. Algorithm 1 and Algorithm 2 show the actions performed
by a generic intermediate node j (child of node i) during the m-th communication
period as a father and as a child, respectively[3]. Clearly, the sink node only executes
Algorithm 1, while any leaf node only executes Algorithm 2.

In the first part of the active interval node j behaves as a parent and talks with its
children (Algorithm 1). Upon wakeup, it schedules the direct beacon transmission
(line 3). To this end, the expected talk interval estimated in the previous communication period is used. During the talk interval node j receives messages from its
children, i.e., data or reverse beacons. Upon receiving a data message, node j updates the statistics on the message inter-reception times and the number of received
messages (lines 9–10) that will be later used to estimate the next talk interval. Upon
receiving a reverse beacon from a child node k, node j realizes that node k requested
it to shift its talk interval by the quantity r_k^{m+1} included in the reverse beacon (line 6).
In case of multiple shift requests from the children, the maximum shift value is considered (line 7). Then, node j estimates the duration of the talk interval for the next
communication period, τ_j^{m+1} (line 12) — based on the obtained statistics and according to the algorithm described in Section 16.4.2 — and derives the starting time of
the next talk interval (line 13). The latter is obtained from the duration of the next

---

[3] In the following, we describe ASLEEP operations in steady state conditions. The initial
schedule is established through a special startup phase, not provided here for the sake of
space.

communication period (γ^{m+1}), the maximum shift requested by child nodes (s_j^{m+1}) and the difference between the current and estimated next talk interval. Finally, node j broadcasts the direct beacon with the new schedule parameters to its children (line 14). This concludes the talk interval with the children.

Algorithm 2. Actions performed by an intermediate node j as a child

---

1 **upon** *wakeup as a child (of node i)*
2   start activity_timer(τ_i^m)
3   **do**
4   send queued data messages to parent node i
5   **if** ($s_j^{m+1} > 0$ **or** $\tau_j^{m+1} - \tau_j^m > 0$) **then**
6   send reverse beacon($s_j^{m+1} + \max(\tau_j^{m+1} - \tau_j^m, 0)$) to node i
7   **until** *activity_timer expires* **or** *direct_beacon received*
8   **if** (*direct beacon*(γ^{m+1}, τ_i^{m+1}, $t_{parent,j}^{m+1}$) *is received from node i*) **then**
9   schedule next wakeup at $t_{child,j}^{m+1} = t_{parent,j}^{m+1}$
10   **else**
11         $\gamma^{m+1} = \gamma^m$
12         $\tau_i^{m+1} = \tau_i^m$
13   schedule next wakeup at $t_{child,j}^{m+1} = t_{child,j}^m + \gamma^{m+1}$
14   start timer($t_{child,j}^{m+1} - t_{now}$)

Then, node j acts as a child of node i (Algorithm 2). First of all, node j schedules the end of the talk interval with its parent (line 2). Then, it sends all queued messages (line 4) and, if needed, a reverse beacon (lines 5–6). When sending the reverse beacon, the intermediate node adds to the maximum shift requested by the children (lines 6–7 of Algorithm 1) its own shift $\tau_j^{m+1} - \tau_j^m$, if greater than zero (line 6). Upon receiving the direct beacon from the parent node i, node j gets the parameters for the next communication period (lines 8–9). If the direct beacon is missed, then node j uses the current parameters also in the next communication period (lines 11–13). In any case, it sets the timer and enters the sleep mode (line 14).

It can be formally proved that, whenever a change has occurred in the talk interval of one or more nodes with the corresponding children, the global network is able to reach a new coordinated and energy-efficient schedule, under the assumptions that: *i*) the clocks of nodes are synchronized; and *ii*) direct and reverse beacons never get lost [2]. Clock synchronization can be achieved through any available clock synchronization protocol [25], and is mitigated by the fact that ASLEEP does not require tight synchronization (values in the order of a few milliseconds are fine). Assumption *ii*) is rather strong and unlikely to hold in practice since beacons can get lost due to transmission errors and/or collisions. However, thanks to mechanisms for beacon protection and schedule prediction, ASLEEP has been shown to be effective even in environments with high message loss rate [4].

16.5 Protocol Implementation and Experimental Testbed

We implemented ASLEEP on Tmote Sky [28] sensor nodes with TinyOS 1.1.15 operating system [7]. The Tmote Sky platform is equipped with the Chipcon CC2420 radio transceiver, which is compliant to the 2.4 GHz IEEE 802.15.4 physical layer and has a maximum bit rate of 250 Kbps. Our protocol used the default CSMA/CA MAC protocol shipped with TinyOS.

According to the case study considered in the project, the experiments were carried out in a vineyard. For convenience, we performed our experiments in a vineyard located in the Chianti area in Tuscany. Of course, the obtained results are general and do not depend on the specific vineyard. To perform experimental measurements we deployed a testbed WSN consisting of 15 sensor nodes in the vineyard (see Fig. 16.2a and Fig. 16.2b). Nodes were placed at about 50 cm from the ground.

(a)   (b)

Fig. 16.2 Experiments site: the vineyard in Vinci (Firenze). General (a) and detailed (b) view.

They periodically sampled the external temperature and humidity, and reported the acquired data to the sink node (node 0). In our experiments we used a laptop as the sink node, for semplicity. In a real deployment the sink node could be the gateway node designed in the previous chapter (Chapter 15), which, then, transmits data to a remote central room. Before starting ASLEEP, we performed an initial time synchronization by using a modified version of the scheme in [21], and built the routing tree by exploiting the MintRoute protocol [29]. Synchronization and routing tree formation are repeated periodically over time in order to cope with node failures and changes in channel conditions. In our experimental analysis we considered the following three scenarios.

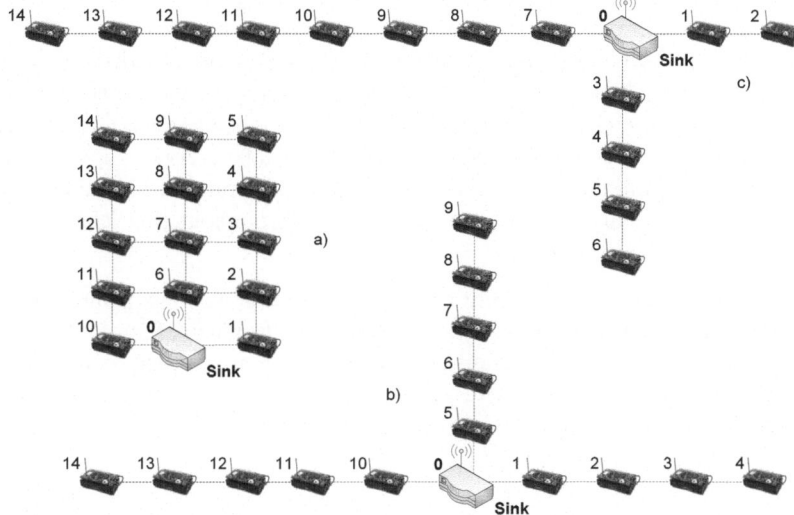

Fig. 16.3 Experimental scenarios: all-in-range (a), multi-hop balanced (b), multi-hop unbalanced (c)

- *All-in-Range Scenario*. All nodes are within the transmission range of the sink (node 0). Sensor nodes are deployed in the vineyard as shown in Fig. 16.3a, at a distance of approximately 10 m from each other, and use the maximum allowed transmission power (so as to reach any other node). Since the transmission range of all sensor nodes is larger than their distance from the sink, a star network topology is expected in this scenario.
- *Multi-hop Balanced Scenario*. Sensor nodes are deployed in the vineyard to form a T-shaped topology with approximately even branches (see Fig. 16.3b). The transmission power of sensor nodes is set to approximately -20 dBm, with a corresponding transmission range of approximately 12 m. Since the distance between neighbouring nodes deployed along the same row is about 10 m, in theory a multi-hop topology should be formed, with routing sub-trees having almost the same depth.
- *Multi-hop Unbalanced Scenario*. Sensor nodes are deployed again to form a T-shaped topology, but sub-trees have now different depths (2, 4, and 8, as shown Fig. 16.3c). The transmission power and the distance between nodes are the same as in the previous scenario. Therefore, in theory, nodes should form an unbalanced multi-hop topology.

For performance comparison, in addition to ASLEEP, we also implemented the following three sleep scheduling strategies.

- *Always-on*. In this scheme there is no sleep schedule: nodes never go to sleep and forward messages as soon as they receive them.

- *TAG-like staggered strategy*. Sensor nodes use a staggered scheme for sleep co-ordination, where the talk interval is *fixed* and *equal* for *all* sensor nodes. The talk interval is set to the value of the communication period divided by the depth of the tree, as in TAG [16]. Throughout, we will refer to such a scheme as *TAG*.
- *Optimal fixed staggered scheme*. In this scheme the talk interval is fixed and equal for all nodes, as in the TAG scheme. However, the talk interval is approximately equal to the minimum value required in that configuration (this value was de-rived by preliminary simulation experiments). Throughout, this scheme will be referred to as *Fixed*.

In a real testbed external conditions change from time to time, sometimes even during the same experiment, and there is no way to control them. In order to improve the statistical accuracy, we replicated each experiment 5 times (each replica was 100 communication-period long). The results presented below are averaged over the en-tire set of 5 replicas. Standard deviations are also reported.

16.6 Experimental Results

In the first part of our analysis we investigated the steady state behaviour of ASLEEP in the three above-mentioned scenarios. We evaluated the performance of the con-sidered protocols in terms of the *average duty-cycle* of nodes at 1-hop from the sink (the most solicited nodes in terms of energy consumption), the *delivery ratio* (i.e., fraction of messages delivered to the sink) and the *average latency* experienced by messages to reach the sink. Table 16.1 shows the parameters used in our experiments. To maintain the experiment duration within reasonable limits, we considered a short communication period (15 s). In a real environment the reporting period is typically much larger, and it is often dictated by the need of maintaining synchronization among nodes. We carried out additional experiments (not presented here) with a communi-cation period of 2 minutes, which substantially confirm the results presented in this section.

When dealing with a real testbed, the actual network topology might be very dif-ferent than the expected one. This is due to the link quality metrics used for building the data gathering tree [29]. For example, we noted that in the all-in-range scenario, nodes located at the upper corners of the grid (see Fig. 16.3a) actually associated with an intermediate node in all experiments, resulting in a routing tree of 2 hops. Since the routing tree impacts on performance, we reported in Table 16.2 the (mean and standard deviation of the) routing tree depth observed in the different scenarios.

Fig. 16.4a compares the average duty-cycle of the different sleep/wakeup schemes in the three considered scenarios. ASLEEP outperforms all other schemes as it is able to adjust dynamically the duty-cycle of each sensor node based on its traffic needs, while the other staggered schemes impose the same talk interval to all nodes belonging to the same level of the routing tree. With ASLEEP and Fixed the average duty-cycle increases when passing from the all-in-range, to the multi-hop balanced, and to the multi-hop unbalanced scenario, as expected. With TAG the trend is just the

Table 16.1 Protocol parameters

Parameter	Value
Communication Period (γ)	15 s
Message rate	1 msg/γ
Message size	28 bytes
MAC frame size	38 bytes
Observation window (L)	10γ
TI time slot (q)	150 ms
Beacon Period (β)	60 ms
TI decrease time threshold (l)	5γ
TI decrease threshold (g)	2q (300 ms)

Table 16.2 Routing tree depth in the experiments (mean and standard deviation under parantheses)

	All-in-range	Multi-hop Balanced	Multi-hop Unbalanced
ASLEEP	2.0 (0)	3.8 (0.45)	5.4 (1.34)
Fixed	2.0 (0)	4.0 (0)	4.0 (0.71)
TAG	2.0 (0)	4.0 (0)	4.8 (1.10)
Always ON	2.2 (0.45)	3.6 (0.55)	4.4 (0.55)

opposite, as in this scheme the length of the talk interval is equal to the communication period divided by the routing tree depth whose average value, for each scenario, is shown in Table 16.2.

We now clarify the impact of the average duty-cycle (i.e., energy consumption) on the network lifetime. We consider the model used in [17] and adapt it to the parameters of the Tmote Sky device [28], which is assumed to be powered with a pair of 3000 mAh AA batteries. For a rough estimate, we only consider the contribution of the radio whilst it is in the transmit or receive states (whose current draw is 19.6 mA), and neglect all other components (i.e., the sensing and processing subsystems). We also consider a communication period of 2 minutes. In the multi-hop unbalanced scenario, the Always-on scheme achieves a network lifetime of approximately 1 month and a half, which is definitely unsatisfactory for a long-term deployment. TAG extends the network lifetime to about 4 months, which is not yet satisfactory. Fixed allows a lifetime of 600 days (20 months), which is quite good, but it requires to know in advance the talk intervals of sensor nodes, which is clearly unfeasible in practice. Finally, ASLEEP is able to extend the network lifetime to approximately 1200 days (more than three years), thus making long-term deployments actually possible.

Fig. 16.4b compares the average message latency introduced by the different sleep/wakeup schemes. Again, ASLEEP outperforms the other staggered schemes. This is because with ASLEEP sensor nodes get shorter active intervals, so that less

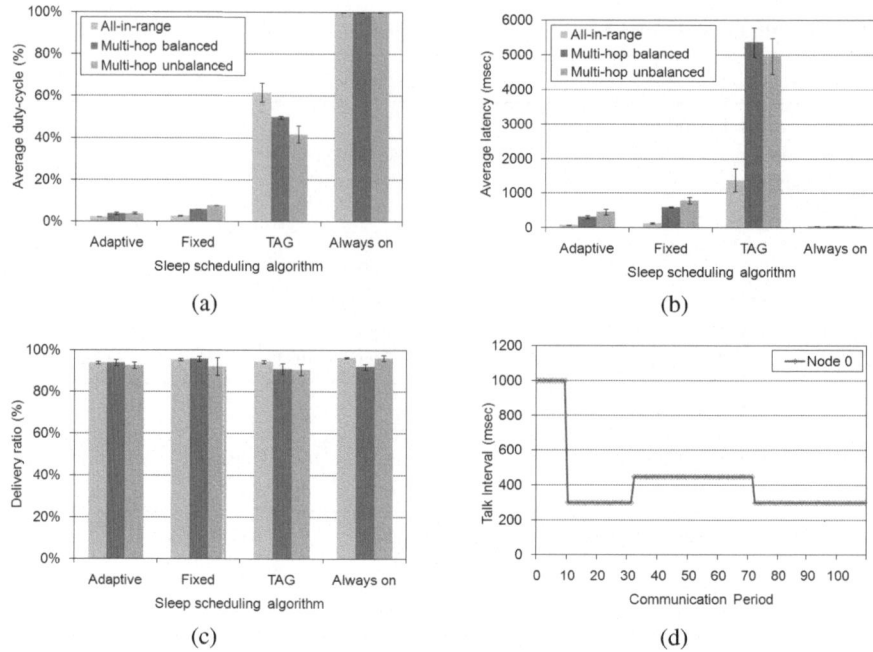

Fig. 16.4 Performance comparison in the three scenarios in terms of Average Duty-Cycle (a), Average Latency (b) and Delivery Ratio (c). Evolution of the Talk Interval over time (d).

time elapses from when the message is sent by the source to when the same message is received at the sink. In particular, ASLEEP introduces an average latency significantly lower than Fixed (464 ms vs. 798 ms in the multi-hop unbalanced scenario), and about an order of magnitude lower than TAG.

Finally, Fig. 16.4c shows the delivery ratio in the three different scenarios. All the sleep/wakeup schemes provide approximately the same delivery ratio (above 90%), and there is no significant difference when passing from one scenario to another. These results confirm that ASLEEP is robust against possible problems related with schedule exchange and maintenance due to beacon losses.

To conclude our analysis, we also analysed ASLEEP in dynamic conditions. We focused on the multi-hop unbalanced scenario and investigated the effect of a variation in the traffic pattern generated by sensor nodes. Initially all sensor nodes generate 1 message per communication period. After the 30-th communication period the rate increases to 3 messages per communication period and, finally, after the 70-th communication period, it reverts back to the initial value. Fig. 16.4d shows the evolution over time of the talk interval shared by the sink and its children as a function of time, in a specific topology (the trend was similar for all other topologies as well). Initially, the talk interval is set to the default value of 1 s. After one observation window ($L = 10$ in our experiments) ASLEEP calculates the first estimate and sets the talk interval to

300 ms. Then, after the 30-th communication period ASLEEP sets the new talk interval duration to 450 ms. The same value is used up to the 70-th communication period, then it is decreased to 300 ms. Hence, we can see that the talk interval is effectively changed by ASLEEP according to the traffic conditions.

16.7 Conclusions

In this chapter we have presented an *Adaptive Staggered sLEEp Protocol* (ASLEEP) for efficient power management in multi-hop wireless sensor networks (WSNs) for vineyard monitoring applications. The proposed protocol staggers the active periods of nodes according to their position in the routing tree. Unlike traditional staggered schemes, however, the proposed approach can tune the active period of each parent node dynamically, depending on the operating conditions experienced by individual nodes. ASLEEP is thus able to adjust the duty-cycle of sensor nodes to their actual needs, thus minimizing the energy consumption. Finally, ASLEEP is conceived as an independent sleep/wakeup protocol operating above the MAC layer, and can thus be used with any available sensor platform.

We implemented ASLEEP on a real testbed deployed in a vineyard, and performed an extensive experimental analysis. The experimental measurements confirmed the previous very promising simulation results, and showed that a significant reduction in both the sensor nodes duty-cycle and the message latency can be achieved with respect to other staggered approaches where the active periods are fixed and equal for all nodes in the network. Assuming that *i*) sensor readings are reported every 2 minutes (which is more than enough for vineyard monitoring); and *ii*) the energy consumption of the CPU and sensors is negligible with respect to that of the radio, our ASLEEP protocol is able to increase the network lifetime to more than 3 years, thus making a long term WSN deployment actually possible.

References

1. Akyildiz, I.F., Su, W., Sankarasubramaniam, Y., Cayirci, E.: Wireless Sensor Networks: a Survey. Computer Networks 38(4) (March 2002)
2. Anastasi G., Conti M., Di Francesco M.: An Adaptive Sleep Strategy for Energy Conservation in Wireless Sensor Networks. Technical Report DII-TR-2009-03, University of Pisa, http://info.iet.unipi.it/~anastasi/papers/DII-TR-2009-03.pdf
3. Anastasi, G., Conti, M., Di Francesco, M., Passarella, A.: Energy Conservation in Wireless Sensor Networks: a Survey. Ad hoc Networks 7(3), 537–568 (2009)
4. Anastasi, G., Conti, M., Di Francesco, M.: Extending the Lifetime of Wireless Sensor Networks through Adaptive Sleep. IEEE Transactions on Industrial Informatics 5(3), 351–365 (2009)
5. Baggio, A.: Wireless Sensor Networks in Precision Agriculture. In: Proceedings ACM Workshop on Real-World Wireless Sensor Networks (REALWSN 2005). ACM (2005)

6. Cao, Q., Abdelzaher, T., He, T., Stankovic, J.: Toward Optimal Sleep Scheduling in Sensor Networks for Rare Event Detection. In: Proc. IPSN 2005 (April 2005)
7. Hill, J., Szewczyk, R., Woo, A., Hollar, S., Culler, D., Pister, K.: System architecture directions for networked sensors. SIGPLAN Not. 35(11), 93–104 (2000)
8. Hohlt, B., Doherty, L., Brewer, E.: Flexible Power Scheduling for Sensor Networks. In: IEEE and ACM International Symposium on Information Processing in Sensor Networks (April 2004)
9. Hohlt, B., Brewer, E.: Network Power Scheduling for TinyOS Applications. In: Proc. IEEE Int'l Conf. on Distributed Computing in Sensor Systems (DCOSS 2006), San Francisco, USA (2006)
10. Intanagonwiwat, C., Govindan, R., Estrin, D.: Directed Diffusion: a Scalable and Robust Communication Paradigm for Sensor Networks. In: Proc. ACM (MobiCOM 2000), Boston, USA (August 2000)
11. Jurdak, R., Baldi, P., Lopes, C.V.: Adaptive Low Power Listening for Wireless Sensor Networks. Transactions on Mobile Computing 6(8), 988–1004 (2007)
12. Keshavarzian, A., Lee, H., Venkatraman, L.: Wakeup Scheduling in Wireless Sensor Networks. In: Proc. ACM MobiHoc 2006, Florence, Italy (May 2006)
13. Li, Y., Ye, W., Heidemann, J.: Energy and Latency Control, in Low Duty-cycle MAC Protocols. In: Proc. IEEE Wireless Communication and Networking Conference, New Orleans, USA (March 2005)
14. Lu, G., Krishnamachari, B., Raghavendra, C.S.: An Adaptive Energy-efficient and Low-latency Mac for Data Gathering in Wireless Sensor Networks. In: Proc. PDSP 2004 (April 2004)
15. Lu, G., Sadagopan, N., Krishnamachari, B., Goel, A.: Delay Efficient Sleep Scheduling in Wireless Sensor Networks. In: Proc. IEEE Infocom 2005 (March 2005)
16. Madden, S., Franklin, M., Hellerstein, J., Hong, W.: TAG: a Tiny AGgregation Service for Ad-Hoc Sensor Networks. In: Proc. of OSDI (2002)
17. Madden, S.: The Design and Evaluation of a Query Processing Architecture for Sensor Networks. UC Berkeley Ph.D. Thesis (2003)
18. Manjeshwar, A., Agrawal, D.P.: APTEEN: A Hybrid Protocol for Efficient Routing and Comprehensive Information Retrieval in Wireless Sensor Networks. In: Proc. International Workshop on Parallel and Distributed Computing Issues in Wireless Networks and Mobile Computing, Ft. Lauderdale, Florida (April 2002)
19. Mirza, D., Owrang, M., Schurgers, C.: Energy-efficient Wakeup Scheduling for Maximizing Lifetime of IEEE 802.15.4 Networks. In: Proc. International Conference on Wireless Internet (WICON 2005), Budapest (Hungary), pp. 130–137 (July 2005)
20. Paruchuri, V., Basavaraju, S., Kannan, R., Iyengar, S.: Random Asynchronous Wakeup Protocol for Sensor Networks. In: Proc. of BROADNETS 2004 (2004)
21. Ping, S.: Delay Measurement Time Synchronization for Wireless Sensor Networks. IRB-TR-03-013, Intel Research Berkeley Lab (2003)
22. Raghunathan, V., Schurgers, C., Park, S., Srivastava, M.B.: Energy Aware Wireless Microsensor Networks. IEEE Signal Processing Magazine 19(2), 40–50 (2002)
23. Rhee, I., Warrier, A., Aia, M., Min, J.: Z-MAC: a Hybrid MAC for Wireless Sensor Networks. In: Proc. ACM SenSys 2005, S. Diego (USA) (November 2005)
24. Schurgers, C., Tsiatsis, V., Srivastava, M.B.: STEM: Topology Management for Energy Efficient Sensor Networks. In: Proc. of the IEEE Aerospace Conference 2002, Big Sky, MT, March 10-15 (2002)
25. Sivrikaya, F., Yener, B.: Time Synchronization in Sensor Networks: A Survey. IEEE Network 18(4), 45–50 (2004)

26. Sohrabi, K., Gao, J., Ailawadhi, V., Pottie, G.J.: Protocols for Self-organization of a Wireless Sensor Network. IEEE Personal Communications 7(5) (October 2000)
27. TinyDB: a Declarative Database for Sensor Networks,
    http://telegraph.cs.berkeley.edu/tinydb/
28. Tmote Sky Platform, MoteIV Corporation,
    http://www.sentilla.com/files/pdf/eol/tmote-sky-datasheet.pdf
29. Woo, A., Tong, T., Culler, D.: Taming the underlying challenges of reliable multhop routing in sensor networks. In: Proc. of the 1st ACM Conference on Embedded Networked Sensor Systems (SenSys 2003), Los Angeles, California, pp. 14–27 (November 2003)
30. Yang, X., Vaidya, N.: A Wakeup Scheme for Sensor Networks: Achieving Balance between Energy Saving and End-to-end Delay. In: Proc. of the IEEE Real-Time and Embedded Technology and Applications Symposium (RTAS 2004), pp. 19–26 (2004)
31. Zhang, Z.: Investigation of Wireless Sensor Networks for Precision Agriculture. In: Proceedings 2004 ASABE Annual Meeting. American Society of Agricultural and Biological Engineers (2004)
32. Zheng, R., Hou, J., Sha, L.: Asynchronous Wakeup for Ad Hoc Networks. In: Proc. ACM MobiHoc 2003, Annapolis (USA), June 1-3, pp. 35–45 (2003)

Chapter 17
Extracting Data from WSNs:
A Service-Oriented Approach

Gaetano F. Anastasi, Enrico Bini, and Giuseppe Lipari

Abstract. This chapter describes the architecture of a middleware layer between low-level sensing devices and higher level software layers, to support the requirements of a software infrastructure for networked enterprises. The development of such middleware layer is an important problem, as demonstrated by the number or research papers and the variety of approaches that can be found in literature. The main goals are to hide the complexity of low-level pervasive technologies, such as Wireless Sensors Networks (WSNs); and to help the higher software layers in managing the heterogeneous real-time data coming from the environment. In this chapter, after analysing the different approaches, we select the Service Oriented Architecture (SOA) design paradigm as the most suitable for allowing a seamless and effective integration of pervasive technologies into the enterprise information systems. We also present SensorsMW, our middleware proposal implemented in the context of the `ArtDeco` project, which is based on some of the many technologies that spin around the SOA world. In particular, our software is a service-oriented, flexible and adaptable middleware that allows applications to configure WSN functionalities and exploit them in the form of Web Services.

17.1 Introduction

The adoption of a middleware in networked enterprises is motivated by the need to connect high-level enterprise management applications with low-level sensing technologies based on tiny and complex physical devices.

Networked enterprises are heavily based on systems that leverage communication and information technologies. These technologies enable, on the one side, inter-organisational cooperation and collaboration; on the other side, they support business processes and strategic decisions. Usually, this kind of support originates from data

Gaetano F. Anastasi · Enrico Bini · Giuseppe Lipari
Scuola Superiore Sant'Anna, Pisa, Italy
e-mail: {g.anastasi,e.bini,g.lipari}@sssup.it

G. Anastasi et al. (Eds.): Networked Enterprises, LNCS 7200, pp. 329–356, 2012.

that constitutes the building block of the analysis of each process. Those data are also fundamental in driving and monitoring the execution of some enterprise tasks or allowing the completion of some others. For these reasons, data must be collected, filtered, merged, and finally made available to decision-makers (or to trusted users) through the enterprise informative system.

Among the various sources of data, the physical environment can be considered as one of the most important. In fact, the production process has usually strict interactions with the environment, which can have a significant impact on the quality of the final product, both directly, in case of outdoor production, and indirectly, by affecting the correct functioning of the factory plant.

Nowadays, many low-cost technologies exist that allow to collect data from the physical world, such as Wireless Sensor Network (WSN) or Radio Frequency IDentification (RFID). RFID and WSN are two fundamental components of pervasive computing. Although they have been developed following different principles, they can be integrated in an unified view as suggested for example by Zhang and Wang [34], or by Ho et al. [17], who observe that RFIDs can be integrated within WSNs just treating them as peculiar kind of sensors.

These technologies, although full of potentiality, are characterised by some distinctive features that must be carefully taken into account. Consider the typical structure of a WSN: it is formed of many sensor nodes that communicate over wireless channels and are spatially distributed for environmental sensing and monitoring. Sensor nodes are usually of very small size and characterised by having few resources: limited amount of energy (they are typically battery-powered with little or no capability of harvesting energy), small communication bandwidth, and reduced computational capabilities. They embed a very simple operating system that provides a few essential features, as event-driven computation and multitasking, basic interaction with sensors, power control, and radio communication.

Due to these constraints, operations in a WSN have to be performed by directly exploiting the underlying hardware components and/or the embedded operating system. This complexity is certainly unacceptable for non-trivial WSN-based applications, which must preferably be designed to be flexible, reusable, and reliable. Thus, the necessity of a middle layer of software arises, i.e. the *middleware*, which lies between the embedded operating system and the application (see Figure 17.1).

Following a well-established approach (Yu et al. [33], Hadim et al. [14], for example), a middleware for WSNs should abstract the underlying system by providing the following components:

1. programming abstractions, that provide high-level programming interfaces with different abstraction levels and programming paradigms;
2. standardised system services, that are exposed through abstraction interfaces and must provide various management functionalities;
3. run-time environments, that can extend the embedded operating system in supporting and coordinating multiple applications to be concurrently run over a single WSN;
4. Quality-of-Service (QoS) mechanisms, that allow to achieve adaptive and efficient utilisation of system resources.

Moreover, middleware for WSNs should obey to certain design principles, that, as it can be argued, on the one hand should address WSN distinctive features, and on the other hand should meet application needs.

The chapter is organized as follows. In Section 17.2 we briefly compare the many existing design approaches for middleware that have been proposed in literature. In Section 17.3 we describe some SOA technologies that constitutes the enabling technologies for SensorsMW, our middleware proposal. Section 17.4 presents the architecture of SensorsMW, whilst Section 17.5 details a case study that has been built to show the effectiveness of the proposed solution. Section 17.6 draws conclusions.

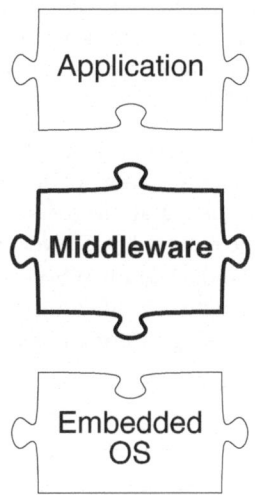

Fig. 17.1 Middleware, the missing piece of software layers for Wireless Sensor Networks

17.2 Common Middleware Approaches

In this section we present some critical aspects that have to be considered when designing middleware for networked enterprises, where data collecting technologies have to be used to fulfil the enterprise informative system. Without loss of generality (see Section 17.1), we will focus on the WSN technology to present these aspects and analysing the approaches that have been presented in the literature for implementing such principles.

17.2.1 Design Principles

Basing on the analyses conducted by Hadim et al. [14] and by Masri and Mammeri [23], the following key principles should be considered when designing WSN middleware. Often, these principles are in contrast with each other and, in that case, appropriate solutions that balance all the aspects are recommended, keeping in mind specific application requirements.

Data-orientation. In WSNs, the main focus is not on the node that produces data but on the data itself, i.e. the application is more interested on data that on the identity of the data source. For this reason, middleware should support this principle of data-centric networking by means of proper techniques of data-extrapolation or specific routing and querying within the network.

Energy-awareness. Energy management is a key issue in WSN, as sensor nodes are typically powered by batteries that could not be easily replaced, especially when many nodes are involved and distributed over wide, and sometimes impervious, areas. Therefore, in order to increase the network lifetime, middleware should be capable of managing properly sensor hardware resources, especially communication bandwidth and processing speed.

Quality of Service support. A WSN should always provide the performance level required by applications, i.e. the required Quality of Service (QoS) level. Unfortunately, WSNs are subjected to many status variations (energy loss, hardware failures, etc.) during their lifetime, and thus the required QoS level could not be provided if proper in-time actions are not taken. Middleware for WSNs should provide QoS support by allowing adaptive changes in the network, affecting determining QoS properties like event notification reliability or sensing information accuracy.

In-network processing. Sensor nodes of a WSN constitute elements of a distributed network, in which the transmitted bits regards not only collected data but also information about the network itself. Examples of this in-network processing, that should be supported by middlewares, are data aggregation and data compression, two techniques that allow to reduce data transmission in the network, respectively by condensing them or by correlating them. The classic example of data aggregation is the distributed evaluation of a maximum value [35].

Scalability and robustness. Scalability and robustness are correlated from a middleware point-of-view, as they both regard performance of the network as a function of his size. On one hand middleware should support scalability by continuing to offer a certain performance level even when the number of network nodes grows, on the other hand middleware should support robustness by tolerating node failures that can reduce network size or change network topology.

Reconfigurability and maintainability. A WSN is usually deployed to be long-time living, as the coverage area can be very wide and impervious, and sensor nodes could not be easily reachable. However, after the initial deployment, new necessities may arise and application may need to change tasks of some nodes or assign them new tasks to perform. For these reasons, middleware should support easy reconfigurability and maintainability of the network with minimal or null manual intervention.

Heterogeneity. Sensor nodes of a WSN can differ each other for their hardware features (such as battery capacity, transmitting power, processing speed, sensing capability, etc.) at deploying time or while the system evolves, due to different tasks each node have to perform. Middleware should support this heterogeneity by adopting a proper task dispatching policy based on the state of nodes.

Real-world awareness. WSNs are used to monitor phenomena that happen in the real world, subjected to time and space laws. In many cases, applications could be interested in knowing when and where a certain datum has been collected or an event happens, and middleware should support them in this sense.

17.2.2 Classification of Approaches

Many middleware for WSN has been developed, trying to face peculiar challenges of this domain. Below we present some representative ones by also giving a classification of existing approaches [14, 23].

Database approach

This approach abstracts the whole sensor network as a relational database, allowing applications to extract sensor data by using traditional methods, such as SQL-like queries. It provides applications with an easy and simple way for dealing with the sensor network but, generally, it lacks of time-space correlation between events.

TinyDB [8] is a project following this approach. It can be considered a tight extension of TinyOS [16], one of the most common Operating System (OS) used for WSNs. TinyDB permits to extract data from a WSN by using SQL-like queries but it does not provide high-level interfaces for QoS configuration and management.

One of the most elaborated project following this approach is SINA [29], in which a sensor network is viewed as a collection of data-sheets, where each data-sheet abstracts a sensor node. Data-sheets are composed of cells, representing an attribute of a sensor node: attributes can express single values, such as energy level, or multiples values, like the change history of a measured value. Moreover, SINA incorporates two robust low-level mechanisms: hierarchical clustering, which consists of grouping nodes into clusters, and attribute-based naming scheme, which allows to facilitate the data-centric characteristics of sensor queries.

Chapter 18 in this book introduces PerLa (http://perla.elet.polimi.it) — a language and middleware for data management in pervasive systems — which adopts a Data Stream management approach based on a SQL-like language. The middleware is responsible for the transparency to the end user of the idiosyncrasies of sets of highly heterogeneous devices by means of their abstraction as "logical objects", which also allows for a seamless run-time integration of new devices into the system [2].

Event-Based approach

Event-based middleware provides an asynchronous approach based on the publish/subscribe interaction model, allowing to decouple event producers from event consumers.

A noteworthy event-based message-oriented middleware is Mires [30], that implements a traditional publish/subscribe solution designed to run on top of TinyOS. Mires is mainly composed of the publish/subscribe service, which allows to query and extract data from the network by advertising the topics available, by maintaining the list of subscribed topics and by publishing messages. Moreover, the publish/subscribe service allows the communication between middleware services: among them, the routing service and the aggregation service can be mentioned, as they can make Mires more energy-efficient.

Application driven approach

This approach focuses on the high level requirements of applications, allowing to adaptively manage the whole network according to application's needs, often expressed as QoS requirements.

MiLAN [15] is a middleware belonging to this category, as it allows applications to specify their QoS requirements and to adjust the network to maximise application lifetime while providing the required QoS. Application requirements are provided by means of specialised graphs, that are used in conjunction with up-to-date network information about available sensors and resources, to determine which set of sensors can satisfy the application QoS needs.

Modular approach

Modular middlewares provide a run-time environment for modular applications running on each sensor node. In this case, applications formed by different tiny modules can be easily transmitted and installed for ensuring a longer life of the network, as sensors can adapt their tasks over time and software updates can be performed with reduced transmission energy loss.

Impala [21] is a middleware belonging to this category, as it enables application modularity and adaptivity, by allowing that software updates can be received via the node's wireless transceiver and dynamically applied to the running system. Essentially, Impala proposes a run-time system that acts as a lightweight event and device manager for each sensor node, providing also mechanisms for adapting the application protocols to different run-time conditions.

Virtual Machine approach

This approach aims to bring the benefits of virtualisation in the WSN domain. In particular, Virtual Machines (VMs) can be used in this context to represent a wide range of programs by using a small set of high level primitives, exploiting the fact that sensor network applications are usually composed of a common set of services and sub-systems, combined in different ways.

Maté [20] is a middleware following this approach, as it is basically a bytecode interpreter to run on motes equipped with TinyOS. Maté has a concise instruction set that comprises different types of instruction, that hide the asynchrony of TinyOS programming by allowing to perform synchronous operations, like sending packets or requesting data from the sensor TinyOS component. Instructions are defined in such a way that up to 24 ones can be encapsulated in a single network packet, allowing to a quickly installation of programs with little network traffic.

Service Oriented approach

This approach aims to exploit the advantages of the Service Oriented Architecture (SOA) design methodology (see section 17.3.1) in WSNs, allowing to build flexible and interoperable systems, in which heterogeneous sensor devices can communicate (with each other and with the rest of the world) in an easy way.

A service-oriented middleware has been proposed by Delicato et al. [7], where sensors functionalities are exposed as services, that are mainly defined as gathered data and operations to be executed on them. This way, flexibility and interoperability are achieved. However, the proposed message exchange pattern between nodes, based on SOAP (see Section 17.3.1), seems not to carefully take into account the stringent resource constraints (e.g. RAM size, CPU speed) of typical sensor nodes.

The SOA architecture proposed by Samaras et al. [28] aims to integrate WSNs into the enterprise information systems, with the purpose of achieving interoperability between WSNs and other enterprise components from different manufacturers. The device level technology used for implementing the SOA paradigm is Device Profile for Web Services (DPWS) [6], a standard that has been designed for industrial automation environments and is fully compatible with the Web Services technology. However, authors recognise high requirements DPWS imposes on device resources and propose a middleware implementing a stripped down version of traditional DPWS. Moreover, SOA messages exchanged by nodes are reduced and compressed by exploiting application specific information.

17.3 Enabling Technologies

The middleware solution presented in this chapter follows the service-oriented approach and thus its design is completely technology-independent. However, for a real implementation, technological choices are required and must be carefully taken: in particular, we consider Web Services for service provisioning and WS-Agreement for service negotiation. Throughout the description of these technologies, we highlight as they are adherent to the SOA principles and constitutes an effective solution for building SOAs.

17.3.1 Web Services

Web Services are distributed systems technologies that aim to provide a common standard mechanism for allowing interoperable integration between heterogeneous systems, so they constitutes the preferred vehicle for implementing Service Oriented Architectures (SOAs). We first describe SOA key principles and then we give a brief overview on technologies constituting the core of Web Services.

Service Oriented Architecture

The concept of SOA has certainly been gaining momentum in Information and Communication Technology (ICT) application area in recent years. SOA is a design methodology that relies on services, that constitute the smallest bricks of software necessary to build distributed applications: services are published, discovered and invoked over a net whilst applications are simply built by putting services together.

The publish/find/bind operations simplify this architectural model, that involves three main parts playing different roles (see figure 17.2): a service provider, publishing and providing services; a registry, containing list of services; a service consumer, seeking for services and requesting them. Besides the involved parts, the service-oriented design methodology, aiming to build open systems composed of heterogeneous and autonomous components, must obey to certain well-known interrelated principles [10, 18] that can be summarised as follows.

High abstraction. This principle regards service interfaces, sometimes called *contracts*, through which service's functionality is exposed: as a service usually represents a business task, it should hide implementation details to capture, at an high-level, its valuable contribution in broad business contexts. For this reason, a service interface is mostly coarse-grained and stateless, reducing communications with other

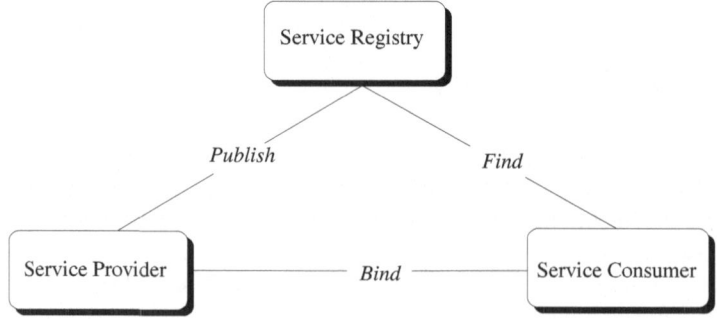

Fig. 17.2 SOA architectural model

agents to few exchanges of standard documents, independent from the specific service implementation.

Loose coupling. This principle refers to the coupling of services, seen as communication entities of SOAs, whose interfaces must be highly abstracted and thus, described by sacrificing precision in favour of interoperability. In fact, in this way services are more adaptable to new requirements, as the implementation of a service may be modified without affecting its users.

Reusability. It is a key concept is service-oriented computing, as a service can be potentially used by other organisations in a different scenario comparing with that the service has been designed for. To be reusable, a service must be highly generic and its functionality must be independent to context and domain of application.

Flexibility. A service can be characterised by requirements or properties (e.g. QoS requirements, security support, computational resources) that should be negotiated at run-time, as the capability of a service to be dynamically configured preserves generality and improves reusability.

Autonomy. A service should be autonomous, in the sense that its computation can only be affected by modifying the input parameters as specified in its interface. Service autonomy increases reliability and fault isolation, allowing to realise distributed computation, in which many autonomous parties working on a team as partners.

Composability. This principle states that services must be used for composition of other services, like many toy bricks, that can be put together to build bigger bricks, or separated each other to be reused in another construction. Service composition is a key features, as it allows to create value from existing parts and should be always possible regardless of size and complexity of the composition.

As many of the techniques used for SOA components (e.g. databases, transactions, software design) are already well-established in isolation, it should be stressed that the main innovation of SOA relies on the architecture, that is capable of putting into cooperation autonomous and heterogeneous components, allowing to build large-scale systems, in which interacting parties could be not only intra-enterprise components, but also inter-enterprise ones, that can be choreographed to achieve cross-enterprise processes.

Core Technologies

Web Services technology is composed by a set of open standards, that aim to reach interoperability among heterogeneous systems, by using XML as standard format of data exchange. Web Services spirit moves away from common distributed object technology to embrace the service-oriented computing paradigm and, by providing a common mechanism for delivering services, it constitutes one of the most suitable implementation choice for SOAs. In fact, Web Services architecture perfectly embodies the typical publish/find/bind model of SOA (see figure 17.2), by exploiting the following core technologies:

- WSDL [4], an XML language that realises the *publish* operation by allowing for describing services through its location, the exposed methods, their bindings to the transport protocol, etc. WSDL also allows for data type definition by using XML Schemas [32].
- UDDI [5], a specification that realises the *find* operation, by providing a mechanism to register and locate Web Services. UDDI is itself a Web Service, providing a set of pre-defined interfaces for client interaction.
- SOAP [13], a messaging protocol that realises the *bind* operation, by allowing to exchange service-related XML messages formatted according to the corresponding WSDL definitions.

The Web Services architectural model perfectly clarifies what are the interactions among the three main agents involved: providers, brokers and consumers. However, for a deeper understanding of how Web Services works, we also consider the inside-out perspective proposed by Vogels [31]. According to this perspective, Web Services comprises four core components:

1. The *service*, a software capable of processing XML documents it receives.
2. The *XML document*, that contains all the application-specific information a consumer sends to the service for processing.
3. The *address*, also called port reference, a protocol binding combined with a network address that allows consumers for accessing the service.
4. The *envelope*, a message-encapsulation protocol that ensures that the XML document to be processed is clearly separated from other system information.

The envelope could be considered optional but it is very useful in practise as it provides a framework for managing message exchanges. SOAP (that, as already said, is the reference messaging protocol for Web Services) realises this partition by dividing a message in two elements:

- The SOAP header, that contains all system information (e.g. security, routing, message handling, etc.).
- The SOAP body, that contains the actual XML documents to be processed.

Although SOAP, WSDL and UDDI represents the basis of Web Services technology, they are not sufficient for building complex applications that exploit Web Services potentiality to its maximum: for this reason, many other standards has been created for enhancing Web Services protocol stack. Among these standards, we mention: **WS-Addressing** [27], for providing transport-neutral mechanisms to address Web services and messages; **WS-Security** [19], for ensuring end-to-end message integrity, confidentiality and authentication; **WS-Reliable Messaging** [11], for ensuring the completion of message exchanges; **WS-Transactions** [25], for coordinating the outcomes of distributed application actions; **BPEL** [1] (Business Process Execution Language), for formally describing overall business processes composed by different Web services.

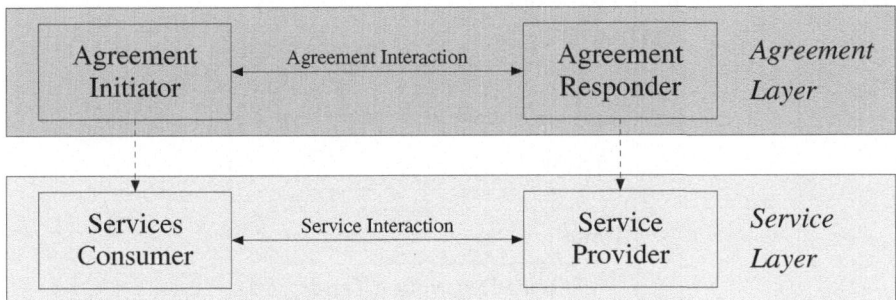

Fig. 17.3 WS-Agreement layered model

17.3.2 WS-Agreement

WS-Agreement [3] is a proposal of the Open Grid Forum for defining and managing Service Level Agreements (SLAs), contracts between service providers and consumers, in which providers describe, on one side, services and related guarantees on provisioning and, on the other side, fruition costs and penalties on guarantee violations.

A system that leverages WS-Agreement must follow the conceptual model illustrated on Figure 17.3, that relies on an architectural division in two layers:

1. the *agreement layer*
2. the *service layer*

The agreement layer provides a Web service based interface that can be used to exploit the functionalities WS-Agreement has been designed for:

- defining a language to specify agreements between consumers and providers;
- defining a protocol to create an agreement;
- defining a protocol to verify agreement compliance at run-time.

Instead, the service layer represents the layer in which the service execution takes place: it is totally independent respect to the upper one and thus services may be provided with interfaces different from Web service-based ones.

As the service layer is application-specific, in the following only operations relative to the agreement layer will be taken into consideration and briefly analysed.

Agreement Structure

In WS-Agreement, a contract (or *Agreement*) is defined according to a reference template, (or *Agreement Template*). An Agreement, represented by an XML document divided in different sections (see Figure 17.4), contains a set of parameters that

Fig. 17.4 Agreement structure in WS-Agreement

describe related services and QoS guarantees to be provided. The key parts of an Agreement can be summarised as follows:

- Name - Field exploited to name an agreement in human-readable way: this field can not be considered as an unique identifier and the attribute AgreementId must be used to this purpose;
- Context - Section exploited to store various information related to the agreement, like the parties involved, the lifetime of an agreement or the template name from which the agreement is created;
- Terms/ServiceDescriptionTerms - Section exploited to describe services related to the agreement. Each Service Description Term (SDT) can describe (fully or partially) one of the provided services and it consists of three parts: a name for the section itself, a name for the described service, the service description in a domain specific language.
- Terms/GuaranteeTerms - Section mainly used to specify guarantees to assure in providing services described in the corresponding SDTs. In this section, the user constraints to respect can be also specified, in conjunction with business values representing the importance of obtaining agreement goals.

An Agreement is based on an Agreement Template, that constitutes a reference guide containing instructions useful to create new agreements. Thus, an Agreement Template has the same sections of an Agreement, plus the following one:

- CreationConstraints - Section exploited to specify configurable fields of the agreement and acceptable values for those fields. These informations can be specified with Item and Constraints fields.

A template is also characterised by an identifier, the TemplateId attribute, used to bind each agreement with the corresponding template.

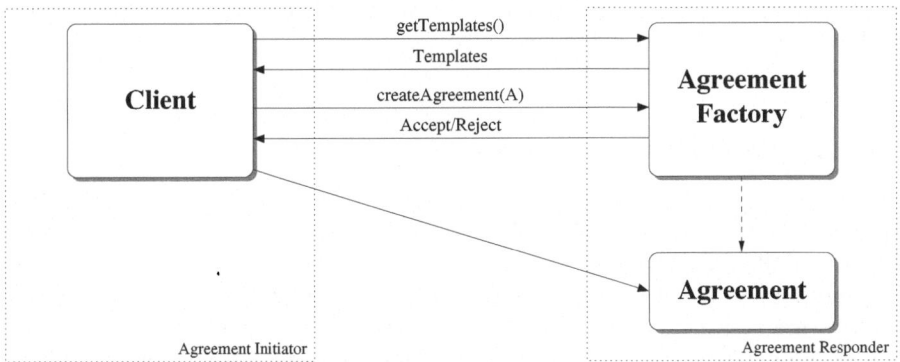

Fig. 17.5 Agreement creation process in WS-Agreement

Agreement Creation

WS-Agreement defines an interaction protocol for agreement creations, in which two parties are involved, the *Agreement Initiator* and the *Agreement Responder*. The initiator begins the interaction by requesting an Agreement Template provided by the responder: the initiator takes the template, fills it with the QoS parameters that wants to request and sends it back. At this point, the responder examines the proposal and decides to accept or reject it, according to its internal management of resources: if an offer is accepted, the responder creates an Agreement and makes available its reference; if the offer is not accepted, an exception is thrown to the initiator, to make him aware of the refusal. This synchronous model of interaction is realised through the methods *getTemplates()* and *createAgreement(A)*, as can be seen in figure 17.5. These methods are exposed by the Agreement Factory component, through a Web Service interface.

Agreement Compliance

WS-Agreement defines a monitoring protocol which has the main purpose of verifying agreement compliance at run-time, in order to monitor the state of agreements, services and guarantees. The state of an agreement can give useful information to determine, for example, if an agreement has been accepted or not. Instead, the state of a service can be exploited to determine if the execution of a service has been started or if it was completed. Finally, the state of the guarantees indicates, for a particular service, if the guarantees negotiated has been violated. All the methods that realise this operation are exposed by the Agreement component of the Agreement Responder. This components also exposes methods to retrieve information about the Agreement and the method *terminate()*, used to terminate an Agreement.

A common reason of agreement termination is the provider's inability to meet agreement objectives, causing the interruption of provided services. To avoid this situation, an extension of the WS-Agreement specification has been proposed [9], in order to enabling the run-time (i.e. during the service provisioning) re-negotiation and modification of QoS guarantees.

17.4 The ArtDeco Middleware

In this section we present SensorsMW, a middleware proposed for the ArtDeco project to allow an easy and seamless integration of pervasive technologies into the informative system of networked enterprises. For this reason, a service-oriented middleware is a natural choice, as it assures both intra-organisational and inter-organisational interoperability, making available precious information to applications, that can take advantage of them in an effortless manner.

Our approach differs from the SOA-related ones illustrated in Section 17.2, as it does not aim to implement a service-oriented middleware directly on sensor nodes, forcing SOA-compatible protocol stacks (like DPWS) in resource constrained devices. In our opinion, this kind of approaches has the major drawback to impose too much complexity in devices that are not enough powerful to transmit and elaborate XML messages. These constraints often lead developers to adopt a-priori knowledge in XML message definition, and thus loosing middleware flexibility. Moreover, usage of web services in resource constrained devices imposes a certain energy and latency overhead (as an example, cost for such implementations has been quantified in the work by Priyantha et al. [26]) that could be unacceptable in some cases.

Instead, the proposed middleware allows high-level applications to exploit data-centric network functionalities and configure a WSN according to their needs. It is thus devoted to expose network functionalities as **services**, besides of the low-level technologies used for programming the WSN.

In fact, the logic that allows to abstract the WSN is concentrated on a powerful gateway, to which the sink node is connected. The traditional technique of backup nodes is used to overcome the single-point-of-failure issue, in case it arises. The gateway solution is not new, for example it has been used by Kansal et al. [12] for building a peer-to-peer infrastructure for sharing sensors through the Internet. However, as their work covers a wide range of sensors, it does not explicitly address typical WSN issues. A gateway-based solution has been also proposed by Moeller and Sleman [24], aiming at integrating WSNs into other existing IP-based networks. However, their work is oriented to ambient intelligence at home, so they do not abstract functionalities of the whole network but only of single sensors. Also, they do not offer a Web Service interface, thus making it difficult to integrate and compose such services.

The following are the main innovative features of the ArtDeco middleware.

Service-orientation It allows for a fruitful exploitation of pervasive technologies in enterprise contexts, by abstracting WSNs as a collection of services.

Flexibility It can be used in many contexts or domains, even when specific network critical issues have to be addressed.

Adaptability It can support well-known low-level techniques or legacy deployments that can be already in-place.

The proposed middleware has been designed keeping in mind the main issues in this domain, as highlighted by its key features, that can be summarised as follows:

- it allows to easily access network-provided data with different grain;
- it allows energy-aware sensing and battery management;
- it supports QoS specification and management;
- it is independent from the network size and topology;
- it allows to reconfigure and maintain the network during its lifetime;
- it supports time and space recognition of network events.

It is also noteworthy to mention that, thanks to its flexibility, the middleware can always embed features that are not explicitly supported by design, as it is completely independent from low-level techniques implemented at the device level.

17.4.1 Architecture

As highlighted in section 17.4, one of the main features provided by SensorsMW is the possibility to configure a WSN according to the needs of client applications. For providing such a feature, we leverage the WS-Agreement framework (see section 17.3.2), that allows an easily creation, management and monitoring of Service Level Agreements (SLAs), i.e. contracts that service providers and service consumers establish for defining characteristics of service provisioning. In this case, SensorsMW acts as a service provider, as it provides services to client applications, that have to negotiate SLAs (also called agreements or contracts in this scope) before consuming services.

The SensorsMW architecture is comprised by four main components. They are illustrated in Figure 17.4.1, that, for simplicity, does not highlight interactions with clients. Such interactions are realised by the ContractsCreator and the Service-Provider components. In the following the components of SensorsMW will be described in details.

ContractsCreator

This component is responsible for interacting with client applications in all the operations that regard contract creation and management. It comprises the following sub-components.

SensorsMWFactory. It interacts with the client in the agreement creation process and is responsible for publishing the agreement templates related to services provided by the system (see section 17.4.2 for details about template specifications). The templates are fulfilled by clients according to their needs and then evaluated by the

system. In case the client proposal can be satisfied, the SensorsMWFactory interacts
with the SensorsMWAgreement in creating the agreement.

SensorsMWAgreement. It realises all the operations related to an agreement, by providing status information for agreements and allowing for an anticipate ending of
them. It appears only if an agreement creation process has been successful completed
by the SensorsMWFactory component. There will be an instance of SensorsMWA-
greement for each contract that is actually in-place.

BrokerAgent. It is responsible for forwarding requests of SensorsMWFactory and
SensorsMWAgreement to the lower levels of the architecture. In particular, it forwards:

- admission requests coming from the SensorsMWFactory when a new contract
  has to be admitted;
- delete requests coming from the SensorsMWAgreement when a contract has to
  be deleted.

The introduction of this sub-component allows to improve responsiveness of the SensorsMWFactory and SensorsMWAgreement sub-components, that have to interacts
with clients, and also allows for decoupling the ContractsCreator and the WSNGateway, that could be deployed in two different physical hosts.

ServiceProvider

This component is responsible for providing services to client applications, in accordance with established contracts. Services are made available by using Web Services

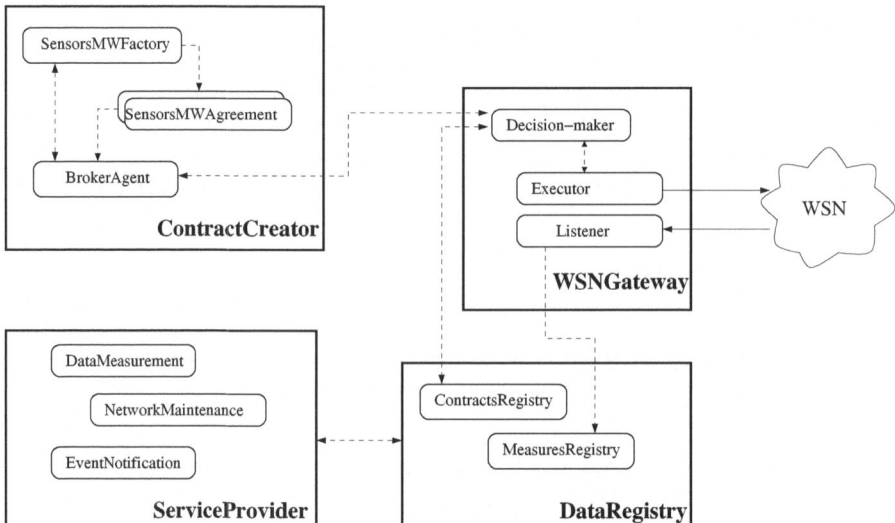

Fig. 17.6 SensorsMW architecture

technologies, that allow, as discussed in Section 17.3.1, an easily integration and interoperability in enterprise systems. In SensorsMW, three main services have been identified as essentials: they exploit the database provided by the DataRegistry component for providing their functionalities, as described in the following.

DataMeasurement. This service allows client applications to obtain presently gathered data, by presenting the identifier of the previously established contract, that contains all the configuration parameters used by the WSN for gathering measurement data related to a certain physic quantity (see also section 17.4.2).

EventNotification. This service allows client applications to receive notifications about events of interest, related to the measurement of a certain physic quantity. Applications can configure events they are interested on, and subscribe to them by means of specific contracts, that will be described in section 17.4.2.

NetworkMaintenance. This service allows client applications to perform network maintenance by measuring and monitoring quantities that are necessary for a proper WSN functioning, like the battery level or the number of active sensors in a certain region. Applications can exploit this SensorsMW features by establishing proper contracts, that will be described in section 17.4.2.

DataRegistry

This component is responsible for managing all the data that have to be persistently stored for the proper functioning of SensorsMW. It comprises the following subcomponents.

ContractsRegistry. This sub-component maintains the registry of all contracts presently established with client applications. Each contract is represented by an unique identifier plus the featuring parameters, that depend on the type of contract (see Section 17.4.2 for a detailed description of such parameters). The knowledge contained in this registry can be used for admitting new contracts and for providing applications with information regarding established contracts.

MeasuresRegistry. This sub-components maintains the registry of measures gathered by the WSN, in accordance with the presently established contracts. A measure is represented in the registry by the following information:

- the identifier of the measured physical quantity,
- the measure value,
- the datum aggregation type,
- the location in which the datum has been gathered,
- the time at which the datum has been gathered,
- the date in which the datum has been gathered.

In order to provide applications with requested data, the knowledge contained in the MeasuresRegistry is leveraged by the ServiceProvider component, that also contains the logic for binding data with contracts and for correlating data in order to respect established contracts: as an example, the ServiceProvider component may aggregate data *a-posteriori* if such in-network processing feature is not available in the WSN.

WSNGateway

This component is responsible for acting as a gateway respect to the WSN, in the sense that all the communications to and from the WSN pass from this component. It comprises the following sub-components.

Decision-maker. This sub-component appears when a new contract has to be admitted and it decides if a service requested by a client with a certain parameter configuration can be provided by the system. This can comprises both an analysis of existing contracts and of the current status of the WSN. When the component takes decision about an high-level request, it communicates the response to the BrokerAgent, that in turn forwards it to the SensorsMWFactory. If the response is negative, the Decision-maker does not take any further action; if positive, it triggers the creation of a new contract in the DataRegistry and interacts with the Executor for triggering tasks for the WSN, in order to fulfil new requirements of applications.

Executor. This sub-component receives commands from the Decision-maker and translates them into a language understandable from the sensor nodes. This level of indirection allows the independence of the admission control logic from the low-level technology used for the WSN programming, to whom the Executor is strictly bound. It is worth to note that, in order to port SensorsMW to another WSN technology, the Executor and the Listener, described later, are the only sub-components that need to be customisable.

Listener. This sub-component receives data gathered from sensor nodes and store them in the DataRegistry. It can be subdivided in two main modules (not highlighted by figure 17.4.1): one is responsible for listening data communications from sensors and it is strictly dependent to the low-level WSN technology, the other one is responsible for binding data coming from nodes with respective locations, formatting measures as specified by the MeasuresRegistry and triggering storage.

17.4.2 Contract Specification

SensorsMW allows applications to configure the WSN according to their needs before service provisioning. In particular, for each kind of service, SensorsMW provides an agreement template that has to be fulfilled by applications in order to create agreement proposals. If an agreement proposal is accepted, a contract is established with the client application, and both parties are obliged to honour it.

The agreement templates provided by the SensorsMW layer have been designed by keeping in mind the following principles:

1. they should be well-structured and easily usable by clients;
2. they should be compatible with the limited hardware resources of sensor nodes (e.g. battery power).

For this reasons, templates are specified by using the WS-Agreement [3] framework and are characterised by a certain time span of validity, that could be renegotiated during the agreement lifetime.

In SensorsMW, we define three different types of services, whose execution parameters can be negotiated by means of templates. Correspondingly, three different kind of contracts have been specified, that can be summarised as follows:

1. *periodic measurement* contract, to periodically measure a certain physical quantity;
2. *event monitoring* contract, to monitor specific events related to quantity measurement;
3. *network management* contract, to control and maintain particular situations related to WSN functioning.

Each kind of contract is characterised by key parameters, that will be described in the following. In particular, as such parameters are negotiated by using templates specified with WS-Agreement, we will concentrate on the Service Description Terms section of each template type (see Figure 17.4), as it is devoted to contain service-related parameters. The time span of validity of a contract is instead stored into the Context section, as it refers to the contract as a whole.

Periodic Measurement Contract

The periodic measurement contract allows to periodically measure a certain physical quantity. It requires specifying some parameters that characterise the WSN behaviour during service provisioning.

For this kind of contract the following parameters have been defined:

- physical quantity to be measured
- time span for which the measurement has to be done
- sampling period
- type of data aggregation
- region to be measured
- QoS level

SensorsMW allows applications to negotiate such parameters by formally describing them with XML Schema [32] elements, that are inserted in a SDT section of an Agreement Template (see Section 17.3.2).

A single SDT can refer to only one physical quantity, as the different quantities that a WSN can measure, can have very different features from one each other.

A possible SDT for an agreement template related to a periodic measurement service can be the following.

```
<wsag:ServiceDescriptionTerm wsag:Name="temperature_measurement"
                             wsag:ServiceName="data_measurement">
  <smw:DataMeasurement xmlns:smw="schemas.sensor_mw">
    <smw:Measure>Temperature</smw:Measure>
    <smw:AggregationPeriod>PT1H10M</smw:AggregationPeriod>
    <smw:SamplingTime>PT10S</smw:SamplingTime>
    <smw:Aggregation>avg</smw:Aggregation>
    <smw:Aggregation>max</smw:Aggregation>
    <smw:Region>
      <smw:Location>North Area</smw:Location>
      <smw:Location>Sensor185</smw:Location>
    </smw:Region>
    <smw:QoSLevel>100</smw:QoSLevel>
  </smw:DataMeasurement>
</wsag:ServiceDescriptionTerm>
```

Referring to the proposed example, the meaning of values associated to each element is explained below.

Measure. It expresses the physical quantity to be measured as enumerate. The example specifies the temperature as the quantity of interest.

AggregationPeriod. It expresses the period for data aggregation, by using the *duration* XML data type [22]. In the example, data are aggregated each 1h and 10min. In the simplest case, this value is equal to the SamplingTime.

SamplingTime. It expresses the sampling period of sensing by using the *duration* XML data type. In the example, data are sampled by sensors each 10 seconds.

Aggregation. It expresses the aggregation mode of data collected in the same location, by using an enumerate data type (possible values could be avg, max, min).

Region. It expresses the list of locations we are interested to monitor.

Location. It expresses the location of interest by using an unique identifier (it could also be a human-readable name).

QoSLevel. It expresses the QoS level to be provided, by using values belonging to the set $\{x \in N: 0 \le x \le 100\}$. A QoS level equal to 100 is equivalent to the maximum quality of service.

It is worth to note that the QoSLevel parameter gives an high level of flexibility to SensorsMW. In fact, it has been introduced for conveying other non-functional parameters besides of those explicitly considered in the contract. In this way, the middleware can address different contexts and clients can specify application-dependent QoS requirements (e.g. minimum coverage area, accuracy of measurements). Thus, the mapping between the values assumed by the QoSLevel parameter and the provided QoS varies according to the application domain.

Depending on the particular service configuration, some parameters could not be negotiated during the agreement phase: as they are bind to the physic quantity to be measured, a different SDT is used for each quantity. Other parameters are instead negotiable and their default values can be modified by applications when presenting an agreement proposal. In particular, for being adherent to WS-Agreement specification, the `CreationConstraints` template section must contain an `Item` element for each SDT parameter that can be modified.

By using the `Item` element, possible values for variable parameters can also be specified, as highlighted in the following example, in which usable values for the `Aggregation` item are limited to `min`, `max` and `avg`.

```
<wsag:Item wsag:Name="AggregationItem">
  <wsag:Location>
    //wsag:ServiceDescriptionTerm[@Name=
'temperature_measurement']/smw:DataMeasurement/smw:Aggregation
  </wsag:Location>
  <wsag:ItemConstraint>
    <xs:simpleType xmlns:xs="http://www.w3.org/2001/XMLSchema">
      <xs:restriction base="xs:string">
        <xs:enumeration value="min"/>
        <xs:enumeration value="max"/>
        <xs:enumeration value="avg"/>
      </xs:restriction>
    </xs:simpleType>
  </wsag:ItemConstraint>
</wsag:Item>
```

This fragment also highlights as restrictions on values can be specified in an Agreement Template by following the XML Schema model.

Event Monitoring Contract

The event monitoring contract allows applications to express an interest in certain events, and, in particular, to monitor specific events related to quantity measurement.

An event monitoring contract has some similarities with the periodic measurement contract, as some of its key features are the same. In particular, the following key parameters have been considered:

- the physic quantity of interest
- the event triggering condition
- the event notification delay
- the data aggregation mode
- the region of interest
- the QoS level

The condition that triggers the event has been specified in the formal definition as an interval on values of the physical quantity of interest, in order to express comparative and equality conditions in the same way.

An agreement template provided by SensorsMW for this service can contain many ServiceDescriptionTerms, where each SDT is relative to a single quantity and can specify a single event triggering condition. A possible SDT section, containing default values for each element, can be the following one.

```
<wsag:ServiceDescriptionTerm wsag:Name="temperature_monitoring"
                             wsag:ServiceName="event_monitoring">
  <smw:EventMonitoring xmlns:smw="schemas.sensor_mw">
    <smw:Measure>Temperature</smw:Measure>
    <smw:MeasurementInterval>
      <smw:LowerBound>20.0</smw:LowerBound>
      <smw:UpperBound>INF</smw:UpperBound>
    </smw:MeasurementInterval>
    <smw:NotificationDelay>PT15S</smw:NotificationDelay>
    <smw:Aggregation>avg</smw:Aggregation>
    <smw:Region>
      <smw:Location>1</smw:Location>
      <smw:Location>3</smw:Location>
    </smw:Region>
    <smw:QoSLevel>100</smw:QoSLevel>
  </smw:EventMonitoring>
</wsag:ServiceDescriptionTerm>
```

Besides already described elements (see section 17.4.2), specific elements for the event monitoring contract are the following.

MeasurementInterval. It expresses the condition that triggers the event as an interval on values of the physical quantity of interest. The event is generated when the measured value falls into the closed interval. Such interval is specified by the LowerBound and UpperBound elements, whose values are interpreted according to the International System of Units (for temperature we consider Celsius temperature). In the example, the event is triggered when the temperature reaches a value greater than 20°C (in fact the upper bound of the interval is $+\infty$).

NotificationDelay. It expresses the granted delay from when an event occurs to when the same event is notified. It is specified by using the *duration* XML data type. The example shows a delay of 15 seconds.

Network Maintenance Contracts

The correct behaviour of a WSN can be compromised by many events, like sensor node failures, battery discharges, node displacements or additions. Thus, this kind of contract allows to configure services for controlling and maintaining a WSN, in order to prevent dangerous events or take proper actions in case they happen.

SensorsMW allows to negotiate both measurement services and event-based services on critical quantities for the WSN maintenance, like energy consumption and the number of sensors in a certain region.

The negotiation of network maintenance services is very similar to that described for periodic measurement services and event monitoring services (please refer to Sections 17.4.2 and 17.4.2), as the number of sensors or the battery level can be treated as quantities to be measured in the network.

As an example, it is possible to establish a contract for monitoring the number of sensors on a region and triggering an event when it is behind a certain threshold by using the following excerpt:

```
<wsag:ServiceDescriptionTerm wsag:Name="sensor_number"
                    wsag:ServiceName="network_maintenance">
  <smw:NetworkMonitoring xmlns:smw="schemas.sensor_mw">
    <smw:Measure>SensorNumber</smw:Measure>
    <smw:MeasurementInterval>
      <smw:LowerBound>20.0</smw:LowerBound>
      <smw:UpperBound>INF</smw:UpperBound>
    </smw:MeasurementInterval>
    <smw:NotificationDelay>PT15S</smw:NotificationDelay>
    <smw:Region>
      <smw:Location>1</smw:Location>
      <smw:Location>3</smw:Location>
    </smw:Region>
  </smw:NetworkMonitoring>
</wsag:ServiceDescriptionTerm>
```

In that case the event is triggered when there are less than 20 nodes in the region formed by locations tagged as 1 and 3.

17.5 A Case Study

In this section a case study is presented to show the effectiveness of the proposed architecture and demonstrate its adaptability and flexibility for building a middleware for networked enterprises.

As a candidate application, we consider the temperature monitoring in a certain region of a vineyard and we study the behaviour of a concrete implementation of SensorsMW, tailored onto TinyOS 2.x.

For the purpose of this case study, we also designed and deployed a WSN testbed with a star topology, in which a node acts as a coordinator and the other ones act as end-devices. This topology has been chosen for the sake of simplicity, as it does not require the use of routing algorithms and can be easily implemented. In any case, other more complex topologies, better suited to particular applications (e.g. the monitoring of a vast vineyard), can be used in conjunction with the proposed middleware, as it does not rely on any low-level technique.

The nodes of the WSN are deployed all around the monitored area and have different tasks according to their category:

- The Coordinator node, connected to a resource-unconstrained machine, is responsible for interfacing the WSN with the SensorsMW architecture. In particular, it receives data coming from nodes and forwards them to the WSNGateway. Also, it receives commands from the WSNGateway and forwards them to proper end-devices.
- An End-Device node is responsible for gathering data from active sensors and sending them to the Coordinator.

For tailoring SensorsMW onto TinyOS, only the Listener and the Executor sub-component (see Section 17.4) of SensorsMW has been modified, by properly adapting the *Listen* and *Send* TinyOS applications.

In this case study, applications can require DataMeasurement services, that can be configured by filling a periodic measurement contract (see Sec. 17.4.2) with the desired values. Applications can require QoS-enabled or QoS-disabled services and the application-dependent QoS parameter is considered the reliability of the measure. A QoS-enabled service can be requested by setting the QoSLevel parameter equal to 100, whilst a QoS-disabled one can be requested by setting QoSLevel equal to 0. When applications require QoS-enabled services, SensorsMW sets the sampling time of nodes in a certain location to the minimum value necessary to gather data at the exact instants of time. Instead, in case of QoS-disabled services, data may be gathered at instants different than required.

As a possible scenario, consider the situation in which no contract has been stipulated and three client applications require a QoS-enabled DataMeasurement service related to the temperature monitoring. Applications configure the service by creating contracts that differ for the parameters described in Table 17.1.

By analysing these parameters, it can be noticed that each application can specify the desired values independently from the other applications. In fact, applications can specify a different sampling period, even in case they choose the same location for monitoring. This capability is highlighted by Figure 17.7, in which the temperature obtained by *app1* and *app3* for location loc1 is plotted as a function of time. Each point in the graph represents a sample of temperature and it can be seen that both applications receive data according to their requirements: *app1* receives data each $30s$ (label agr1-loc1) whilst *app3* receives data each $10s$ (label agr3-loc1).

It is worth to note that agr3 has been created while agr1 is in place and they both require data from the same location (i.e., the same sensors). The fact that samples are correctly received highlights as the system is capable to reconfigure itself for transparently support application requirements.

Table 17.1 Application requested parameters

Application	Agreement	Sampling (s)	Location
app1	agr1	30	loc1, loc3
app2	agr2	20	loc2
app3	agr3	10	loc1

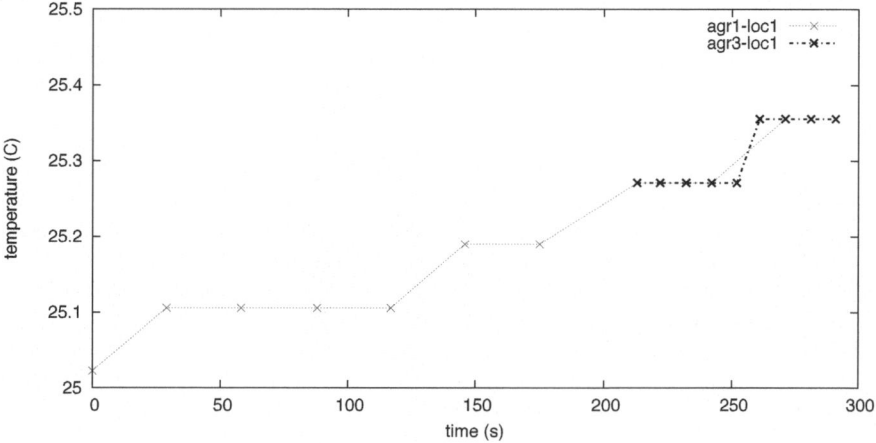

Fig. 17.7 Varying of temperature for location loc1

What happens behind the scene is that, when *app3* requires a sampling of 10*s* for location loc1, the sampling time of nodes in that location changes from 30*s* (value imposed by agr1) to 10*s*, in a manner that both *app1* and *app3* can obtain data gathered at the required time instants. Then, the ServiceProvider component is responsible for analysing the contract established by each application and for providing data following their requirements.

This scenario is particularly suited for highlighting as SensorsMW can transparent support different requirements of applications, even when consuming services related to the same areas of a WSN.

17.6 Conclusions

In this chapter we have described SensorsMW, a service-oriented approach for the middleware of a wireless sensor network. Any client that is willing to use the services of the sensor network must establish a connection with the service manager. The manager grants or deny the service depending on the resource availability.

References

1. Andrews, T., Curbera, F., Dholakia, H., Goland, Y., Klein, J., Leymann, F., Liu, K., Roller, D., Smith, D., Thatte, S., Trickovic, I., Weerawarana, S.: Business process execution language for web services. Specification version 1.1 (2003), http://download.boulder.ibm.com/ibmdl/pub/ software/dw/specs/ws-bpel/ws-bpel.%pdf

2. Schreiber, F.A., Camplani, R., Fortunato, M., Marelli, M., Pacifici, F.: Perla: A data language for pervasive systems. In: Proc. PerCom, pp. 282–287 (2008)
3. Andrieux, A., Czajkowski, K., Dan, A., Keahey, K., Ludwig, H., Nakata, T., Pruyne, J., Rofrano, J., Tuecke, S., Xu, M.: Web Service Agreement Specification (WS-Agreement) (2007), http://www.ogf.org/documents/GFD.107.pdf
4. Chinnici, R., Weerawarana, S., Moreau, J.J., Ryman, A.: Web services description language (WSDL) version 2.0 part 1: Core language. W3C recommendation, W3C (2007), http://www.w3.org/TR/2007/REC-wsdl20-20070626
5. Clement, L., Hately, A., von Riegen, C., Rogers, T.: UDDI version 3.0.2. OASIS specification, OASIS (2004), http://www.uddi.org/pubs/uddi_v3.htm
6. Driscoll, D., Mensch, A.: DPWS version 1.1. OASIS specification, OASIS (2009), http://docs.oasis-open.org/ws-dd/ns/dpws/2009/01
7. Delicato, F., Pires, P., Pinnez, L., Fernando, L., da Costa, L.: A flexible web service based architecture for wireless sensor networks. In: Proceedings of 23rd International Conference on Distributed Computing Systems Workshops, 2003, pp. 730–735 (2003), doi:10.1109/ICDCSW.2003.1203639
8. Madden, S.R., Franklin, M.J., Hellerstein, J.M., Hong, W.: Tinydb: An acquisitional query processing sys- tem for sensor networks. ACM Trans. Database Syst. 30(1), 122–173 (2005)
9. Di Modica, G., Regalbuto, V., Tomarchio, O., Vita, L.: Enabling re-negotiations of sla by extending the ws-agreement specification. In: IEEE International Conference on Services Computing, SCC 2007, pp. 248–251 (2007), doi:10.1109/SCC.2007.55
10. Domingue, J., Fensel, D., GonzAlez-Cabero, R.: Soa4all, enabling the soa revolution on a world wide scale, pp. 530–537 (2008), doi:10.1109/ICSC.2008.45
11. Fremantle, P., Patil, S., Davis, D., Karmarkar, A., Pilz, G., Winkler, S., Ümit Yalçinalp: Web Services Reliable Messaging (WS-ReliableMessaging) Version 1.1. OASIS specification, OASIS (2007), http://docs.oasis-open.org/ws-rx/wsrm/200702/wsrm-1.1-spec-os-01.pdf
12. Grosky, W., Kansal, A., Nath, S., Liu, J., Zhao, F.: Senseweb: An infrastructure for shared sensing. IEEE Multimedia 14(4), 8–13 (2007), doi:10.1109/MMUL.2007.82
13. Gudgin, M., Hadley, M., Mendelsohn, N., Lafon, Y., Moreau, J.J., Karmarkar, A., Nielsen, H.F.: SOAP version 1.2 part 1: Messaging framework (2nd edn.) W3C recommendation, W3C (2007), http://www.w3.org/TR/2007/REC-soap12-part1-20070427/
14. Hadim, S., Mohamed, N.: Middleware for wireless sensor networks: A survey. In: First International Conference on Communication System Software and Middleware, Comsware 2006, pp. 1–7 (2006), doi:10.1109/COMSWA.2006.1665174
15. Heinzelman, W., Murphy, A., Carvalho, H., Perillo, M.: Middleware to support sensor network applications. IEEE Network 18(1), 6–14 (2004), doi:10.1109/MNET.2004.1265828
16. Hill, J., Szewczyk, R., Woo, A., Hollar, S., Culler, D., Pister, K.: System architecture directions for networked sensors. In: ASPLOS-IX: Proceedings of the Ninth International Conference on Architectural Support for Programming Languages and Operating Systems, pp. 93–104. ACM, New York (2000) doi: http://doi.acm.org/10.1145/378993.379006

17. Ho, L., Moh, M., Walker, Z., Hamada, T., Su, C.F.: A prototype on rfid and sensor networks for elder healthcare: progress report. In: E-WIND 2005: Proceedings of the 2005 ACM SIGCOMM Workshop on Experimental Approaches to Wireless Network Design and Analysis, pp. 70–75. ACM, New York (2005) doi:
    http://doi.acm.org/10.1145/1080148.1080164
18. Huhns, M., Singh, M.: Service-oriented computing: key concepts and principles. IEEE Internet Computing 9(1), 75–81 (2005), doi:10.1109/MIC.2005.21
19. Lawrence, K., Kaler, C., Nadalin, A., Monzillo, R., Hallam-Baker, P.: Web Services Security: SOAP Message Security 1.1 (WS-Security 2004). OASIS specification, OASIS (2006), http://docs.oasis-open.org/wss/v1.1/
20. Levis, P., Culler, D.: Maté: a tiny virtual machine for sensor networks. In: ASPLOS-X: Proceedings of the 10th International Conference on Architectural Support for Programming Languages and Operating Systems, pp. 85–95. ACM, New York (2002) doi:
    http://doi.acm.org/10.1145/605397.605407
21. Liu, T., Martonosi, M.: Impala: a middleware system for managing autonomic, parallel sensor systems. In: PPoPP 2003: Proceedings of the Ninth ACM SIGPLAN Symposium on Principles and Practice of Parallel Programming, pp. 107–118. ACM, New York (2003) doi: http://doi.acm.org/10.1145/781498.781516
22. Malhotra, A., Biron, P.V.: XML schema part 2: Datatypes second edition. W3C recommendation, W3C (2004),
    http://www.w3.org/TR/2004/REC-xmlschema-2-20041028/
23. Masri, W., Mammeri, Z.: Middleware for wireless sensor networks: A comparative analysis. In: IFIP International Conference on Network and Parallel Computing Workshops, NPC Workshops 2007, pp. 349–356 (2007), doi:10.1109/NPC.2007.165
24. Moeller, R., Sleman, A.: Wireless networking services for implementation of ambient intelligence at home. In: 7th International Caribbean Conference on Devices, Circuits and Systems, ICCDCS 2008, pp. 1–5 (2008), doi:10.1109/ICCDCS.2008.4542655
25. Newcomer, E., Robinson, I., Little, M., Wilkinson, A.: Web Services Atomic Transaction (WS-AtomicTransaction) Version 1.2. OASIS specification, OASIS (2009), http://docs.oasis-open.org/ws-tx/wstx-wsat-1.2-spec-os.pdf
26. Priyantha, N.B., Kansal, A., Goraczko, M., Zhao, F.: Tiny web services: design and implementation of interoperable and evolvable sensor networks. In: SenSys 2008: Proceedings of the 6th ACM Conference on Embedded Network Sensor Systems, pp. 253–266. ACM, New York (2008) doi: http://doi.acm.org/10.1145/1460412.1460438
27. Rogers, T., Hadley, M., Gudgin, M.: Web services addressing 1.0 - core. W3C recommendation, W3C (2006),
    http://www.w3.org/TR/2006/REC-ws-addr-core-20060509
28. Samaras, I.K., Gialelis, J.V., Hassapis, G.D.: Integrating wireless sensor networks into enterprise information systems by using web services. In: SENSORCOMM 2009: Proceedings of the 2009 Third International Conference on Sensor Technologies and Applications, pp. 580–587. IEEE Computer Society, Washington, DC (2009) doi: http://dx.doi.org/10.1109/SENSORCOMM.2009.96
29. Shen, C.C., Srisathapornphat, C., Jaikaeo, C.: Sensor information networking architecture and applications. IEEE Personal Communications 8(4), 52–59 (2001), doi:10.1109/98.944004
30. Souto, E., Guimaraes, G., Vasconcelos, G., Vieira, M., Rosa, N., Ferraz, C., Kelner, J.: Mires: a publish/subscribe middleware for sensor networks. Personal Ubiquitous Comput. 10(1), 37–44 (2005) doi: http://dx.doi.org/10.1007/s00779-005-0038-3
31. Vogels, W.: Web services are not distributed objects. IEEE Internet Computing 7(6), 59–66 (2003), http://dx.doi.org/10.1109/MIC.2003.1250585

32. Walmsley, P., Fallside, D.C.: XML schema part 0: Primer second edition. W3C recommendation, W3C (2004),
http://www.w3.org/TR/2004/REC-xmlschema-0-20041028/
33. Yu, Y., Krishnamachari, B., Prasanna, V.: Issues in designing middleware for wireless sensor networks. IEEE Network 18(1), 15–21 (2004), doi:10.1109/MNET.2004.1265829
34. Zhang, L., Wang, Z.: Integration of rfid into wireless sensor networks: Architectures, opportunities and challenging problems. In: International Conference on Grid and Cooperative Computing Workshops, pp. 463–469 (2006) doi:
http://doi.ieeecomputersociety.org/10.1109/GCCW.2006.58
35. Zhao, J., Govindan, R., Estrin, D.: Computing aggregates for monitoring wireless sensor networks. In: Proceedings of the First IEEE International Workshop on Sensor Network Protocols and Applications, 2003, pp. 139–148 (2003), doi:10.1109/SNPA.2003.1203364

Chapter 18
Extracting Data from WSNs: A Data-Oriented Approach

Fabio A. Schreiber, Romolo Camplani, and Guido Rota

Abstract. The PerLa language and the related middleware have been developed to ease the task of querying heterogeneous devices in pervasive systems. This paper presents, in a detailed way, some of the main features of the PerLa language by showing how it can be applied to the wine production process.

18.1 Introduction

The wine production process requires the cooperation of many different "technologies" and expertises, which work in a pipelined manner: from the "wine design", performed by the oenologist, of the blending and timing for a quality wine, to the grape cultivation in the vineyard, up to the wine delivery to the consumer table, as discussed in chapter 17 and chapter 21.

As we can see in Figure 18.1, sensors and portable computing devices play an important role in each of the process steps; however the different environments and scopes require very different devices to gather and send data to the production control system, each with its own format and protocol.

Many different data are to be collected from sensors in the vineyard in order to control both the grape ripening conditions and the possible parasites attacks or on the barrels to control the wine ageing; even more differences among the information needed by the vineyard workers on their PDAs and on the RFID tags to be attached

Fabio A. Schreiber · Romolo Camplani
Politecnico di Milano, Dipartimento di Elettronica e Informazione, 34/5
Ponzio, 20133 Milano, Italy
e-mail: {schreiber,camplani}@elet.polimi.it

Guido Rota
Politecnico di Milano, Dipartimento di Elettronica e Informazione, 34/5 Ponzio,
20133 Milano, Italy
e-mail: guido.rota@gmail.com

G. Anastasi et al. (Eds.): Networked Enterprises, LNCS 7200, pp. 357–373, 2012.

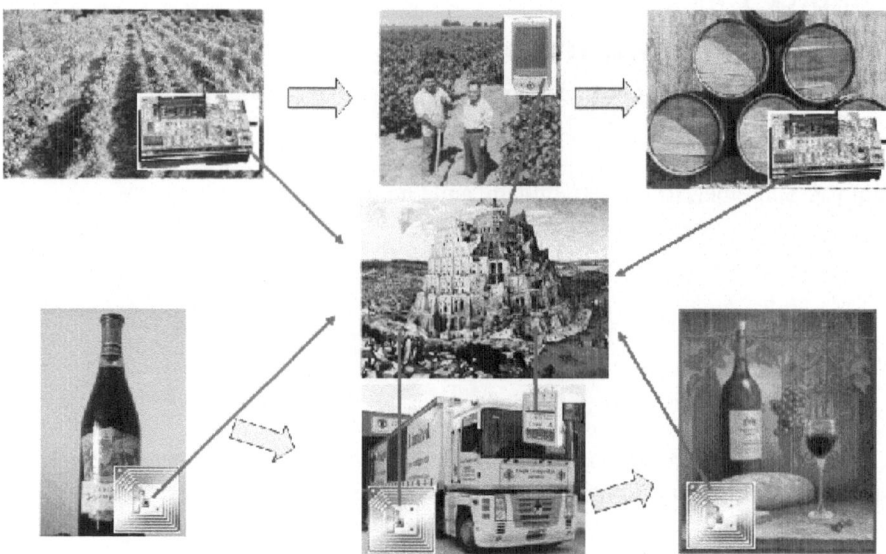

Fig. 18.1 The wine production and delivery process

to the pallets or to each bottle. In this chapter we outline the winemaking monitoring process (see Figure 18.2), and we shall discuss how we overcome the challenges
raised by the different sensing devices that support this process.

In this chapter we introduce *PerLa* - a language and middleware for data management in pervasive systems - by showing how its features can be applied to the winery
scenario. In section 18.2 we introduce the monitoring requirements. In section 18.3,
the main feature of the PerLa language are introduced by means of set of queries
relevant to the wine production monitoring processes, while in section 18.3 a more
formal presentation of the language is introduced.

18.2 The Vinification Monitoring Process

The task of controlling the vinification process begins in the vineyard. Humidity and
temperature have a strong influence on the final quality of the wine, as they directly influence the grape's maturation process. Keeping these parameters under strict control
is therefore an activity of paramount importance. The monitoring system, by means
of sensors deployed in the vineyard, is able to provide continuous readings of humidity and temperature, as well as fire warning alarms should one of the parameters of
interest exceed the ranges considered safe by the oenologists. Furthermore, to make
full use of the resources available in the field, any relevant data sensed by the *PDAs*

provided to the vinery workers is automatically integrated with the readings coming from the stationary sensor nodes.

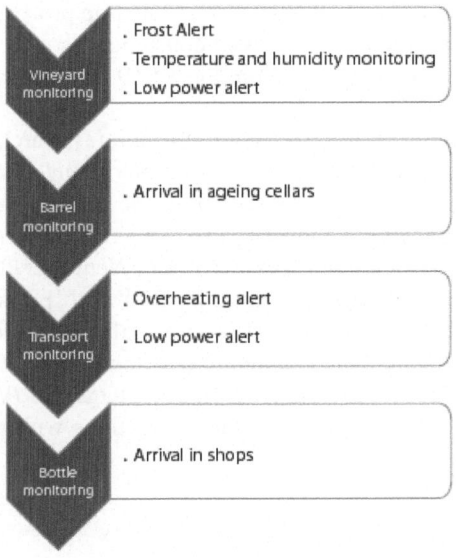

Fig. 18.2 Wine production process monitoring - Workflow

The wine production monitoring process doesn't stop in the vineyard. Just after the grapes are harvested, and the fresh wine is made, the ageing process begins. The age and the peculiar character of the wooden barrels, the humidity of the ageing cellar, and unwanted temperature spikes are just few of the several factors involved in the development of wine flavour. The oenologists are required to constantly monitor all these aspects throughout the entire ageing period, to ensure the correct environmental conditions needed to mature the desired wine savour are maintained. The winemaking monitoring system is employed to chronicle the cellars where a wine barrel is stored during the ageing period. To this purpose, *RFID tags* are to be attached on every barrel, and the *RFID Readers* installed in the cellars will be used to trace the different locations where the wine is stored. This information can then be crossed with the data sensed by the cellar's legacy environmental control systems to obtain a complete monitoring of the wine maturation process.

To provide a detailed account of the entire production activity, it is essential to monitor the transportation phase as well. Since the flavour of wine can be easily spoilt by a sudden change of temperature, every pallet of bottles is to be provided with a thermal sensor during shipment. The wine will then be marked accordingly in the eventuality of an overheating. Moreover, by means of *GPS receivers* installed on the trucks, the monitoring system can assess if the current situation requires additional surveillance (e.g. when the payload is stationary in a sunny area), and ensures that the temperature is sampled with adequate frequency.

All the information gathered during the vinification is stored in a database for future evaluation. The complete account of the production process of a bottle of wine, made available to winemakers and the final consumers as well, can be retrieved by means of the identification code stored in the *RFID tag* located under the label of each bottle.

In the remainder of this chapter we will briefly outline the architecture of the aforementioned database and provide a short description of *PerLa* language and middleware, the system employed to collect data from the sensing devices. The decision to adopt *PerLa* is the results of a thorough analysis of current state of the art technologies for *Pervasive Systems* and *Wireless Sensor Networks* [1]. TinyDB [2], DNS [3], Cougar [4], Maté [5], Impala [6], Sina [7], DsWare [8], MaD-WiSe [9], Kairos [10], GSN [11], SWORD [12] and *PerLa* were evaluated to determine which one met the requirements of the winemaking monitoring system. *PerLa* has been chosen by virtue of its SQL-like declarative query language and *Plug & Play* node addition system, which greatly simplified the interactions with the sensing network devices.

18.3 PerLa: System Description

While in chapter 17 a service oriented approach is presented, *PerLa* is a language and a data processing middleware for *Pervasive Systems*, developed to mask the idiosyncrasies of the nodes employed in complex sensing networks based on the Database approach. The pervasive systems, exposed as a Database by *PerLa*, can be queried through a declarative language with SQL-like syntax. This feature allows application developers to release themselves from the burden of managing the peculiar behaviour of different sensing devices. Collecting data from a pervasive system abstracted by *PerLa* can be as easy as writing a typical database query. Moreover, the *PerLa* language [13][14][16] is composed of special syntactic statements designed to fully exploit the capabilities typical of pervasive sensing networks. Various examples of these statements will follow in the remainder of this chapter.

All nodes present in the sensing network are abstracted by the *PerLa middleware* as proxies called *FPC* (Functionality Proxy Component). These components have common and homogeneous interfaces, and are used by *PerLa* queries to access the data gathered from the network nodes. No knowledge of the node's hardware and computational characteristics is needed to perform a *PerLa* query. Moreover, by means of the FPC abstraction, the language is not tied to any particular type of sensing device.

PerLa support for pervasive systems extends to node developers as well. The addition of new sensing devices in an existing network is facilitated by a *Plug & Play* connection system, i.e. a runtime factory that generates all the software components needed to query new sensor nodes. The information required to automatically assemble a device driver are stored in an XML file drafted by the node developer. This file, dubbed *Device Descriptor*, details all the node's characteristics in terms of data structures, protocols of communication, computational capabilities, and behavioural patterns. The descriptor can be sent by the device itself upon startup or directly injected by the user (e.g. for RFIDs or other dumb devices). All running queries will automatically make use of every new node added in the system.

Results generated by *PerLa* queries are automatically stored in a relational database. This features allows third party software (see chapter 13) to easily access the data collected from the sensing network (see chapter 15 and 16).

PerLa middleware is being continuously updated. At the present time, a series of ongoing projects is aiming at expanding the system by adding an intelligent power management, new context-aware query statements and context-management features [15][17] and a virtualization layer to fully exploit the computational capabilities of the network nodes.

18.3.1 PerLa: Integration in the Winemaking Monitoring Process

As the reader may have already realized from the introductory sections of this book, the wine production monitoring system makes use of data coming from a wide variety of heterogeneous sensing devices. *RFID* readers and tags, *PDAs*, legacy environmental control systems and ad-hoc sensors are just a few examples of the different nodes that support the winemaking control system. Developing a custom-made application merely to administer and query scores of heterogeneous sensing devices would not be an effective solution, since even the slightest change in the data acquisition network would surely entail a substantial rewrite of the system. Moreover, the resulting product could not be reused in any other domain of application.

All these considerations led to adopt *PerLa* as a middleware to decouple the core monitoring application from the Pervasive System used as a data source. The information retrieved from the sensing network is stored in the relational database described in chapter 9, and then analysed and processed by the wine production control system. A simplified version of this database schema is shown in Figure 18.3. While a detailed description of the PerLa language and its EBNF formal description can be found in [14], in the remainder of this section we shall describe the *PerLa* queries used for the monitoring process, and we shall use them to introduce the semantics of the *PerLa* language and the major software components of the *PerLa* middleware.

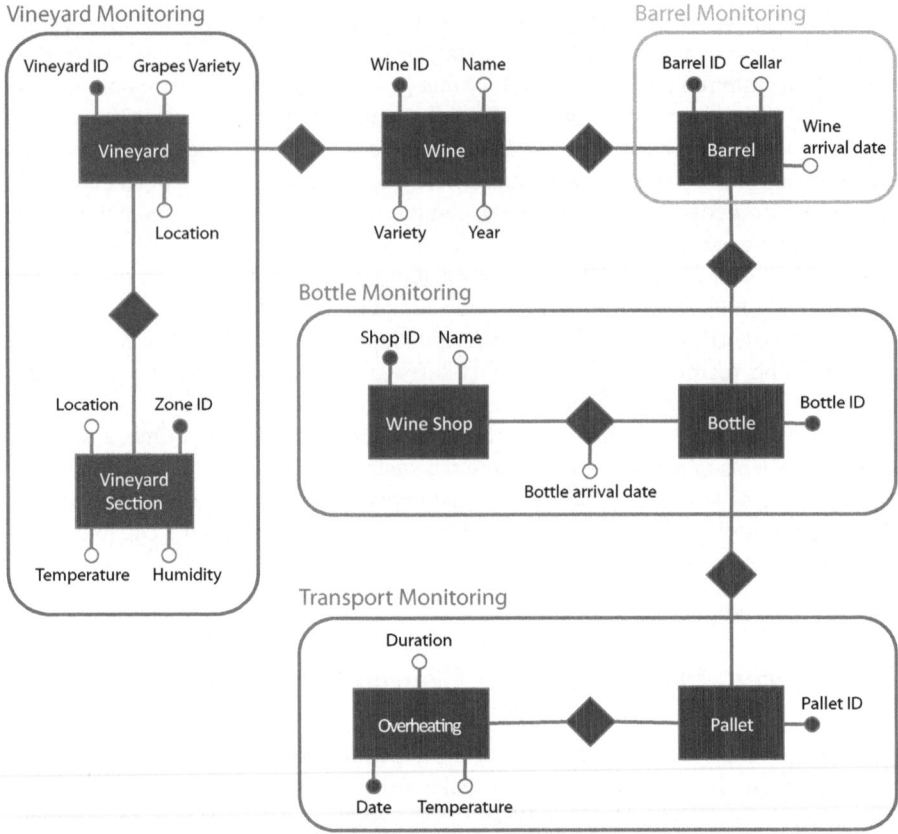

Fig. 18.3 Wine production process monitoring - Database

```
1    CREATE OUTPUT STREAM Monitoring
2        (nodeId ID, temperature FLOAT, humidity FLOAT, locationX FLOAT, locationY FLOAT) AS
3    LOW:
4        EVERY ONE
5        SELECT ID, temperature, humidity, locationX, locationY
6        SAMPLING
7            EVERY 1 m
8        EXECUTE IF EXISTS (temperature) AND is_in_Vineyard (locationX, locationY)
9            REFRESH EVERY 10 m
```

Fig. 18.4 Vineyard monitoring query

18.3.1.1 Query A: Vineyard Monitoring

The purpose of this query is simple: collect temperature and humidity from all the nodes located in the vineyard.

The first 2 lines of Figure 18.4 contain the declaration of an *OUTPUT STREAM*, one of the two data structures available in *PerLa*. A *STREAM* is fundamentally an unbounded table, designed to be used mostly as an output data structure. As can be seen in the second line of Figure 18.4, every record of a *STREAM* is composed of a fixed set of fields, each of which has an identifier (nodeID, temperature, ...) and a type (ID, FLOAT, ...). In addition to the fields declared by the user, every record is provided with a native *TIMESTAMP* field.

The keyword *AS* (Figure 18.4, second line) is then used as a shorthand to indicate that the results of the query have to be stored in the previously declared data structure.

At the third row, the body of the query begins. *PerLa* supports two different types of queries:

- **Low Level Queries:** define the behaviour of a single sensing or actuation device. *Low Level Queries* allow the user to:

  - Set the sampling mode of the nodes
  - Execute SQL operations on the data sampled from the *Pervasive System*, (e.g. filtering, aggregations, ...).

- **High Level Queries:** perform SQL operations on *STREAMS* generated by *Low Level Queries* or other *High Level Queries*. Since the data extracted from the sensor nodes is stored in a relational database, the winemaking monitoring application does not use this type of queries extensively. An example of a *High Level query* is to be given in Fig.18.8.

Low Level Queries and *High Level Queries* are respectively identified by the keywords *LOW* and *HIGH*.

The *EVERY* clause (fourth line) specifies the execution condition of the Low Level *SELECT* statement. In this case, an event-based approach is chosen using the keyword *ONE*. As a result, the selection is scheduled to run every time a sample is gathered from the device. The *EVERY* clause, in addition to the event-based behaviour (*EVERY ONE, EVERY 2 SAMPLES*, ...), supports a time-based semantics as well (*EVERY 20 m, EVERY 2 h*, ...).

The selection clause (Figure 18.4, line 5) is then introduced by the keyword *SE-LECT*. The semantics of this clause is identical to its SQL counterpart, with just one difference: the data source of *PerLa* queries is not a database table, but a group of sensing devices instead. Depending on the peculiar characteristics of the nodes that compose the sensing network, same types of information may be collected by means of different techniques. The location of the devices employed in the vineyard monitoring, for example, can either be read from memory (if the node is known to be stationary) or sampled from a *GPS* receiver (if the device is a *PDA* assigned to a field worker). Despite these differences, the same selection statement (*SELECT locationX FLOAT, locationY FLOAT*) correctly retrieves the location from both devices.

The *SAMPLING* keyword (Figure 18.4, line 6) defines which sampling mode is required to collect the data requested in the *SELECT* clause. Two different semantics are available:

- **Time based:** the sampling frequency is set explicitly by the user. This mode is chosen by means of the keyword *EVERY*, followed by the desired time interval (e.g. 1 m, 10 m, 3 h, ...).

- **Event based:** the sampling operation is performed upon the occurrence of an event. The event based semantics, further discussed in one of the following queries, is activated with the keyword *ON EVENT* followed by the list of events chosen to trigger the sampling.

In the instance of Figure 18.4, this clause simply specifies a sampling interval of 1 minute.

PerLa queries may also be used to describe other aspects of the data gathering process, empowering the user with the ability to greatly influence the operational behaviour of the single devices. It is worth mentioning that no knowledge about the hardware and software features of the sensing nodes is required when writing *PerLa* queries, even when dealing with clauses that govern the lowest layers of a pervasive system. We will delve further into this topic while describing the forthcoming queries.

The *EXECUTE IF* clause (Figure 18.4, line 8) is used to determine, by means of a logical expression, the set of devices on which the low level query is to be deployed. The statement of Figure 18.4 is set to be executed on all nodes located in the vineyard equipped with at least a temperature sensor. The vigilant reader may have noted that the selection clause demands a humidity reading, whereas the *EXECUTE IF* clause doesn't require the presence of the corresponding sensor. Whenever this query is executed on a device lacking the humidity transducer, a *NULL* value will be returned instead of the missing reading.

The *REFRESH* clause (line 9) forces the *EXECUTE IF* condition to be reassessed every 10 minutes, in order to update the list of devices capable of running the query. This allows the *PDAs* employed by the winemakers to be included when they access the vineyard. Should the *REFRESH* clause be omitted, the *EXECUTE IF* condition would be evaluated only once.

18.3.1.2 Query B: Frost Alarm

The second query, shown in Figure 18.5, is designed to monitor the environmental conditions that may induce frost in the vineyard. This phenomenon is known to show itself when temperature is near 0°C and the amount of water vapour in the air is significantly high. In this implementation of the query, the frost alarm is raised by adding a record in the *OUTPUT STREAM* whenever the temperature decreases below 5°C and the humidity in the air is over 75%.

```
1    CREATE OUTPUT STREAM FrostAlarm (nodeId ID, ts TIMESTAMP) AS
2    LOW:
3         EVERY ONE
4         SELECT ID, TIMESTAMP
5         HAVING (AVG (temperature, 10 m) < 5) AND (AVG (humidity, 10 m) > 0.75)
6         SAMPLING
7              EVERY 1 m
8         EXECUTE IF EXISTS (temp) AND EXIST (humidity) AND is_in_Vineyard (locationX, locationY)
```

Fig. 18.5 Frost alarm

This particular behaviour is enforced by means of the selection statement of a *Low Level Query* (Figure 18.5, lines 4 and 5). The *HAVING* clause, which, in contrast with standard SQL language, works both for aggregates and single values as well, filters the data sampled by the network nodes and discards all the inappropriate records. The predicate used by this selection clause imposes a condition on both temperature and humidity readings, which evaluates true only when the vineyard is at risk of freezing.

Note that the frost condition is evaluated on the average of the last 10 minutes samples in order to remove incidental noise. This is accomplished by using *PerLa* aggregate operators. Differently from their SQL counterpart, these operators have two parameters: the value on which the operation has to be carried out and the number of samples that are to be used to compute the aggregate. This difference from standard SQL aggregates is due to the peculiar nature of the input data sources. The information collected from pervasive system nodes is a continuous stream of data. Specifying to which extent the aggregation operation should be performed is therefore mandatory, otherwise the computation would never terminate due to the potentially infinite data set. The number of samples on which the aggregate is calculated can be specified either as an explicit number (e.g. *AVG(temperature, 25 SAMPLES)*) or as a time duration (e.g. *AVG(temperature, 10 m)*).

Since air humidity is fundamental to forecast the frost phenomenon, the *EXECUTE IF* clause (Figure 18.5, line 8) mandates the presence of the corresponding transducer on all the nodes involved in the execution of this query. By contrast, this sensor was tagged as optional in the first query (Figure 18.4, line 8).

18.3.1.3 Query C: Devices with Low Battery

The state of a sensing network managed by *PerLa* is abstracted as a set of records collected from a group of potentially virtual sensors. Therefore, non functional information regarding the network nodes such as battery status, processor speed, software or hardware revision, etc. can be accessed by means of a plain *PerLa* query. No special statements or clauses are needed for this purpose.

The instance of Figure 18.6 exploit this feature to list the identifiers (*IDs*) that belong to wireless devices with low residual battery charge.

```
1   CREATE OUTPUT STREAM LowPoweredDevices (sensorID ID) AS
2   LOW:
3        EVERY ONE
4        SELECT ID
5        SAMPLING
6            EVERY 24 h
7            WHERE powerLevel < 0.15
8        EXECUTE IF deviceType = "WirelessNode"
```

Fig. 18.6 Low Powered Devices

In contrast with the examples shown up to this point, the query of Figure 18.6 employs the *SAMPLING* clause to determine whether a node's battery is nearing exhaustion (line 7, *WHERE powerLevel < 0.15*). Previous examples (e.g. Figure 18.5) relied entirely on the *SELECT ... HAVING* syntax to filter records; the usage of the *SAMPLING ... WHERE* construct produces the following effects:

- the discarded records are not processed by the *SELECT* statement
- the discarded records do not trigger the execution of the *SELECT* statement when the execution condition is event based (e.g. *EVERY ONE, EVERY 2 SAMPLES,* ...)
- the records that does not fulfill the *WHERE* criterion are discarded by the sensor node itself; no data is transmitted over the network.

It is worth mentioning that the *SAMPLING ... WHERE* and *SELECT ... HAVING* constructs are not interchangeable, even though most trivial *PerLa* queries can be written using either of them. The *HAVING* clause is meant to be used when the filtering condition involves aggregate operations or other functions that require two or more records to be computed. The *WHERE* clause is a better choice if the filtering operation can be performed evaluating a single record. The latter approach can lead to a significant performance improvement, since the discarded values are not processed by the *SELECT* statement. Therefore, the *WHERE* clause should be preferred over the *HAVING* one whenever possible.

```
1   CREATE OUTPUT STREAM NumberOfLowPoweredDevices (counter INTEGER) AS
2   HIGH:
3       EVERY 24 h
4       SELECT COUNT(*)
5       FROM LowPoweredDevices(24 h)
```

Fig. 18.7 Number of low powered devices

The number of devices that need a battery replacement may be computed with the query of Figure 18.7. The statement, whose syntax closely mirrors standard SQL, is a simple *High Level Query*. Like the foregoing *Low Level Queries*, the selection statement activation condition is expressed via the *EVERY* clause. In this instance, the query is run every 24 hours (Figure 18.7, line 3). The only operation required to reckon the number of devices in need of a new battery is a simple *COUNT(*)*. The raw data regarding the battery conditions of the network nodes are retrieved from the *LowPoweredDevice STREAM*, which is constantly updated by the query shown in Figure 18.6.

One of the major differences among SQL queries and *PerLa High Level Queries* lies in the *FROM* clause. As can be seen at line 5 of Figure 18.7, the input *STREAM* name is complemented with a duration (namely *Window Size*). This information determines how many records are to be processed whenever the *SELECT* statement is activated. The *Window Size* can be specified either as a time interval or a number of records.

18.3.1.4 Query D: Pallets Out of Temperature

High temperatures can yield devastating effects on wine flavour. Even if protracted for a short amount of time, the exposure to a warm environment may irreversibly change the product's typical character and savour. These alterations are to be prevented at all costs to avoid wine depreciation and to increase customer satisfaction. To this purpose, the query in Figure 18.8 is employed to signal every thermal shock experienced by the wine during shipment. Considering that the thermal sensors employed in this application are powered by an autonomous battery, this query has been designed to minimize unnecessary data communications.

```
1    CREATE SNAPSHOT TrucksPositions (linkedBaseStationID ID) WITH DURATION 1 h AS
2    LOW:
3        SELECT linkedBaseStationID
4        SAMPLING
5            EVERY 1 h
6            WHERE is_in_CriticalZone (locationX, locationY)
7        EXECUTE IF deviceType = "GPS"

8    CREATE OUTPUT STREAM OutOfTemperatureRangePallets (palletID ID) AS
9    LOW:
10       EVERY 10 m
11       SELECT ID
12       SAMPLING EVERY 10 m
13           WHERE temperature > 18
14       PILOT JOIN TrucksPositions ON baseStationID = TrucksPositions.linkedBaseStationID
```

Fig. 18.8 Pallets out of temperature

The shipment monitoring system is composed of these main elements:

- A temperature sensor node with low-range radio transmitter installed on every pallet of wine
- A *GPS* receiver installed on every truck of the shipment fleet
- A *Base Station*, i.e. a special network node used to relay the data from *GPS* and temperature sensors to the monitoring headquarters. Every truck is provided with a single base station.
- The *PerLa* query of Figure 18.8

Since under normal circumstances the trailer's air cooling system is considered adequate to safeguard the integrity of the wine, the continuous temperature monitoring is activated only when the shipment travels across a critical location. This behaviour is achieved through the *PILOT JOIN* execution condition.

With *PILOT JOIN*, the decision to include a node in the execution of a query is based on the content of a support data structure. Therefore, the output of any sensing device can be employed to trigger the activation of a query on other network nodes. By contrast, the *EXECUTE IF* only allows the use of attributes gathered from a single device to determine whether that node could take part in a query or not.

The example in Figure 18.8 makes use of the *PILOT JOIN* clause to activate a monitoring query on those pallets that are traversing a critical location. To obtain this behaviour, a first subquery (lines 1 to 7) is run to gather the *IDs* of the *Base Stations* installed on trucks considered at risk of overheating. Then, the second subquery (lines 8 to 14) is used to collect the identifier of the pallets whose temperature exceeded the safe threshold (line 13). The *PILOT JOIN* (Figure 18.8, line 14) ensures that the scope of the monitoring query is limited to the bottles inside a critical location.

The first subquery, in contrast with all the preceding examples, is used to insert records in a *SNAPSHOT* table (Figure 18.8, first line). This type of data structure is intended to hold data for a limited time only, known as *SNAPSHOT DURATION*. Upon expiration, the *SNAPSHOT* is purged and filled with new records. A *PILOT JOIN* execution condition, if used in combination with a *SNAPSHOT* table, is evaluated when the content of the data structure is refreshed (conditional semantics). Were the same *PILOT JOIN* used with a *STREAM*, the evaluation of the execution condition would be triggered by every new record added in the data structure (event based semantics).

18.3.1.5 Query E: Cellar Monitoring

The wine monitoring process does not stop once the product is delivered. To ensure that flavour and aroma are preserved, the wine has to be stored in a controlled environment, where unexpected variations in temperature, humidity or light are kept under strict control. All wine cellars must therefore provide some sort of controlled environment, either by nature or by means of a climate control systems. To complete the chain of trust between the winemaker and the final buyer, the wine dealers must be able to provide the list of cellars where every bottle in their catalogue has been stored. The query of (Figure 18.9) is specifically designed for this purpose.

```
1  CREATE OUTPUT STREAM BottlesEnteredInCellar (bottleId ID, ts TIMESTAMP) AS
2  LOW:
3        EVERY ONE
4        SELECT ID, TIMESTAMP
5        SAMPLING
6            ON EVENT lastReaderChanged
7            WHERE lastReaderId = "CellarX_ReaderId"
```

Fig. 18.9 Cellar monitoring

Since every bottle is labelled with an *RFID*, the content of every wine cellar can be easily monitored by installing a tag reader at the entrance door. The instance in Figure 18.9 is then executed to maintain a log with the *IDs* of all the wine bottles that crossed the threshold of the storing room.

In this situation, a continuous monitoring of the *RFID reader* would not make sense. Hence the decision to use an event based sampling technique. As explained in the description of the first query, the event based sampling mode triggers a reading from a sensing device upon the occurrence of one or more events. In the query of Figure 18.9, the sampling operation is set to be performed whenever the *RFID tag* of a wine bottle is sensed by a new *RFID reader* (i.e. the bottle has been moved to a new cellar). The *WHERE* clause of line 7 is additionally specified to monitor one cellar at a time.

The information produced by this query, when crossed with the data collected by the cellar's legacy environmental control system, provides a comprehensive chronicle of the wine storage phase. The final consumer can exploit this information to determine the exact conditions endured by the wine bottle during its entire lifetime.

It is worth mentioning that this query can be easily adapted to monitor the wine barrels introduced in the ageing cellars.

18.4 PerLa Language Digest

We conclude this chapter with a brief digest on the *PerLa language*, intended to give the reader a broad vision over the most important language statements and features [14].

18.4.1 Data Definition

PerLa language provides the user with two distinct table types:

- **STREAM:** a table composed of an unbounded number of records. *STREAMS* represent *PerLa's* main data structure.
- **SNAPSHOT:** a set of records generated during a specified time period (*SNAPSHOT DURATION*). When the given duration expires, the *SNAPSHOT* is cleared and filled with new records.

PerLa Data Definition statements are introduced by the keyword *CREATE*, followed by the identifier of the desired data structure. Both *STREAMs* and *SNAPSHOTs* are a homogeneous collection of records. The record structure declaration, defined as a list of identifiers and data types, is therefore mandatory. *PerLa* tables can be additionally tagged with the keyword *OUTPUT* if their content is to be shown to the user who submitted the query.

A *STREAM* or *SNAPSHOT* definition is usually followed by a *PerLa* query, introduced via the keyword *AS*, which is executed to generate the data structure content.

18.4.2 Low Level Queries

Low Level Queries are used to access the data produced by the sensing network. By means of this type of statement, *PerLa* users can define which information is to be gathered from the pervasive system , set the sampling mode of the sensing devices, select the network nodes on which execute the query, filter data and perform simple *SQL* operations. *Low Level Queries* are identified by the keyword *LOW*.

Sampling

The *SAMPLING* clause is used to specify the behaviour of every single sensing device. Two different semantics are available:

- **Time based:** the sampling operation is performed periodically, following user indications. The sampling period can be specified as a fixed duration (*SAMPLING EVERY 3 m*), as a numeric expression (*SAMPLING EVERY (1000 / temperature) s)*) or as a conditional statement (*SAMPLING IF powerLevel > 50 EVERY 1 s ELSE EVERY 15 s REFRESH EVERY 1 h*).
- **Event based:** the sampling operation is triggered by the occurrence of one or more events (e.g. SAMPLING ON EVENT highTemperature, lowTemperature).

The *SAMPLING* clause may also be employed to perform basic filtering operations (SAMPLE EVERY 1 m WHERE temperature > 50).

Data Managing

PerLa Low Level Queries are provided with a *Data Managing Section*, introduced by the *SELECT* keyword, through which users define the exact output of the query. Legit *SELECT* statements can make use of constants (SELECT 3, "Temperature"), data sampled from sensors (SELECT temperature, humidity, powerLevel), numerical expressions (SELECT $0.55 * (farenheitTemp - 32)$) or aggregate operations (SELECT timestamp, AVG(temperature, 10 m)). Two different activation semantics are available:

- **Time based:** query results are computed periodically (EVERY 10 m SELECT ...)
- **Event based:** query results are computed when a determined number of records is available (EVERY 5 SAMPLES SELECT ...)

SELECT statements can be complemented with a *HAVING* clause to specify advanced filtering conditions.

Execution Conditions

This optional section is used to define which sensing devices can partake in executing a query. Execution conditions are expressed through the following clauses:

- **EXECUTE IF:** allows the definition of a simple predicate, which is evaluated to determine whether a node can take part in a query or not. Execution conditions set via the *EXECUTE IF* clause are reassessed periodically if the *REFRESH* keyword is specified.
- **PILOT JOIN:** provides the user with the ability to declare sophisticated execution conditions based on a support query.

Termination Conditions

The optional *TERMINATE AFTER* clause may be employed to declare the lifetime of a *Low Level Query*. Queries can be set to stop at the end a specified time period (e.g. *TERMINATE AFTER 1 h*) or after the selection statement has been executed an established number of times (e.g. *TERMINATE AFTER 10 SELECTIONS*).

18.4.3 High Level Queries

High Level Queries are designed to perform data manipulation operations over *STREAM* tables. Syntax and semantics of this type of statement is closely related to standard SQL. There are, however, two major differences:

- **Activation conditions:** the *High Level* selection statements can be activated periodically (*EVERY 10 m*) or by the insertion of a new record in a *STREAM* (*EVERY 3 SAMPLES IN <StreamName>*)
- **FROM clause:** when using *STREAM* tables, the user is required to define a duration window that identifies how many records are processed during selection

High Level Queries are identified by the keyword *HIGH*.

18.4.4 Actuation Queries

Actuation Queries are employed to set parameters on network nodes. This type of statement is mainly used to drive mechanical and electronic actuators or to modify software variables. *Actuation Queries'* syntax is composed of the keyword *SET*, which introduces the parameter to set, and the keyword *ON*, used to list the nodes interested by the query.

18.5 Final Remarks

The PerLa middleware has been adopted in a prototypical deployment for vineyard monitoring. In particular, we deployed half a dozen nodes, each one composed by a ZigBee-compliant Jennic JN39R131 mote and endowed with humidity, temperature and luminosity sensors. For the tests, we fixed the sampling rate to 1Hz (which is considerably high for the chosen scenario).

Despite the considered testbed is a significant case study for our system, further improvements may be only obtained by studying the scalability of the entire systems. In particular, we want to focus on the scalability both in terms of number of nodes per network and of number of served networks.

Acknowledgements. We thankfully acknowledge the work of Ing. Marco Fortunato and Ing. Marco Marelli, who pioneered the design and development of *PerLa*, and the contribution of Ing. Diego Viganò, who assisted us in the development of the system.

References

1. Akyildiz, I.F., Su, W., Sankarasubramaniam, Y., Cayirci, E.: Wireless sensor networks: a survey. IEEE Communications Magazine 40(8), 102–114 (2002)
2. Madden, S., Franklin, M.J., Hellerstein, J.M., Hong, W.: TinyDB: an acquisitional query processing system for sensor networks. ACM Trans. Database Syst. 30(1), 122–173 (2005)
3. Chu, D., Popa, L., Tavakoli, A., Hellerstein, J.M., Levis, P., Shenker, S., Stoica, I.: The design and implementation of a declarative sensor network system. In: SenSys 2007: Proceedings of the 5th International Conference on Embedded Networked Sensor Systems, pp. 175–188. ACM, New York (2007)
4. Yao, Y., Gehrke, J.: The cougar approach to in-network query processing in sensor networks. SIGMOD Rec. 31(3), 9–18 (2002)
5. Levis, P., Culler, D.: Maté: a tiny virtual machine for sensor networks. SIGPLAN Not. 37(10), 85–95 (2002)
6. Liu, T., Martonosi, M.: Impala: a middleware system for managing autonomic, parallel sensor systems. In: PPoPP 2003: Proceedings of the Ninth ACM SIGPLAN Symposium on Principles and Practice of Parallel Programming, pp. 107–118. ACM Press, New York (2003)
7. Srisathapornphat, C., Jaikaeo, C., Shen, C.-C.: Sensor information networking architecture. In: ICPP 2000: Proceedings of the 2000 International Workshop on Parallel Processing, p. 23. IEEE Computer Society, Washington, DC (2000)
8. Li, S., Son, S.H., Stankovic, J.A.: Event Detection Services Using Data Service Middleware in Distributed Sensor Networks. In: Zhao, F., Guibas, L.J. (eds.) IPSN 2003. LNCS, vol. 2634, pp. 502–517. Springer, Heidelberg (2003)
9. Amato, G., Baronti, P., Chessa, S.: Mad-wise: programming and accessing data in a wireless sensor network. In: Proceedings of the International Conference on Computer as a tool EUROCON 2005 (2005)
10. Gummadi, R., Kothari, N., Millstein, T., Govindan, R.: Kairos: A macroprogramming system for wireless sensor networks. In: Proceedings of the Twentieth ACM Symposium on Operating Systems Principles SOSP 2005 (2005)
11. Aberer, K., Hauswirth, M., Salehi, A.: Global sensor networks. School of Computer and Communication Sciences Ecole Polytechnique Federale de Lausanne (EPFL), Tech. Rep. LSIR-REPORT-2006-001 (2006)
12. http://webdoc.siemens.it/CP/SIS/Press/SWORD.htm
13. Schreiber, F.A., Camplani, R., Fortunato, M., Marelli, M., Pacifici, F.: Perla: A data language for pervasive systems. In: Proc. PerCom, pp. 282–287 (2008)
14. Perla Home Page, http://perlawsn.sourceforge.net/
15. Bolchini, C., Curino, C.A., Orsi, G., Quintarelli, E., Rossato, R., Schreiber, F.A., Tanca, L.: And what can context do for data? Communications of ACM 52(11), 136–140 (2009)

16. Schreiber, F.A., Camplani, R., Fortunato, M., Marelli, M., Rota, G.: Perla: A language and middleware architecture for data management and integration in pervasive information systems. IEEE Transactions on Software Engineering, doi:10.1109/TSE.2011.25
17. Schreiber, F.A., Tanca, L., Camplani, R., Viganó, D.: Towards autonomic pervasive systems: the PerLa context language. In: Electronic Proceedings of the 6th International Workshop on Networking Meets Databases (Co-located with SIGMOD 2011), Athens, pp. 1–7 (2011),
http://research.microsoft.com/en-us/um/people/srikanth/
netdb11/netdb11papers/netdb11-final4.pdf

Chapter 19
Optimal Design of Wireless Sensor Networks

Marcello Mura, Simone Campanoni, William Fornaciari, and Mariagiovanna Sami

Abstract. Since their introduction, Wireless Sensor Networks (WSN) have been proposed as a powerful support for environment monitoring, ranging from monitoring of remote or hard-to-reach locations to fine-grained control of cultivations. Development of a WSN-based application is a complex task and challenging issues must be tackled starting from the first phases of the design cycle. We present here a tool supporting the DSE phase to perform architectural choices for the nodes and network topology, taking into account target performance goals and estimated costs. When designing applications based on WSN, the most challenging problem is energy shortage. Nodes are normally supplied through batteries, hence a limited amount of energy is available and no breakthroughs are foreseen in a near future. In our design cycle we approach this issue through a methodology that allows analysing and optimising the power performances in a hierarchical fashion, encompassing various abstraction levels.

19.1 Introduction

When envisioning applications that directly interface with the physical world through use of wireless sensor networks (WSNs), some low-level design aspects have to be taken into account already while drafting the application at a high abstraction level. In fact it becomes necessary to consider some requirements and constraints that go

Marcello Mura
ALaRI - Faculty of Informatics - University of Lugano (CH)
DEI - Politecnico di Milano (IT)
e-mail: muram@usi.ch

Simone Campanoni · William Fornaciari · Mariagiovanna Sami
DEI - Politecnico di Milano (IT)
e-mail: {campanoni,fornacia,sami}@elet.polimi.it

G. Anastasi et al. (Eds.): Networked Enterprises, LNCS 7200, pp. 375–395, 2012.
© Springer-Verlag Berlin Heidelberg 2012

well beyond the "typical" information technology ones: just as an example, a few points may involve:

- The physical environment where the sensor network will be deployed. Dimensions are not the only relevant point; topography, presence of physical obstacles, even such aspects as availability of sunlight may have an impact on design or even on feasibility itself;
- Requirements of sensor deployment (e.g., critical positioning of some given types of sensors);
- Technological choices (e.g., are there commercially available devices capable of sensing specific phenomena? Are such sensors capable of being ported into a sensor network in terms of costs, dimensions, energy consumption?);
- The "survival" requirements for the sensor network. If (as it is usual) battery-operated nodes are envisioned, how long do we expect the network to operate? Is this goal compatible with technological choices as well as with processing and transmission loads?
- How harsh is the physical environment and how critical is survival of an individual node and of the network as a whole?

Only after such basic questions have been answered it is possible to state whether (given the available technology) an efficient and effective deployment will be possible and what is necessary in order to achieve it. To this end, the application designer must be able to rely on a full design flow, based on suitably developed methodologies and tools, that will ultimately support the low-level design but that will also provide a first set of indicators guiding high-level decisions. More specifically, the designer should be able to:

- Verify the feasibility in terms of basic technologies;
- Take preliminary decisions (e.g. , communication protocol, programming model, etc.);
- Evaluate costs, power consumption, capacity of survival of a first draft design and check them against the specifications;
- Identify the aspects for which optimisations might be particularly relevant;
- Provide low-level specifications for the final network design.

Coming to the specific application envisioned within ArtDeco, recurring to sensor networks in fine-grained monitoring of high-quality cultivations has great potential relevance, as it allows bringing integrated adoption of ICT techniques from the initial production to the final commercial support in a sector (the so-called primary sector) where such overall impact has not yet been fully exploited. Remote sensing techniques, adopted by a few very large actors, are rather costly, do not provide 24-hours coverage and afford limited precision with respect to micro-climate aspects and to some ground-level sensed data. Use of sensor networks would overcome such limitations and make advanced monitoring available to a much larger community of users.

The experiment envisioned for ArtDeco represents a spectrum of applications in which sensor networks must be deployed in a well-studied and carefully tuned way, rather than by random distribution of highly redundant numbers of nodes. Position of the individual nodes is related to physical characteristics of the environment, and this will in turn impact on network operation. Cost needs to be minimized. Analysis and optimisation must thus be performed for the specific network well before actual deployment to support deployment decisions.

The initial design phases of a WSN require modelling and simulation tools meeting a number of challenges: as different aspects of the system need to be analysed, different modelling approaches may be more suitable and different components of the system may be targeted (often attention is on the radio section, but processors and sensors as well may need specific attention). Hence, identifying a single reference tool satisfying the requirements of a vast majority of designers is difficult.

Various approaches to deal with this task have been proposed, and related simulation tools have been made available (see, e.g., [1], [4], [14], [15]); nevertheless, custom-built simulators may need to be realized in specific instances (see, e.g., [7] where the necessity of modelling the node together with its harvesting section did not allow the use of a standard network simulator).

A major research focus is represented by optimising the location of the sensors to maximize their collective coverage of a given region. This challenge has been tackled using several approaches, such as integer programming or greedy heuristics to incrementally deploy the sensors. Adaptive techniques considering scenarios where multiple sensors are needed, accounting for a possible real-time deployment have been proposed.

Functional and non-functional requirements of a WSN may be very strict and must be tackled at design time. Power-aware design and operation of nodes as well as of network has received in particular much attention. Several power-aware protocols and routing algorithms have been presented (see, e.g., [3], [6]), the ultimate goal being that of increasing battery and network life.

A further factor that may have great relevance in some application classes is the capacity of prompt sensing of, and reactions to, particular events. The planning of WSNs for such applications is a off-line activity and requires to:

1. specify the characteristics of the events to be discovered;
2. select a proper set and type of sensors to enable the capturing of such events;
3. embed the sensors in the environment in a way to ensure the capturing of the desired events while optimising some design goals.
4. estimate energy consumption of proposed solutions so as to evaluate their viability in terms of network lifetime and eventually define optimisation policies

The approach described in this chapter is first of all to make sure a priori, with a good confidence, that there exists a feasible solution to the sensing problem with the accuracy required by the application. Then, by exploiting the capabilities of the SWORDFISH optimisation engine, the WSN is refined according to design constraints and users goals. After a preliminary network has been drafted, power modelling is tackled following a hierarchical approach (see [12]) that allows creating technology independent models at a high abstraction level and subsequent estimation and comparisons when low level technology-dependent models are inserted. Node and network level simulations based on such approach allow to refine the various choices (from node selection to network organization) and to perform possible power-related optimisations.

This chapter presents the contributions given by the authors to the design flow of a WSN, concerning both the initial phase of node-level and network-level design in view of a specific application and the subsequent steps focusing on power-oriented simulation and optimisation. The chapter is organised as follows. The concept of multi-level WSN design methodology is introduced first (section 19.2); design tools developed for sensor-level, node-level and network-level design are presented in the subsequent sections 19.3 to 19.6. The methodology devised for modelling power consumption at node and network level is then discussed in sections 19.7 to 19.9. A case study is discussed in section 19.10.

19.2 Multi-level Design

One of the objectives of the multi-level design methodology is to create a design flow for WSN offline planning, which is scalable with the application complexity (see Table 19.1). To this purpose, the first step is to identify a proper set of sensor-position pairs, considered optimal to capture the desired behaviour of the WSN.

Table 19.1 Abstraction layers of the design space exploration

Level	Activities
Sensor	Selection and positioning of sensors set; Sensitivity analysis
Node	Aggregation of sensors onto some nodes; Sensitivity analysis
Network	Identification and positioning of gateways; Protocol selection

This initial solution is the baseline for any architectural design space exploration. The next optimisation step is related to possible aggregation of the previously identified sensors set onto nodes, so as to balance network cost with effectiveness and performance of the WSN. The outermost layer accounts for complex sensor networks or for networks deployed in rugged terrains, where some of the nodes have to manage hierarchies of sensors/subnets.

We consider here all of the three levels mentioned in Table 19.1. The support to such system-level design is provided by a modular framework, called SWORD-FISH. The SWORDFISH framework has a graphical user interface to describe the actors (sensors, network, events, and environment) and the design goals of the systems (properties of the network and target optimisation parameters), whose roles are explained below and depicted in Figure 19.1.

Environment Editor. This module allows defining a simplified representation of the environment where the WSN will be deployed, with graphical views of the associated physical parameters (e.g., temperature, humidity, 3D-spatial

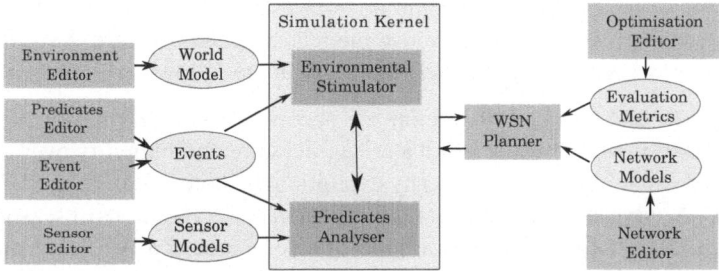

Fig. 19.1 Coarse-grain architecture of SWORDFISH

representation, obstacles) and the possibility to specify constraints such as position and type of some sensors, if relevant for the users.

Event Editor.   The purpose of this editor is to support the description of the events to be captured in terms of variation of some physical parameters to be sensed, along with their timing characteristics. The models are flexibly implemented via plugins.

Sensor Editor.   It allows obtaining the analytic representation of the sensing nodes, which is a modelling of the relation existing between the sensed physical parameters and the signal produced. The model of the node includes additional information like cost, type of sensors, energy consumption, accuracy, speed, etc.

Network Editor.   In addition to the node features, a model of the available connection channels among nodes is specified. In general, this model can cover both wired and wireless links, although in our first implementation we focused on wireless only.

Predicate Editor.   This editor allows the user to specify via a logic formula the properties to be verified when a given event occurs. This is of paramount importance to verify that a WSN is actually capable to properly react when an event is recognized, or, dually, to select the proper set of sensors to recognize the events. This is a concept more abstract and powerful then a simple measurement.

Simulation Kernel.   It is the engine which, based on a simulation of the event occurring, modifies the configuration of the world model accordingly. This allows feeding the sensor node models with the real (location aware) data of the world, including their dynamics. Hence, both the physical parameters of the environment and the events to be monitored can be jointly modelled and verified by the Predicate Analyser (see Figure 1).

Optimisation Editor.   It is an editor allowing the designer to specify and tune the goal functions and the formal model of the network properties/constraints.

Planner.   This is the main module for both verification and network design. It allows formally verifying that a given WSN is able to capture a set of events as well as to support the building and optimisation of the overall network according to the selected policies and goals.

Being the overall software toolsuite organized as a set of plug-ins, possible replacement or improvement of the above models is straightforward. SWORDFISH architecture aims at allowing the users to deal with the following problems:

Verification: the goal is to determine the occurrence of a set of events (e.g., fire in a defined region, presence of water, temperature and humidity over a certain threshold for a time window, etc.) by exploiting the potential of a given WSN

Sensitivity Analysis: evaluation of the impact of some variations of sensors, environment and network properties, onto the performance of a WSN. Examples are fault tolerance w.r.t. sensors and network errors, effect of sensor ageing or displacement of their location, influence of the observation time, etc.

Design/Planning: given a set of events and some constraints/goals, the task is to discover the optimal sensor network capable to identify the events while maximizing a user-controlled goal function.

Based on the application requirements, the first step for the user is formally defining the events to be captured and possibly some optimisation goals/constraints (Predicate, Event and Optimisation editors). Network properties and sensor behaviour can be also specified (Network and Sensor editors), if the default settings are not considered suitable. According to the model of the environment (specified using the Environment editor), the events are then *fired* to get a profiling of the evolution of the physical parameters corresponding to the events. Such results are then used as a testbench to compare the sensing capabilities of alternative WSNs. The Predicate analyser and the selected optimisation goals are extensively used by the Network Planner to explore the design space. Useful information for optimisation can be gathered by analysing the sensitivity of the network over the variation of parameters like observation time or sensor accuracy.

19.3 Sensor-Level Design

As said before, the model of the environment is 3-D, so that each point is represented using (x, y, z) coordinates belonging to a user-defined grid. Before starting the exploration of the WSN design space, there are three preliminary steps to define the purpose of the network, the benchmark and the hardness of recognizing physical parameters corresponding to an event.

The first activity is the definition of an overall Sensing Goal (SG) for the WSN, that is a multi-value logic formula composed of some predicates (implemented via plug-ins), each corresponding to an event. For example Water(x, y, z, magn, trend) is a plug-in modelling the presence of water in the point (x, y, z), starting from a given magnitude and with a specified trend over the time. A predicate is an instance of Water applied to a specific point. A catalogue of plug-ins (e.g., Fire, Water, Humidity) is available, and its extension is straightforward. An example of sensing goal is the following.

$$SG = Water(0, 1, 2, 20, const) \wedge Water(3, 3, 5, 10, const)$$

Such SG means that the goal of the WSN is to discover the concurrent presence of two events, namely, having a certain amount (20 and 10, respectively) of water in two points (0, 1, 2),(3, 3, 5) of the environment.

The second step consists in characterizing changes in the environment whenever the events occur, i.e. the identification of a testbench to evaluate the WSN performance. To this purpose, based on the (user defined) sampling rate of the environment simulator, a profiling stage is triggered by firing each of the defined events, namely running the related plugins. At the end, for every (x,y,z), and for every predicate of SG, all the data patterns are obtained.

The other two problems the designer has to face concern the types of sensor to be chosen and their best positioning, in order to maximize their capacity to recognize the events, i.e. maximizing the SG. The former point impacts mainly on the feasibility of designing a WSN capable of recognizing the events encompassed by the SG. The latter is related to the dissemination of sensors in order to enhance the possibility of satisfying the composing the SG, i.e. improving the systems performance.

SWORDFISH is a very fast tool, able to provide results within seconds of computation, so as to actually enable sensitivity analysis. First of all we ensure that a solution to the SG can exists, using a proper set of sensors that is incrementally built up and significantly optimised by sharing sensors among the set of (specified in the SG) to be verified. Then, this set of candidate sensors are placed in the environment taking into account the information coming from a configurable *hardness* function. In such a way it is guaranteed that a WSN formally satisfying the SG with a quasi-optimal cost will be obtained, with runtimes in the order of a few seconds.

As far the positioning of the sensors is concerned, we defined a *hardness* function modelling the difficulty in evaluating in a given point (x,y,z). A formal definition of hardness and confidence can be found in [2]. The hardness is a function linking: (i) the data patterns obtained during the initial profiling (depending on the type of sensors); (ii) the difficulty to recognize the event predicates Pr (those composing the SG) within the time frame of a profiler sampling rate; (iii) the confidence to infer the truth of Pr based on the speed of variation over the time of the above data patterns obtained during the initial profiling. To represent how a given sensor is actually capable to capture its target events from a position, a proper metric has been defined, called confidence [2], that is comparing the hardness in one point, with its maximum value.

The optimisation strategy uses some default heuristic (alternatively, some taboo conditions,such as e.g. a maximum number of sharing may be imposed). Additional features like the cost of sensors or the requirements to achieve multiple coverage of predicates to enhance fault tolerance/reliability of the WSN response may be considered.

As mentioned above, the placement of sensors is based on the use of the Hardness grid obtained by adding the contribution of each of the predicates that the sensor has to evaluate. In such a way, the identified position will be optimal in the sense of reaching the minimum Hardness total value.

19.4 Cost Model

The sensor-level design, carried out within SWORDFISH, produces a set of (sensor, position) pairs tailored to optimise cost-effectiveness and capability to fulfil the sensing goal of the WSN. On the other hand, realistic design and deployment typically require simplifying both node hardware and network architecture , by exploiting boards hosting multiple sensors. This constraint necessarily modifies the optimal positioning of sensors, with the risk of side effects on the desired WSN behaviour.

Depending on the application, three different cases can be envisioned: new ad-hoc boards are realized for the application, use of off-the-shelf boards already existing on the market, and customization of boards, e.g. by adding daughter boards to create gateways or to add specific sensors.

Based on market availability of sensing modules and the results emerging from the application scenarios defined in ARTDECO research project, we found reasonable to adopt a general model of monetary cost for each board (node). We observed that there exist a variable cost which is related to the type and number of sensors in a linear manner and a processing cost that is logarithmic, due to the typical price trends of CPU and micro-controllers.

To consider different suppliers, we partitioned the available sensors into classes, to capture their relative cost, instead of considering the absolute values. Concerning the cost of the network, we assume a constant value depending on the protocol for wireless connections (typically built-in in commercial nodes).

Furthermore, some influence of the network topology should be considered in the case of some gateway nodes, managing hierarchies of sensors patches, were identified. In such case, there is an additional cost related to the wired connection or the use of other long-range radio communication standards and modules.

19.5 Node-Level Design

The clustering of the set of sensors identified by SWORDFISH is a multi-stage process, including the following main activities : compatibility analysis between all the possible pairs of sensors, identification of the boundaries of the clustering problem (worst and best case), and generation and evaluation of the candidate solutions.

Compatibility. Initially, the user (e.g., by accepting default settings) has to provide taboo conditions, by specifying constraints on the possible clustering of different sensors onto the same board. Based on these information, an Interference Graph $G=<N,E>$ is built, where nodes n are sensors and an edge e between two nodes represents a possible sensor interference to be avoided.

From the interference graph, the complementary compatibility graph $G'=<N,E'>$, gathering all feasible solutions, is built. (Note: any possible clustering of sensors is a clique of the compatibility graph, since all sensors hosted by the same board must be compatible with each other). All the maximal cliques of the compatibility graph G'

is computed next; since this is recognized to be a NP-hard problem, some heuristics are adopted.

Coverage. At the end of the previous step we obtain a partitioning of the compatibility graph in cliques clustering the maximum number of compatible nodes. The design space spanning between the two boundaries cases so identified contains a number of possible solutions that is exponential with the sensor cardinality. Suitable heuristics are introduced to extract a set of (not necessarily maximal) cliques, allowing the optimiser to consider solutions possibly less homogeneous but characterized by a lower board-level cost.

Comparison of Solutions. This step takes into account the candidate WSNs from the Pareto standpoint. The task of the *Pareto Efficient Solution Clustering Algorithm* (PESCA) is to find out a solution to the multi-objective clustering problem, considering two metrics: cost and functional quality, i.e. its performance.

The cost of a solution (set of boards) is evaluated through the cost model described in Section 19.4, that is depending on the number and type of sensors associated with each partition. Concerning the performance, the quality of a solution is computed by exploiting the Hardness functions of the event i covered by the sensor j belonging to the same board. The hardness of the WSN is evaluated, and its minimum corresponds to a point where the positioning of the board is optimal. This new location, which is shared by all the sensors on the same node, is the best to ensure that all the events associated with the sensors can still be captured after clustering. The solution so discovered is a Pareto efficient solution.

19.6 Network Design

In the previous sections, node level analysis was considered. It is anyway useful to move further our perspective, and to include network dimension in the analysis/optimisation phase. Problems such as network topology, gateways placing, definition of a suitable communication protocol or tuning of an existing one must be addressed considering the particular application.

Fine-grained monitoring of a high-quality vineyard may require deploying a consistent number of nodes, especially if the area to be monitored is wide. If the monitoring area is small or at least more uniform with respect to the measured parameters, networks can be smaller and simpler. It is possible to group the network topologies in two categories.

Star Topology. The star topology is the simplest network topology we consider; a central coordinator is responsible of orchestrating network activities so as to collect information from sensor nodes. All the nodes in the network have a communication link only with the coordinator. In the particular case of the precision agriculture system of Donnafugata it is convenient to use such a schema when the distance between nodes is small enough that there exists a radio link between all the nodes and a central coordinator. Given current legislation limits[1] the range should be in the order of hundred of meters to few Kilometers; considering power requirements and current technologies the range should be reduced to tens of meters to one hundred meters for most applications.

Multi-hop/Mesh or Hierarchical Topologies. There may be cases in which extension of the deployment area of the WSN (or the location topographical characteristics) does not allow the use of the star topology. It may be then necessary to send messages with data from sensor nodes stepping through a set of intermediate nodes before reaching their final destination (i.e. multi-hop). Alternatively, it is possible to have local star networks with a gateway nodes, while the information is locally exchanged using a star topology, gateways communicate between them (usually through a wider range network) forwarding or possibly aggregating information coming from the node.

The problem then requires a design space exploration phase that allows identifying the appropriate solution for the particular application class targeted. The design space can be seen as made of multiple dimensions; as an example, choosing or tuning the communication protocol involve evaluation of a multiplicity of parameters; moreover topological choices as optimal partitioning of the network or selection of coordinators, gateways or sinks should be considered.

While in the case of medium/small size WSN it is possible to identify the optimal solution, as the network grows heuristics must be used producing near optimal solutions. In this phase, the positions of the nodes, evaluated in the previous phases, are taken as an input to calculate the optimal routing through the network and the parameter configurations while keeping the nodes positions fixed.

The output of this phase is a set of Pareto-efficient solutions, specifying: the type of the nodes, the partitioning of the network (in terms of communication links), the overall throughput of the WSN, and the cost of the WSN.

19.7 Power Modelling Methodology

The previously described design phase tackles functional requirements and network cost in terms of components and placement so as to reach a first feasibility assessment. A second main problem remains to be tackled, namely, dealing with energy efficiency and power limitations. Power consumption cannot be navely inserted in the cost functions used above, as it is related to network operation; the previous choices obviously impact on it, and ultimately the power analysis phase may lead to modifying either choices for nodes or network topology or possibly even network protocol. A further optimisation phase then has to be carried out, based on suitable power-related modelling and simulations at node and network level.

---

[1] The maximum power considering the band and the modulation for 802.15.4 networks is 13dBm.

We approach power modelling for WSN in a classical top-down way, starting from abstract choices such as the adopted protocol, the generic node and network structure, etc.. An abstract, implementation-independent model thus derived then allows both validating the high-level concepts and proceeding through subsequent design space exploration steps, by identifying critical points, possibly suggesting optimisations that still comply with the initial solutions, evaluating and comparing from the energy perspective the alternative implementations, supporting feasibility decisions for a specific application.

The modelling style chosen is that of StateCharts [5], a well-known and widely adopted formalism that allows us to efficiently model hierarchy and concurrency and to detail operation of Logical Activities. At the highest abstraction level, the application is modelled through a (set of) FSMs representing the behaviour of the system; no implementation choice is evident here. Operation of this FSM corresponds to activation of lower-level, concurrent FSMs representing with finer detail solutions that have been adopted in design; consider a very simple example, referring to the Donnafugata case study and to the individual node model. The end-user and the designer initially reach an agreement on the node operation, from which the top-level FSMs representing the nodes operation are derived (see figure 19.2 A)). No information on actual low-level actions, technologies, protocols etc. is as yet present. Operation of this FSM can be seen as a path in the state diagram, traversing suitable nodes in suitable order; reaching a state in this FSM actually activates one or more lower-level FSM, that represent an (intermediate) implementation. When creating and activating the lower-level FSMs, design detail is added and choices are made (e.g., concerning the transmission protocol), but technological detail may still be absent (see figure 19.2 B)). For example, one of the the top FSMs includes a sense state, which at lower level activates a path on the FSMs coordinating the behaviour of the individual sensors and of the microprocessor, even though no technological detail on any of these components is as yet introduced (as an example the number of sensors is yet not relevant at this abstraction level). Iteratively, design detail is added by creation and activation of lower-level, concurrent FSMs. Stepping further down, execution of the chosen protocol is modelled as an activation of elementary machines representing the behaviour of all the different components of a node, and finally adding detail to such machines until the physical implementation (bottom) level is reached, where actual values (in particular, concerning energy and power) are introduced to annotate the bottom FSMs. A similar procedure holds at the network level, where the initial deployment reached through use of SWORDFISH may be taken as the starting point.

Defining a possible high-level use of the nodes implies subsequent FSM activations through the whole hierarchy and finally provides power consumption estimation for that use mode; modelling a different use mode will just result in different firing of lower-level FSMs, while exploring different choices will again lead to different firing sequence and/or to different bottom-level power annotations.

As already said, at the bottom level the technological details are taken into account so that quantitative information may be added in correspondence of a specific implementation. Once more, we start from an abstract (implementation-independent) model representing providing functional but not technological detail, afterwards

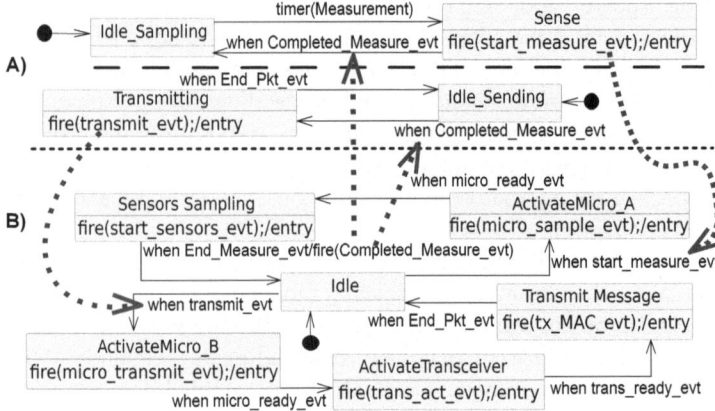

Fig. 19.2 A simplified example showing the mechanism of hierarchical modelling

annotating it with implementation-specific information. To this end, in order to analyse the nodes energy consumption we identify first the different sources of consumption, performing both an architectural and a functional breakdown. Keeping the model
implementation-independent means that, rather than detailing a specific hardware
architecture, execution of a protocol on an abstract node architecture is modelled.
Power consumption is related to both the functional sections of the architecture (i.e.,
transceiver, processing unit, sensors etc.) and to the activity performed by it. This follows the concept of *Logical Activities (LAs)* introduced in [12] that leads to estimating
the total energy consumption as the sum of the energies corresponding to the sequence
of activities performed by the node (the implicit linearisation assumption has been
experimentally proved to grant acceptably accurate results). The abstract architecture
includes the communication standard and models the related timing information as
well, either by adopting definite values (whenever strict timing constraints are given)
or by inserting *Temporal Parameters (TPs)* whenever flexibility of timing is allowed.

Starting from the abstract architecture, an *Implementation Independent Model* is
built; this model can then be characterized for a specific platform obtaining an *Implementation Specific Model*. In this final modelling, quantitative information is associated with the components of the (purely qualitative) abstract model derived before.
Correspondence to operation of actual platforms validates both Implementation Specific and Implementation Independent models.

The methodology comprises the following phases:

- **Abstract Model:** Starting from the protocol and the application, LAs and TPs
  are identified.
- **Implementation Independent Model:** the FSMs model of the system is built
  based on the Abstract Model,. *LAs* are mapped onto states (and when necessary
  to transitions) and *TPs* are inserted as timeouts on the corresponding transitions
  of the FSMs.

- **Implementation Specific Model:** By performing a series of measurements, the implementation-independent model is characterized for one or more given platforms. In particular, in this phase the power values corresponding to the *LAs* and the times corresponding to the *TPs* isolated in the previous phases are evaluated.
- **Experimental Validation:** The Implementation Specific Model is evaluated by comparing its predictions for a set of activities of the nodes to the measured consumptions of the node. (Iteration of some previous phases may be requested to achieve a more accurate model).

19.8 Hierarchical Modelling

The low-level node model consists of independent FSMs associated, respectively, with radio, microprocessor and sensors. These FSMs are then suitably activated and coordinated by the higher-level operations required by protocol execution and ulti- mately by the application[2].

19.8.1 Bottom Layer Machines

To better clarify the LA concept, let us refer to an example: if we are considering the transceiver activities, we may isolate different sources of Power Consumption, e.g. *Reception, Transmission, Idle, Low-Power*. We extract such LAs in an abstract way by analysis of the communication standard.

Identification of LAs represents the basics for creation of the bottom layer machines. We only annotate power consumption in lowest-level state machines; in such FSMs there is a one to one mapping between States and LAs. Subsequently, design- ers using the model thus created can easily see the projection of high-level actions onto activations of suitable paths in the low-level FSMs, and thus possibly identify critical points as far as power consumption is concerned. In figure 19.3 a simplified sets of FSMs suitable for the three main sections (i.e. transceiver, microprocessor and sensor) are shown.

19.8.2 Higher-Level Models

The bottom layer FSMs are driven by models at higher levels so as to compose the set of activities performed by the system. By the "event broadcast" mechanism of the StateCharts formalisms, the various layers of the system can model the behaviour of the node. Hierarchical modelling allows representing the high-level aspects of the

---

[2] The underlying philosophy is similar to that adopted in the case of instruction-level power modelling for microprocessors (e.g. [13]).

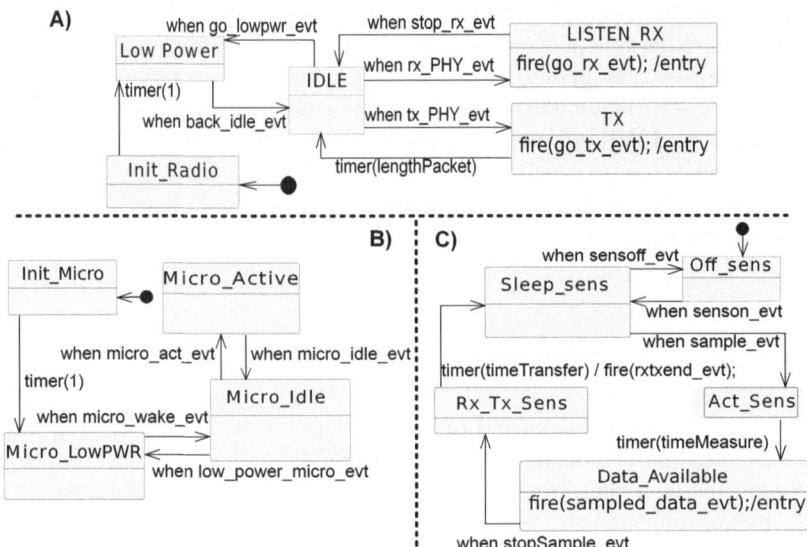

Fig. 19.3 The bottom level State Machine for the Transceiver (A), the Microprocessor Unit (B) and a Sensor (C) are shown in this figure

envisioned application with minimal or no reference to the underlying technology (i.e. technology independence).

As already hinted, at the highest abstraction level, the application (the "scenario") is modelled by an FSM that will activate the lower-level FSMs through the suitable hierarchy. At any abstraction level, FSMs operate concurrently. Considering now the hierarchy involved, on the radio side the scenario machine directly operates through a sequence of FSMs representing, as an example, scheduling of packets, networking MAC layer etc, down to the lowest-level radio model. In the same way, where the microprocessor is concerned, the sequence of FSMs corresponds in order to the scheduling and the processing of the MAC layer, to processing and storing of the sensed data, etc. The case is much simpler for the sensors section, where basically only the lowest-level FSM exists and is directly activated by the scenario FSM.

In any case, the events notified by higher level machines are not strictly bound to a particular implementation, or even to a particular choice on lower levels (e.g. choice of the MAC layer of the communication protocol etc.) but can be reused in the design space exploration phase, activating alternative lower-level FSM sets or the same FSMs in an alternative way.

In Fig. 19.3 A) a reference application running on a wireless node is modelled. The node samples a physical quantity with periodicity defined by ; the measure is added to a data packet and when a number of measures defined by are collected, the packet is sent. As the scenario starts the sensor is turned on (through senson_evt); when lower level machines notify an event communicating that the node has been correctly associated with a network (association_completed_evt) , the sampling procedure begins.

In Fig. 19.3 B) a simplified MAC layer is presented. In the top part a schematic representation of the scanning procedure is modelled; after completing scanning and association (association_completed_evt) the node synchronizes with the chosen coordinator and enters the macro-state *beaconed_network*. Operation in this macro-state is periodically reactivated through a timer so as to mimic the beacons periodic structure;

The microprocessor is restored and it wakes up the radio after a time interval - depending on the duty cycle of the network - before the beacon is received (beacon_evt notified) so as to account for listening tolerances. After the end of the beacon packet, the node enters in the macro-state corresponding to the contention access period. When the node is in the Active state it notifies it to the scheduling layer. We emphasised the Txbeaconed state so as to show how it drives the lower layer models by causing a change of state in the radio FSM model.

Further features, such as, e.g., channel arbitration, can be easily introduced even at a later time as intermediate layers in the hierarchy.

Messages are produced by the application, but nothing guarantees that the MAC layer is in a consistent state to send such packets when created. Therefore a scheduling FSM, as sketched in figure 19.3 C) is necessary. Such FSM on one side collects the requests for sending packets from the application an on the other monitors the capability of the node to transmit packet so as to notify the event triggering transmission in the appropriate moment.

Experiments carried out through a measurement campaign [9] show that results of the model approximate real consumption within an interval of less than five percent. Moreover all the main components of the nodes are included and this allows giving a real estimation of the node consumption and may help isolating some inefficiency of the protocol.

19.9 From Node Models to Network Models

In general, to allow dealing with the complexity of systems with increasing dimensions, the node models used for network simulation are more simplistic than those for node level simulators. Simplification of the model may anyway lead to missing some important aspects for evaluation and optimisation.

On the other hand node level models allow devising only some possible optimisation steps for design of the network as a whole, so that network level models are certainly needed. It is possible to emulate network level simulations by feeding various node models with synthetic data and extrapolating the performance of various solutions, but data collected in this way may not be realistic. The use of network level models gives higher credibility to the evaluation as the node model is fed with more realistic input streams.

We propose a framework for automatic generation of power simulators starting from the protocol-level model designed through our methodology. Our approach consists in generating executable C++ code, using an appropriate simulation library,

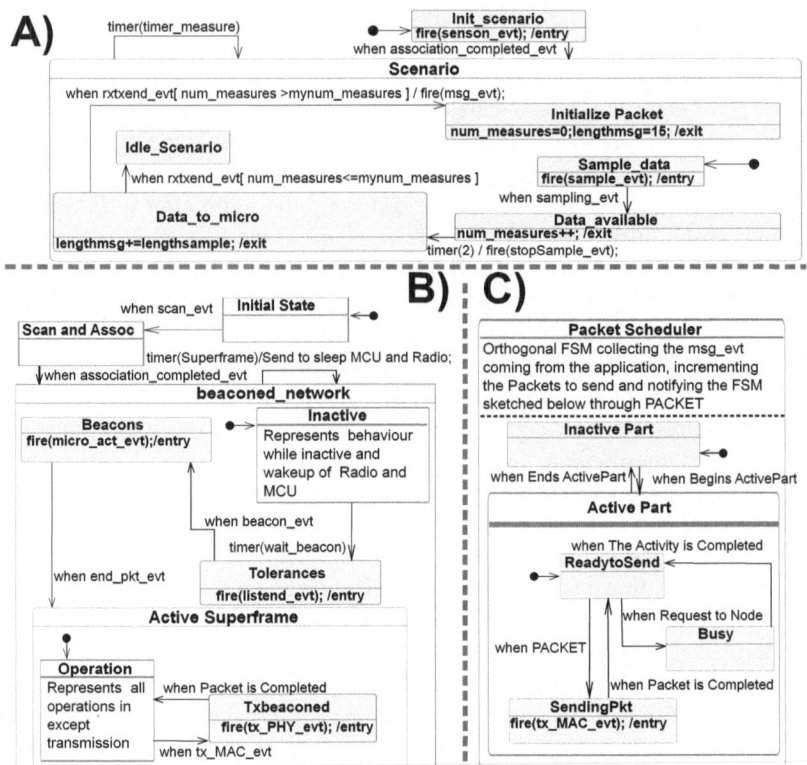

Fig. 19.4 The simplified FSMs representing the real-world application scenario A),A simplified representation of an 802.15.4 like MAC layer B) and a scheduler interconnecting Application and MAC C)

starting from the representation of the models through StateCharts formalisms [8], [10], [11]. This simulation framework has been extended to support simulation of network level models.

Multi-Instantiation of the node models is the basic step that allows building of network models. Just as in the Object-Oriented paradigm classes can be instantiated into objects, we use instantiation of models into simulation objects. Coexistence in a network simulation scenario of instances of different models may be necessary; as an example in the case of a sensor network based on IEEE 802.15.4, different models may correspond to coordinator and simple devices. The network models are built using the nodes models so that the same level of detail is kept.

As the simulation objects are executed concurrently in a common scenario, a communication pattern between various simulation objects is necessary; to this end, appropriate interfaces suitable for allowing communication between objects are defined in the various models. As the nodes are composed in a network scenario, it becomes necessary to separate a local and a global context; for example, an event that brings the transceiver of a node in receiving state is local to that node, while an event that

Table 19.2 Optimal position and type of sensors

Sensor	Position (x,y,z)	Type	Sensor	Position (x,y,z)	Type
s_0	$(0,0,0)$	Pressure	s_1	$(0,1,0)$	Temperature
s_2	$(1,1,0)$	Water	s_3	$(2,1,0)$	Water
s_4	$(2,2,0)$	Temperature	s_5	$(1,0,0)$	Pressure
s_6	$(3,2,0)$	Water			

synchronizes the network has to be processed by all the participating nodes and has global scope. Solutions dealing with this issue depend also on the network topology adopted. Our tool supports creation of network models for the two basic networking topologies presented in section 19.6 (essentially, the ones taken into account when devising the ArtDeco experiment).

19.10 Case Study

Synthetic applications and some real use cases extracted from the ARTDECO project are here used in order to provide an evaluation of the algorithms and approaches presented within this chapter.

19.10.1 Node-Level Optimisation

To highlight the importance of a quantitative tradeoff when moving toward realistic deployments, let us consider an application setup with the following sensing goal SG:
$SG = Pressure(0,0,0) \land Temperature(0,0,0) < 30 \land Water(1,1,0) \land$
$Temperature(1,1,0) > 20 \land Water(2,2,0) \land Temperature(2,2,0) > 20$

For the sake of clarity, the environment is open space, the sensor model is ideal, the time window is set to 1 second and there are no taboos specified.

In general, each board hosting sensors includes the following sections: PCB/ package; power supply and energy management; radio (RX/TX); control/processing Unit (CU); connectors/Interfaces; one or more sensors. Based on our experience in realizing PCB-level embedded systems and on market availability of sensing modules, we found reasonable adopting the model 19.1 for the cost of each board (node).

$$NodeCost = Const + K * log(N) + \sum_{j=1}^{SensorTypes} SC_j * NumS_j \qquad (19.1)$$

Where, for each board, N is the overall number of sensors, $NumS_j$ is the number of sensors of a given type j, SensorTypes is the number of possible types of sensor, and SC_j is a cost of a sensor of type j. More details can be found in [2].

The other parameters of the cost are Const=12.5, K=0.5 and all the sensors have the same $SC_j = 1$, no matter their type. The output of SWORDFISH is a set of 7 sensors (see Table 19.2). Starting from this configuration, PESCA (see Section 19.5) computes the following two cliques with the max cardinality: $\{s_0, s_1, s_2, s_3, s_4, s_5\}, \{s_4, s_6\}$. Then the covering of G' is performed using the subgraphs: $\{s_0, s_1, s_2, s_3, s_4, s_5\}$ and $\{s_6\}$. Due to space limits, the entire set of solutions generated and evaluated is not reported. In this example there is only a single solution in the Pareto frontier, which is the following.

$$Sol_1 \quad Board0=\{s_0, s_1, s_2, s_3, s_5\}, \quad position=(1,1,0),$$
$$Board1=\{s_4, s_6\}, \quad\quad\quad\quad position=(2,2,0),$$
$$Total\ cost=34.8, \quad\quad\quad\quad hardness=41$$

Should we consider a different technology with Const=2.0 instead of 12.5, the solutions populating the Pareto frontier become those depicted in Table 19.3. It worth nothing that these solutions require more boards w.r.t. the previous one, as a consequence of the reduction of the board model fixed cost. Concerning the "quality" in terms of performance, the hardness (badness) of all the solutions is better (lower) than Sol_1. This behaviour is reasonable, since the more board are used, the closer to the optimal output of SWORDFISH are the sensors.

Table 19.3 Pareto solutions for Const=2

	Sol_2	Sol_3	Sol_4
$Board_0$	$\{s_0, s_5\}(0,0,0)$	$\{s_1, s_2, s_3\}(1,1,0)$	$\{s_4, s_6\}(2,2,0)$
$Board_1$	$\{s_0, s_5\}(0,0,0)$	$\{s_1, s_2,\}(1,1,0)$	$\{s_3, s_4, s_6\}(2,2,0)$
$Board_2$	$\{s_0, s_5\}(0,0,0)$	$\{s_1, s_2,\}(1,1,0)$	$\{s_4, s_6\}(2,2,0)$
$Board_3$			$\{s_3\}(2,1,0)$
Total cost	14.8	14.8	16.5
Hardness	78	78	61

The quantitative analysis produces a significant value added for the designer when the tradeoff is not so "obvious". In such a way, the driver may be not only the cost, but also the capability of the WSN to fulfil the initial application requirements.

19.10.2 Power Estimation of Selected Configuration

The modelling and simulation methodology illustrated in section 19.7 can be profitably used for evaluating power performance of WSN deployments so as to optimise parameter tuning or to design solutions aiming at improving standard operation. In situations where the positions of sensor nodes is carefully engineered before actual deployment - as e.g. in the precision agriculture experiment of Donnafugata where appropriate node positioning is essential for reaching sensing goals - the solution we propose allows accurate evaluation of power consumptions in pre-deployment phase. This is essential for choosing the most suitable solution inside the design space of the foreseen application.

The simulator was used in a preliminary phase to evaluate, and possibly optimise power consumption for a WSN with application-specific requirements and topology such as the ones adopted in the Art-Deco case. Ten nodes send periodically measures (in packets of about 50 bytes) to a base station (the network topology is a star). From such general high level description of the application it is possible to design a model to simulate the data collection scenario (based on the IEEE 802.15.4 protocol).

Our investigation was on one side meant at identifying appropriate transmission periods (impacting on network duty cycle). Beaconed operation was chosen and a design space exploration on duty cycle of the network was performed[3]. The Beacon Order (BO) determines the temporal distance between two successive active parts, while the Superframe Order (SO) determines the length of such active parts. Therefore these two parameters together define the duty cycle of the network. Simulation were carried for 400 superframes[4].

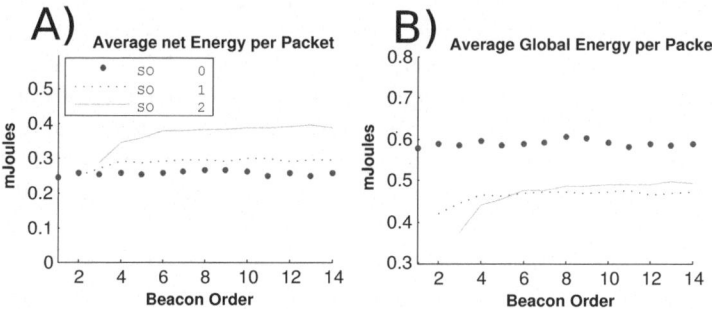

Fig. 19.5 Results of the simulation campaign, the energy was measured on an interval corresponding to 400 superframes, traffic was adapted to duty cycle of the network. The average net energy does not consider the contributions due to infrastructure communication but only the energy spent for transmitting packets (including possibly retransmissions).

In the graph A) of the figure 19.5 the average net energy that is necessary for transmitting a packet is shown, the graph suggests that power performance is better when SO is smaller. On the other hand graph B) representing the average global energy per packet (i.e. including the energy spent for infrastructure communication) shows that using SO 2 is optimal up to BO 5 then SO 1 is optimal. This can be explained considering the fact that if the superframe is longer more packets are sent in the same superframe and consequently the number of beacon messages that are received per application packet is smaller.

---

[3] It was established that the global throughput on the network was 50 kbps (5kbps per device) times the duty cycle, therefore by varying the duty cycle a different load was reached.

[4] The number of Superframes was chosen so as to guarantee reliable results. The temporal length depends on the BO parameter.

These results helped in devising that an adequate data transmission interval should be in the range of few minutes, considering the power requirements of the platform. Moreover they show that having an excessive inter-beacon distance tends to concentrate the transmissions at the beginning of the active part and therefore boosts the channel contentions and consequently power consumptions, therefore guiding towards the optimal choice of parameters.

19.11 Conclusions

In this Chapter we outlined the links existing between the application requirements and the constraints related to the environment in which the wireless sensor network will be deployed and the technology of the nodes. Particular attention has been paid to the task of modelling the *functional goal* of the sensor network and on deriving from that a set of guidelines and constraints to drive the following phases of node level and network level design and modelling for optimisation of energy related issues. It has been shown that the pure availability of a HW technology for the node and of some SW layer to support the distribution of applications it is not sufficient to provide a viable answer to basic questions regarding i) the feasibility of the project and ii) the fulfilment of severe design/application constraints. To cope with such needs, particular emphasis has been devoted to describe the main activities, actors and figures of merit involved in the different stages of a comprehensive design flow and on the crucial impact of the modelling of the node and network behaviour on the quality of the final results. The focus has been mainly kept at system level, while providing appropriate references to move more in depth into the covered topics.

References

1. Buck, J., Ha, S., Lee, E.A., Messerschmitt, D.G.: Ptolemy: a framework for simulating and prototyping heterogeneous systems, pp. 527–543 (2002)
2. Campanoni, S., Fornaciari, W.: Node-level optimization of wireless sensor networks, pp. 1– 4 (2008)
3. El-Hoiydi, A., Arm, C., Caseiro, R., Cserveny, S., alii: The ultra low-powerwisenet system. In: Proc. DATE 2006, vol. 1, pp. 1–6 (2006)
4. Fall, K., Varadhan, K.: The ns Manual (formerly ns Notes and Documentation). The VINT Project 16 (2006)
5. Harel, D.: Statecharts: A visual formulation for complex systems. Sci. Comput. Program., 231–274 (1987)
6. Heinzelman, W., Chandrakasan, A., Balakrishnan, H.: Energy-efficient communication protocol for wireless microsensor networks. In: System Sciences
7. Moser, C., Brunelli, D., Thiele, L., Benini, L.: Real-time scheduling for energy harvesting sensor nodes. Real-Time Syst. 37(3), 233–260 (2007)
8. Mura, M., Paolieri, M.: Sc2: State charts to system c: Automatic executable models generation. In: Proceedings FDL 2007, Barcelona, Spain (2007)

 9. Mura, M., Paolieri, M., Fabbri, F., Negri, L., Sami, M.: Power modeling and power analysis of IEEE 802.15.4: a concurrent state machine approach. In: Proc. CCNC (2007)
10. Mura, M., Sami, M.G.: Code generation from statecharts: Simulation of wireless sensor networks. In: Euromicro Symposium on Digital Systems Design, pp. 525–532 (2008)
11. Negri, L., Chiarini, A.: StateC: a power modeling and simulation flow for communication protocols. In: Proc. FDL, Lausanne, Switzerland (2005)
12. Negri, L., Sami, M., Macii, D., Terranegra, A.: FSM–based power modeling of wireless protocols: the case of Bluetooth. In: Proc. ISLPED, pp. 369–374 (2004)
13. Sami, M., Sciuto, D., Silvano, C., Zaccaria, V.: An instruction-level energy model for embedded vliw architectures. IEEE Transactions on CAD 21(9), 998–1010 (2002)
14. Varga, A., Hornig, R.: An overview of the omnet++ simulation environment. In: Proc. Simutools (2008)
15. Zeng, X., Bagrodia, R., Gerla, M.: GloMoSim: a library for parallel simulation of large-scale wireless networks. ACM SIGSIM Simulation Digest 28(1), 154–161 (1998)

Chapter 20
Enabling Traceability in the Wine Supply Chain

Mario G.C.A. Cimino and Francesco Marcelloni

Abstract. In the last decade, several factors have determined an increasing demand for wine supply chain transparency. Indeed, amalgamation, fraud, counterfeiting, use of hazardous treatment products and pollution are affecting the trust of consumers, who are more and more oriented to consider the so-called "credence attributes" rather than price. Thus, consumers demand detailed information on the overall process from the grape to the bottle. In this chapter, we present a system for traceability in the wine supply chain. The system is able to systematically store information about products and processes throughout the entire supply chain, from grape growers to retailers. Also, the system manages quality information, thus enabling an effective analysis of the supply chain processes.

20.1 Introduction

Winemaking has a very long tradition in Italy: Etruscans and Greek settlers produced wine in the country long before the Romans started developing their own vineyards in the 2nd century BC. Until the mid-1980s, wine production was not generally of a high standard and, indeed, much table wine was cheap and of very poor quality. In the last years, however, consumers have become to consider traditional determining factors such as price less important than other qualities, called credence attributes. This has lead the Italian wine industry to go through a series of reforms aimed at introducing strict quality controls. Thus, the standard of the production has risen to a level whereby Italian wines can now compete at international level with French wines. Consumers however request certifications for these credence attributes and for this reason traceability is gaining more and more importance in characterizing

Mario G.C.A. Cimino · Francesco Marcelloni
Dipartimento di Ingegneria dell'Informazione: Elettronica, Informatica, Telecomunicazioni,
University of Pisa, Largo Lucio Lazzarino 1, 56122 Pisa, Italy
e-mail: {m.cimino,f.marcelloni}@iet.unipi.it

G. Anastasi et al. (Eds.): Networked Enterprises, LNCS 7200, pp. 397–412, 2012.

production quality [7, 10]. Further, food traceability became a legal obligation within the European Union [22]; similar requirements for traceability systems are present in the United States and Japan [26, 18]. On the other hand, traceability is becoming an essential management tool for improving production efficiency. Indeed, traceability enables an effective process control and allows generating reliable risk assessment models, for identifying various factors that cause quality and safety problems [27]. Finally, traceability can play important roles in promotion management and dynamic pricing, with more dynamic and agile planning approaches. Traceability data can also provide instantaneous decision-making responses to variations in the supply chain. Nonetheless, enabling traceability in complex supply chain is not trivial, due to the high number of activities and actors. Further, companies generally outsource operations and leverage global sourcing.

Traceability is defined as the ability to follow a product batch and its history through the whole, or part, of a production chain from raw materials through transport, storage, processing, distribution and sales (called chain traceability) or internally in one of the steps of the chain, for example the production step (called internal traceability) [19]. Traceability of products has been introduced since the 1990s [13, 6] and is still under investigation by scientific and industrial bodies [4, 7, 10, 15]. A number of traceability systems, technologies and standards have been developed to carry out supply chain traceability and internal traceability, with different business objectives [2, 3, 9, 12, 17, 20, 25]. Nevertheless, only large enterprises, which are characterized by a tightly aligned supply chain and supported by a considerable use of information and communication technology, employ very efficient and fully automated traceability systems [10]. On the contrary, small enterprises only rarely implement traceability and, when they do, they add the traceability management to their normal operation, decreasing the efficiency and increasing the costs. Thus, today, a considerable challenge is to develop agile and automated traceability platforms for communities of small-scale enterprises [21]. On the other hand, just these enterprises are typically involved in the different activities of a wine supply chain.

In the automation of supply chain traceability, some standards and technologies gained a leading role [2]. In particular, radio-frequency identification (RFID) [24] and Electronic Product Code (EPC) global (EPCglobal) [9] are considered to be the most appealing sensing technologies and paradigms, respectively, for supply chain traceability. Further, in the vision of "The Internet of Things" [16], promoted by the Auto-ID Labs network [1], a global application of RFID allows all goods (bottles, casks, kegs, etc.) to be equipped with tiny identifying devices. Also, a globally distributed information system, made of networked databases and discovery services, allows managing an "Internet of Physical Objects" to automatically identify "any good anywhere".

The need to share data in this globally distributed information system requires the adoption of some coding standard which is agreed by all parties and allows them to communicate with each other, so as to ensure the continuity of the traceability throughout the chain. To this aim, the most promising coding system is certainly the GS1 (formerly EAN.UCC) system [12], a specification compliant with the EPCglobal Architecture Framework (EPC-AF) [9]. The EPC-AF is a collection of interrelated standards for hardware, software, and data interfaces (EPCglobal Standards), together with core services (EPCglobal Core Services).

Although standardized identification technologies and data carrier middleware are today mature, tracing items in a production chain, across different-scaled enterprises and through the full process scope, is an inherently expensive design task. Indeed, the various approaches proposed in the literature are often designed for specific good categories, and are characterized by the need of a top-down design approach for each supply chain. This approach usually produces some specific form of application middleware. However, general enterprise solutions are more difficult and more costly to develop, because they often need to be tailored to different applications. On the other hand, the wine supply chain is complex and fragmented, with distant suppliers and different demanding customers. Further, only the largest companies have significant technology requirements. Finally, there is also a myriad of other support companies that provide materials, transportation, storage and other services that are also impacted by traceability [11].

Companies vary greatly in their technical capabilities: from phone, fax and paper based transactions, through robust e-commerce, bar code, and other internal systems. Thus, their ability to identify products, and perform tracking and tracing activities is directly related to their technical skills [8].

To overcome these issues, a wine supply chain traceability system with a high level of automation is discussed in this chapter. In particular, the chapter is organized as follows. Section 20.2 presents a set of traceability requirements for the wine supply chain. Section 20.3 is devoted to the representation and the management of traceability information, whereas Section 20.4 details the behavioral model of the system in terms of transactions. The architecture is discussed in Section 20.5. Finally, Section 20.6 draws some conclusions.

20.2 Traceability Requirements in Wine Supply Chain

In 2003, GS1 co-established the Wine Traceability Working Group, joining representatives of international wine trading companies from France, Germany, South Africa, United Kingdom and United States. Further, industry peers in Argentina, Australia, Chile, New Zealand, Spain, and other wine regions, have collaborated with the Working Group on building a traceability model that has global applicability. In particular the Working Group defined a reference wine supply chain [11], which has been employed as reference in our framework to assess the fundamental requirements of wine traceability. Fig. 20.1 shows this scenario by highlighting the main actors of the supply chain. Each actor is responsible for specific activities which have to be traced so as to enable supply chain traceability. In the following, for each actor, we describe these activities and the corresponding data which have to be collected to make traceability effective.

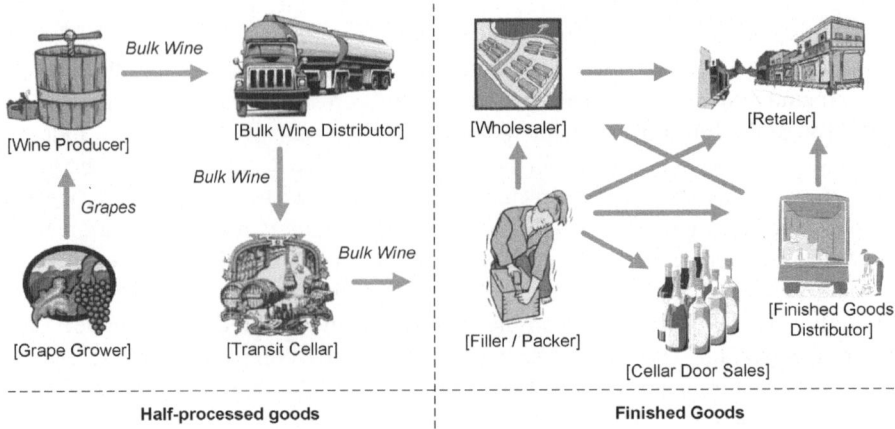

Fig. 20.1 A representative scenario of wine supply chain

Grape Growers are responsible for the production, harvest and delivery of grapes. Growers should record, for each plot of vines, details about the location, type and care of the vines, annual production record, origin and chemical content of water used for cleaning and irrigation, and the annual treatment [11]. Further, for each receipt of treatment products from suppliers, growers should record the supplier's details, a description of the product received, as well as applicable batch numbers. Each plot of vines is identified with a location number, which is allocated by the grape growers. The growers supply, with each delivery, the location number of the plot from which it comes and the date of picking, so that the receiving wine producers can link the related details to the wine made from these grapes.

Wine Producers are responsible for the production, manufacture and/or blending of wine products. Wine producers should record where, in the winery, grapes or juice were stored and must keep accurate records for the large number of procedures and operations performed to transform juice into wine. The wine producer is responsible for identifying each production run with a batch number. Further, for each receipt of additives from suppliers, the wine producers should record supplier's details, receiving date, a description of the product received, as well as applicable batch numbers.

The *Bulk Distributor* is responsible for receipt, storage, dispatch, processing, sampling and analysis of bulk wine. The wine is usually pumped into transport containers such as road tankers or barrels. When the wine arrives at the "tank farm", the bulk distributor checks the receiving documents, records all the information including the amount of received wine and takes samples for tasting and analysis. If the wine is rejected, the wine returns to the source, otherwise, two distinct processes are performed: (i) storage and dispatch of bulk wine without any blending or any other processing; (ii) storage, blending of different wines and dispatch of the new bulk blend. The bulk distributor sends batches of wine to the transit cellar. Identification is handled for the bulk distributor and the bulk wine container. To ensure forward tracking, it is essential to record references of the delivery items and to link these to the recipient.

The *Transit Cellar* is responsible for the receipt, storage, dispatch, processing, sampling and analysis of bulk wine. The transit cellar receives bulk wine from bulk distributors in different kinds of containers. Each of these containers is identified with a proper code. The transit cellar sends batches of bulk wine to the filler/packer. Each container sent is identified with a unique number, and with the associated quantity of wine (litres). In order to maintain accurate traceability throughout the chain, it is necessary that the transit cellar records the item and batch numbers, as well as the identifier of each dispatched item. To ensure forward tracking, it is necessary to record the global identifiers of the shipped items and link these to the location of the recipient.

The *filler/packer* is responsible for the receipt, storage, processing, sampling, analysis, filling, packing and dispatch of finished goods. The filler/packer receives containers of bulk wine from the transit cellar, and also "dry goods" in contact with wine (bottles, caps, corks, etc). Each of the containers of bulk wine and logistic units of dry goods are identified with a proper batch number. During this stage, the wine is poured into different kinds of containers, such as bottles, bags, kegs or barrels, and a lot number is allocated to them. A link between these components (bulk wine, finished product) should be maintained. The next step is the packaging into cartons and pallets and the dispatch of these cartons and pallets (identified with a lot number) to the finished goods distributor. The lot number must be linked to the batch(es) of bulk wine used to fill the bottles. To ensure forward tracking, it is necessary to record the global lot number of the shipped items and link these to the location number of the recipient.

The *finished goods distributor* is responsible for the receipt, storage, inventory management and dispatch of finished goods. The finished goods distributor receives pallets and cartons from the filler/packer and dispatches them to the retailer. These trade items are identified with lot numbers. To ensure forward tracking, it is necessary to record the global lot number of the shipped items and link these to the location number of the recipient.

The *retailer* receives pallets and cartons from the finished goods distributor and picks and dispatches goods to the retail stores. The container number of an incoming pallet is recorded and linked to the location number of the supplier. The retailers keep a record of the container number and the lot numbers of the components of the pallets and cartons they receive. The retailers sell consumer items (bottles, cartons) to the final consumer. These items are identified with a number allocated by the brand owner.

This brief description of the wine supply chain has highlighted that all the processes from the grape grower to the consumer can be traced by associating appropriate identifiers with the traceability entities managed by the single supply chain actors and, for each identifier, creating a record with all the information required about the entity. Each actor of the supply chain is therefore responsible for recording traceability data corresponding to specific entities. Further, each actor has to create the links between identifiers which identify correlated entities. For instance, the filler/packer has to link the lot number of the bottles to the batch number which identifies the bulk wine used to fill the bottles. This link enables forward and backward traceability. The identifiers are physically associated with the traceability entities. To this aim, we can use both RFID tags and bar codes. Typically, RFID tags are used in the first stages of the wine supply chains for speeding up the logistics operation. Currently, in the last stage, which involves bottle traceability, bar code is still preferred to RFID tag. However, in the next future, it is likely that also in this stage RFID tags will replace bar codes.

In the next section, we will introduce a simple data model which allows enabling traceability in the wine supply chain.

20.3 Traceability Information Representation and Management

The data model must be general enough to represent the variety of traceability items which are managed within a wine supply chain (for instance, grapes, vines, tanks, bottles) and also the activities which have been performed on these items at different stages of the supply chain. Thus, the data model has to provide a means to univocally identify traceability items and activities, and to record information about items and activities, and their relations. Further, a traceability system for the wine supply chain has to take additional data on quality features explicitly into account. For example, during the storage of the wine in the bulk distributor it is important to monitor temperature and humidity.

Each item is identified by a *global identifier*, which has to be unique within the supply chain. To avoid a centralized administration of the identifiers, we adopt a solution inspired to the approach used in the GS1 [12] standard. Each actor is assumed to be uniquely identified in the supply chain by an *actor identifier*. Moreover, an actor is allowed to freely associate an identifier (*traceable entity identifier*) with each traceable entity (i.e. either an activity or an item) the actor is responsible for. If an actor manages several distinct items, the item identifier may consist of the item type identifier and one progressive number. The only constraint we impose is that the identifier is unique within the amount of items managed by the actor. The global identifier is composed of the *actor identifier* and the *traceable entity identifier*.

We adopt the data model we introduced in [2]. Fig. 20.2 shows this data model. Here, classes are grouped into two distinct UML packages: *Traceability* and *Quality*. The former contains the entities that allow tracing and tracking the product path. The latter contains the components related to item quality. The *TraceableEntity* is an abstract class that models the basic characteristics of the two entity types involved in traceability: items and activities. The field *TraceableEntity.id* implements the traceable entity identifier. The association *is managed by* enforces a traceable entity to be always associated with a responsible actor. This constraint guarantees the univocal identification of the traceable entity, as described above. Further, *TraceableEntity* is also associated with *Site*, which holds its own unique identifier: i.e., each item is placed in one site. Thus, at each stage of the supply chain, the traceability system is

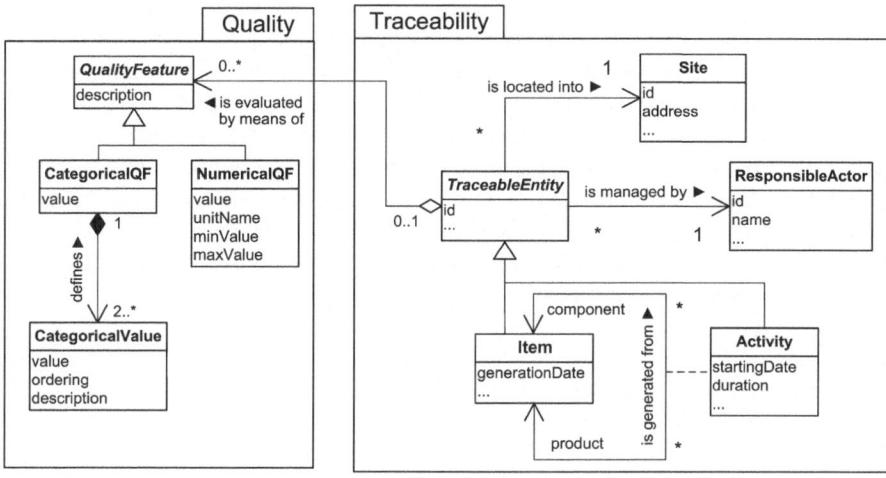

Fig. 20.2 UML class diagram of the traceability data model

able to retrieve the information about the site where the item has been processed or stored. Both *Site* and *ResponsibleActor* are characterized by a number of attributes that summarize all the information required for traceability. The association *is generated from* states that each item may be generated from zero or more items (zero in the case of an initial item). The generation is ruled by an activity.

Fig. 20.3 shows an example of the objects used to record an activity: a filler/packer purchases a red wine cask from a transit cellar, and carries it to her/his storehouse by a truck. The input and the output items of the activity are definitely the same cask. However, transit cellar and filler/packer typically identify the cask in a different way. Further, transit cellar and filler/packer are, respectively, responsible for the output and the input items. Therefore, for traceability purposes, input and output items are different. Thus, several different instances of class *Item* can correspond to a unique physical item (the same cask in the example).

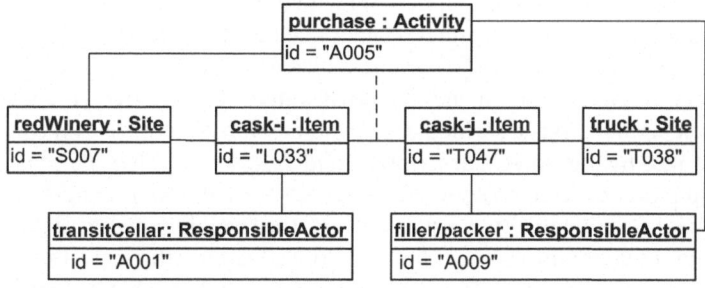

Fig. 20.3 Objects involved in recording the actual execution of a simple activity

In Fig. 20.4, a UML sequence diagram describes a possible message exchange within a purchase activity. We refer to a distributed model with no central tracking management. Here, the actor responsible for an activity is also responsible for recording and managing the relation between input and output items. The transit cellar communicates the global identifier of the input item to the filler/packer, who is in charge of binding such an identifier to the other corresponding identifier for the output item. This association allows both item tracing and item tracking. Typically, the global identifier is attached as barcode or RFID tag to the item. Thus, part of the communication consists of reading item identifiers (by means of appropriate devices) at successive supply chain actors.

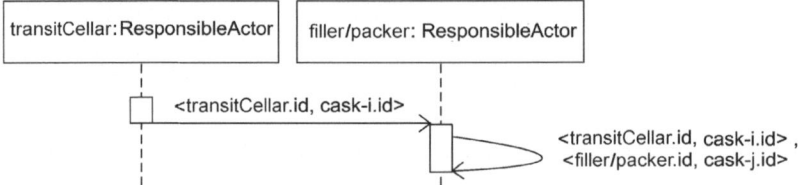

Fig. 20.4 Sequence diagram of a purchase activity, in a distributed model

In order to retrieve the history of an item, each actor of the supply chain has to communicate with its trading partners. In fact, legally, the requirement [22, 26, 18] for traceability is limited to ensure that businesses are at least able to identify the immediate supplier of the item and the immediate subsequent recipient (one step back-one step forward principle), with the exemption of retailers to final consumers. The data exchange must of course be carried out in a secure and reliable way.

Quality requirements often play a crucial role in modern business process management, and thus they deserve particular attention in the corresponding traceability systems as well [23]. The ISO 9000 standard [14] defines quality as the totality of features and characteristics of a product or service that bear on its ability to satisfy stated or implied needs. To meet quality requirements, we introduced the *Quality* package shown in Fig. 20.2. This package contains the abstract class *QualityFeature* (QF), which includes a description of the feature itself and a collection of methods to set and retrieve feature values. Values can be either categorical or numerical. *CategoricalQF* and *NumericalQF* concrete classes implement features that can assume, respectively, categorical and numerical values. *CategoricalQF* contains a set of *CategoricalValue* objects, which define the possible values. A *CategoricalValue* is characterized by the value, a description, and an ordering value. This last item can be used whenever ordered categorical values are needed. *NumericalQF* is qualified by the value, the unit name (for instance, "Kg" for "weight" quality factor), and the minimum and maximum values. This class organization allows dealing uniformly with different quality features. Fig. 20.5 shows an example of object diagram that describes the quality features "color intensity" and "rating" associated with item cask-i of wine. Color intensity can assume numerical values in the interval 1-10. Rating

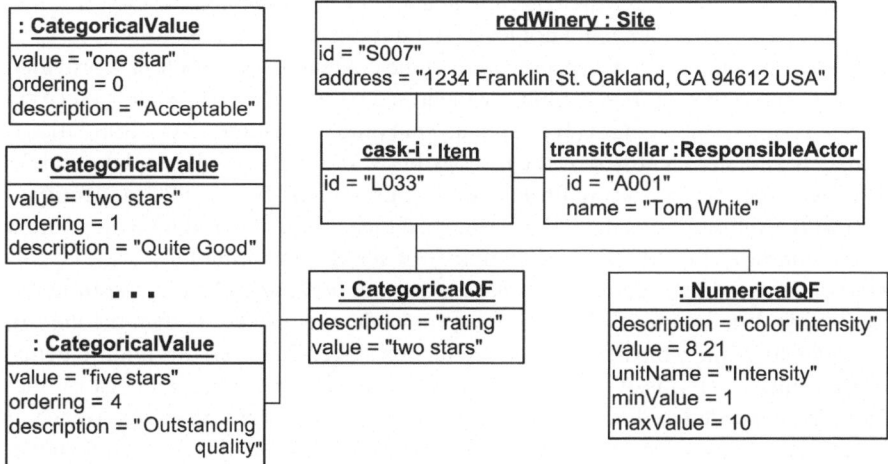

Fig. 20.5 Example of objects related to quality features

takes the wine excellence into account. Here, excellence is evaluated by employing a 1- to 5-star rating system [5].

20.4 A Transactional Model for Process Tracking

The full comprehension and monitoring of what actually happens along the supply chain requires not only a precise data model for the involved assets, but also a clear understanding of the item temporal progression towards successive stages in the supply chain. In a nutshell, a simple formal characterization of the item "history" is needed for the investigation on the actual requirements of the overall traceability system. The key observation is that the item progression is determined by *activities over it*, and thus its behavior can be described recording the activities that a generic item may undergo.

A transactional model describes the way in which the system can use transactions in message flows to accomplish certain tracking tasks and tracing results. From a tracking perspective, each activity that terminates correctly generates some item, and for each generated item a proper business transaction is recorded by the traceability information system. A business transaction is an atomic part of work that can be associated with the activity. For instance, from an activity with N output items, a set of N independent transactions can be tracked. A single transaction cannot be decomposed into lower level independent tracking pieces of information. A business transaction is a very specialized and very constrained semantics designed to achieve product state alignment when needed by third parties. As a transaction, it must succeed or fail, from both a technical and business protocol perspective. If it succeeds from both perspectives, it can be designated as a piece of the item history. If it fails from any perspective, it should not leave any trace of its existence.

In the following, an exact specification of the content of a transaction is provided. Let us suppose that an *item* is globally identified by the responsible actor ID (Axx), the site ID (Sxx), the item ID (Ixx), and the generation date-time (Dxx). Similarly, an activity is globally identified by the responsible actor ID, the site ID, and the activity ID (Txx). Indeed, considering further constraints, it could be possible to identify an item with a subset of this data. For instance, let us consider a product with a simple production process consisting of a number of serial transformations, with no fork and join of activities, such as *fermentation, aging, packing* and *transport* of home-made wine. If a unique RFID tag is used for each transformation, then the item ID is enough to identify the item at each production stage. However, this requirement is very expensive in terms of tags. If a unique tag is used for the entire item history, then date-time is needed to distinguish the item at different processing stages. Hence, in each transaction, the item ID and the date-time are supposed to be necessarily known. The pair (Ixx,Dxx) allows identifying an item in a specific stage of the supply chain, even if the RFID tag is re-used after the item has been sold. To follow the production path, when a new tag is applied to the output item, it is important to keep track of the input item ID [2].

It is worth noting that the times recorded in transactions can play a crucial role in the tracking of items. For this reason, a clock synchronization mechanism among the distributed units has been realized. More specifically, each SU is periodically synchronized with a global Internet time clock service, whereas each TU is automatically synchronized with the related SU during the daily start-up. This two-level synchronization process allows a sufficient precision. Indeed, the actual precision needed to determine an ordering between production activities is very coarse with respect to the clock technology available on digital devices.

Together with the item, some contextual information is fundamental to support a series of tracing processes, which need to be connected with the real world at a business level. For instance, when some contamination event occurs, it is important to know *who* and *where* to investigate, and also further features of the item itself. Hence, in a general traceability model transactions have to contain at least the input/output items, their site and their responsible actor.

All the possible transactions can be represented by using the following two patterns.

a) providing-acquisition. Fig. 20.6 represents a scenario of providing-acquisition of an item. At the instant $D1$, the actor $A1$ provides the actor $A2$ with the item $I0$, which was stored at the site $S0$. At that moment, $A1$ could not know the site in which $A2$ will store the item, and then, in her/his vision, that site is denoted by $S?$ (unknown site). This is usual, for instance, if the two actors belong to different companies, or if some module has not been properly configured. In this case, the transaction will have an undefined output site (transaction $TR1$ in Fig. 20.6).

Similarly, at the instant $D2$, the actor $A2$ acquires the item $I0$ and stores it in its own site $S3$. However, he cannot know where the item was previously stored. Again, in this case the acquisition transaction will have an undefined input site (transaction

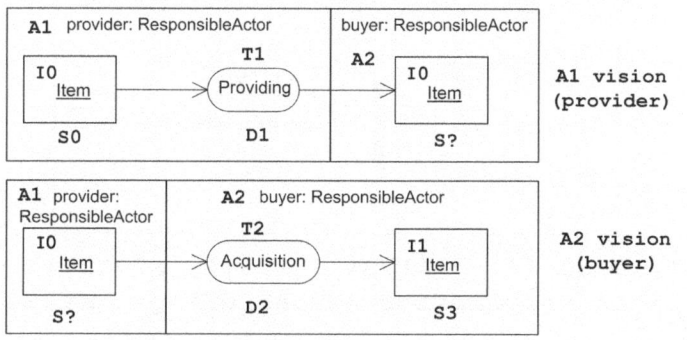

TR1: [A1,S0,I0] -> *providing* (T1,A1,D1) -> [A2,S?,I0]
TR2: [A1,S?,I0] -> *acquisition* (T2,A2,D2) -> [A2,S3,I1]

Fig. 20.6 A scenario of the *providing-acquisition* transactions

TR2 in Fig. 20.6). Note that, in Fig. 20.6, the item is identified by two different RFID tags before and after the acquisition, i.e., *I0* and *I1*, respectively. On the other hand, if the RFID tag is kept, *I0* will be equal to *I1*. Note how, starting from the input item of the transaction *TR2* (i.e., [*A1*, *S?*, *I0*]), and replacing its actor (i.e., *A1*) with the actor in the output item (i.e., *A2*), it is possible to derive the output item of the transaction *TR1* (i.e., [*A2*, *S?*, *I0*]). This means to identify the transaction *TR1* with some data available in the transaction *TR2*, i.e. a step backward in the tracing back. If more than a transaction with the same output item is available, the transaction *TR1* closest in time to *TR2* is considered (i.e., with *D1* such that *D1* is closest to *D2*). Vice versa is also valid for a step forward (tracing forward).

b) transformation. In the case of processing activities that are internal to a company, a group of *N* items can be transformed into a group of *M* items, via splitting, merging, moving, processing, etc. This activity can be represented as a series of *M* transformations of *N* items into an item, having the same items as input. Fig. 20.7 describes a scenario with three input items. Here, at the instant *D3*, the actor *A3* performs the activity *T3*, taking as inputs the three items, *I0*, *I1* and *I2*, and giving as output the item *I4*. The input items were stored at the sites *S0*, *S1* and *S2*, respectively, and owned by the actors *A0*, *A1* and *A2*, respectively. The output item is stored at the site *S4*, and owned by the actor *A4*. Note that, in this transaction, tracing back and forward are simpler to perform with respect to the providing-acquisition transaction, because sites are known.

As an example, let us consider a simplified wine supply chain. The starting point of the supply chain is the harvesting of wine grapes (from nature, in our simplified setting). In the first place, this can be accomplished using mechanical harvesting or traditional hand picking one. Subsequently, during fermentation, yeast interacts with sugars in the juice to create ethyl alcohol. Fermentation may be done in stainless steel tanks, in an open wooden vat, in a wine barrel and even in the wine bottle itself. Hence, during the aging of wine, complex chemical reactions involving sugars, acids

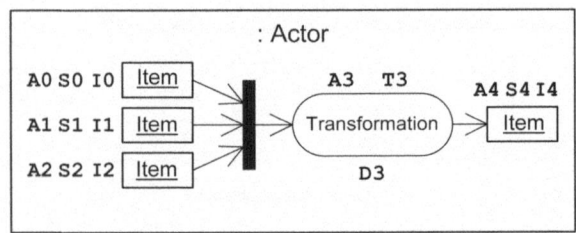

```
TR:  {[A0,S0,I0],  [A1,S1,I1],  [A2,S2,I2]}
     -> transformation (T3,A3,D3) -> [A4,S4,I4]
```

Fig. 20.7 A scenario of the *transformation* transaction

and tannins can alter the aroma, color, mouth feel and taste of the wine, in a way that may be more pleasing to the taster.

This simple supply chain can be modeled as depicted in the UML communication diagram shown in Fig. 20.8. For simplicity, we have supposed that harvesting, fermentation, aging, packing and transport are performed by the same supply chain actor. Actually, this is typically the case especially for high quality productions. We have denoted this actor as *wine maker*. In the figure, *nature, wine maker, shop* and *customer* are the different *ResponsibleActors*, and they interact according to given activities, possibly producing new items. The activity ordering is specified by the numbers associated with the shown procedures.

At the beginning, the *wine maker* performs an acquisition from the *nature* (harvesting) and creates a new item. Then the *wine maker* performs two transformations (fermentation and aging): each transformation produces a new item. Finally, the *wine maker* provides (transport) the shop with the wine and generates a new item. The *shop* performs an acquisition (buying), which produces a new item. When the *shop* provides (sale) the wine to the customer, it creates a new item. The *customer* comes after the last responsible actor of the supply chain: he/she does not create any item because his/her acquisition has not to be traced.

As highlighted in the last example, the tracking process along the chain production possibly generates and manages a huge amount of data records. In order to allow tracing procedures to remotely retrieve such data, a pervasive architecture is needed. This aspect is detailed in next section.

20.5 An Architectural View of the System

Let us consider more specifically the architectural view of the traceability system. Fig. 20.9 shows a deployment diagram containing different kinds of units. The proposed traceability system comprises different *Tracking Units (TUs)* equipped with RFID or code bar readers. A *TU* gathers data and transmits them to a *Storing Unit (SU)*. *SUs* are in charge of keeping local production data, supplied by *TUs*, according

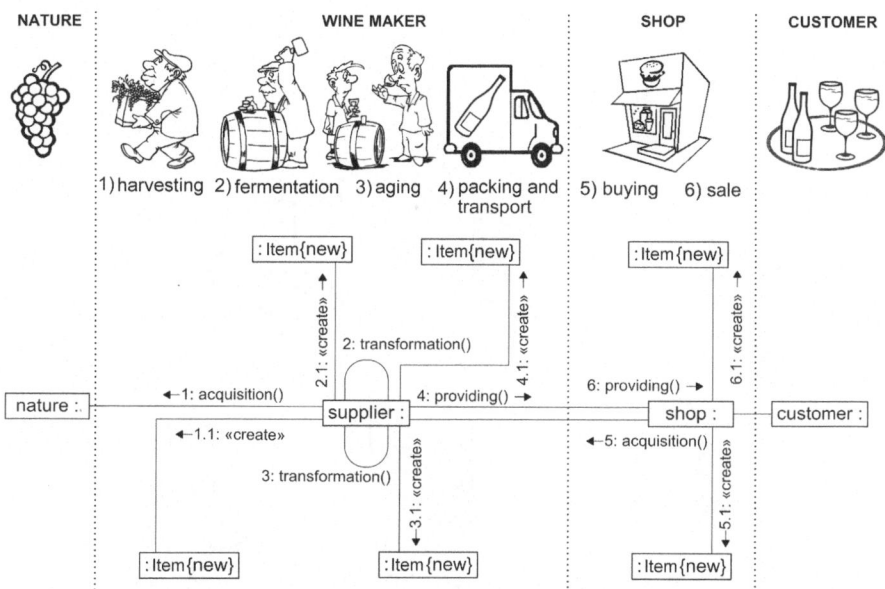

Fig. 20.8 Communication diagram for a simplified traceability system in wine production

to some criteria. *Analysis Units (AUs)* steer business process analyses and harvest data supplied by *SUs* in terms of pieces of a global tracing problem. *TUs* can be hosted by a mobile device (e.g., PDA or smart phone equipped with an RFID reader), or fixed device (e.g., bank reader, door gate reader). Further, *TUs* allow data harvesting supported by user agents, because *TUs* are self-configured on the basis of the local context. More specifically, there are some *Context Units (CUs)*, which are able to provide a local business process context. Indeed, *CUs* and *AUs* are strictly related to each other. For a given business analysis, an amount of data need to be collected, and this process can be guided configuring the *TUs* via *CUs*. Furthermore, *CUs* contain also the definitions of the quality features used by *AUs*. Thus, for instance, when quality attributes such as color intensity and rating have to be inserted, the *TU* is automatically configured by the corresponding *CU* so as to show appropriate interface widgets. Finally, there are some lookup services for *SU*, accomplished by *Registry Units (RUs)*. The traceability system is based on a distributed architecture in which data is managed according to a "pull" model [2]. In the pull model, at the *tracking stage*, data is stored at the site where it was generated. At the *tracing stage*, an *AU* actively requests a particular analysis from the system. Hence, *SUs* wait for a pull request to reconstruct an item history. When a pull request arrives, only related tracing data is collected and returned to the *AU*. According to the service-oriented paradigm, the communication between *SUs* and *AUs* relies on an asynchronous message-centric protocol, which provides a robust interaction mechanism among peers, based on the SOAP/HTTP stack. On the other hand, the communication between *TU* and the other units can be proficiently achieved using a more efficient and lightweight XML-RPC/HTTP based interaction.

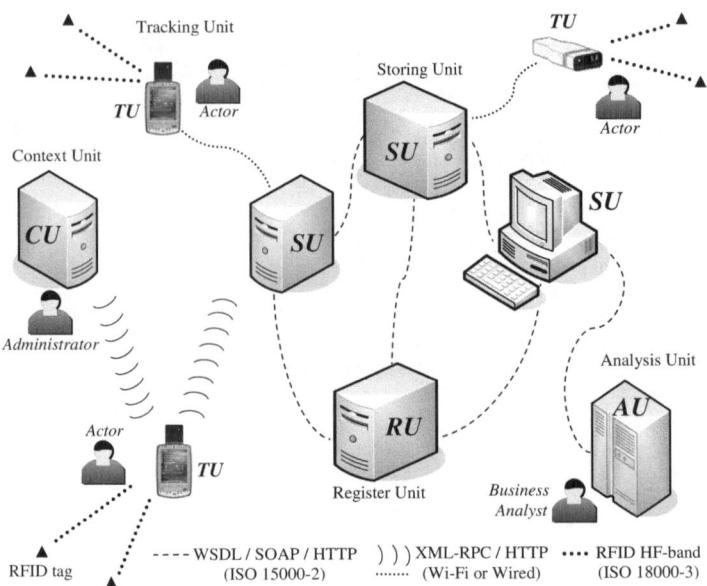

Fig. 20.9 An overall Deployment Diagram of the Traceability System

20.6 Conclusions

In this chapter, we have proposed a solution for wine chain traceability that relies on a general model and a pervasive and mobile architecture, employing RFID technologies. After a business and technological overview, encompassing wine supply chain requirements, key properties of a data representation model have been pointed out. Hence, a transactional view of process tracking has been provided, together with the discussion of the application of the system to a simplified example. Finally, the detailed architecture has been discussed.

The system has been realized considering a real wine supply chain in Tuscany, made of more than a hundred small (family) grape growers, four medium-large wine producers, three fillers/packers and a large wholesaler (a consortium). In terms of processes, such supply chain comprises 20 different types of production activities, 37 types of quality features and 14 types of sites. The participating enterprises are characterized by different levels in technological competence, economic resources, and human skills. In this setting, the system has been oriented to support the following goals: (i) to reduce the time and effort needed to execute every-day transactions; (ii) to significantly lower the rate of errors that are currently caused by replicated data entries and manual interventions; and (iii) to reduce the software maintenance and usability cost.

References

1. Auto-ID Labs, http://www.autoidlabs.org (accessed 2010)
2. Bechini, A., Cimino, M.G.C.A., Marcelloni, F., Tomasi, A.: Patterns and technologies for enabling supply chain traceability through collaborative e-business. Information and Software Technology 50(4), 342–359 (2008)
3. Bertolini, M., Bevilacqua, M., Massini, R.: FMECA approach to product traceability in the food industry. Food Control 17(2), 137–145 (2006)
4. Bevilacqua, M., Ciarapica, F.E., Giacchetta, G.: Business process reengineering of a supply chain and a traceability system: A case study. Journal of Food Eng. 93(1), 13–22 (2009)
5. Broadbent, M.: Michael Broadbent's Vintage Wine. Websters Int. Publishers, London (2002)
6. Cheng, M.L., Simmons, J.E.L.: Traceability in manufacturing systems. International Journal of Operations and Production Management 14(10), 4–16 (1994)
7. Donnelly, K.A.-M., Karlsen, K.M., Olsen, P.: The importance of transformations for traceability - A case study of lamb and lamb products. Meat Science 83, 68–73 (2009)
8. European Federation of Wine and Spirit Importers and Distributors (EFWSID), Voluntary Code of Practice for Traceability in the Wine Sector (2001)
9. EPC Global, http://www.epcglobalinc.org (accessed 2010)
10. Gandino, F., Montrucchio, B., Rebaudengo, M., Sanchez, E.R.: On Improving Automation by Integrating RFID in the Traceability Management of the Agri-Food Sector. IEEE Transactions on Industrial Electronics 56, 2357–2365 (2009)
11. GS1 Working Group, Wine Supply Chain Traceability, GS1 Application Guideline, Brussels, Belgium (September 2008)
12. GS1 Traceability, http://www.gs1.org/traceability (accessed 2010)
13. ISO, ISO 8402:1994, Quality management and quality assurance, http://www.iso.org
14. ISO, ISO 9000 family of standards (2006), http://www.iso.org/iso/en/iso9000-14000/index.html
15. ISO, ISO 9001:2000 and ISO 9001:2008, Quality management systems - Requirements, http://www.iso.org
16. International Telecommunication Union, The Internet of Things, ITU Internet Reports, Geneva, Switzerland (2005)
17. Jansen-Vullers, M.H., van Dorp, C.A., Beulens, A.J.M.: Managing traceability information in manufacture. Int. Journal of Information Management 23(5), 395–413 (2003)
18. Ministry of Agriculture, Forestry and Fisheries of Japan, Guidelines for Introduction of Food Traceability Systems (March 2003)
19. Moe, T.: Perspectives on traceability in food manufacture. Food Science and Technology 9, 211–214 (1998)
20. Mouseavi, A., Sarhadi, A., Lenk, A., Fawcett, S.: Tracking and traceability in the meat processing industry: a solution. British Food Journal 104(1), 7–19 (2002)
21. Opara, L.U.: Traceability in agriculture and food supply chain: A review of basic concepts, technological implications, and future prospects. Journal of Food, Agriculture and Environment 1(1), 101–106 (2003)
22. Regulation (EC) n. 178 of the European Parliament and of the Council of January 28, 2002, ch. V (2002)
23. Food Standards Agency of Actored Kingdom, Traceability in the Food Chain - A preliminary study (2002), http://www.foodstandards.gov.uk/news/newsarchive/traceability

24. Roussos, G.: Enabling RFID in retail. IEEE Computer 39(3), 25–30 (2006)
25. Sasazaki, S., Itoh, K., Arimitsu, S., Imada, T., Takasuga, A., Nagaishi, H., Takano, S., Mannen, H., Tsuji, S.: Development of breed identification markers derived from AFLP in beef cattle. Meat Science 67, 275–280 (2004)
26. U.S. Food and Drug Administration, regulation 21CFR820, Title 21: Food and drugs, subchapter H: Medical devices, part 820 Quality system regulation, (revised April 1, 2004), http://www.accessdata.fda.gov/scripts/cdrh/cfdocs/cfcfr/CFRSearch.cfm?CFRPart=820
27. Wang, X., Li, D.: Value Added on Food Traceability: a Supply Chain Management Approach. In: IEEE International Conference on Service Operations and Logistics, and Informatics (SOLI 2006), Shanghai, China, June 21-23, pp. 493–498 (2006)

Part V
Case Study

Chapter 21
Putting It All Together: Using the ArtDeco Approach in the Wine Business Domain

Eugenio Zimeo, Valentina Mazza, Giorgio Orsi, Elisa Quintarelli,
Antonio Romano, Paola Spoletini, Giancarlo Tretola,
Alessandro Amirante, Alessio Botta, Luca Cavallaro, Domenico Consoli,
Ester Giallonardo, Fabrizio Maria Maggi, and Gabriele Tiotto

Abstract. This chapter summarises the results achieved by the ArtDeco project and presents the overall approach for developing the models that the large-scale middleware infrastructure designed during the project uses to drive the behaviours of the information systems of a sample networked enterprise operating in the domain of wine production. This domain is sufficiently wide to cover many business aspects that highlight the ability of the proposed infrastructure to adapt its behavior to the evolving execution context by reacting to unexpected events with a very limited human intervention.

The same approach has been experimented also in different domains, including fashion industry and value added services for TLC providers.

Eugenio Zimeo · Giancarlo Tretola
University of Sannio
e-mail: {zimeo,tretola}@unisannio.it

Valentina Mazza · Elisa Quintarelli
Politecnico di Milano
e-mail: {vmazza,quintarelli}@elet.polimi.it

Giorgio Orsi
University of Oxford
Politecnico di Milano
e-mail: giorgio.orsi@cs.ox.ac.uk

Antonio Romano
Scuola Superiore S. Anna di Pisa
e-mail: a.romano@sssup.it

Paola Spoletini
University of Insubria
e-mail: paola.spoletini@insubria.it

Alessandro Amirante · Alessio Botta · Luca Cavallaro · Domenico Consoli ·
Ester Giallonardo · Fabrizio Maria Maggi · Gabriele Tiotto
Students of the GII Ph.D. School 2008, L'Aquila

G. Anastasi et al. (Eds.): Networked Enterprises, LNCS 7200, pp. 415–452, 2012.

21.1 Introduction

With the advent of globalization, companies typically operate capturing the trends of the market and delivering products that fit the needs of customers. Moreover, value chains should be faster to react to changes by using information systems that collect and process a large amount of data from the business environment in order to satisfy the needs that emerge from the customers and to respond to anomalies that occur in the environment.

In the wine industry, the value chain typically consists of all the enterprises and consumers that participate to the overall business process, starting from wine design and ending with wine sales. The enterprises interact with the aim of improving their profits and reducing costs by producing the right amount of high-quality wine and by exploiting at the best the available resources in the virtual organisation.

Each enterprise runs *internal processes*, which aim at improving the benefits for the specific organisation, and participates to *external processes*, which are generated by the interaction of each enterprise with the other ones belonging to the network.

While wineries' internal processes are typically orchestrated, as in every other manufacturing organisation, due to their well-known structure, external processes may be orchestrated (every supplier works for the wine producer and interacts with it - see Fig. 21.1.a) or choreographed (suppliers have their own independent market and interact among them by forming a peer network - see Fig. 21.1.b), depending on the business models adopted.

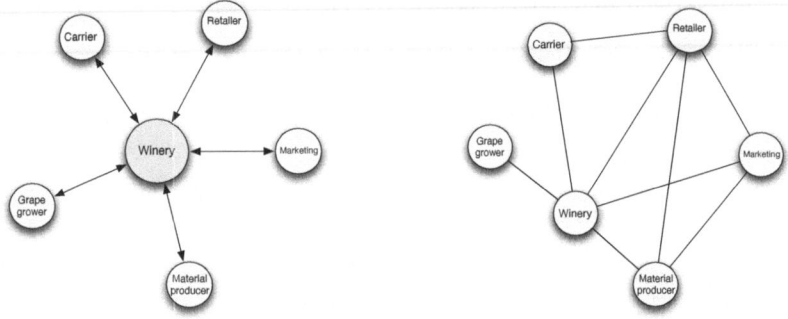

Fig. 21.1 (a) Centralised networked enterprise; (b) Decentralised networked enterprise

While the former model is mostly adopted by specialised enterprises, the latter is better for not specialised ones since it is much more flexible and tolerant when the market is unstable (the enterprises can organise themselves in networks to create new business opportunities). However, in both cases contextual events can alter a predefined business process so that it is able to evolve towards the successful termination in the new running conditions.

The structure of orchestrated processes can be changed as result of a centraslised decision, triggered by internal or external events. On the other hand, the structure of decentralised processes can be altered as a result of a decentralised decision that leads to a new equilibrium point of the networked enterprise to align the business to the changing market conditions (since in this case, the external goal could be not known a priori).

Handling these events and changing the behavior of enterprises belonging to a network are not simple tasks (for related efficiency and cost dependency) without an adequate IT support that overcomes the current technologies adopted by supply chains. The technologies and the methods developed in the ArtDeco project could be a promising way to provide the desired IT support to help enterprises in improving their businesses and to survive in the global market.

The overall ArtDeco system is composed of three main layers (see figure 21.2): *business process*, *application*, and *logical* layers. The high-level layer is in charge of handling business processes, crossing different enterprises, through their functional, non functional requirements, and active and reactive behaviours. The second layer represents the information systems distributed among the value chain partners that are able to execute workflows by providing data from different sources: *data bases*, *data warehouses* and *wireless sensor networks*. The lower level layer is responsible for abstracting the physical resources with the aim of enabling concrete actions and interactions with them.

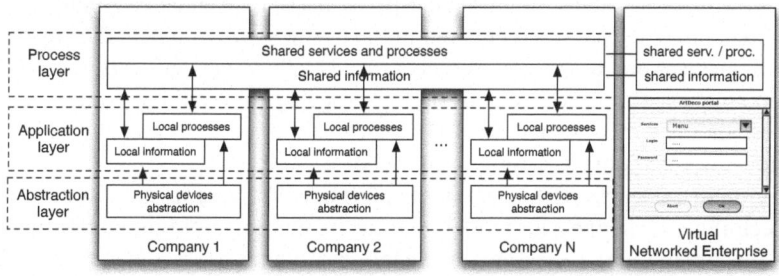

Fig. 21.2 Networked Enterprise: the vertical view shows different abstraction processes whereas the horizontal view highlights the owners of the different information systems. The enterprises involved in the network are federated through the registration in a business registry

The basic paradigm to implement the *Process Layer* is Service Oriented Computing (SOC). With this paradigm, the value chain may become more automated basing the collaboration among participants on interactions among IT systems through the invocation of service operations, coordinating and controlling complex processes in the form of distributed applications by means of workflow technology. However, basic SOC is not sufficient to ensure the desired degree of autonomy since human beings are responsible to perform most of optimizing activities in more and more complex workflows.

Context awareness significantly enhances workflow technology by allowing business processes to adapt themselves to their environments exploiting the opportunities and avoiding the problems that might emerge. However, this implies: (1) an increasing complexity of the analysis and the conception phases of processes' lifecycle; (2) handling a large amount of data due to the need for analyzing processes' data during the execution; (3) run-time actions to dynamically improve and adapt the business processes.

A further step to handle workflow complexity is performed by *autonomic computing* (see chapters 6 and 7). It has been already applied to automatically manage software systems by assigning to them the responsibility to self-manage their resources and services, in order to stay operative and efficient when the external conditions change.

Self-management is typically based on the MAPE [7] cycle: *Monitor* the resources, *Analyze* the collected data, *Plan* the intervention, *Execute* the proper actions. Monitored events can be originated internally to the organisations by physical or virtual (human driven) sensors, or externally as the result of the actions of other enterprises or by other sources of the execution context.

This chapter discusses how do ArtDeco tools and methods can be used for developing the models that drive the behaviours of the information systems of a sample networked enterprise in the domain of wine production. The approach requires enterprises re-engineering their business and organisational models to work in a networked environment (see Part 1 of this book).

The information systems of these enterprises need to be enhanced with new technology to handle the changes that occur in the business environment during the execution of business processes, opportunely sensed by these new infrastructures. Moreover, domain experts need novel design approaches and validation techniques to model business processes that are able to survive to variations and anomalies arising in the business context, as proposed in Part 2.

Working in a open world [4] and with continuously changing conditions requires novel techniques to define data-models that can be successfully exploited in heterogenous environments and in a variety of execution contexts. Moreover, new techniques to extract knowledge from the business environment are needed, as described in Part 3. This knowledge regards whatever kind of information that can be collected by the middleware infrastructure and that can be usefully exploited to suggest a proper control onto the enterprises' choices.

A particular kind of information is the one coming from the physical environment where the business processes take place. This information can be observed by one or more networks of wireless sensors that are disseminated in the physical space and whose data are propagated to the information systems for their analysis finalised to take proper decisions.

Driving the actions of the higher-level software infrastructure requires wireless networks to be programmed with expressive and powerful constructs that are able to instruct networks to deliver only useful information by reducing at minimum the possibility of faults due to excessive power consumption or other kinds of unexpected events, as discussed in Part4.

The remaining part of the chapter is organized as follows. Section 21.2 describes and models the scenario used in this chapter to show the role of each technology or methodology developed in the ArtDeco project. Section 21.3 presents an integrated view of the middleware architecture and how applications for it can be modelled, programmed, executed and verified. Section 21.4 reports on lessons learnt. Finally, Section 21.5 concludes the chapter.

21.2 Application Scenario and Modelling

The scenario presented in chapter 2 is detailed in this section with the aim of identifying some demonstration cases of the results produced in the ArtDeco project. The focus of the scenario is around a hypothetical winery[1] that operates as main contractor in a networked enterprise.

The winery uses IT technologies in order to increase *wine quality* and to reduce *production and distribution costs*. It interacts with other companies that provide materials, transportation, storage and distribution.

The whole value chain is able to adapt its business processes to the evolution of the environment with the aim of satisfying the requirements defined during the *wine design* phase. In particular, wine quality depends on all the phases of the wine production: *grapes cultivation*, *wine maturation* and *distribution*. However, the major contribution to quality depends on the basic ingredient of wine: the *grapes*.

The winery handles several vineyards that are monitored during the cultivation to provide the agronomist, the oenologist and the information systems with sufficient data for avoiding irrecoverable damages to the grapes, ensuring their high quality, and controlling the amount of grapes useful for producing the desired wine. However, if critical damages occur, the winery is able to replace the harvesting with an order from other wineries of specific kinds of grapes, with a competitive price, that ensure producing the amount of wine that the winery is able to sell (data that can be extracted from the statistics of the winery or from the market trend).

The winery is also interested in monitoring wine during storage in cellars since this information helps the oenologist and the system to monitor wine aging. Moreover, wine temperature during transportation should be monitored to reduce losses of quality and consequent returns.

In addition to information needed for ensuring wine quality, the winery is also interested in reducing the final cost of a wine bottle in order to increase sales for medium-quality wines or to improve the saving for high-quality wines. To this end, many inefficiencies should be eliminated by revising the organisational models of the winery and its cooperation with other enterprises that are involved in the business scenario. For example, transportation of wine could be optimised by coordinating the planning of several wineries that are interested in delivering the final product to the same distribution or selling point.

---

[1] The information about wine life-cycle reported in this chapter is inspired by discussions that we had with two sicilian wine producers: Donnafugata and Planeta. The scenario depicted is a collection of fragments (not complete) of real business processes that highlight the main innovations obtained as a result of the ArtDeco project.

To achieve the desired level of flexibility and control, the winery adopts both traditional (legacy) and novel IT technologies to handle its internal processes and the interaction with the external environment. This way, the application logic of existing IT systems is integrated by using workflow management systems and accessed by human operators through a portal, which enables the sharing of both information and services inside the winery and outside with other enterprises that belong to the same business network.

21.2.1 BPMN Modelling

All the IT systems of the winery are handled in a uniform way through a set of processes (each one modelling the wine life-cycle for each year). In the following, we mainly refer to one of this process (named *main process*) that characterises the life cycle of the current year (nevertheless, interactions with other processes related to other years could be useful in some cases). The human operators involved are *agronomist, oenologist, farmer, worker, warehouseman, quality manager*, etc..

The main process can be decomposed in some sub-processes (see Fig. 21.3) that are related to the main phases of wine life-cycle: *cultivation, production* and *distribution*. This process is executed and controlled by a workflow management system that is able to interact with the environment though several peripheral systems: physical and logical sensors, and physical and logical actuators.

Before starting such a process, information from the marketing is needed to estimate the kind and the related amount of wine to produce for different geographic areas, which are the commercial target of the winery. This marketing information represents the input to derive the kind and the amount of grapes to cultivate in the first part of the main process.

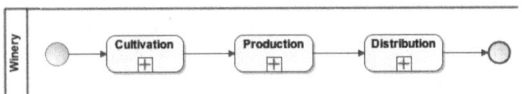

Fig. 21.3 Main process of the winery

The *cultivation* phase is handled by *grape growers* that can be internal to the winery or external organisations. Anyway, this phase is organised according to a process composed of sub-processes that reflect the phases of the grape life-cycle: *vegetative rest, sprouting, flowering, accretion*, and *harvesting*.

Each one of this sub-process is split into as many sub-processes as there are vineyards to manage and are assigned to operators. The activities of these operators are organised on daily basis and can be changed as consequence of unexpected events. For example, during *vegetative rest*, processes to be performed daily may be those shown in Fig. 21.4.

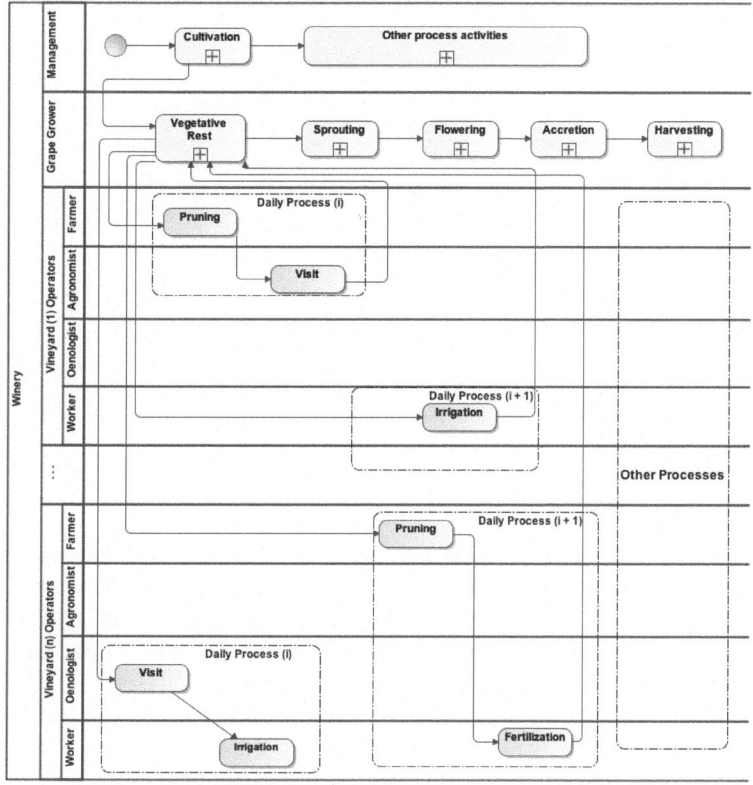

Fig. 21.4 Cultivation sub-process

The events that can occur during the different phases of the grapes life-cycle and some possible recoverable actions are reported in Table 2.1. The events are tied to observable phenomena that are monitored though specific sensors.

The values monitored are assigned to high-level parameters that are used to identify some critical conditions, such as: (1) *hailstorm has destroyed the grapes of the vineyard*, (2) *rain does not allow for harvesting*, (3) *presence of powdery mildew*. For every phenomenon, a possible corrective action can be defined by using reactive rules. For example, in case (1), the system could solve the problem by buying grapes from other producers. In case (2), the system delays or suspends the harvest, and in (3) it plans the intervention of the agronomist.

During the *production* phase, the measured parameters can be used to assess changes in the mix of grapes used in previous years. A wide array of sensors monitors the environmental conditions of cellars. Also the fermentation phase is controlled with periodical chemical analysis that allows for assessing how the composition of the wine is evolving. When the wine is in bottles, monitoring humidity and temperature is important to ensure a good preservation: wine, in fact, must not suffer of temperature fluctuations higher than 5 degrees, and humidity fluctuations higher than 5%.

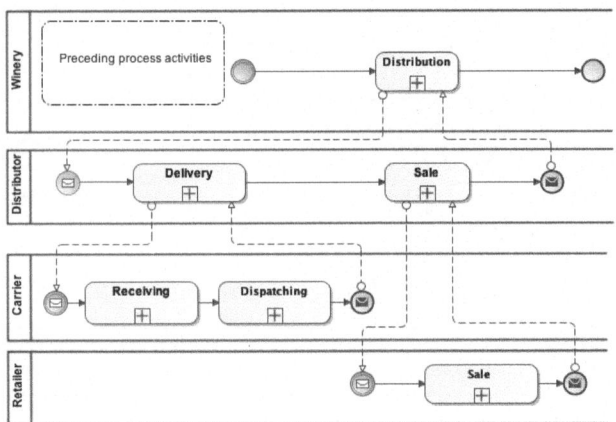

Fig. 21.5 Distribution sub-process

As a consequence, during the *distribution phase*, these two attributes may be used to substantially influence the choice of the carrier performed by the winery. These parameters, in fact, are measured and exploited for evaluating the quality of the previous subcontractors' performances. Temperature and humidity of the package during transportation are recorded in data loggers for subsequent analyses. Fig. 21.5 shows the distribution sub-process. It is aimed to discover and to select a carrier for delivering the wine bottles, on the basis of the quality of the delivery service. During the delivering activity, temperature and humidity are measured and stored. Process performance is assessed and used to subsequently update the carrier reputation.

21.3 Middleware Architecture and Programming

Due to the complexity of the interactions and the need for handling unexpected events, existing technology for managing business processes are no longer sufficient for networked enterprises working in the global market. Therefore, we consider having a new technology substrate in each enterprise that contributes to create a virtual information system able to drive the actors of the whole network.

The complex mix of information systems adopted at application layer is hard to manage without a support from the system itself. In this direction, autonomic computing, data integration and, in particular, the techniques and the technologies discussed in the previous chapters represent promising solutions to handle such kind of complexity.

It is worth noting that the results proposed in the previous chapters of this book contributed to address some scientific issues that represent the enablers of a *new-generation middleware*, which will be described and detailed in the next sub sections. However, several additional aspects (*security, privacy, transactions*

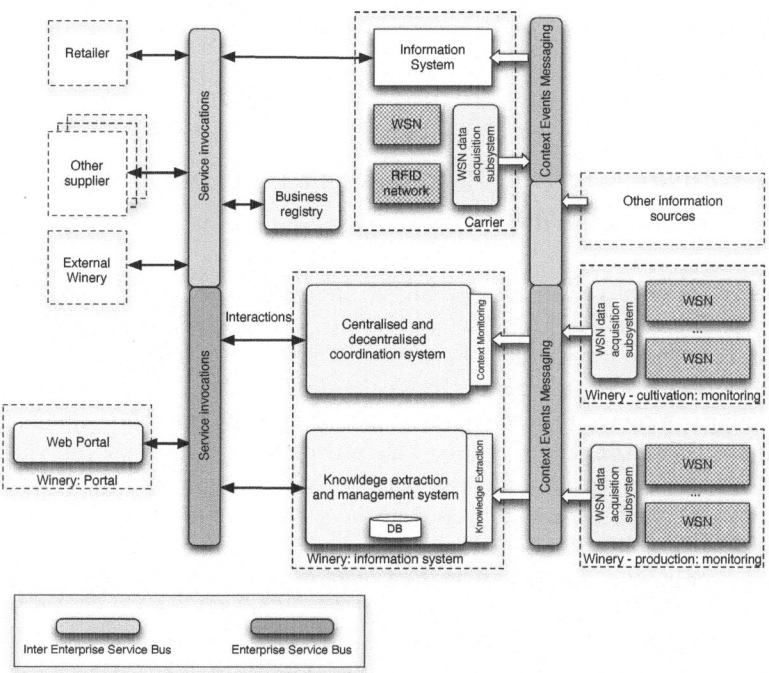

Fig. 21.6 Architecture of the ArtDeco System deployed in the networked enterprise

management, *performance*, and *scalability*), which make it possible to adopt such a middleware in real deployments, have not been addressed during the ArtDeco project but they will be addressed in future work.

The main components of the current design of the middleware infrastructure are (see figure 21.6): (1) the autonomic sub-system for centralised and decentralised co-ordination of actions; (2) the knowledge extraction and management sub-system; (3) the wireless sensors networks and the related middleware for data acquisition.

These components are connected by exploiting some buses: an *enterprise bus* is used for connecting components inside enterprises and an *inter-enterprise bus* is used for connecting components belonging to different organisations. These buses are also structured in order to enable two different kinds of interactions: *event-driven*, to handle events from the business environment, and *invocation-based*, to ask for the execution of specific services provided by the middleware components. The current implementation of these buses is only for proof of concepts; therefore, a significative work could be devoted in the future to enhance existing enterprise service buses with the features identified during the ArtDeco project.

Developing applications for the ArtDeco middleware requires the definition of internal processes and their relationships with external information systems, global processes that aims at finding new business collaborations, data acquisition and management from different sources. In the next sub-sections, the three main subsystems shown in Fig. 21.6 are detailed and discussed as concerning their roles and programming features.

21.3.1 Handling Internal Processes and Their Interactions with External Systems

The main components of the sub-system "centralised and decentralised coordination system" presented in Fig. 21.6, are (see Fig. 21.7): (a) a *workflow engine* able to orchestrate automated services and human activities; (b) a *manager* used to observe and analyse external events in order to change the structure and the behavior of the current running workflows; (c) a *configurator* that is used to design - before the execution or automatically during the execution - the processes, as well as the rules and constraints used for driving the autonomic behavior of workflows; (d) a set of *autonomic components* - named *selflets* - that emulate the behavior of the networked enterprise as regarding the ability of handling specific needs by using a cooperative approach. The first three components belong to the *Semantic and Autonomic Workflow Engine* (SAWE) described in chapter 7. Moreover, this section addresses the problem of the verification of autonomic workflows as regarding the ability to intercept conflicting rules and constraints when they are added to the knowledge base of the WfMS.

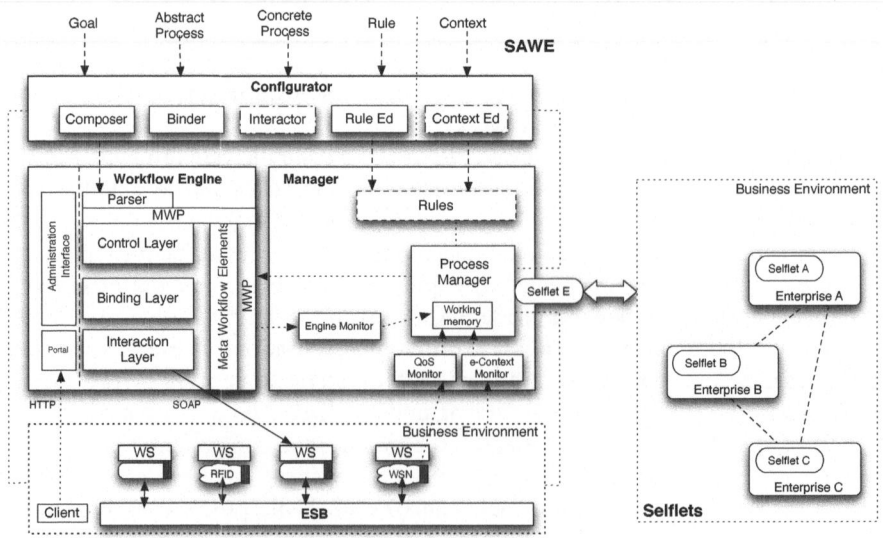

Fig. 21.7 Coordination system architecture

21.3.1.1 Centralised Management of a Networked Enterprise

In this sub-section, we discuss some examples of autonomic workflows by using an orchestrated view. The workflows are mainly referred to the internal processes of the winery even though some considerations are reported as regarding the interactions with other enterprises belonging to the network. These interactions typically involve the conversation among the information systems of different enterprises by using web services as enabling technology.

As reported in chapter 7, programming an autonomic workflow requires the definition of two separate sets of instructions: a typical flow of control, following an imperative model, and an event-driven behavior, based on a declarative language. In the following, the two parts of some sample workflows related to the scenario presented in Section 21.2.1 are illustrated.

Programming the Workflow Engine

Workflow programming in this section will regard mainly the cultivation and distribution processes described by the BPMN diagrams reported in Section 21.2.1. In particular, we consider an example in which the winery owns three vineyards for cultivating the grapes. The UML Activity Diagram depicted in Fig. 21.8 shows the explosion of the cultivation phase in three concurrent workflows, one for each vineyard to manage.

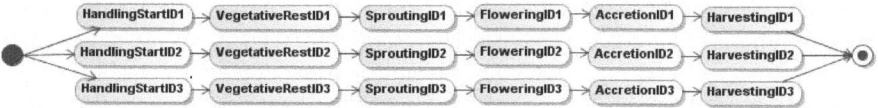

Fig. 21.8 Activity diagram representing the cultivation process

During each phenological phase, low level processes are used to manage daily tasks to be performed. These processes may be designed in advance, defining a pool of processes, covering a week or a month or a different period of time. The grape grower's management system has the task of identifying, day by day, the process to be started for each vineyard cultivated. In Fig. 21.9 an example of daily process is shown.

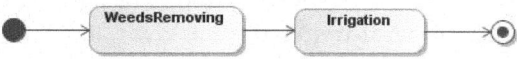

Fig. 21.9 Activity diagram representing an example of daily process

Starting from the process above, SAWE is able to change its structure according to actions triggered by possible events occuring during every day.

As regarding the Distribution phase, we consider a simple process composed of the two activities depicted in Fig. 21.10. During this phase, critical delays could occur when the carrier fails and the company cannot replace it. The notification of this unavailability to SAWE may be handled by discovering another qualified carrier from the shared registry of services. If this search successes, the process can perform a re-bind to ensure the process continuation.

Also in this phase, some exceptional events can be captured and handled by chang-ing the structure of the planned process. In particular, when the carrier fails, an event triggers the start of a discovery process to find another qualified carrier from a shared registry of services. SAWE is also able to continue towards a succesful termination also in the presence of a discovery failure (as reported in the following subsection).

Fig. 21.10 Activity diagram representing the distribution process

Programming the Manager Behavior

After the canonical phase of defining workflows derived from the business processes described in BPMN, the next step is the definition of rules. Rules allow for handling internal (to the WfMS) or external events during workflows execution. In particular, according to the MAPE cycle, events from the observed sources (external or internal contexts) are *monitored* and *analysed* by a class of rules defined as *monitoring rules*. The analysis is used to assert in the WfMS's knowledge base a new fact, which repre-sents a synthesis of the context changes. Hence, a different class of rules, named *man-agement rules*, *plans* a proper reaction by identifying and *executing* possible changes to the current structure of the running workflow, making it able to continue towards a succesful termination.

Events listened by the manager are generally defined context events [8]. They can be classified depending on the observation side: with reference to the manager, these events can come from the engine, from the services used in the running process or from the environment where the process runs. We name these three class of events: *engine events*, *QoS events*, and *e-context events*, respectively. In the following, we will refer to only two classes of events (the same considerations are valid for QoS events): e-context and engine events.

E-context events are generated each time a measure of a physical quantity is collected and stored in the WfMS's KB. Examples of physical quantities consid-ered in this chapter are: Temperature, WindSpeed, Light, Humidity, WindDirection, Pluviometer. They are related to the domain rules specified in Table 2.1. These ones, in fact, are examples of monitoring rules, that can be formalised by using a specific language, named SAWE policy language (SPL), as reported below:

```
on (Temperature > 30) if ((Humidity > 75) && (Wind > 9)) assert PowderyMildew;
on (Pluviometer > 1) assert Rain;
on (Pluviometer > 20) assert HeavyRain;
```

The first rule checks for the condition that may cause the emerging of the *pow-dery mildew* disease. The second rule is used for monitoring rain in the vineyard, which may be an obstacle for a certain type of activities and, at same time, may be the opportunity for saving costs avoiding any activity. The third rule is used for identifying *heavy rain*, which, if is coupled with *hailstorm*, can cause heavy damage to the vineyard.

The same approach can be followed when measures are referred to virtual sensors that are able to collect more complex and aggregate measures. An example of such a sensor is a feedback from a human being related to an observation of the external environment.

The rule reported below uses a physical quantity, named GrapesQuality, monitored by a virtual sensor to assert that the quality of the vineyard identified by ID1 is not able to produce high-quality grapes and therefore harvesting is not possible.

```
on (GrapesQuality.<ID1>==LOW) assert NotHarvesting.<ID1>;
```

The monitoring rules typically are used to generate higher level events, possibly coupled with management rules. These rules exploit meta-operations to change the current structure of the running workflow (see chapter 7 for the complete semantics of the meta-operations), as exemplified below:

```
on (PowderyMildew) add Spraying;
on (Rain) drop Spraying;
on (Rain) drop Irrigation;
on (HeavyRain) add AgronomistVisit;
```

Powdery mildew insurgence is contrasted using appropriate chemical substances to be sprayed in the vineyard (thus, an activity - Spraying - is added to the workflow). In another situation, rain may be an obstacle for spraying while, at same time, it is an opportunity that may be exploited for avoiding to perform irrigation. Heavy rain requires particular attention and call for planning an agronomist visit to supervise the phenomenon and, if it is the case, introducing in the system information that may cause the generation of an event signaling *low quality* for some grapes.

As regarding the example on grapes quality, the *id* of the vineyard interested by low quality is obtained by exploiting specific info provided by the agronomist. This info is used to extract statistical information on grapes variety and amount from the database and consequently the harvest activity is replaced by buying.

```
on (NotHarvesting.<ID1>) replace Harvesting.<ID1> with Buying.<ID1>;
```

The angular brackets are used to refer to an implicit information used to correlate the replacing and the replaced activities.

Management rules have to be verified against a set of high level constraints, expressed in SPL. The following constraints can be used for supporting the autonomic adaptation of the cultivation process, in order to avoid inconsistency at run-time. They allow for reacting to events and, at same time, to guarantee workflow correctness. The symbol ->! means mutual exclusion between the two involved activities and <-> means dependence between the involved activities.

```
AgronomistVisit ->! Irrigation
OenologistVisit <-> Harvesting
Buying ->! Harvesting
```

The first constraint states that it is not possible to have, in the same handling process (that spans for a day), an agronomist visit and an irrigation activity. The second constraint states that during harvesting process, an oenologist visit is needed for assessing grapes quality. The last constraints suggests that buying of grapes is permitted only if harvesting is not possible (for a whole vineyard in the example).

In the Distribution process, it is possible that a carrier is not available, i.e. the static binding defined at the beginning of the process is not satisfied and/or there is no service in the registry able to satisfy the needed functionality. If this happens, a binding failure occurs. The related event is propagated to the management level for its processing by firing the following management rules.

```
on (BindingFailure) compose <BindingFailure>.service
on (ComposingSuccess) invoke <ComposingSuccess>.service
```

The first rule is used to request the composition of the failed service to external systems supporting the workflow execution: the semantic functional description of the service is used for defining the service that must be composed.

The second rule is used to receive the result of the composition, as performed by the external system and to invoke the composed service. This alters the executing process, replacing the failed activity with another one that is in charge of invoking the received composition. Also in these cases, the angular brackets refer to implicit information used to correlate the events and the actions (in the example, the service description used for the composition is extracted from the triggering event of type BindingFailure).

It is worthy noting that also the composition may fail. In this case, another level of autonomic action could be performed (not described in this example but available in SAWE): *re-planning* of the remaining process, from the current state to the goal. This recovery action is able to find an alternative flow of activities able to reach the process objective. This kind of failure recovery is not always possible. It depends a lot on the kind of workflow executed and on the availability of equivalent services.

When every failure mechanism provided by SAWE fails, a notification is sent to the human manager.

21.3.1.2 Decentralised Management of Networked Enterprises

During the execution, the workflow requires the invocation of a service for a delivery activity. Such activity could be accomplished either by a delivery company, or by some other wineries belonging to the network of enterprises subscribed in the service registry. It is possible that no service is available in the registry and as a consequence, no binding is found for that activity and a binding failure occurs.

In the Artdeco architecture, different components could support the workflow engine in order to offer an autonomic behaviour; in particular the workflow of the winery (and the workflow of the others enterprises belonging to the network) can be bound to a SelfLet (see Chapter 6) able to react to changes in the surrounding environment and make decision on the basis of the sensed context.

Reasoning on the sensed context means managing information about the internal and/or the external execution environment. In particular, a SelfLet could access to the information regarding other enterprises belonging to the network. Such information could be related to a particular service offered for a certain time by an enterprise and, due to its nature, not possible to find in the registry. In fact, while the workflow is mainly thought to access to stable information valid for a long period of time (generally stored in a registry), the SelfLets are conceived to be more suitable to observe and to react to the continuously changing environment in which the network evolves; in particular, the SelfLets are able to access to information having a "temporary" validity obtained reasoning on the current status of the enterprises composing the network.

Since the workflow is not able to identify any service satisfying its needs, it generates an event requesting the intervention of the SelfLet. It receives the description of the problem and reacts to the event: on the basis of the gathered information it can decide, which is the best solution optimizing cost and efficiency requirements. For example, it can decide to contact a delivery company or another winery and can inform the workflow of the result of the decision. Thanks to this, the workflow of the winery is able to continue the execution without failures.

While a binding failure, occurring during the execution of a traditional workflow, implies a human intervention to decide the next step or the abort of the process, in the Artdeco workflow the binding failure suspends the failed task while the other activities currently executing can continue without problems. In such a way, the workflow engine can access to external resources in order to overcome the failure; due to this, the workflow switches in the "composing" mode and communicates with the other components (i.e. SelfLets) able to accomplish the needed task.

Due to some constraints on its delivery, the SelfLet can decide to invoke the service of a delivery company to rent a truck; on the basis of the analysis of the internal knowledge, or of observation of the external environment, the SelfLet can decide to advertise the possibility to be a carrier itself. In fact, it could realize that it has enough space on the truck rent for the delivery, to satisfy some other possibly incoming delivery requests or even, observing the external environment, it could discover that one (or more) company has a delivery request having suitable requirements (in term of volume and date of the delivery, starting and destination place) that it could be able to meet. Then, it could notify this new condition to all the network. The notification of the new behaviour (carrier) could happen either by means of a publication in the service registry, or, even, sending the event to all the other companies in the network. In the following figure the SelfLet behaviour related to the described situation is shown.

When the SelfLet receives the event from the wokflow, it evaluates the delivery requirements and search for a suitable external company to which assign the delivery. It could search for a delivery company, or share the truck with a company belonging to the network. Whenever the delivery is assigned to a delivery company, if the available space is enough, the SelfLet could advertise to the network its capability to be a carrier. While the notification of the event towards all the registered component is enabled by a publish-subscribe mechanism, the publication in the

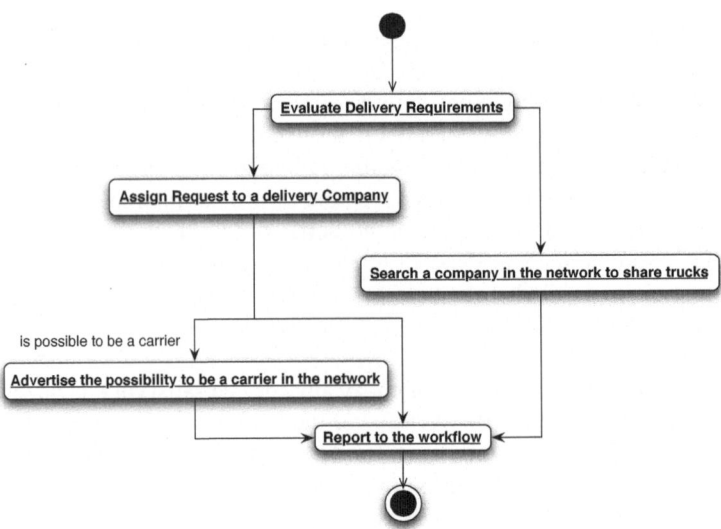

Fig. 21.11 SelfLet behaviour

registry, could happen only if the new service is *"stable"* and usable, as is, for a certain time; in such case it could be used as a new service in the chain and, thanks to this, it could be needed its publication in the registry. In particular, also the registry could be interested in receiving such kind of events, but the publication in it could respect certain policies: a logic is needed to determine, on the basis of the lifetime of the new service, if and how the service could be discoverable in the registry. Moreover it could happen that, trying to identify a possible enterprise able to share the delivery truck, the SelfLet observes that no single enterprise could perform its delivery; in fact it could be possible that the delivery could be partially accomplished by more than one winery. It may happen, in fact that more companies could be potentially used as carrier, since satisfying the requirements for data and starting/destination places, but they do not have enough space for the whole delivery. In such a case, the SelfLet could combine all the available spaces of the suitable potential carriers in order to detect the best combination of carriers able to fulfil its delivery task.

21.3.1.3 Dynamic Verification of Autonomic Compositions

Changing a running workflow can cause consistency failures. Therefore, correctness verification takes an important role in the case of dynamic adaptation of workflows. Correctness of an autonomic workflow may be defined as the ability of the workflow to reach its goal also in the case of dynamic adaptations. While run-time verification of the satisfiability of some properties, related to specific constraints posed by the structure, the semantics and business rules of a workflow, has been discussed in chapter 7, in this section a different kind of verification is analysed. It regards the identification of violations potentially introduced when a new rule is added to the knowledge base (we refer to this approach as on-line verification). This verification is performed by using model checking [3], an automatic technique for formal verification of reactive systems behavior.

The main violation analysed is the deadlock; however, the same approach could also be used to verify the correctness of the workflow against other constraints defined by the user (such as business rules). By adopting this approach, all the possible execution traces, i.e., the whole state space of the system, are analysed. Therefore, the identified violations could refer to only a subset of the complete set of traces. This suggests to not removing potentially conflicting rules from the knowledge base but to move the verification to runt-time. On the other hand, if the current set of rules satisfies the desired properties, run-time verification is not needed.

On-line verification has been used for analysing the cultivation handling phase, considering also the adaptation rules. The model checking has been performed using the following steps:

1. *Modelling*: target system is formally described, using the verification tool formalism;
2. *Specification*: temporal logic is used for specifying system properties;
3. *Verification*: the model is checked against the property using the engine of the model checker.

The case study workflow has been evaluated using two different Model Checking approaches. The first one is the symbolic model checker NuSMV [2], derived from SMV (Symbolic Model Verifier). The model has been realized using Petri nets for representing the workflow, as described in the XPDL*, and the possible adaptation, as described in the management rules, since Petri nets are easily translated in the checker input language. Then, the branching temporal logic CTL has been used for stating the correct behavior that we want to guarantee with the specification of the workflow and its adaptation rules. The automatic verification of the model has given a positive result: the workflow and the policy rules are correct against the properties.

The second approach has used the model checker Spin [6]. In this case the process has been modeled using a finite state machine with a state for each phenological phases of the grapes. Additional states has been used for modeling alarm states, which describe critical situation to be handled, and disastrous states, which describe states in which the system goes when a critical situation has not been solved.

The external events, which occur with the time evolution, have been introduced using a dedicate process (Nature) that simulates the changing of season. The automaton and the external components are easily mapped in Promela, the input language of Spin. The overall behavior has been then checked using Spin and the obtained result has been that the workflow is able to handle correctly the events that may emerge during cultivation. Further details on the representation of the systems in both the approaches and some example of the verified properties can be found in chapter 8.

The verification process has been iterated online in order to consider the alteration of the KB during workflow enactment. In such a view, verification of correctness must be performed when the rules are asserted or retracted in the KB. The proposed approach employs model checking, in particular the use of Spin, to evaluate correctness of the autonomic workflow at runtime, interfacing it with the workflow that runs. This would allow to check the correctness of the workflow every time the KB is modified.

In order to have a complete automatic execution and evolution of the workflow, the online use of model checking has been embedded in the workflow and the modeling phase of the target system has been automatised using a translation mechanism that transforms the XPDL process together with the rules and the constraints on the activities into Promela, the input language of Spin.

The translation process consists mainly of four functions:

- The function `variableGenerator(XPDL)`, using information from the XPDL process and the rules, generates the variables needed for the model. In particular, the possible states of the workflow, that consist of the states actually described in the XPDL file together with the states that can be added by the rules, are encoded assigning to each of them a unique identifier and a boolean value, contained in the array (`active_states` , that denotes if a state has to be executed or not. Analogously, all the possible events are represented by a variable that can assume different values depending on the different possible occurring events.
- The function `environmentGenerator(Rules)`, using the events in the rules, randomly generates the events that trigger the ECA rules, together with two special values that simulate "no event" and the special event "end of activity".
- The function `rulesTranslator(Rules,XPDL, Constraints)` encodes the rules and the constraints into two communicating processes. The first one simulates the rules, each with a branch in a conditional structure. The condition is given by the event and when the considered event occurs, the corresponding consequence simulates the performing of the action, after having checked with the other process if the constraints are not violated.
- The function `processTranslator(Rules,XPDL)` generates a process that simulates the overall structure of the workflow, also taking into account the possible new states that can be created by the rules.

Using this functions, every time the KB is modified a new Promela model is generated ad is checked against predefined safety and liveness properties, that once verified can be assumed as guaranteed for the new process. In this way it should be possible to verify, at run time and automatically, if the alteration of the adaptation rules cause the workflow to be still correct or to become not correct.

Consider as example the simple process represented in Fig. 21.9, composed by the sequence of the two states `WeedsRemoving`and `Irrigation`, together with the following simple rule:

```
on (PowderyMildew) add Spraying;
```

and the following constraint:

```
Spraying <-> WeedsRemoving
```

The workflow together with the rule and the constraint will be translated[2] into five communicating processes: environment , rules , constraints , procWorkflow and the special process init , that initializes all the involved variables and runs the other processes. First, the process environment uses the following non deterministic choice to simulates external events:

```
if
    ::event=0;
    ::event=1;
    ::event=2;
fi;
```

where the value 0 means that no event occurred, 2 simulates the end of the current activity and 1 encodes the event PowderyMildew . The above process communicates with the others, with the only exception of init , using a randezvous interaction via Promela channels. The process rules analyzes the occurred event and, if it is the case, performes the required action. For example the rules in our example is encoded as

```
if
    ::e==1-> in!1,3; in?x,1;
    if
        ::x!=0->memo=status; status=4; active_states[4]=1;
    ::else;
    fi;
    ...
fi;
```

The condition of the if checks if the rule is triggered; if it is the case, the process communicates with the process constraints to check if some constraint is violated (in!1,3 , where 1 represents the rules that we want to apply and 3 the identifier of the process to communicate with); it evaluates if the rule violates the constraint, i.e. it checks the array active_states to ensure that the position in the array corresponding to the state WeedsRemoving is set to 1 . If it is the case, it sends on the channel the value 1 , otherwise the value 0 . The process rules reads from the channel (in?x,1) and, only if the property of the constraint holds, it applies the rule adding the needed state.

The process procWorkflow simulates the overall structure of the workflow. The duration of each activity, i.e. the permanence in a given state, is bounded, but not fixed. Once in a state, the process checks from events in the environment: if the event is 2 the process changes state as suggested in the workflow, if it is 0 it remains unchanged, while if it is a different value it checks if some rule is applicable. After a bounded number of interaction with the environment, if event has always assumed the value 0 , it changes state according to the workflow structure.

As already said, the obtained communicating processes can be checked against pre-defined properties. If they hold, the workflow meets the requirements, otherwise a violation may occur. Notice that this information can be used in a double

---

[2] Notice that the translator implements an optimized Promela code, while in the following example the proposed translation does not consider any optimization of the code such that the translation is more intuitive for the reader.

way: indeed, if the violated property is critical, the workflow execution needs to be interrupted and a designer intervention is required, but, if it is not the case, the property can be monitored online to check if the sequence of events at runtime is a sequence that will violate it.

The proposed technique requires a new translation and verification phase every time the KB is changed. In the future, we aim to improve this mechanism using incremental techniques that do not need to verify again the whole process but only concentrates on the differences between the new and the old specification.

21.3.2 Extracting Knowledge from Heterogeneous Sources

According to the global system architecture presented in Fig. 21.6, we now describe the "Knowledge Extraction and Management" sub-system and the relationships between its components.

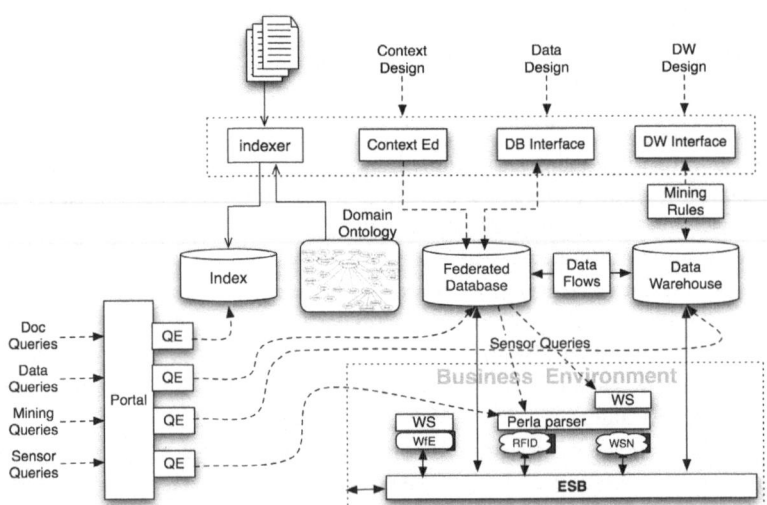

Fig. 21.12 Data and knowledge base architecture

The sub-system consists mainly of three modules: (i) an ontology-based indexing mechanism for accessing Web documents that may be of interest for the quality control unit (e.g., the quality-manager); (ii) a federated relational database that provides a virtual integrated-schema for various information sources such as the operational and analytical databases of the enterprise as well as the database extensions for managing the Web documents and the sensors data; (c) a data warehouse materialising the result of interesting pre-computed queries for supporting precise analytical tasks ranging from common sales analysis to more innovative tasks such as product-quality assessment using the sensors data collected during the product's life-cycle and customers taste prediction based on data collected from wine blogs and web-sites.

21.3.2.1 Semantic Indexer

This component can be exploited by the Quality Manager or by the Marketing (during wine design) to search for specific documents from a shared repository. Search is supported by semantic annotations that link documents words with the concepts of a domain ontology. This way, the keywords used during search can be significant also when they have not a direct match with the words of the searched documents. Ontology navigations make it possible to find semantically equivalent concepts and consequently to find further significant keywords for performing the search. This approach makes it possible to retrieve a larger set of documents from the repository.

To perform semantic search, three phases are needed: (1) text extraction from heterogenous documents; (2) text annotation with concepts of a domain ontology; (3) semantic indexing of documents with concepts. Only at this point, queries can be issued. In the following, the three phases are described with reference to an example in the context of the wine scenario.

The *Text Extractor* module gathers text from several document formats (PDF, HTML pages, etc.). The extracted text, as well as the original document file (if any), are stored into the Document Repository. For each word of the documents, the engine selects the related concept of the Domain Ontology. Thus, the Conceptual Index provides an abstract view of the documents, permitting users to search for concepts. A classic TF-IDF scheme is applied, in order to generate the documents' concept vectors, stored into the Conceptual Index.

The concepts-words mappings generated during the indexing procedure are then arranged as an index and stored into the Conceptual Index. As an example, consider the sentence "The bottle with red label contains Barolo, a red wine". The human expert associates words and concepts, producing "The bottle/WINE-BOTTLE with red label/WINE-LABEL contains Barolo/BAROLO-WINE, a red/RED-WINE wine/WINE", where WINE-BOTTLE, WINE-LABEL, BAROLO-WINE, RED-WINE, and WINE represent concept IDs (see chapter 9 for details).

Now, if the system has been trained to recognise concepts in the following set WINE-BOTTLE, WINE-LABEL, BAROLO-WINE, RED-WINE, WINE and the document set is composed of D_1 ="Barolo is a small village where good wine is produced" and D_2="Barolo is a red wine", the system extracts the following sets of concepts C_1=WINE, WINE and C_2=BAROLO-WINE, RED, WINE. The TF-IDF calculates the weight $w_{c,d}$ associated to each concepts c, for each document d; thus, the following vectors are generated: V_1=[0, 0, 0, 0, W_{WINE,D_1}] and V_2=[0, 0, $W_{BAROLO-WINE,D_2}$, $W_{RED-WINE,D_2}$, W_{WINE,D_2}].

The *Query Engine*, exploiting both the Conceptual Index and the Domain Model, permits to formulate concept-based queries on the document collection. Searching for a given concept, the system finds every mapped word, and calculates a ranked list of documents. Keyword based queries are composed of a sequence of words, con-

```
WINE(appellation, category, vinification)
REF-COMP(appellation,tech-name, min_%, max_%)
GRAPEVINE(tech-name, name, variety)
HARVEST(h-id,vineyard-id, note)
CELLAR(cl-id, material, type, vineyard-id, harvest-id)
BARREL(bl-id, wood, type, cellar-id)
COMPOSITION(bottle-id,barrel-id, wine_%)
BOTTLE(b-id, bott-date, harvest-date, price, appellation,lot-id)
LOT(l-id, pkg-date, pkg-type, return-date, sale-date, cr-id, cs-id)
CUSTOMER(c-id, name, address)
VINEYARD(v-id, hectares, fraction, municipality, district, region, zone)
ROW(r-id,vineyard-id, plant-date, tech-name, phenological-phase)
PHENOMENON(vineyard-id,phenomenon,date, type, notes, emergency-plan)
DAMAGE((vineyard-id,row-id,name,date), analysis)
SENSOR-BOARD((sb-id), coordinates, act-date, obj-id)
SENSOR(s-id, type, model, meas-unit)
MEASURE-DATA(date,time,sensor-id, value)
```

Fig. 21.13 Relational Schema for the Wine Use-case

nected by either AND or OR boolean logic operators. As an example of AND query, if the user issues Q="Barolo wine", the system indexes the text and produces the following sequence of concepts: $V_Q=[0, 0, W_{BAROLO-WINE,Q}, 0, W_{WINE,Q}]$. Then, it compares the vector V_Q against V_1 and V_2, and finds the nearest document.

21.3.2.2 Federated Database

As it is commonly done in modern information integration systems, the data structures supporting the Knowledge Extraction and Management system are engineered by following classic conceptual and logical design flows consisting of:

- Conceptual Design of the mediated global schema, also called integration schema.
- Logical Design of the mediated global schema.
- Logical Design of the mappings between the global schema and the data-source schemas.
- Conceptual Design of the data warehouse for the analysis of facts of interest.
- Logical Design of the data warehouse.

In addition to the steps above, we also need to design the necessary structures to allow context-aware querying. This requires two additional steps in the design methodology, namely:

- Conceptual Design of the context-model for quality assessment.
- Logical Design of the context-aware views.

The first step is the design of the mediated schema, which acts as a global view over the enterprise's information legacy.

The schema models all the information related to the wine production process and the structures needed to store sensors data collected for all the objects of interests (e.g., cellars, vineyards, barrels and bottles). Moreover, we collect also data about atmospheric phenomena that interested certain areas of the vineyard, since they could have affected the quality of the wine, along with information about the returned lots of products that are usually a starting point for the quality assessment process. The relational schema that we adopt for the wine case-study is shown in Fig. 21.13.

21.3.2.3 Creating the Data-Mart for Quality Assessment

As already said, an interesting analysis to perform is related to the correlation between atmospheric phenomena and the quality of the wine. In order to study this relationship we decided to design a data-mart that combines the data coming from sensors with the data related to the returns of wine lots. The data mart has been designed following the methodology of Golfarelli and Rizzi [5]. In particular, we started from the design of the attribute trees for the identification of the relevant analysis dimensions and measures, then we produced the conceptual schemas for the various facts and translated them into the corresponding relational schemata. An example of attribute tree for the grapevine damages is represented in Fig. 21.14.

The next step is the identification of the facts of interest and the conceptual design of the related data marts for the data warehouse. We identified three facts of interests for the quality manager namely:

- Returns of goods
- Sales
- Grapevine damages (e.g., plant diseases and atmospheric accidents)

Up to now, the design process we presented does not enable context-aware querying of the global schema nor of the data-marts. We now describe the last design step leading to the creation of the contextual views. The methodology followed is the one presented in chapters 13 and 13 of the book.

21.3.2.4 Designing Context Relevant Areas

For the sake of explanation, we consider a subset of the contextual model shown in Part 3 that is relevant for the quality management and the agronomist processes (see figure 21.15). In particular, when considering the quality manager, we are interested in modelling the threats to the vinegrapes such as the diseases and the atmospheric disasters (e.g., flashfloods, fires, droughts and hailstones) that might affect the quality of the final product. Notice that the atmospheric phenomena affects only the vinegrapes, thus they are interesting only during the cultivation phase. However, since the same threat might have different effects in different moments of the cultivation phase, we need a parameter to distinguish among the different moments of the cultivation (i.e., the phenological phases). The other phases represented in the context model are interesting for the analysis of the sensors data, in order to identify possible problems in the production chain that might affect the quality of the wine.

Moreover, we are interested in modelling also the information that are relevant for the agronomist, during the harvest phase, which include, besides the informa-

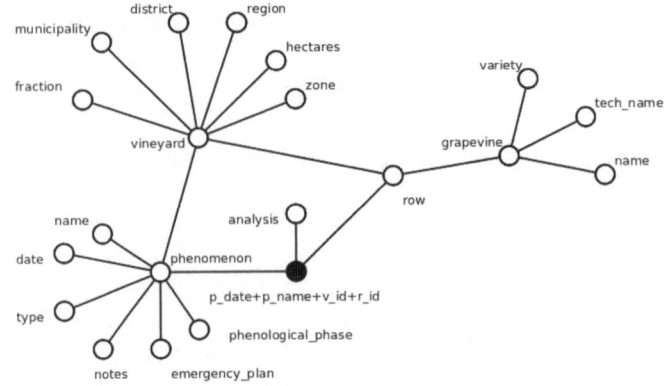

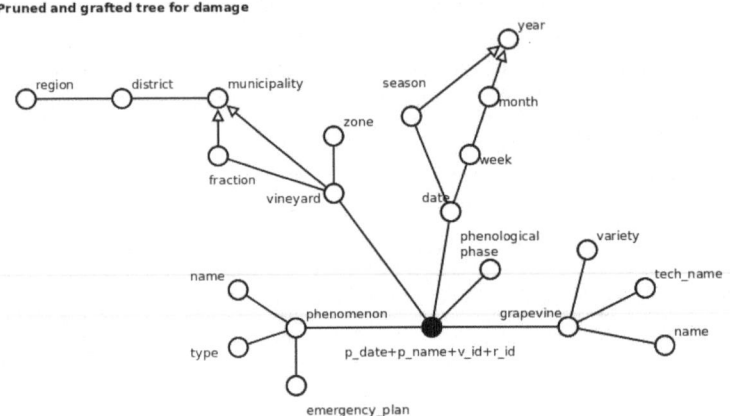

Fig. 21.14 Attribute Tree Diagram for Damage Analysis with Pruning and Grafting Phases

tion related to vineyards and grapevines, the data collected by sensors placed in the vineyards. Other phases represented in the context model, and in particular the cultivation and ageing phases, are interesting for the agronomist as well; indeed she/he can analyse sensors data that are placed in the vineyards, in order to identify possible problems related to the cultivation in different periods of year.

By leveraging on the context model, for each value of each dimension, we define a view over the database (or the data warehouse) that consists of all the data that might be of interest for that value (i.e., the relevant area). Figure 21.16 shows the definitions in Relational Algebra of the views associated to the atmospherical events and to the quality-manager and agronomist roles.

Once the all the views corresponding to the dimension values have been defined, the methodology described in Part 3 allows us to combine them to obtain the view associated to a given context. Consider for example, the situation of the quality

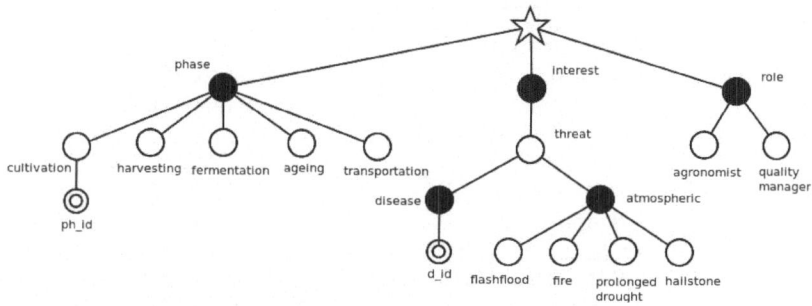

Fig. 21.15 Context-Tree for the Wine Use-case

manager interested in the analysis of data coming from sensors during the transportation phase. The associated contexts C_1 and C_2 will be:

$$C_1 = < transport, *, quality\text{-}manager >$$
$$C_1 = < harvest, *, agronomist >$$

Where the symbol '*' means that we do not care about the value of the interest dimension. Given the views definitions given above, the resulting view for the context C_1 and C_2 is shown in Figure 21.17. This view can be easily implemented as an SQL view over the WineDB database and used whenever the quality manager is analysing the sensors data collected during the transportation phases.

21.3.3 Monitoring Physical Environments with Wireless Sensors

The main components of the "WSN data acquisition sub-system" presented in Fig. 21.6 are (see Fig. 21.18): (a) a *WSN configurator*, able to configure the sensor network with negotiable parameters; (b) a *function-oriented interface* used to query the network through operations that return data coming from the sensors; (c) a *data-oriented interface* used to query the network with a declarative language that exhibits a more expressive semantics.

Wireless Sensor Networks (WSNs) can be used in different phases of the wine business process. Typically, by using WSNs data are acquired from the external physical environment and made available to the enterprise's information system. In the grapes production phase WSNs are used to monitor environmental conditions that can affect the cultivation process and impact on the final product quality. Information made available during this phase are used by agronomists to decide if and when to start possible management actions (see Section 3.1). In the wine production phase, as some high quality wines require a long aging process at controlled temperature and humidity, WSNs can be used to monitor the environmental conditions in the cellar in a very punctual way. Finally, during the wine distribution phase low-cost wireless sensors can be attached to single bottles, or pallets, to monitor

$\mathcal{R}el$(flashfood) $= \{\sigma_{name='flashfood'}$PHENOMENON, $\pi_{type,value,measunit,date,time}$SENSOR
$\bowtie$(MEASURE_DATA$\bowtie$($\sigma_{name='flashfood'}$PHENOMENON)),
VINEYARD$\bowtie$($\sigma_{name='flashfood'}$PHENOMENON, ROW
$\bowtie$(DAMAGE$\bowtie$($\sigma_{name='flashfood'}$PHENOMENON)\}

$\mathcal{R}el$(fire) $= \{\sigma_{name='fire'}$PHENOMENON, $\pi_{type,value,measunit,date,time}$SENSOR
$\bowtie$(MEASURE_DATA$\bowtie$($\sigma_{name='fire'}$PHENOMENON)),
VINEYARD$\bowtie$($\sigma_{name='fire'}$PHENOMENON, ROW
$\bowtie$(DAMAGE$\bowtie$($\sigma_{name='fire'}$PHENOMENON)\}

$\mathcal{R}el$(drought) $= \{\sigma_{name='drought'}$PHENOMENON, $\pi_{type,value,measunit,date,time}$SENSOR
$\bowtie$(MEASURE_DATA$\bowtie$($\sigma_{name='drought'}$PHENOMENON)),
VINEYARD$\bowtie$($\sigma_{name='drought'}$PHENOMENON, ROW
$\bowtie$(DAMAGE
$\bowtie$($\sigma_{name='drought'}$PHENOMENON)\}

$\mathcal{R}el$(hailstone) $= \{\sigma_{name='hailstone'}$PHENOMENON, $\pi_{type,value,measunit,date,time}$SENSOR
$\bowtie$(MEASURE_DATA$\bowtie$($\sigma_{name='hailstone'}$PHENOMENON)),
VINEYARD$\bowtie$($\sigma_{name='hailstone'}$PHENOMENON, ROW
$\bowtie$(DAMAGE$\bowtie$($\sigma_{name='hailstone'}$PHENOMENON)\}

$\mathcal{R}el$(disease) $= \{\sigma_{name=\$d\_id}$PHENOMENON, VINEYARD
$\bowtie$($\sigma_{name=\$d\_id}$PHENOMENON, ROW$\bowtie$
DAMAGE$\bowtie$($\sigma_{name=\$d\_id}$PHENOMENON)\}

$\mathcal{R}el$(harvest) $= \{$GRAPEVINE, ROW, VINEYARD,
$\pi_{type,value,measunit,date,time}$MEASURE_DATA $\bowtie$SENSOR
$\bowtie$(SENSOR_BOARD$\bowtie$VINEYARD)\}

$\mathcal{R}el$(fermentation) $= \{$GRAPEVINE, TUB $\pi_{type,value,measunit,date,time}$MEASURE_DATA
$\bowtie$SENSOR$\bowtie$(SENSOR_BOARD$\bowtie$TUB)\}

$\mathcal{R}el$(ageing) $= \{$GRAPEVINE, ROW, VINEYARD,
$\pi_{type,value,measunit,date,time}$MEASURE_DATA$\bowtie$SENSOR
$\bowtie$(SENSOR_BOARD$\bowtie$BARREL),
$\pi_{type,value,measunit,date,time}$MEASURE_DATA$\bowtie$SENSOR
$\bowtie$(SENSOR_BOARD$\bowtie$BOTTLE)\}

$\mathcal{R}el$(transport) $= \{$BOTTLE, LOT, $\pi_{type,value,measunit,date,time}$MEASURE_DATA
$\bowtie$SENSOR$\bowtie$(SENSOR_BOARD$\bowtie$BOTTLE
$\bowtie$ $\pi_{type,value,measunit,date,time}$MEASURE_DATA$\bowtie$SENSOR$\bowtie$
(SENSOR_BOARD$\bowtie$LOT)\}

$\mathcal{R}el$(agronomist) $= \{$GRAPEVINE, ROW, VINEYARD,
$\pi_{type,value,measunit,date,time}$MEASURE_DATA$\bowtie$SENSOR
$\bowtie$(SENSOR_BOARD\}

$\mathcal{R}el$(qlty-manager) $= \{$WINE, BOTTLE, LOT, CUSTOMER,
$\pi_{type,value,measunit,date,time}$MEASURE_DATA$\bowtie$SENSOR$\bowtie$
(SENSOR_BOARD$\bowtie$ $\sigma_{c\_id<>null}$LOT)\}

Fig. 21.16 Contextual Relevant Areas for the Wine Use-case

periodically the external temperature and avoid to expose wine to adverse environmental conditions that could compromise the wine quality. In the following we will describe, in detail, how WSNs are used in the three above-mentioned scenarios.

$\mathcal{R}el$(harvest, _, agronomist)	= {GRAPEVINE, ROW, VINEYARD, $\pi_{type,value,measunit,date,time}$MEASURE_DATA$\bowtie$SENSOR$\bowtie$ (SENSOR_BOARD$\bowtie$VINEYARD)}
$\mathcal{R}el$(transport, _ ,qlty-manager)	= {BOTTLE, LOT, $\pi_{type,value,measunit,date,time}$MEASURE_DATA $\bowtie$SENSOR$\bowtie$(SENSOR_BOARD$\bowtie$ $\sigma_{c\_id<>null}$LOT)}

Fig. 21.17 Contextual Views

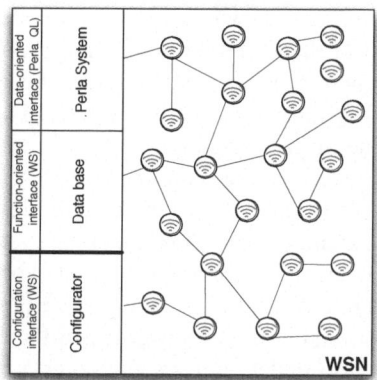

Fig. 21.18 Wireless sensor networks subsystem

21.3.3.1 WSN Middleware

The middleware represents the physical interface between the WSN and the rest of the enterprise's information system. To allow an efficient interaction between application processes and sensor nodes, two different middleware solutions have been developed in the project, which take alternative approaches. The former solution, described in chapter 17, relies on a service-oriented paradigm. The middleware layer allows an application process to establish a service agreement with the underling WSN and hide low level details, such as programming interface and embedded OS (see the module *Configurator* in Fig. 21.18).

In the winery, three different kinds of service have been considered and, correspondingly, the following three service contracts have been defined.

- Periodic Measurement contract: is used to periodically measure a certain physical quantity (e.g., temperature);
- Event Monitoring contract: is used to monitor specific events (e.g., Temperature exceeding or the Battery Voltage go down a given threshold);
- Network Management contract: is used to manage and control the WSN behaviour.

The above schemas are used for both adaptation of the behavior of the sensors to the environmental conditions and for reducing the power consumption of the sensors. Both are important aspects to be considered when using sensor networks for monitoring outdoor scenarios such as vineyards. The first goal was to implement a mechanism to automatically reconfigure a WSN. The scenario envisages a sensor that periodically acquires data related to the brightness level it senses, and sends such data to the base station. We supposed that a year is divided in two different periods, namely the summer and the winter, and that during the winter the rate at which the sensor collects and sends data can be reduced to the half of the corresponding rate during the summer. To this end, we implemented a function that automatically sets the sampling rate (the rate at which the sensor collects and sends data) during the winter at the half of the sampling rate during the summer. Furthermore, we implemented another function, which allows to set the sampling frequency from the base station, by sending an asynchronous message. The payload of the message contains the desired value of the sampling frequency, and the sensor has to extract that payload and appropriately set the sampling rate.

The nesC language has been utilized for the activities on sensor network programming with sensors running the TinyOS operating system. In the activities, they worked towards the implementation of two new in-network functions.

The second goal was related to the reduction of the power consumption. This is a very hot topic and a critical issue whenever it is difficult to replace the batteries of the sensors. The scenario under analysis envisages a couple of sensors sending packets to each other. In this scenario, when a sensor is not sending or receiving packets, it has to 'go to sleep' for a certain time by switching off the radio (that is the most power-consuming functionality). To implement this behavior, we exploit the LowPowerListening functionalities. LowPowerListening interfaces allow to set the local node's radio sleep interval which define the time interval in which the radio is switched off. After this interval the radio is turned on for receiving check intervals to perform Clear Channel Assessment according to the 802.15.4 standard.

This feature has been used for the temperature monitoring in a certain region of a vineyard by using a middleware, tailored to TinyOS 2.x and featured by a simple QoS admission control module. For the sake of simplicity, the WSN testbed is characterized by a network star topology, in which a node acts as a coordinator and the other ones act as end-devices. The nodes are deployed all around the monitored area and have different tasks according to their category:

- The *Coordinator* node, connected to a resource-unconstrained machine (i.e. a pc), is responsible for interfacing the WSN with the higher level architecture through the WSNGateway. The coordinator receives data coming from the end-devices and forwards them to the WSNGateway. Also, it receives commands from the WSNGateway and forwards them to proper end-device nodes. In the current implementation, commands concern activation and deactivation of sensors on a node, and setting/changing of the sampling periods.
- The *End-Device* node is responsible for gathering data from active sensors and sending them to the coordinator node.

Every end-device node is a composed by:

- Crossbow MicaZ a 2.4 Ghz mote module used for enabling low-power, wireless sensor network, featuring the:
- ATMEL ATmega128, a low-power CMOS based on the AVR enhanced RISC architecture.
- Chipcon SmartRF CC2420, a single-chip 2.4 GHz IEEE 802.15.4 compliant RF transceiver.

- Crossbow BelosB mote platform including a suite sensor with light, temperature and humidity sensor, featuring the:

  - TI MSP430, a powerful 16-bit RISC CPU.
  - Chipcon SmartRF CC2420, a single-chip 2.4 GHz IEEE 802.15.4 compliant RF transceiver.

- Crossbow Sensor Board MDA100CB, a sensor board that provides a precision thermistor, a light sensor/photocell and general prototyping area.

Moreover, the TelosB platform is equipped with a Voltage sensor that allow monitoring the battery level and using this information for estimation of battery decrease as a function of the sampling time. On every node, the TinyOS 2.0 embedded operating system runs along with the application program written in the NesC language. In particular, the Coordinator node runs the BaseStation TinyOS application to manage data packets coming from the network and send command packets to the network. Every end-device runs an application program composed by several TinyOS interfaces linked together to provide the following features:

- dynamic adjustment of sensor sampling rates.
- power saving through the asynchronous low power listening (LPL) strategy.
- dynamic sensor activation/deactivation.

The alternative solution, described in chapter 18, takes a data-oriented approach and regards the entire WSN as a database [9]. The WSN can thus be queried just as a traditional database. A specific middleware layer and a language to install queries and extract data from a WSN have been designed and implemented in the project. Specifically, the following three different types of queries can be executed.

- Low Level Queries are used to access data produced by sensor nodes. Both periodic and event-triggered data extraction paradigms are supported;
- High Level Queries are used to perform data manipulation operations;
- Actuation Queries are used to set parameters at sensor nodes (e.g., for modifying software variable).

In the following, we will assume the presence of two different types of devices:

- DA which embeds temperature and humidity sensors.
- DB which embeds only a pressure sensor.

The sensors are configured as logical objects with the following attributes:

- LA: id, device_type, zone, temp, hr
- LB: id, device_type, zone, bar

where id is a unique identifier for the logical object, zone is a static attribute configured at deploy-time and temp, hr and bar represent the sensed values. For example if we want to continuously sample the temperature and the humidity over the entire vineyard and put the results in a stream called MEASURE_DATA. We suppose that sampling frequency is of 1 sample every 300 seconds. The correspondent PERLA query will result as follows:

```
CREATE [OUTPUT] STREAM MEASURE_DATA AS
LOW:
EVERY ONE
SELECT id, temp, hr FROM LA
SAMPLING
EVERY 300 s
```

If we want to filter the previous stream in order to generate an alarm condition about rime (where rime occurs with a condition of temperature <5 C and humidity >75%), we have the following query:

```
SELECT id FROM MEASURE_DATA
WHERE temp < 5 AND hr > 75
```

Sometimes it is not necessary to activate all the sensors at the same time. For example, if we want to monitor a group of adversities with a restricted group of sensors, we might start by monitoring the common magnitudes and then start the other sensors when certain alarm conditions are reached. If we want to monitor the average temperature in a given zone of the vineyard and we suppose to start the sampling of pressure conditions on the same zone only if the temperature raises up to 30 Celsius, we'll have the following query:

```
CREATE [OUTPUT] STREAM ZONE_TEMP AS
SELECT id, temp, zone
GROUP BY (zone)
HAVING AVG(Temp) > 30
WHERE zone = zone_id

CREATE [OUTPUT] STREAM PRESSURE AS
LOW:
EVERY ONE
SELECT bar FROM LB
PILOT JOIN ZONE_TEMP ON ZONE_TEMP.zone=LB.zone
```

21.3.3.2 WSN Deployment and Data Communication

Data acquired by sensors are transferred, through a wireless Gateway node, to the Data Collection Server, from where they can be accessed - even remotely - to support appropriate decisions.

As anticipated in Section 3.1, the following physical quantities need to be monitored in a vineyard to control the grapes cultivation process: temperature, leaf witness, soil moisture, wind speed and direction, and rain intensity. Therefore, sensor nodes must be equipped with appropriate sensors. Specifically, all sensor nodes in the network include sensors for temperature, leaf wetness and soil moisture, so as to achieve a fine-grain monitoring of the corresponding physical quantities. In addition, there are one pluviometer and one anemometer for the entire network. Technical characteristics of sensors used in our testbed are shown in Table 1.

Type	Manufacturer	Model	Range	Accuracy	Power	Out Interface
Temperature	TX Instruments	TMP275	–20 C to +100 C	±0.5 · C (max)	2.7V to 5.5V	Two-Wire SI
Moisture	Sensirion	SHT11	0 to 100% RH	pm3.5% RH	30µW	Two-Wire SI
Leaf wetness	Decagon Devices	LW05-rc Leaf W.	Not available	Not available	2.5VDC@2 mA	Linear 250 -1500 mV
Pluviometer	AANDREAA	3864	max 200mm	±2%	30µA	SR10 Protocol
Wind Speed & direction	COMAI	t008 TVDV-I	0 + 50 m/s	±0,25 m/s (0+20m/s) ±0,7 m/s (>20m/s)	Max 14mA @ 16VDC	Linear 4-20mA

Fig. 21.19 Sensors used for vineyard monitoring

Sensor nodes are deployed on plants in the vineyard - at a distance of 10-20m - so as to form a regular grid. A single-hop or a multi-hop network can be used, depending on the size of the monitoring area. In both cases data acquired by sensors are transferred, through short-range wireless communication, to the local Gateway that is in charge of collecting data and transferring them to the Data Server. To this purpose, the Gateway node uses a long-range radio link. The Gateway is also responsible for receiving control messages from the Data Server and forwarding them to all sensor nodes, or a subset of them (e.g., for changing the configuration parameters). Irrespective of the network topology (i.e., single-hop or multi-hop), communication must be energy efficient as sensor nodes are powered by batteries and have a limited energy budget. Therefore, specific energy conservation techniques must be implemented to prolong the network lifetime [1]. Solutions for efficient energy management in single-hop and multi-hop WSNs have been discussed in chapters 15 and 16, respectively. The proposed approaches rely on the concept of duty cycle, i.e., sensor nodes switch off their radio when there is no message to receive or transmit. Duty cycle is a very effective solution for sensor nodes as they have a limited network activity and use short-range communication. However, this is not sufficient for the Gateway node as (i) it manages an amount of traffic much larger than any other node in the network, and (ii) it uses long-range radio communication to transmit data to the Data Server, which consumes a lot of energy. Therefore, the Gateway also includes an energy harvesting system, based on solar cells, to scavenge energy from the external environment. A detailed description of the Gateway node and its integrated energy harvesting system is given in chapter 15. Many of the concepts presented above for the vineyard WSN also apply to the cellar WSN used for monitoring the wine aging process. In this application scenario the relevant environmental data to be monitored are temperature and humidity. To this end, sensors already integrated in sensor nodes are used, without any need for specific external sensors. Sensor nodes are deployed strategically in some specific points of the cellar to provide a fine-grain monitoring of the sensing area. As in the vineyard WSN, both a single-hop or multi-hop topology can be used - depending on the size of the sensing area - and similar energy management techniques are used for energy conservation at sensor nodes. As above, data acquired by sensors are transferred to the Data Server through the Gateway node. Finally, wireless sensors are also used in the wine distribution process to monitor and record the external temperature at which wine bottles are exposed during the travel from the winery to the final vendor (humidity and possible shocks may be also of interest). This is achieved by means of low-cost wireless data loggers attached to single bottles (or, better, to pallets), that are able to measure periodically the external environmental conditions. Measured values are recorded in a memory and can be retrieved, at the destination side, through a Personal Digital Assistant (PDA) including an RF reader. The PDA has the same role of the Gateway node in Figure 1, i.e., data acquired by sensors are collected at the PDA and, then, transferred to the Data Server. In our testbed we used the MTSens temperature data logger whose technical characteristics are summarized in Table 2 . Similar data loggers also exist for monitoring humidity and possible shocks.

Parameter	Unit	MTSens
Temperature Range	Celsius	-10 to +60
Temperature Resolution	Celsius	1
Temperature Accuracy	Celsius	1
Time Accuracy	%	0.006
Data Logging Capacity		Up to 400 violations
Acquisition Rate		From 1 min to 24 h
Data Retention	Years	40
RF Interface (ISO Std. 15693)	MHz	13.56
Size	cm	8.56 x 5.59
Duration	Months	4

Fig. 21.20 Technical characteristics of MTSens data loggers

21.3.4 ArtDeco Prototype Integration Layer

An integration layer has been developed to integrate the different tools and systems prototyped during the ArtDeco project. Even though, this layer is not a core sub system for the research activities of the project, it is important for showing the benefits of the integration of the different technologies developed during the project.

The section mainly deals with (1) a publish/subscribe middleware used for modelling the execution context and for collecting data from the business environment and for delivering them to the interested information systems' components; (2) a portal server that offers the ability to rapidly integrate different Web-based applications in a uniform and controlled framework.

21.3.4.1 Connecting the External Context with the Information Systems

As Fig. 21.6 shows, the data and events generated by data sources (introduced in sections 21.3.2 and 21.3.3) must be delivered to the information systems for processing (see the *context events messaging* subsystem). This requires an integration layer that reduces the effort for connecting heterogeneous data sources and ensures the delivery of the needed data. Typically, such a layer is built upon an enterprise bus that ensures the needed decoupling among the components. However, since in the project this layer is not a fundamental aspect to address, existing buses for system integration have been used for building an integrated prototype.

Independently of the bus adopted, a specific connector is useful to model the external context and to suggest the parameters to monitor by the workflow management system and the other information systems. Modelling the context means to have a representation of the variables that characterise the environment to observe and defining how and when to monitor these variables.

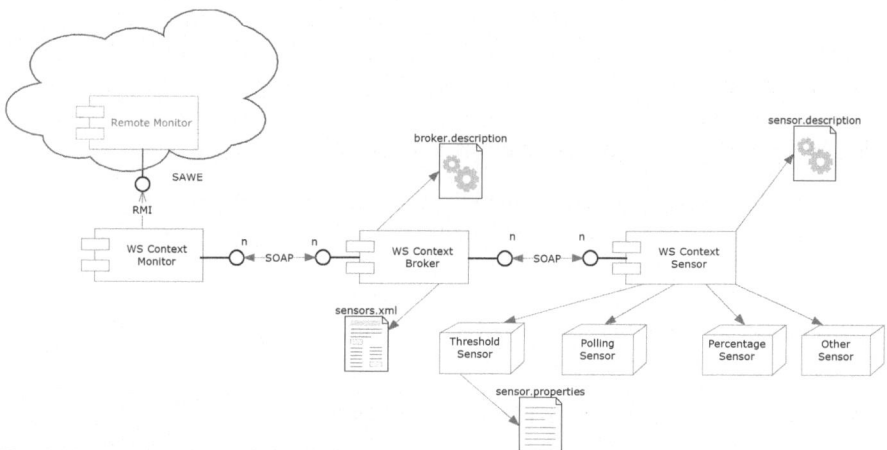

Fig. 21.21 Context monitoring architecture

A framework for sensing the context has been developed, according to the architecture shown in Fig. 21.21. The framework is conceptually composed of three components (to be specialised in order to get the desired behavior):

1. *Sensor*: is the component that measures the value of a parameter and that produces related events. Events can be raised due to one of the following circumstances:

   a. *Polling*: the event is produced in pre-defined temporal intervals;
   b. *Threshold*: the event is produced in case of overcoming of one determined threshold;

    c. *Variation*: the event is produced when the parameter varies of a certain percentage in comparison to its preceding value;

2. *Monitor*: is the component that deals with the received events and notifies them to the upper layers;
3. *Broker*: is the intermediary component that receives events from sensors and notifies them to the monitors. Its use is optional, but strongly recommended, since it promotes the loose-coupling (in fact a Monitor may not know the addresses of all the Measurers from which it is interested to receive events) and it allows for the aggregation and the processing of the received events.

The interface for subscribing is the same for every component, so that it is possible for a Monitor to subscribe to a Sensor, to a Broker or to both and it is possible to define chains of Brokers before reaching a Monitor. A Broker may subscribe more than one Sensor and a Sensor can be subscribed by more than one Broker. The same behavior may be used also between Monitors and Brokers. As the monitoring, also sensing can be built in a 'recursive' way.

For using the framework, it is needed to use a Broker as 'responsible' of every context area. The Broker has the knowledge of all the Sensors of that determined area and acts as intermediary between the Monitor and them. The Broker uses a semantic layer to get the description of an observable area and the parameters available. The framework is specialized through the definition of three objects.

1. *Event*: the object that defines the event; it contains all the useful fields to the qualification of the same event (used by the monitoring rules described in section 21.3.1);
2. *Measure*: the object that represents the measure;
3. *EventListener*: it defines the specific behavior of the monitor for the received events.

The parameters belonging to a context are handled by the Context Broker. For each parameter it is possible to specify whether it is negotiable or not. In the case of negotiable parameter, the Context Broker negotiates a model for data retrieving through the use of WS-Agreement. That agreement will then be used by the sensor of the parameter to take measurements and to send events.

21.3.4.2 Portal

The portal represents the access point for human-oriented interactions with the information systems of the sample networked enterprise analysed in this chapter. Through the portal, several actors are able to cooperate, by exchanging information generated during the different phases of wine production. Some actors belong to the winery whereas other ones operate in different organisations belonging to the same network. By sharing some data related to the wine production processes, the whole value chain becomes more efficient and effective, as regarding the ability of reaching the business targets in a faster and cheaper fashion.

Through the portal, the actors working in the winery can receive daily tasks to perform and can send data obtained by monitoring the vineyards. They can follow the running processes by accessing the desired information in a contextualised fashion, so avoiding wrong queries and reducing noise, since only useful information is retrieved in any context.

The portal was developed by leveraging on an open source product as portal server. It has been used to integrate some Web applications and a CMS in a uniform view. Users can access to the portal after their authentication that identifies their role and propose to them a contextualised view for accessing to data.

21.4 Lesson Learnt

Business environments involve increasingly complex and changeable processes today. To keep and improve their competitiveness, companies need innovative methodologies to efficiently and effectively manage their Business Processes (BPs). The autonomic approach can be considered, in this scenario, as a valuable support for the Business Process Management (BPM) lifecycle automation. BPM represents a systematic approach to incrementally define, analyse and innovate BPs, which can be applied to address intra- and inter-organisational issues. The BPM lifecycle starts with a modelling phase to define user requirements describing a BP as a model. Afterwards, the BP is implemented (in the configuration phase) and executed. Finally, the BP execution is monitored and the gathered data is analysed to identify improvement actions to be applied to the original model. On the basis of the updated model a new cycle can start. Operating in this manner, a BP can increasingly evolve.

Based on organisational and business models for the development of new systems, the autonomic methodology allows the structure and the organisation of a process to be concisely and accurately represented. This guarantees an in-depth comprehension of the process, without ambiguity, and allows the stakeholders to share information about its peculiarities.

Another crucial issue in the BPM life cycle is process automation i.e. the translation of a process model (describing a BP from the domain expert point of view) into an executable process. To support process automation, the proposed methodology provides for the translation of abstract requirements, described as a process model, into an executable process through successive transformations of the original model. The process model is enriched with implementation details and more detailed process models are generated which can be easily converted into executable code. In this way the autonomic methodology also represents an effective support to the configuration phase in the BPM life cycle providing a workflow design approach which produces progressively concrete models directly executable by information systems.

Moreover, companies working in a highly competitive and ever changing environment also have to make their BPs adaptable. It often happens that existing BPs become inadequate due to the natural changeability and competitiveness of the business world. To support flexibility in BPM, the autonomic workflows are implemented according to the 'sense-and-respond' paradigm. From this point of view an autonomic workflow provides for the execution of the normal flow together with reactive rules able to change the business processes when unexpected events occur. Using this approach BPs execution is tailored according to different context factors and is able to fulfill the critical issue of process adaptability in an automated and easy way .

The autonomic approach also includes techniques for process monitoring generating a set of data (e.g. event logs) which can be analysed to re-design the original model and improve it. For this purpose process mining techniques could be a valuable instrument to be used within the proposed methodology. In this context, over the last decade a variety of techniques and algorithms have been proposed for mining process models from event logs, showing that information contained in the logs can be used to formalize or improve process models. Proposed methods have shown that event logs can be used to construct from scratch models underlying an automated process (i.e. process discovery), or to identify discrepancies between event logs generated by an automated process in place and a predefined process model representing its formal definition (i.e. conformance testing). Both of these approaches to process mining can effectively meet the evolutionary nature of the 'sense-and-respond' paradigm.

21.5 Conclusion

This chapter presented an approach for programming the information systems of a value chain though the help of a middleware substrate designed and developed in the framework of the ArtDeco project. The approach has been experimented in the context of wine production and distribution, starting from process information gathered from domain experts. It impacts on modelling and programming aspects (such as service description, process modelling, workflow, data bases) and information systems used to run the whole processes (such as WfMS, DBMS, WSN, etc).

Standard business process management languages have been extended in order to program two orthogonal and concurrent dimensions: normal control flow and context events handling. This involves the need for using a novel approach for programming the workflows that implement the business processes in order to obtain an autonomic behavior at run-time. To this end, reactive actions are driven by context events that are observed by a specific workflow management system, extended with a context handler that gathers these events from heterogenous sources, such as sensors used during cultivation and wine production, or QoS estimates that suggest to use a supplier instead of another one.

Workflow data or queries produced by domain experts (agronomists, oenologists, etc.) are retrieved from data bases by using a context-aware approach that reduces the amount of information that is not particularly useful for specific application contexts. Real-time data are collected by wireless sensors disseminated in the business environment and programmed following autonomic principles in order to improve the capacity of WSNs to survive as long as possible. Moreover, data mining techniques and semantic search have been shown as promising techniques to obtain useful information to take proper decisions in different phases of the whole process of wine production and distribution.

A significant part of the chapter has been devoted to the integration problems among the different components of the whole architecture. The middleware proposed in this chapter represents a conceptual framework for the integration. However, it could be significantly enhanced in the future by providing it with some additional non-functional features, such as: security, privacy, transactions management, performance, and scalability.

Acknowledgements. We thank Gaetano Anastasi, Andrea Cameli, Romolo Camplani, Angelo Furno, Ilaria Giannetti, Eugenio Marotti, Marina Polese, Emilio Sellitto, and Quirino Zagarese for their valuable contribution in developing some additional components that helped the integration of the tools developed in the ArtDeco Project.

The technologies described in this chapter have been partially experimented by the Ph.D. students who attended the GII (Gruppo Italiano di Ingegneria Informatica) school organised in the context of the project. The content of the school was designed around three groups of students, each one composed of about eight people characterized by different skills (software engineering, data base, computer networks, artificial intelligence). The two weeks of the school were structured in theoretical and practical sessions. During the theoretical sessions, students were provided with the main concepts related to the methodologies and technologies developed in the framework of the ArtDeco project, whereas during the practical sessions students were involved in the experimentation with the results of the project though a set of exercises framed within a coherent *project work*. The winner group was stimulated to give some feedbacks about the experimented technologies that have contributed to improve the project results and to write some sections of this chapter.

References

1. Anastasi, G., Conti, M., Di Francesco, M., Passarella, A.: Energy conservation in wireless sensor networks: A survey. Ad Hoc Networks 7(3), 537–568 (2009)
2. Cimatti, A., Clarke, E., Giunchiglia, E., Giunchiglia, F., Pistore, M., Roveri, M., Sebastiani, R., Tacchella, A.: NuSMV 2: An OpenSource Tool for Symbolic Model Checking. In: Brinksma, E., Larsen, K.G. (eds.) CAV 2002. LNCS, vol. 2404, pp. 359–364. Springer, Heidelberg (2002)
3. Peled, D., Pelliccione, P., Spoletini, P.: Model checking. In: Wiley Encyclopedia of Computer Science and Engineering. John Wiley & Sons, Inc. (2008)
4. Ghezzi, C.: The challenges of open-world software. In: WOSP, pages 90 (2007)
5. Golfarelli, M., Rizzi, S.: Data Warehouse Design: Modern Principles and Methodologies. McGraw-Hill, Inc., New York (2009)
6. Holzmann, G.: Spin model checker, the: primer and reference manual. Addison-Wesley Professional (2003)

7. Huebscher, M.C., McCann, J.A.: A survey of autonomic computing—degrees, models, and applications. ACM Comput. Surv. 40(3), 1–28 (2008)
8. Polese, M., Tretola, G., Zimeo, E.: Self-adaptive management of web processes. In: WSE 2010, pp. 423–428 (2010)
9. Schreiber, F.A., Camplani, R., Fortunato, M., Marelli, M., Pacifici, F.: Perla: A data language for pervasive systems. In: IEEE International Conference on Pervasive Computing and Communications, pp. 282–287 (2008)

Author Index